AUSTRALIAN SCIENCE

in the making

Australia
1788-1988

This publication has been endorsed by the
Australian Bicentennial Authority to celebrate
Australia's Bicentenary in 1988.

TERRE DE DIÉMEN : ÎLE MARIA.

TOMBEAUX DES NATURELS.

(Courtesy of the Baillieu Library, The University of Melbourne)

AUSTRALIAN SCIENCE
in the making

edited by

R. W. Home

Published in association with the
Australian Academy of Science

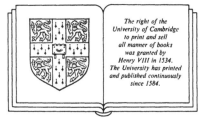

The right of the
University of Cambridge
to print and sell
all manner of books
was granted by
Henry VIII in 1534.
The University has printed
and published continuously
since 1584.

Cambridge University Press
Cambridge

New York   New Rochelle   Melbourne   Sydney

CAMBRIDGE UNIVERSITY PRESS
Cambridge, New York, Melbourne, Madrid, Cape Town,
Singapore, São Paulo, Delhi, Mexico City

Cambridge University Press
The Edinburgh Building, Cambridge CB2 8RU, UK

Published in the United States of America by Cambridge University Press, New York

www.cambridge.org
Information on this title: www.cambridge.org/9780521396400

First published in 1988
Paperback edition 1990

A catalogue record for this publication is available from the British Library

Library of Congress Cataloguing in Publication Data
   Australian science in the making.
   ISBN 0 521 35556 7.
   ISBN 0 521 39640 9 (Pbk.).
   1. Science — Australia — History. 2. Science — Australia
   — Social aspects. I. Home, R. W. (Roderick Weir). II.
   Australian Academy of Science.
509.94

ISBN 978-0-521-35556-8 Hardback
ISBN 978-0-521-39640-0 Paperback

CONTENTS

LIST OF ILLUSTRATIONS

Introduction

The year 1988 marks the 200th anniversary of the European settlement of Australia. Long before that, the Aboriginal people who inhabited the land had developed their own modes of understanding the Australian environment and the natural processes occurring within it. Prior to the arrival of the first white settlers, scientific exploration of the continent had been started by visitors from Europe, most notably Lieutenant James Cook and the scientists who accompanied him aboard the barque *Endeavour* during its exploration of the eastern Australian seaboard in 1770.

Following white settlement, a steady trickle of visiting scientists came from Europe to investigate Australia's unfamiliar flora and fauna, its rocks and soils and sky, its native peoples. A few of the settlers also took an interest in such things. It was many years, however, before Australian scientists ventured beyond investigating their local surroundings to tackle more general questions that were in the mainstream of scientific study elsewhere. Except for a few remarkable but isolated individuals, it took many more years before they could even begin to claim parity, as they do today, with fellow scientists in the leading centres of Europe, North America and Japan.

The present volume has been brought together, under the aegis of the Australian Academy of Science, to mark the bicentenary of white settlement. Its theme is 'Man's attempts to understand Nature in an Australian environment'. The book does not claim to be comprehensive in its coverage of the history of science in Australia. Instead, it takes up a number of more restricted questions bearing on the central theme.

Science of course recognizes no national boundaries. Scientific advance depends on successful cross-fertilization of ideas and techniques. Scientists everywhere depend on a free flow of information. They need to know about the work of fellow scientists at home and abroad, and they need to have their own work incorporated into the accepted body of scientific knowledge: this has been the case ever since modern science first emerged, in a few countries of western Europe, during the seventeenth century. At first glance, therefore, a volume confined to the history of science in one country would appear to be unnaturally constrained. If science is one, and universal, there can be no distinctive French or Japanese or Australian science worth studying as such.

Yet though scientific discourse is international, its practice is at the same time socially bound. Individual scientists inevitably work within the confines of the societies in which they live as well as of the scientific disciplines to which they contribute. The most challenging task facing the historian of science today is to delineate, in any particular case, the interlocking intellectual, social and economic strands that shaped the work of individuals and groups and thus determined events, and very often the social strands will be best defined within a particular linguistic or geographical or national context. Hence, even if 'Australian science' is, strictly speaking, a phrase that lacks content, to study the history of science as it has been practised in Australia remains an intellectually coherent thing to do.

Indeed, it is imperative that such studies be undertaken. Science has in modern times become a powerful social and economic force, the effects of which are apparent in every facet of daily life. To study the place of science in social and economic life and the way in which this has changed as science itself has changed in the course of the past 200 years is thus to focus on one of the central historical questions of our time. From this point of view, the history of Australian science is a vital part of the general history of the nation.

Despite such an obvious truism, until relatively recently all too few scholars have paid serious attention to the subject. In a scientific age, Australians need to develop a better understanding of their nation's scientific past. The publication of the present volume will, it is hoped, help to engender a wider appreciation of this at the same time as it opens up a number of possible new lines of historical investigation.

Yet the history of Australian science is not just the history of science in Australia, so that interest in these topics should not be confined to Australians. Australia is but one of many 'new' nations to which western science was initially quite foreign, but which in recent times have developed, more or less successfully, a scientific culture of their own. Readers more familiar with developments elsewhere are likely to find instructive parallels and in some cases contrasts in the Australian experience.

On the one hand, because the incoming white settlers simply pushed aside or, all too often, exterminated the original inhabitants of the land, we do not find in Australia as we do in, say, China and Japan, an interweaving of traditional culture with western scientific values. Rather, as several of the chapters that follow make clear, science developed in Australia — as it did in other 'settler' societies such as Canada, Argentina and New Zealand — as part of the cultural baggage that the settlers brought with them from their European homelands. On the other hand, in both kinds of 'new' environment, settler society or traditional one undergoing westernization, the same questions arose. These concerned, above all, the maintenance of connections with the scientific heartlands of western Europe, the structuring of relationships with the scientific leaders there, and the small size of the local scientific population and the inhibiting effect on the work of individuals of the resulting lack of regular contact with other researchers working on similar problems. There was the same concern, as well, with developing local supporting institutions (libraries, museums, publishing houses, universities, learned societies), a climate in which scientific work was encouraged, linkages with a growing technical and industrial base and, in general, with developing scientific self-sufficiency.

From the point of view of the formerly imperial powers, yet other questions spring to mind — concerning, for example, the place of science in the rise and fall of imperial hegemony in the modern world, the stability (or instability) of cultural and intellectual forms translated to new environments, the impact of discoveries made in the colonies on science at the metropolitan centre. Here, too, the Australian experience is likely to suggest answers of considerably wider relevance.

Inevitably, therefore, a central theme of a book such as this becomes the changes that have occurred during the past 200 years in the relationship between scientists working in Australia and those in the leading scientific centres in Europe. In this regard, some important ideas have been developed by historians in recent years.

In particular, the American historian Donald Fleming has suggested, on the basis of a comparison of the history of science in Australia, Canada, and the United States, that the science that was practised in these three countries until relatively recent times shared certain features that could reasonably be described as 'colonial'. He notes, in particular, what he calls 'the phenomenon of absentee landlordship' among nineteenth-century naturalists, 'the dominion of European students of natural history over their collaborators on the periphery'. Naturalists in the new territories, anxious to secure recognition from Europe, sent descriptions and specimens to leading authorities such as the Hookers, father and son, at Kew Gardens in London; but it was to the latter that the higher task fell — that of collating the particulars thus supplied them in order to reach some more general understanding. Fleming insists, however, that this is not to be seen as 'a limitation clamped upon a subject race of investigators by their overlords'. On the contrary,

> For the colonial investigator himself, natural history was the ideal refuge from the more perilous enterprise of embarking upon theoretical constructions by which he would be pitched into naked competition with the best scholars of all countries. To be a forager for Linnaeus or correspondent of the Hookers might be an identity in science purchased by bondage to the local and particular; but it was also a shelter against the more bracing winds that would promptly blow upon any man who tried to grapple with undifferentiated Nature in physics.[1]

In addition, Fleming argues, 'the practical associations of natural history were greatly enhanced by its appearing to be the intellectual aspect of pioneering'; as a result, natural history reconnaissance became 'an acceptable style of scientific endeavour in the new societies themselves'.

Fleming also points to the habit of colonial investigators in Australia and Canada completing their training 'at home' rather than in the colonies, with the ablest then finding positions in Britain; of talented young British scientists finding their initial appointments in colonial universities but returning to positions in Britain at the first plausible opportunity; and of colonial scientists continuing to look to election to a British institution, the Royal Society of London, as the ultimate accolade of their work. 'Colonialism in science owed much of its tenacity', he suggests, 'to academics of the second rank, Englishmen who did not succeed in rebounding from exile, native colonials who had no choice but to return home after their studies abroad because nobody tried to detain them'. These men were intellectuals, nevertheless, who felt alienated by the indifference of the communities in which they found themselves to the life of the mind. In compensation, 'they called in the old world to redress the psychological balance of the new, kept their affiliations with Britain in repair, and recovered the sense of belonging to a larger community in which colonial values did not prevail'. In Australia this situation persisted, Fleming implies, at least until the Second World War.

These challenging ideas have been elaborated by another American historian, George Basalla, into a diffusionist model, which he claims has general application, in which science is seen as spreading from its original home in western Europe to other parts of the world.[2] Basalla delineates three different phases in this process, one of which parallels Fleming's 'colonial' science. In his scheme, however, it is preceded by a period in which such science as was done in the new land was done by visitors who, after sampling the new environment, returned to Europe and published their results there, and is followed by one in which local scientists 'struggl[ed] to create an independent scientific tradition'. Moreover, Basalla portrays the relationship between metropolitan scientific centre and colonial periphery in more mechanistic and less psychological terms than Fleming.

According to Basalla's schema, the second, 'colonial science' phase begins when local residents — whether settlers or native citizens — take up scientific pursuits. It is 'dependent science' in the sense that its practitioners look to an external scientific culture in the European heartland for their advanced training, for books and laboratory equipment, for the publication of their work, for recognition. In the succeeding phase there is a conscious attempt to establish local alternatives so that a scientist's major ties will come to be 'within the boundaries of the country in which he works'.

A number of criticisms of the Basalla schema have been published, many of which have drawn heavily on the Australian experience.[3] In particular, Basalla's claim that his model applied universally has been widely challenged, as has his assumption that there was a single western scientific ideology waiting to be diffused into new scientific territories. The variety of historical experience has been much richer and more complex, it has been argued, than his model allows. Attempts to apply the model in detail to particular cases have led each time to the conclusion that, for one reason or another, it cannot be done in any straightforward way. More generally, Basalla has been accused of neglecting the essentially interactive nature of the scientific links that develop between metropolis and province, whereby the science practised at the centre can be powerfully influenced by inputs — usually of new information — from the colonies. He has also been criticized for treating science in isolation and failing as a result to take proper account of the political, economic and social forces that have brought about the changes he describes and that maintain the relationship of scientific dependency even after formal political ties are cut.

He has failed, in short, to take proper account of the ultimately political character of science itself within such a framework of dependency.

Despite such reservations, however, there remains a fairly general consensus that, even in the absence of overt political links, the science practised in a developing country can often properly be described as colonial. Yet what that label implies, beyond the existence of a relationship of dependency of some kind on the science of a metropolitan centre or centres, is a matter of controversy. The precise nature of the relationship seems to vary from country to country and also over time.

In the Australian case, there is universal agreement that after an initial period (corresponding to Basalla's Phase I) in which the only science done was merely an extension of metropolitan science, carried out by temporary visitors from the metropolitan centres, there ensued a period of colonial science characterized, at least initially, by the colonial scientist-collector sending his materials 'home' to Kew Gardens or the British Museum for analysis and description by metropolitan savants. The basis of scientific decision-making and authority remained firmly in Europe; Australian workers could not aspire to become recognized authorities even on Australian materials. Moreover, with no facilities for providing even the most rudimentary training in science — these came only with the founding of the first universities in Sydney and Melbourne in the 1850s — the Australian colonies remained entirely dependent, scientifically speaking, on imported skills. Furthermore, the number of trained people in the country long remained too small to form self-sustaining institutions or to maintain locally based scientific publication outlets.

The same relationship of dependency existed even in sciences such as astronomy and geophysics, where Australia's location in the southern hemisphere gave it peculiar advantages for certain kinds of observing. Here, too, it was not until after 1850 that sufficient Australian resources could be mustered to sustain a significant programme of locally directed scientific work. Until then, Australia served merely as a convenient fixed platform for temporary observatories established under direct British aegis, whether on Dawes Point at Sydney Cove in 1788, at Parramatta under Governor Brisbane in the 1820s, or at Hobart in the 1840s.

The pattern of Australian scientific work changed dramatically during the latter half of the nineteenth century. Though scientific exploring remained important, this now tended to be carried out under local rather than European auspices. Collectors continued to scour the bush for specimens but, as time went by, more and more of the material collected found its way into newly founded Australian museums of natural history. Furthermore, the notion, perhaps first expressed by Leichhardt in the 1840s, that type specimens and other reference or unique materials ought to remain in or at least return, after description, to Australia, came to be increasingly widely held.[4]

The second half of the century saw the establishment in the various colonies of a range of other scientific institutions besides museums, all of them now under local control. The founding of universities in Sydney and Melbourne has already been mentioned; Adelaide and then Hobart later followed suit, as did Brisbane and Perth shortly after the turn of the century. Initially, only a liberal arts course centred on the classics and mathematics was provided but the sciences were also strongly represented in the curriculum from the outset. In time, separate laboratory-based degree courses in science and engineering were introduced and also, in the larger universities, degrees in medicine.

In addition, a majority of the colonies established observatories during this period. Government astronomers, appointed to run them, became dominant figures in colonial scientific circles and were usually responsible not just for astronomical work (including their colony's time service) but also for routine geomagnetic and meteorological observing and for making the fundamental geodetic determinations for government survey purposes. In Queensland and Tasmania, where observatories were not established, government meteorologists were appointed instead. Other sciences, too, received support from newly formed colonial legislatures. Several colonies appointed government botanists, of whom Ferdinand von Mueller in Victoria is by far the best known. With rumours of gold in the air, New South Wales appointed a government geologist, albeit only temporarily, in 1850. Victoria set up an excellent Geological Survey in 1852, and most of the other colonies later followed suit. Public health officials were appointed and likewise government analysts to undertake assays and to oversee the quality of water and food supplies.

These developments, together with the increasing numbers of doctors, school teachers and other professional men required to service the waves of immigrants attracted by the gold rushes and their aftermath, led to a rapid growth, especially in the more populous colonies, in the numbers of residents with scientific interests and expertise. Viable scientific societies became a possibility for the first time and were formed one by one in most of the colonies during the middle third of the century. Each colony, sooner or later, came to have a 'Royal Society' patterned after the London model, with its own journal, which became the basis of exchange agreements with scientific societies elsewhere. Such societies provided both suitable outlets for local scientific work and a mechanism whereby local workers could keep in touch more satisfactorily with what was being done in other parts of the world. The growth of public reference libraries in the different colonial capitals also helped in the latter regard.

Nevertheless, Australian science continued to be quintessentially 'colonial' in character, remaining largely observational and descriptive in style rather than experimental and laboratory based, and being concerned almost exclusively with local questions rather than with topics of more general relevance. Furthermore, scientists in the different colonies remained more or less isolated from each other. For intellectual support and encouragement they generally looked not to their fellow colonists but to the scientific community 'at home' in Britain, and if they did venture on to topics of more than local concern, it was to the English journals rather than their local Royal Society transactions that they sent their work for publication — naturally so, because that was where the relevant readership was, not in Australia.

An important change in attitude can be discerned during the 1880s. It was during this decade, for example, that the universities in Sydney and Melbourne dramatically expanded their commitment to science. (The much smaller University of Adelaide followed suit 15–20 years later.) Talented new professors and lecturing staff, mostly from Britain, were appointed and new laboratories were constructed. Soon, the first research students appeared. Furthermore, many of the new professors were already active researchers, with a mind to keep it up; and their interests were by no means confined any longer to local questions. Publications by them and their students began to appear with increasing frequency in the international journals.

The impact of a number of science-based technological advances also first began to be felt during this period. The first telephone services were introduced in the larger cities, as well as the first municipal electrification systems. Chemical and metallurgical industries mushroomed. Major improvements in public health services followed the general acceptance of the germ theory of disease. Soon afterwards, science began to be applied in the countryside as well, through the establishment of scientific services within colonial departments of agriculture.

In addition, Australian scientific workers were now somewhat less isolated than they had been. The introduction of steamships and the opening of the Suez Canal had halved the time of the journey from Europe. The resulting improvement in the mail service made it possible for the first time for scientific workers in Australia to keep reasonably up to date with the international journal literature in their field and even to take part in some cases in scientific debates in the pages of journals such as *Nature*. Within Australia, improvements in transport helped to bring scientists in the different colonies closer together and made possible the formation of the first inter-colonial scientific organization, the Australasian Association for the Advancement of Science (AAAS), founded in 1888. The Association at first met roughly annually, later at approximately two-year intervals. Its meetings quickly established themselves as the highlight of the Australian scientific calendar, offering those attending welcome opportunities to exchange opinions and establish a basis for subsequent correspondence. They remained so until the growth of specialist societies, especially in the years after the Second World War, challenged the association's hegemony.

The Australian scientific community remained small, however, and the imperial connections continued strong. Even in 1939, Australian scientists tended to see themselves and their work very much within the context of a larger British scientific network. Travelling scholarships such as the 1851 Exhibition science research awards (established in 1891) and the Rhodes scholarships (established in 1904) strengthened the links by taking many of Australia's best young science graduates to England for further training. A significant proportion, including some of the best of them, did not return.

Several chapters in this book explicitly take up questions concerning Australia's scientific links with Britain and, in almost all of the chapters, such questions are present implicitly. Indeed, they can scarcely be ignored, and much work remains to be done adequately to explore their ramifications for the history of Australian science. Scientific work, the production of new scientific knowledge, is an inherently social process if for no other reason than that such knowledge needs to be certified by the wider scientific community of the day. Certification, however, implies authority, something that was long denied the small and scattered groups of Australian scientific workers. For example, when data on the southern skies collected at Sydney's Parramatta Observatory were found to disagree with those recorded by Sir John Herschel during a visit to the Cape of Good Hope, the immediate (and subsequently justified) assumption among the world's astronomers was that the Sydney observations must be in error.[5] Likewise, though there had been many reports from Australia that monotremes laid eggs as well as suckled their young, it took a telegram from a visiting British authority, W.H. Caldwell, to convince the zoologists of Europe.[6] The story of colony and metropolis is thus, in science as in so much else, a story of a struggle for authority. Meanwhile, the strong political and cultural ties with Britain ensured that the authorities to whom Australian scientific workers looked were British rather than, say, French or German.

A major contributing factor to the maintenance of power relations of this kind was the chronic isolation of Australian scientific workers. Active scientists need to interact constantly with their fellows, exchanging information, opinions and ideas and drawing encouragement from such contacts. They need such exchanges to publicize their work and to offer it up for certification. In most scientific fields, however, such interaction was long impossible in Australia because there were too few scientists and those few were too widely scattered among the major population centres. The formation of viable scientific societies in the various colonies in the second half of the nineteenth century was a first step towards facilitating interchanges, but it usually remained the case that, at best, only a tiny handful of active workers could be mustered in any particular scientific field, even in Melbourne or Sydney. The establishment of AAAS created opportunities for occasional meetings with larger groups of specialists. Being relatively infrequent, however, such meetings scarcely constituted an adequate solution to the problem. For the most part, individual Australian scientists were thrown back on their own devices. Lacking the support of informal networks of intellectual exchange such as their peers in the metropolitan scientific centres enjoyed, their only recourse was to do the best they could, alone, and then submit their work directly to formal scrutiny by the metropolitan authorities.

In these circumstances, it is small wonder that most Australian scientists continued to focus their attention, as Fleming has noted, on questions arising from the local natural environment. Here, they could retain (or establish) a measure of intellectual control; occasionally, as von Mueller did, they could even themselves gain the status of authorities in their special fields.

Once they took up more general questions, as W.H. Bragg did in 1903–4 when he embarked upon a major investigation of the ionizing radiations emitted by radioactive substances, they found themselves almost literally on their own. Through attending the AAAS congresses regularly, Bragg had already built up a network of friends and correspondents within Australia's nascent community of physical scientists. Once fairly launched on his study of radioactivity, he exploited this network by sending out drafts of his paper for comment. The responses he received offered encouragement and some useful criticism. None of his correspondents were researching similar problems, however, and hence no-one was able to go beyond this and provide detailed, penetrating feedback such as Bragg now needed. Where friendly exchanges with fellow physicists had previously sufficed, these were no longer enough. Bragg was now suffering, as have a number of other scientists working in Australia who have reached the first rank in their particular fields, from a more refined form of isolation, 'the isolation of the élite'.[7] Until relatively recently, if an Australian scientist reached the top rank in his or her speciality, there was virtually no chance that there would be another researcher within striking distance within Australia who was equally knowledgeable in that speciality. Yet researchers working at this level seem to feel, quite acutely, a need for close and regular personal contacts with other leading researchers in their field. Bragg's response to the new situation in which he found himself was all too characteristic of scientists working 'on the periphery' — he sought a job in England. As he wrote to Ernest Rutherford, the unquestioned leader of research in his field: 'I would be glad to go to England for many reasons: you must not mind my saying that one of these is to be near people like yourself'.

During the 1914–1918 war, Australian scientists, like their compatriots from other walks of life, flocked to support the allied cause. Scientific work was not a reserved occupation, and many scientists simply joined the fighting services. In some cases, however, their special skills were recognised by the authorities. For example, many Australian chemists were recruited to go to England to help to develop the munitions industry there, while Australian geologists, working as miners under the leadership of Sydney's Professor Edgeworth David, performed remarkable service in the trench warfare on the western front.

For many of the Australians involved, the war brought with it a heightened sense of their Australian-ness. In its aftermath, several new and consciously Australian institutions claiming nation-wide coverage and authority were founded — the Australian National Research Council, the Australian Chemical Institute (later the Royal Australian Chemical Institute) and the Institution of Engineers, Australia. Nevertheless, throughout the 1920s most Australian scientists continued to see themselves within a larger, imperial framework.[8] The popular vision of an integrated imperial economy in which Britain possessed the factories while the Empire supplied the raw materials and markets for the finished goods implied that Australian scientists would concentrate on fields relating to agriculture and mining, whereas the British would concentrate on sciences such as physics. The overwhelming emphasis on agricultural research in the early years of the Council for Scientific and Industrial Research (CSIR), formed in 1926, in part reflected this doctrine, though it also reflected the more parochial political and economic circumstances within Australia at the time that had led to the council's creation.[9]

The economic collapse of the early 1930s brought an end to imperialist dreams of this kind. Already, however, local needs had been working against them. CSIR's charter explicitly envisaged the organization's undertaking research that would assist manufacturing as well as the agricultural sector, especially through the establishing of physical and engineering standards. An early and highly successful involvement in radio research led the way in non-agricultural research. Then during the late 1930s, as war clouds gathered again and Australia began at last to build up its manufacturing capability, major CSIR divisions of Aeronautics and Industrial Chemistry were founded, and also the long awaited National Standards Laboratory.

The Second World War had a much more dramatic impact on Australian science than had the First. With invasion threatening and traditional British sources of supply cut off, Australia was forced to look to its own resources for essentials that had always previously been imported. As the existing science-based industries such as munitions and electronics expanded, they demanded more and more scientifically trained staff. University scientists worked to create new industries where none had existed before in fields such as pharmaceuticals and optical components. Physicists and engineers were recruited in large numbers to work on a new invention of strategic importance — radar. By war's end, it was clear that Australian science had undergone an irreversible change in line with the general industrialization of the nation's economy. The number of scientists working in all fields had greatly increased, and the demand for their services did not decline with the coming of the peace. Moreover, no longer did Australian science look quite so automatically to England for leadership and research opportunities. For some, the United States had become an enticing alternative, but others looked forward to Australia making its own, independent contribution to the new, scientific age that seemed to have been ushered in with the explosion of the first atomic bombs and the promise of 'atoms for peace'.

The post-war period saw a continued rapid expansion of Australian scientific institutions. CSIR, reconstructed in 1949 in the wake of a savage and unprincipled political attack as the Commonwealth Scientific and Industrial Research Organization (CSIRO), remained pre-eminent as it spawned more and more divisions. The majority of these conducted research in areas related to primary industry, but those relating to manufacturing industry also grew in numbers and strength.[10] No longer, however, did the organization confine itself to the applied research envisaged by its creators. Instead, it was urged that the ideal for CSIRO was something closer to a 50–50 'mix' of applied and basic research. University research likewise expanded, at first as a result of the creation of the Australian National University (ANU) in Canberra, later with the injection of Commonwealth funds on a large scale into the state university system. The commitment of the universities to post-graduate education grew steadily following the introduction of the PhD degree in the late 1940s. The increased funding made available by the Commonwealth government in the 1960s led to the development of substantial research schools in several of the state universities as well as at ANU, and a general enhancement of research activity throughout the system. As a result, young scientists no longer needed to go abroad to complete their training. The local scientific community had at last become self-sustaining.

Moreover, there were many more job opportunities for science graduates outside CSIRO and the universities than there had ever been before. The Department of Defence emerged as a major employer of research scientists, as did the Australian Atomic Energy Commission, established in 1955. In the area of medical research, several small existing privately or semi-privately funded research foundations, the best known of these being Melbourne's Walter and Eliza Hall Institute of Medical Research, expanded dramatically. Together with the new John Curtin School of Medical Research at the ANU, they earned the nation an enviable reputation in this field that was crowned in the early 1960s by the award, in quick succession, of Nobel Prizes for medicine and physiology to two Australian scientists, F.M. Burnet and J.C. Eccles. In some other fields, most notably radio astronomy, Australian scientists found themselves in the vanguard of world research. In almost all fields, they could now hold their own.

During these same years, the impact on Australian society of new, science-based technologies became ever more pervasive. Revolutionary changes in communications[11] and, more recently, the rapid spread of computer technology are but the most visible signs of a thorough-going invasion of science into all aspects of daily life. This has been accompanied by — indeed, has been dependent on — a rise in the level of scientific and technical skill in the general population, much of it provided, as in earlier periods, by waves of immigrants entering the country. The degree of scientific and technical sophistication of Australia's manufacturing and service industries is now incomparably higher than it was in 1945. The primary sector, too, has been transformed and its productivity dramatically enhanced under the impact of modern science.

Yet in comparison with most other countries, in Australia scientific research remains confined to a disturbing degree to public institutions. Though Australian science has attained a high level of achievement, Australian industry has failed to keep pace. The longstanding tendency of the nation's manufacturers to purchase the results of foreign industrial research rather than investing in such research themselves, and to limit their horizons to import-replacement manufacturing rather than looking to export markets, has left their companies vulnerable and ill-equipped to meet foreign competition in a manufacturing environment increasingly dependent in the 1980s on the exploitation of new scientific discoveries and techniques. It remains a moot point whether Australia can build a modern science-based industrial economy or whether, in Australia, science will remain on the margins of the nation's economic life.

The chapters that follow open windows on to various aspects of Australia's scientific past. Many of the events described may be unfamiliar to the average reader but they are of considerable historical interest nonetheless. Chapters have been selected in an attempt to provide breadth of coverage both chronologically and in terms of subject-matter. The decision was made at an early stage in the planning, however, that treatment of a relatively small number of topics in some depth was to be preferred to skimming lightly over a greater number. As a result, many readers will doubtless be pained to discover that their own favourite subject has been overlooked. It is to be hoped that publication of this volume will spur them to take up their pens for themselves. In far too many cases, almost everything remains to be done.

'Science' often tends to be equated with 'modern western science', in which case the history of Australian science would begin with the first European contacts. There are, however, other ways of viewing nature's workings besides the one that evolved in western Europe. People have been living in Australia for at least 40 000 years and, as L.R. Hiatt and Rhys Jones argue below, they learned long ago to classify the objects making up their immediate environment on the basis of a close and detailed knowledge of their properties and behaviour. Many of the categories they enunciated were similar to those later adopted by European scientists, others were not; in either case the Aborigines' systems of classification gave them a remarkable intellectual mastery of their surroundings. Sometimes, non-Aboriginal botanists or zoologists have successfully used knowledge gained from the Aborigines as the basis of their own investigations; in places like Arnhem Land the detailed knowledge that the Aborigines have of the regional flora remains ahead of that of modern science. Aboriginal natural knowledge is, however, embedded in a totemic religious system, an explanatory structure very different from the structure that developed in European science and one that has very different implications as far as the conceptualization and control of nature are concerned. Hiatt and Jones see a tendency for totems to figure as objects of contemplation and argue that, in its more contemplative aspects, Aboriginal totemism represents an attempt to epitomize the structure of the cosmos, based, as was Plato's philosophy, on a notion of archetypes or ideal forms.

It is doubtful whether the first European scientists visiting Australia, convinced as they were of the superiority of both their science and their religion, would have seen any merit at all in Aboriginal totemism, even if it had been accessible to them. In fact, cultural preconceptions and barriers of language and custom long made Aboriginal patterns of thought as inaccessible to those arriving from Europe as the European mode of thinking was to the Aborigines. As Miranda Hughes' account of the interchanges between a group of Tasmanian Aborigines and the French scientists who visited them briefly in 1802 makes clear, mutual incomprehension was the inevitable outcome even when, as in this case, the visitors had come with the best of intentions and explicit instructions to evaluate the native societies they encountered by their own internal standards rather than by European ones. As Hughes indicates, the difficulties of interpretation that the French confronted in studying the Tasmanians are unavoidable in anthropological research. All that one can do is recognize their existence and work to mitigate their effects. As Hughes' discussion of the rival accounts of the Tasmanians prepared by Baudin and Péron reveals, some investigators succeed in doing this more than others.

Hughes' account is valuable from another perspective, too. Histories of 'white' Australia inevitably emphasize the British roots and continuing British connections of the various colonies. The contributions of other European nations, though often significant, are all too frequently overlooked. As far as the scientific exploration of the continent was concerned, the French were particularly active; indeed, the various French sea-borne expeditions, of which the one led by Baudin was probably the most successful of all, achieved a great deal more than most of the rival British expeditions. The massive collections of Australian materials assembled on these voyages still survive in France and remain a major scientific as well as historical resource. Hughes' chapter serves to remind us that 'white' Australia was not an exclusively British preserve.

Sybil Jack's concern is not, however, with the assembling of collections of Australian materials for study by the scientists of Europe, but with the growth of scientific consciousness among the early white settlers in Australia. The structure of society and the conditions of life in the early colonial period were, she argues, generally unfavourable to science. The attitude of the governing classes was crucial, but for the most part the government did not encourage the pursuit of science. There was no substantial leisured class, some members of which might take up the subject, and neither were there sufficient professional men with scientific interests. Jack challenges the accepted, more positive assessments of early Australian science. Throughout the period to 1850, in her view, science made very little headway in establishing itself in Australia's white settler community.

The story is very different in the years after 1850, when colonial science rapidly increased in strength and international standing. Two chapters focusing on different but overlapping periods consider in detail the relationship between metropolis and province in nineteenth-century science. Robert A. Stafford describes the manifold lines of influence whereby the British geologist Sir Roderick Murchison promoted the geographical exploration, scientific investigation and settlement of the Australian continent for the greater glory of the British Empire. He shows at the same time how Murchison's defence of his own geological theories, developed in a quite different part of the world, long shaped the development of Australian geological understanding. Murchison emerges as one of the greatest of all patrons of imperial science, rivalling or even surpassing in influence Robert Owen at the British Museum and the Hookers, father and son, at Kew Gardens.

The Hookers play a major role in A.M. Lucas' account of the career of nineteenth-century Australia's greatest scientist, Ferdinand von Mueller. Lucas shows, on the one hand, how Mueller drew support and intellectual sustenance from the Hookers and other leading British botanists and men of influence and, on the other, how he himself dispensed patronage to collectors and others who assisted him throughout the Australian colonies. Mueller rendered thanks to his patrons and encouraged his proteges by naming species or geographical features in their honour. As an active explorer and, later, patron of exploration, and as the country's pre-eminent taxonomic botanist, he had every opportunity to dispense favours of this kind. Once himself elected to learned and scientific societies, he also had the opportunity, of which he made extensive use, to honour friends and acquaintances by nominating them, too, for membership. Not only, then, did the patronage system link metropolis to colony in nineteenth-century science, it also here played an important role in generating support for science within the colonial environment itself.

The chapter by Ian Inkster and Jan Todd is likewise concerned with the support for science in the Australian colonies in this period and, more generally, with the place of science in colonial society. Inkster and Todd highlight the 'utilitarian and localized profile' and 'pragmatic empirical emphasis' of most Australian science during the second half of the 19th century. Few Australian scientists, they suggest, took on the wider intellectual and cultural responsibilities of their European counterparts.

Inkster and Todd stress the importance for the scientific enterprise of larger movements in the economic and social structures of the various colonies. The boom of the 1850s and the accompanying rapid increase in population, especially in Victoria, the subsequent narrowing of the resulting economic gap between Victoria and New South Wales, the growing prosperity of South Australia, the boom of the 1880s and the depression and subsequent recovery of the 1890s all directly affected the level of colonial scientific activity. So too, they argue, did the infrastructure of cultural institutions, especially mechanics' institutes, that generated support for science within a broad spectrum of the colonial population. Within the institutions of science itself, it was not until the 1880s that the preoccupation of colonial scientific societies with 'pragmatic issues of development' began to give way to 'a more consistent offering of original scientific papers'. Finally, they argue on the basis of a number of case studies that in the period from the mid-1890s to around 1910, there was a significant change in the nature of Australian economic development, and that this was based on efficiency gains deriving from specific applications of scientific knowledge. Science came to be recognised at about this time, they suggest, as a potential asset rather than a mere commodity.

Two other chapters consider developments in Australian science in its colonial heyday in the second half of the 19th century. Barry W. Butcher focuses on a public dispute between the founding professor of Australia's first medical school, George Britton Halford, and the famous British naturalist, Thomas Henry Huxley. He shows that although this was ostensibly over certain points of detail in the comparative anatomy of monkeys, gorillas and Man, it was in reality part of a much larger 19th-century debate over Man's place in Nature. Abstruse questions of taxonomy are shown to be no mere matters of fact but to be inextricably bound up with the taxonomist's more general philosophical stance.

In the wake of the publication of Darwin's *Origin of Species*, the debate between Halford and Huxley and their respective supporters could perhaps have occurred anywhere in the then scientific world. However, Butcher shows how, in certain respects, its course was mediated by the colonial environment in which Halford was working. In particular, he points to the very different reactions of metropolis and province to Halford's contributions to the discussion. In the small colonial scientific community, the newly arrived professor achieved a great success by defending a position that had broad support from local social and intellectual élites. He did so, however, only at the expense of his scientific reputation internationally. Arguments developed for and applauded by a colonial audience were found seriously wanting when read by his scientific peers in Britain. Butcher does not speculate on why this should have been so, but the question is worth asking, nevertheless, especially when one recalls that almost all the leaders of Melbourne intellectual life at the time had themselves but recently arrived from Europe. There are important implications here for our understanding of the nature of scientific authority and of the way in which this is achieved.

George Bindon and David Philip Miller take up one of the questions raised in the chapter by Inkster and Todd, namely the conditions governing the successful application of science to industrial production in a peripheral economy such as that of late 19th-century Australia. They do so by considering the remarkable case of the growing influence of a cadre of chemists within the Colonial Sugar Refining Company (CSR), the Australian company that came to dominate the sugar industry in the south-west Pacific region during this period. The relative paucity in Australia in more recent times of industry-based or industry-financed research and development has become a matter of notoriety. Bindon and Miller show how, in the case of CSR in the late 19th century, an 'aggressive innovative strategy' based on the work of the company's chemists became an integral part of the company's approach and in the process transformed the economic circumstances not just of CSR but of the entire Australian sugar industry. The science involved was not in any sense fundamental research; indeed, much of the work was routine testing and analysis. What was significant was the way in which scientific techniques were appropriated in a systematic way to bring about major improvements in the efficiency of the production processes employed by the company. Chemists skilled in the requisite techniques were initially imported from Europe without the company being really aware at first of their potential. Later, they were trained within the company. Strikingly, there was almost no input from Australia's institutions of higher education or from other support structures of late 19th-century Australian science identified by Inkster and Todd.

This study by Bindon and Miller raises a number of questions of long-term significance about Australian science in its relations with industry. Was CSR unique in launching a successful programme of industrial research and innovation at this period? It certainly appears to have been. Why was this so? Why did other sectors of Australian industry then and subsequently fail to create equally innovative enterprises? And why was there, even in the case of CSR, so little interaction with the nation's universities and other institutions in which scientific knowledge and manpower were supposed to be created? There is a striking contrast here with what happened in Germany and the United States, in particular.

Finally, this chapter brings home the importance, in studying the history of science, of not confining one's attention to the 'high flyers' of scientific research. Especially in the case of nations on the scientific periphery where genuine high-flyers are relatively scarce, doing so has degenerated all too often into hagiography of the second-rate. The work of Bindon and Miller shows that important insights for the wider history of the nation can flow from adopting a less élitist stance.

The same may be said, in respect of a later period, of the chapter by Hugh Hamersley, which charts the rise during the 1920s and 1930s of new, medically orientated career opportunities for physicists in Australia. These arose in connection with the spread of radiotherapeutic methods for the treatment of cancer. Hamersley shows how the development of the new techniques led to a recognition that the services of specialist physicists were required to oversee standards of dosage and the physical measurements associated with treatment, and how, in the process, the Commonwealth was able to expand its influence in comparison with the states in this new area of public health responsibility.

By tracing the sorry history of the research undertaken under the aegis of the University of Sydney's Cancer Research Committee, Hamersley also highlights some of the difficulties confronting any attempt to establish a major scientific research programme in Australia at this period. The techniques of scientific research are not easily learned from books. They include a substantial element of craft knowledge that is best learned through apprenticeship to an established top-rank researcher. Unfortunately, none of the University of Sydney professors who unexpectedly found themselves responsible for administering huge sums of money contributed by the public for cancer research had had that kind of experience, which at the time could only be acquired by going overseas. The same problem would have arisen at most other Australian centres at that time. Though those involved doubtless did their best, they were out of their depth. Major commitments were made in regard to the investigations to be pursued where a more research-wise group would almost certainly have adopted a more sceptical and cautious approach; and, as the research programme expanded, there was no-one who could adequately supervise the work or provide the leadership that the mostly young and inexperienced researchers who had been engaged to do it required. By the time someone who had had experience of high-level research, the young physicist V.A. Bailey, joined the group, the major decisions had already been made. Bailey's doubts were over-ridden and, soon afterwards, he severed all links with the work. Within a few years, the entire research programme disintegrated, leaving almost nothing to show for the huge sum of money invested in it.

There is a striking contrast between this story and that of Melbourne's Walter and Eliza Hall Institute of Medical Research, which is a principal focus of the chapter by F.C. Courtice. During the same years in which Sydney's Cancer Research Committee floundered, the Hall Institute went from very modest beginnings, with vastly fewer funds at its disposal than were available in Sydney, to become a significant centre of research with a growing international reputation. The difference cries out for explanation, which seems to lie in the different calibre of scientific leadership in the two cases. Whereas in Sydney the research programme was directed by a committee, no member of which had worked for any length of time in an active research environment, at the Hall Institute the direction was for many years in the hands of a single, strong individual, C.H. Kellaway, who had himself worked in several of the world's leading research centres in his field. Kellaway exercised direct personal control over the Institute's research programme, he recruited some outstanding young scientists, including F.M. (later Sir Macfarlane) Burnet, and he made sure that in due course they too had an opportunity to gain experience in leading laboratories in Britain. Under Kellaway's leadership and later that of Burnet, the Hall Institute went from strength to strength, culminating in the award of a Nobel Prize to Burnet in 1960 for his work, done at the Institute, on acquired immunological tolerance.

The rise of the Hall Institute is, however, but part of the wider theme of Courtice's chapter, which charts the growth of Australian medical research from the days of J.T. Wilson and his school in Sydney in the 1890s to the 1960s. During that time, Courtice argues, Australian medical science achieved national independence. In earlier days, as he shows, Australian medical scientists were crucially dependent on their links with leading researchers in England for both their advanced training and, more generally, to help them maintain contact with the leading edge of research in their respective fields. Wilson himself and most of his group came from Britain and all eventually returned there. One of the group, C.J. Martin, in due course became a key figure linking Australian and British workers and research institutions. Later, Kellaway's teacher, Sir Henry Dale, played a similar role. Independence came in the years after the Second World War with the expansion of a number of previously small institutes, including the Hall Institute, and the foundation of the John Curtin School of Medical Research at the ANU. Now, for the first time, Australia could provide sustained opportunities for front-rank research in several fields of medical science. Instead of promising young Australian medical scientists having to go overseas to learn the most advanced techniques, Australian institutions began attracting outstanding young researchers from other parts of the world. Burnet's Nobel Prize and that awarded to J.C. Eccles three years later for his work on the transmission of nerve impulses not only rewarded outstanding individual achievements but also provided public recognition of Australia's new-found standing in this field.

The chapter by R.L. Burt and W.T. Williams charts the rise to maturity of a very different category of Australian scientific work, dealing with the controlled introduction of new plant species into Australia. Their concern, however, is not so much the emergence of a self-sustaining high level of Australian research in this field — though this is implicit in the story they tell — as the application of increasingly scientific methods to the age-old question of the transfer of plant and animal species between different parts of the globe. In particular, they contrast the more-or-less undisciplined importation of new plant species into Australia in the period prior to 1930 with the increasingly systematic effort made thereafter to identify, and then establish in Australia, useful and much-needed cultivars that did not at the same time threaten to become pests in the way that, earlier, blackberry and prickly pear had done.

Burt and Williams see the formation in the late 1920s of a Plant Introduction Section as one of the original units within CSIR's Division of Economic Botany (soon re-named Plant Industry) as an important step, which for the first time provided some co-ordination of Australian activities in this area. However, it is to the years immediately after the Second World War that they date the major transformation of the field. Previously, species had been accepted more or less randomly and their utility then investigated. Now, prior ecological and soil surveys enabled specific requirements to be laid down for new plant types, which were then sought out on plant-hunting expeditions sent in increasing numbers to other parts of the world. Detailed classificatory and recording schemes essential to such work were developed. More recently, Australian scientists have participated in the establishment of genebanks for various cultivated plants. Though the ownership of genetic resources has in recent years become a matter of considerable international sensitivity, Australia's record, Burt and Williams argue, of making germplasm freely available from its collections, has been good.

This is not the only chapter in which the Second World War appears as a major turning point in the history of science in Australia. The chapter, 'Science on service, 1939–1945', focuses on the war years themselves and the impact of the war on Australian science, and points, in particular, to a shift in the balance of the nation's scientific power that occurred at this period. Prior to the war, the agricultural sciences had dominated the Australian scientific scene, most obviously in the emphasis given to them within CSIR. The growth of Australian manufacturing industry that began in the 1930s and accelerated dramatically during the war led to a great expansion of activity in the physical sciences, even as agriculturally orientated research was held temporarily in check. After the war, the situation did not revert to its pre-war state; on the contrary, the major shift that had occurred in the distribution of Australia's scientific resources was maintained in the peace.

During the war, it is argued, much of the work that was done by Australia's scientists was not front-rank research but was of a fairly routine problem-solving kind associated with the build-up, under war-induced 'hot-house' conditions, of the nation's manufacturing capacity. In many cases, the scientists in their laboratories themselves became directly involved in production. The numbers of people engaged in scientific work expanded rapidly. Young graduates (or even undergraduates) were pressed into the work without prior training in research, on the assumption that this could be provided later. Large new physical-science laboratories mushroomed, especially within CSIR, devoted for the time being to essential war-related work but in advance of any clear idea of how they would relate to the needs of Australian industry once peace returned. Hence, on the one hand, the war led to a rapid rise in Australia's level of scientific and industrial sophistication, but on the other, it both created short-term problems that had to be resolved once it was over and opened up major new issues of long-term science policy.

The chapter by Woodruff T. Sullivan, III, describes in detail how one of the biggest new CSIR divisions, the Radiophysics Laboratory, made the transformation to peace-time conditions and, in the process, became a world leader in a brand-new field of scientific research, radio astronomy. Sullivan surveys the Laboratory's exciting achievements in the first few years of radio astronomical research and provides an evocative picture of the way the work of the radio astronomy group was organized by its leader, J.L. Pawsey, with the support of the Chief of the Laboratory, E.G. Bowen.

The remarkable and continuing success of the Australian radio astronomers during this period raises in a particularly striking way the question of the conditions under which science in a 'new' country such as Australia can come to achieve parity with the established science of the Old World. For in this case Australia unquestionably did achieve parity — indeed, it probably did better than that. Sullivan points to certain structural features of post-war Australian science that led to the large group of radio physicists and engineers that had been brought together for the war-time radar project staying together afterwards rather than dispersing, as happened in the United States and Britain. As a consequence, the Australian radio astronomy group was for some time considerably larger than any other in this field. The old problem of the isolation of the élite, of the outstanding individual scientist remote from his peers, thus not a problem here. Also, many technicians and a lot of the most advanced electronic equipment of the day were left over from the radar work.

Isolation was still, however, a consideration. Sullivan describes in some detail the measures that Pawsey and Bowen took to ameliorate the problem and to make the Australian work known. At the same time, they had enough confidence in what they were doing to establish a policy of publishing most of the group's results in Australian journals rather than the international scientific journals traditionally used by aspiring scientists on the periphery. Interestingly, Macfarlane Burnet adopted a similar publication policy, at about the same time.[12] For the first time, Australian scientists had attained a sufficient degree of authority in their respective fields to make this declaration of scientific independence.

The chapter by S.C.B. Gascoigne is another story of Australian science attaining independence and maturity. Gascoigne extends the story of Australian radio astronomy beyond the point, in the early 1950s, where Sullivan leaves off; but he does so within the context of a broad survey of Australian astronomy as a whole in the post-war period.

In optical as well as in radio astronomy, the Second World War emerges as a major turning point; and in this field, too, though somewhat more slowly than in the case of radio astronomy, Australian work came to rank consistently with the best in the world. The key was the acquisition of suitable instruments, namely the Mount Stromlo Observatory's 74-inch reflecting telescope and, later, the 150-inch Anglo-Australian Telescope constructed on Siding Spring Mountain, NSW. Radio astronomy, too, gradually became more and more the preserve of the large instrument, and the Parkes radio telescope in particular. As the scale of the instrumentation increased, the telescopes concerned came to be treated as national facilities, open to any researcher who could present a suitable proposal. The whole chapter vividly conveys the excitement of contemporary astronomy and the important role that Australian workers have come to play in this field.

The Anglo-Australian Telescope operates under the control of the Anglo-Australian Telescope Board, one of many new scientific institutions established in Australia in the post-war era to oversee the expenditure of Commonwealth Government funds on scientific research. Prior to the Second World War and for a number of years after it, CSIR/O was the locus for virtually all Commonwealth scientific activity outside the Department of Defence and the rather specialised interests of the Department of Health. When the government needed advice on scientific matters, it naturally turned to its Council for Scientific and Industrial Research, this being one of the purposes for which the Council had been set up in the first place. With the proliferation of Commonwealth scientific agencies in the post-war period and also the increasing commitment of the Commonwealth to supporting scientific research in the universities, CSIRO came to be seen in a very different light, as merely the largest of the government agencies competing for the limited public funds available for science. Its privileged position as government scientific adviser came under increasing challenge from scientists outside the organization who pressed for the formation of a more broadly representative Science Advisory Council.

The final chapter in this volume, by Ron Johnston and Jean Buckley, describes the evolution of Australia's science policy system in response to this growth and profileration of institutions. A further factor they identify is an increasing public scepticism during the 1960s and 1970s about the benefits supposed to flow from untrammelled scientific research. They see a steady expansion of bureaucratic control over science from about 1965 onwards, leading to a seemingly endless series of inquiries, the formation of a Commonwealth Department of Science — the demise of which in July 1987 came too late to be considered by them — and eventually, in 1979, the passage of legislation establishing the Australian Science and Technology Council (ASTEC), which had already existed in interim form for several years.

Contrary, however, to the hopes of those who had first pressed, over 20 years before, for the formation of a body like ASTEC, neither it nor the other elements of science policy bureaucracy that have been established have served to represent the interests of scientists to government. Together with the massive growth of government expenditure on research that has occurred in the post-war years has come an increasing desire on the part of government to control and direct more of that expenditure into commercially exploitable fields of research. Here, it is felt, in the development of new, science-based technologies, lies the key to the restructuring of the Australian economy.

On all fronts, Australian science has gained remarkably in strength since 1945 and has on any reasonable criterion at last established its independence. Many individual Australian scientists do outstanding work that ranks at least on a par with that of the leading scientific nations. Unfortunately, Australia's industrial base has not developed at the same pace. Australia is still producing far too few technically trained people, and far too many of the nation's managerial class are woefully ignorant of matters scientific. These problems, rather than any relating to science itself, are likely to restrict the role of science in Australia's short-term economic future. If they can be resolved, the range of technical expertise and of expertise in working at the frontiers of knowledge that is now available within the Australian scientific community will be found to be a priceless national resource.

* * *

This volume had its origins in the deliberations of the Bicentennial History of Science Committee established by the Australian Academy of Science in 1981 with the enthusiastic support of the then President of the Academy, Dr L.T. Evans, FAA, FRS. It has profited greatly from the support and encouragement of members of the committee and especially the committee's chairman, Professor J.M. Swan, FAA. Other members of the committee were H.C. Bolton, L.A. Farrall, R.W. Home, J.W. McCarty, C.B. Schedvin (until September 1982) and D.F. Waterhouse, FAA, FRS. From the outset, it was hoped that the project would lead not just to the production of a worthwhile book but more generally to a heightened interest in studying the history of science in Australia. To that end, two highly successful conferences were held at the Australian Academy of Science, one in August 1982 with an organizing committee chaired by H.C. Bolton, the other in February 1985, organized by Ian Inkster and David Philip Miller. Financial support for the first of these was provided by the James Kirby Foundation. In addition, a *History of*

Australian Science Newsletter was launched with financial support from the Academy, and a general invitation was issued to scholars at home and abroad to submit manuscripts with a view to their being included in the volume. These manuscripts were then subjected to formal refereeing. In the event, the task of selecting contributions for inclusion in the volume proved a difficult one and several valuable essays had to be turned away.

Vital secretarial support for the project has been provided by staff of the Department of History and Philosophy of Science at the University of Melbourne, especially by Lynne Padgham and Renae Stoneham. Valuable assistance has also been provided from time to time by officers of the Australian Academy of Science, in particular Peter Vallee and Rosanne Clayton. A large number of busy people who must remain anonymous gave freely of their services as referees of the various papers submitted. To all of these, named and unnamed, to Robin Derricourt and Marjorie Pressley of Cambridge University Press, and above all to John Swan, I record my grateful thanks.

Notes

1   Donald Fleming, 'Science in Australia, Canada, and the United States: Some comparative remarks', *Proceedings of the 10th International Congress of the History of Science, Ithaca, 1962*(Paris, 1964), 179-96; p. 182.
2   George Basalla, 'The spread of western science', *Science*, 156 (1967), 611-22.
3   For example, Roy MacLeod, 'On visiting the "moving metropolis": Reflections on the architecture of imperial science'. *Historical Records of Australian Science*, 5(3) (1982), 1-16; Ian Inkster, 'Scientific enterprise and the colonial "model": Observations on Australian experience in historical context', *Social Studies of Science*,15 (1985), 677-704; and several of the essays in Nathan Reingold and Marc Rothenberg, eds, *Scientific Colonialism: A Cross-Cultural Comparison* (Washington, D.C., 1987).
4   Leichhardt to Gaetano Durando, 20 May and 27 September 1846; *The Letters of F.W. Ludwig Leichhardt*, ed. M. Aurousseau (London, 1968), III, 868-72 and 906-8. Also Sally Gregory Kohlstedt, 'Natural heritage: Securing Australian materials in 19th-century museums', *Museums Australia*, (December 1984), 15-22.
5   F. Baily to J.F.W. Herschel, 18 January 1836; Royal Society of London, Herschel correspondence, vol. 3, f. 131.
6   Kathleen G. Dugan, 'The zoological exploration of the Australian region and its impact on biological theory', in Nathan Reingold and Marc Rothenberg, eds, *Scientific Colonialism*(op. cit., n.3), pp. 79-100; esp. pp. 88-95.
7   R.W. Home, 'The problem of intellectual isolation in scientific life; W.H. Bragg and the Australian scientific community, 1886-1909', *Historical Records of Australian Science*, 6(1) (1984), 19-30.
8   Cf. Barry W. Butcher, 'Science and the imperial vision: the Imperial Geophysical Experimental Survey, 1928-1930', *Historical Records of Australian Science*, 6(1) (1984), 31-43.
9   G. Currie and J. Graham, *The Origins of CSIRO: Science and the Commonwealth Government, 1901–1926* (Melbourne, 1966); C.B. Schedvin, 'Environment, economy, and Australian biology, 1890-1939', *Historical Studies*, 21 (1984), 11-28, reprinted in Nathan Reingold and Marc Rothenberg, eds, *Scientific Colonialism* (op.cit., n.3), chap. 5.
10   *Historical Directory of Council for Scientific and Industrial Research and Commonwealth Scientific and Industrial Research Organization, 1926–1976* (Melbourne, 1978).
11   Ann Moyal, *Clear across Australia: A History of Telecommunications* (Melbourne, 1984).
12   Frank Fenner, 'Frank Macfarlane Burnet, 1899-1985', *Historical Records of Australian Science*, 7(1) (1987), 39-77.

PART I:
Early days

Aboriginal conceptions of the workings of nature

L.R. Hiatt and Rhys Jones

PART I

> These cosmogonies belong to the ancient world — a world peopled so sparsely that nature was not as yet overshadowed by man . . . Nature hit you in the eye so plainly and grabbed you so fiercely by the scruff of the neck that perhaps it really was still full of gods . . . This ancient world ended with Rome, overpopulation put a stop to it.
>
> Boris Pasternak, *Doctor Zhivago*

The distant ancestors of the Aborigines colonized the Australian continent at least 40 000 years ago. To do this required the crossing of oceanic water barriers between 60 and 100 kilometres wide even at times of glacial low sea levels.[1] The colonization of Australia constitutes the oldest firm evidence we have of substantial watercraft and of the human capacity to cross the sea. Recent archaeological field research in New Ireland has shown occupation in a cave there with a basal date of 32 000 years, showing that two other water barriers, namely the Vitiaz Strait between New Guinea and New Britain and the St George's Channel between the latter and New Ireland had also been crossed during this primary colonizing process.[2] Modern man was already established out on the largest islands of the western Pacific at the same time that his relatives, the Cro Magnon population, were rapidly replacing Neanderthalers in Europe. Indeed the two events are probably linked, both being manifestations of a fundamental change in human history from about 40-50 000 years onwards associated with the appearance and rapid geographic spread of biologically modern humans.[3] New Guinea and Australia formed parts of the same land mass. The flat and ancient Australian plate with its laterized and deeply weathered soils extended to the foot of the New Guinea cordillera, the latter formed of recent, mineral-rich volcanics and under a regime of nearly constant rainfall. Later, during the height of the last ice age from 24 000 to 12 000 years ago, the floor of Bass Strait was also exposed, with a dry land bridge to the glaciated highlands of Tasmania.

The first colonists, coming from the coastal swamps of South East Asia, would have found the tropical Australian–New Guinea shores familiar to them in several essential elements as regards their subsistence. The molluscan fauna of the two regions share most genera, as do the fish. Important food plants such as the tubers of yams (*Dioscorea*), taro (*Colocasia*), and arrowroot (*Tacca*); the seeds of rice (*Oryza*) and cycads; and the fruits of such trees as *Terminalia, Canarium, Syzygium* and various palms are found both in tropical Australia/New Guinea and in South East Asia.[4] However, the land fauna of the new continent would have been totally different, with its marsupials, monotremes and reptiles. Also, people extending their range southwards into the heart of the Australian continent would have to cope with an increasingly alien flora. Such trees as Eucalypt, Casuarina and Acacia shaded grasses with edible seeds, such as *Panicum* or *Nardoo*, but the technology to grind and to utilize these had to be learnt.

Research over the past twenty-five years has revolutionized our knowledge of the prehistory of Australia. In 1961, there was but one securely dated piece of evidence of human presence at 10 000 years ago. Now we can demonstrate that occupation had occurred throughout every major ecological zone of the continent at least by about 25 000 years ago. These areas ranged from the highland valleys of New Guinea by 30 000 years ago; the tropical lowlands of both New Guinea and the savanna plains of northern Australia by 30 000 years at least; the rivers and lakes of western NSW back to 40 000 years; the south-west tip of the continent by the same date and even the very heartland arid core in the mountain ranges west of Alice Springs by 20 000 years ago. In Tasmania, the glacial low-sea-level bridge gave access to people who promptly took it, leaving evidence of their occupation at about 20 000 years ago even in caves overshadowed by the valley glaciers of the extreme south-west.

This occupation of all of the major ecological zones of the continent implies that the Aboriginal ancestors had mastered the key problems for subsistence. These centred on the utilization of plants and the hunting or capturing of animals. In addition there would have been the need to organize these activities on a seasonal basis, and to have systems for the alleviation of the effects of droughts or other natural stresses. To hunt, to dig and to carry required the manufacture of artefacts of wood and of fibre. To cut these required tools of stone, the suitable materials for which were usually restricted to isolated outcrops. The first colonists used fire both as the essential element of the hearth — as warmth, to cook, and as the focus of domestic society — and also as a major tool to transform the environment through firing the country in a systematic way in hunting or to clear the ground for easier travel. The early phase of human colonization saw the rapid extinction of more than a third of the existing land fauna, in particular the so-called 'giant marsupials' including the rhino-sized Diprotodontids, Procoptodons and Protemnodons and larger cousins of still existing species. All of this evidence shows that, even by 25–30 000 years ago, the essential economic adaptations to the Australian continent had been made. We can assume that these people had a detailed and sophisticated knowledge of the properties and behaviour of plants and animals, and they knew where to get specialized resources such as fine stone or woods from which to manufacture their technology.

Direct archaeological evidence

On the shores of the now extinct Lake Mungo, at 30 000 years ago, people, probably women, dug fresh-water mussels (*Velesunio ambiguous*) from the mud and fished for golden perch (*Plectroplites*), using either nets or movable woven barriers.[5] The integration of these lake-edge foods with those of the interior scrubs, such as small marsupials and emu eggs, indicates a planned economic system. In some highland New Guinea cave sites, now-extinct large marsupials such as Protemnodon were hunted.[6] In the extreme south-west of the continent at the cave of Devil's Lair, the animals killed were kangaroos and wallabies, their charred and smashed bones discarded on the edges of camp hearths.[7] Equally, in the valleys of south-west Tasmania, overlooked by glaciated peaks, red-necked wallabies were hunted in large numbers and their carcasses brought back to snug limestone caves to be cooked and eaten.[8] Here also there is evidence for the utilization of bracken fern, microscopic pieces still adhering to stone tools.

At such antiquity, the main surviving artefacts are those of stone, and archaeologists, while seeming to be obsessed by such tools, nevertheless can deduce remarkably accurate scenarios as to their prehistoric usage. To cut wood, meat and fibre, sharp flakes need to be produced from suitable rocks. These must be capable of conchoidal fracture, which will result in the formation of sharp flakes. Such rocks in Australia include silicified quartzites, as well as finer chalcedonies and cherts. In the Mungo site mentioned above, the stone tools had been manufactured from silcretes obtained some 30 kilometres from the main habitation sites. In caves on the escarpment edge of western Arnhem Land, the oldest tools, dating back to about 30 000 years, were also made from fine quality highly silicified quartzites, obtainable only in a few places on the exposed cliff faces; and also pink cherts probably located in a small area (Barramundi Creek), up to 50 kilometres from the sites where they were discarded.

There were also more exotic raw materials, which, while being of minor importance in the total tool kit, nevertheless indicated that their makers appreciated their superior qualities and were prepared to go to highly localized quarries or to obtain them through trade. Late Pleistocene assemblages along the coast of Western Australia contain numerous artefacts made from an Eocene fossiliferous chert.[9] The distribution of these artefacts extended over an area of 40 000 square kilometres, and in time from about 35 000 years ago to 5 000 years ago. Then they stopped being used. The explanation is that the source itself was located west of the present coastline and was drowned in the late and post-glacial rise of sea level. The actual deposits of chert have been deduced geologically from the archaeological data.

In south-western Tasmania, from about 17 000 to 14 000 years ago, a few flakes and tools were made from a high-quality true glass. This is now called Darwin glass and was originally formed as an 'impactite' when a huge meteor hit the earth and ploughed a crater one kilometre wide. Such was the dissipation of the energy of impact that parts of the surface rocks were melted and the resulting glass blown outwards to the west in a shock wave. These pieces of glass were discovered by the Aboriginal people of western Tasmania during the last ice age. With the warming of the post-glacial climate, and the flooding of Bass Strait, this region became abandoned by people and eventually clothed in dense rain forest. Darwin glass was first discovered by modern science in 1908, in the form of little pieces scattered on the eastern faces of the West Coast Range, found through mining activities. The actual Darwin crater was found only in 1974, perhaps 17 000 years after its original discovery.[10]

In coastal sites in north-western Tasmania, a new raw material made its appearance some 1500–2000 years ago. This was a silicified spiculite, that is, rock of organic origin from the silica spicules of prehistoric sponges. The source was discovered in December 1985 by bulldozer drivers cutting the survey tracks for a proposed road. The find was confirmed by archaeologists a few months later.[11] However, the Aborigines had found this stone in the depths of rain forest, 2000 years earlier, and had exported it to most of their domestic sites of the region.

Although most artefacts from the distant past are made from stone, in special circumstances artefacts of bone, wood or even fibre can be preserved. A spectacular find of this sort was made in a swamp, called Wyrie, in the south-eastern coast of South Australia. Here, during peat-digging operations, several wooden artefacts were recovered, and subsequent excavations showed them to be 10 000 years old.[12] Tools included finely carved barbed wooden harpoon heads from hard *Casuarina stricta* wood, and several long sections of boomerangs. These had fully carved aerofoil sections, some with a slight twist of angle between one arm and the other. These were totally sophisticated returning boomerangs and represent the oldest evidence in the world for a practical application of knowledge of aerodynamics. Probably these were designed to imitate hawks, in order to get ducks to rise into nets set at the outlet points from the swamps.

Mankind is not to be measured only by his belly. The cremation ceremonies of the young girl at Mungo, about 26 000 years ago, represent the oldest evidence for cremation in the world. The Mungo 3 human was laid in a grave scattered with red ochre. In the Lindner Site, in Deaf Adder Gorge (Northern Territory), large pieces of industrial-quality haematite were found, with clearly defined grinding facets, and dated to between 25–30 000 years ago.[13] On the walls of the Arnhem Land cliffs are some motifs in red ochre actually bonded to the rock. Mineralogical examination shows the skin to be a complex silicate mineral.[14] Ion-probe analysis indicated this to consist in some cases of a complex polyhalite, perhaps formed under the colder drier conditions of the last ice age, with salt-laden winds, blowing across the exposed Arafura Plains, and bonding the art on the Arnhem rocks. The motifs show men and women hunting animals, some now extinct; all are bonded beneath a silica sheen.

The concept of the taxon

To understand how Aborigines classified the natural world, it is necessary to combine a working knowledge of natural sciences with skills in linguistics and social anthropology. The enterprise is a co-operative effort between research workers and Aborigines. On the Aboriginal side, some people are especially interested in their own knowledge of the world and how to transmit this to another cultural milieu. Some of the most systematic ethno-biological work carried out in recent years has been in Arnhem Land. Contemporary Arnhem Landers have a high retention of their traditional knowledge and are fluent in their own languages. Some older people are not able to speak English. Many groups still hunt and gather for their food and manufacture many elements of their technology from natural sources.

Let us take as an example the Gidjingarli-speaking people of the Blyth River. In general, Gidjingarli names for plants and animals in their environment correspond to scientific species. Thus there are different names for jabiru, brolga, magpie goose, burdekin duck, grass whistle duck, black duck, pelican, and so on. All these birds form part of the diet. However, many small birds, especially those seen fleetingly in rain forest patches, are given a collective term *badaitja*, the best translation being 'small bird'. *Badaitja* have no economic value. However, some birds that are not part of the normal diet do have proper names. These birds play important roles in totemic song cycles; their habits, or characteristics such as distinctive plumage or song, are depicted on bark or body paintings, or enacted in dance and song.

For the economically most important animals, the Gidjingarli subdivide within the species. Thus in the case of both the agile wallaby (*M. agilis*) and the kangaroo (*M. antilopinus*), the male and female are given separate names. The Gidjingarli know perfectly well that they are of the same biological species. In the case of the barramundi (*Lates calcarifer*), different stages in its life-cycle are given different names. The silvery coloured young fish moving in from the sea at the beginning of the wet season are called *anamutjala*, whereas the older, larger and dark coloured fish speared or trapped on the estuarine wetlands at the end of the wet season are called *djanambal*.[15] Again the Gidjingarli realize they are dealing with the same species; indeed, their knowledge of the life-cycles of this and other estuarine/wetland fish is profound and accurate.

The most systematic investigation of the relationship between Aboriginal taxa and Linnean species was done by Meehan,[16] using the Gidjingarli classification of mollusca as her case study. All molluscs collected, including those from storm beach debris, were identified as belonging to a total of 106 formal Linnean species. Of these, ninety-seven were assigned names by the Gidjingarli who placed them into fifty-four Gidjingarli taxa. In twenty-five cases, a taxon corresponded to a species; in fifteen cases to two species; in eleven cases to three species. There was a general taxon *lugaluga* into which eleven species were lumped — these were all non-edible species that formed part of the beach debris and had been blown up from deep water as dead shells. The relationship between taxa and species can be illustrated by reference to the Arcidae in the table on page 6.

Similar tables can be constructed for the Veneridae and Mytilidae families.[17] In one case a single species, the oyster *Crassostrea amasa,* was assigned to two distinct taxa on the basis of ecological habitats. Oysters that grow on the roots of mangroves are called *an-guldjaraba,* whereas those on ocean-fronting rocky outcrops are *waiyanaka.* The favourite shell fish to eat, and also the one with the greatest role in songs, is the bivalve *Tapes hiantina,* called *diyama.* This is subdivided by the Gidjingarli into two types based on different patterns of decorations on their shells and named after different ducks, the shell patterns reminding the Gidjingarli of plumage patterns.

At a higher level than species, there are some terms referring to classes or sub-classes of animals. Thus fish are all grouped under the general term *djidjidja,* and there is another term for all turtles and tortoises. Among the mollusca and crustacea there are two terms describing their flesh. One is *kaparra* and refers to the flesh of bivalves that have habitats on the open sea

Relationships between nine species of Arcidae and one species of
Cucullaeidae and five Gidjingarli shellfish taxa[18]

Family and species	Gidjingarli taxa				
	ngandipurda	*an-galidj-awurrigiya*	*gunagulumba*	*gumunka*	*gunamil-amilawa*
Arcidae					
Anadara aliena	x				
Anadara crebricostata	x				
Anadara desparilis	x				
Anadara gubernaculum	x				
Anadara granosa		x			
Mesocibota luana	x				
Arca imbricata			x		
Arca multivillosa			x	x	
Trisidos yongei					x
Cucullaeidae					
Cucullaea labiata	x				

shore. The other is *ngarl* (literally meaning 'tongue'), which refers to the flesh
of many animals including bivalves from mangrove or fresh water habitats, all
gastropods, crustacea such as crabs and prawns, and some insects such as
witchetty grubs. The Gidjingarli make no distinction between birds and
mammals, referring to the flesh of both as *mindjak*, meaning meat; though
they do have a subdivision into 'bloody' (red) and 'clean' (white) meats. Thus,
unlike medieval Europeans, they correctly classify sea mammals, such as
dolphins and dugongs, into the same group as the land mammals on the basis
that both are *mindjak*.

The Gidjingarli language assigns all objects in the universe to one or other of
four noun classes, signified by the prefixes *an, djin, man,* and *gun*. Although
the classification is not rigorous, there is a tendency towards an association
between *an* and masculinity, *djin* and femininity, *man* and edible plants, and
gun and non-edible matter. Almost all animals, including fish and shell fish are
classified as either *an* or *djin*. While most edible plants are in the *man* group,
some are either *an* or *djin*. Many trees are in the *gun* group, as are such things
as timber or sticks and branches. However, within this classificatory system
there are some interesting anomalies and nuances. For example, a few
molluscs are in the *gun* group because of their habitat on rocks at low or
subtidal levels or inside rotting tree trunks.

The Gidjingarli see each taxon as an immutable entity, enshrined in the
totemic religious system. Many are said to 'have a song', that is, they form part
of the great song cycles, which relate them to the other elements of the panoply
— land and people. Some taxa have secret ('inside') names used only in
ceremonies and usually with restricted access as to their true meaning. This
contrast between inner, religious and esoteric on the one hand and outer,
secular and practical on the other is a common feature of Gidjingarli thought
and practice.

Plants

In recent years, research workers have drawn up plant lists for various areas of
Arnhem Land and Cape York, and through the medium of several different

Aboriginal languages.[19] Among the Gidjingarli, Meehan and Jones collated a list of some seventy-five plant taxa, each related to a plant with some useful function either as a food or a raw material, growing in that environment. A similarly sized list, referring to essentially the same species, was made by Altman from Gunwinggu informants living inland some 50 kilometres south-west of the Gidjingarli. Both these environments consist basically of savanna woodland and estuarine/riverine communities with small patches of vine thicket.

Along the escarpment edges of Kakadu, another 100 kilometres to the west, are extensive pockets of floristically complex monsoon forests, which contain a great diversity and number of plant species. The inhabitants speak the Mayiali language and have a corresponding richness of taxonomic vocabulary to describe their environment. In a study carried out recently by the botanist Russell-Smith,[20] four elderly Aboriginal people were able to identify no less than 420 plant species and assign names to them. For each species, there was a detailed knowledge of its season of flowering, its habitat, whether or not parts were edible or poisonous and what could be made from it as a raw material. In several cases, the species have not yet been scientifically described. It must be remembered that botanical knowledge in this region is still at the pioneering stage, with even one of the dominant trees, *Allosyncarpia ternata,* being formally described only in 1979. Yet the Mayiali word for it, *anbinig,* has presumably been in use for centuries at the very least.

There has been a tradition for pioneering botanists to work with Aborigines on the classification and properties of plants ever since the days of Robert Brown; and this process still continues, both at the level of systematics and in the growing field of nutritional studies. Through experimentation over a long period, Aborigines have learnt many of the nutritional and pharmaceutical properties of plants — which ones have poisons to be careful of or to be removed. Systematic nutritional analyses of these food plants result in the documentation of their calorific values, protein content, minerals, vitamins and so on. In some cases there have been surprising results such as the high calorific content of *Pandanus* nuts[21] and the high vitamin C content of *Terminalia ferdinandiana.*[22]

Most edible tubers in northern Australia taste slightly bitter when raw, and some are poisonous in that state. The Gidjingarli describe tubers that are fit to eat, or that are 'sweet', as *man-bala;* those that are potentially edible but bitter as *man-baitjarra;* and those that are poisonous as *man-erra.* The same suffixes can be used to describe a good or calm man as *an-bala,* a fierce one with a strong temper as *an-baitjarra* and an evil man as *an-erra.* Bitter tubers are cooked in earth ovens, which breaks down the toxic compounds (usually oxalates) and makes the vegetable taste 'sweet'. The round or 'cheeky' yam, *Dioscorea bulbifera,* is diced using a hole in the outer whorl of a land snail, soaked and then baked.[23]

A major carbohydrate food source is a bread made from nuts of *Cycas media.* This is the one truly toxic plant utilized by northern Aborigines. In its raw state it has a poison that affects the central nervous system and causes severe vomiting. It also contains cycasin, a natural carcinogen of great potency used in modern cancer research. To remove these toxins, the nuts are de-husked and left to dry in the sun, ground into a paste, fermented in fresh water ponds and baked. Analysis shows that this removes all of the cycasin.[24]

Many plants were used for their real or imagined pharmacological properties. Some were used internally as infusions for the treatment of gastric complaints and diarrhoea, others as poultices for wounds (e.g. from fighting or from cat-fish spines). Linaments were prepared to be rubbed on the skin to alleviate respiratory troubles, arthritis or burns.[25]

In north-eastern Arnhem Land some fifty-seven medicinal species have been recorded. Almost 200 years ago, a young medical student, François Péron, with the Baudin expedition to Australia and Tasmania, recommended that special notice be taken of the plants used medicinally by the various native people to be encountered, who 'could again show us substances no less precious than quinine; nearer than us to nature, forced by the need to alert all their senses, is it surprising that they have been better served by instinct than we have by our methods, however scholarly they may be?'[26] It is fair to say that pharmacological research on the active ingredients in Australian plants used medically by Aborigines is still in its infancy.

Landscapes and the seasonal cycle

Arnhem Landers have an integrated view of their environment, seeing associations of plants, animals and soil types within the context of a changing seasonal cycle. There are six main seasons,[27] characterized by temperature, rainfall and the direction of the main trade winds, grouped around the concepts of the full wet season, the early dry and late dry.

The Gidjingarli define their environment in terms of named entities, which are similar to the land systems of the CSIRO classification.[28] Thus the *malpi* is the open woodland, dominated by the stringy barks *E. tetrodonta* and *E. miniata,* with understories containing the edible palm *Livistonia humilis* and cycads. Central to each concept are the various foods contained within it — honey from tree trunks, kangaroos, the various nuts and also seasonal markers, such as flowering plants, that indicate the readiness of various foods. Other land units include the *kapal* or estuarine black-soil plains, the *djaranga,* inland fossil Quaternary dunes with their fringing groves of vine thickets, the *madua* or sea beach, and so on.

Over the seasonal cycle, people locate themselves in different places at different times of the year according to a regular pattern. Thus, during the full wet season, people are pinned down on their coast beaches, with the grass tall and green and much of the inland flood plains under water. Then, with the coming of the dry season, more excursions are made inland to the vine thickets to get yams until eventually the plains begin to dry out and water becomes restricted to large swamps with their abundant supplies of swamp plants and magpie geese. Finally, the late dry season is often a period of stress with little water, plant foods hard to find and most animals thin. The budding clouds along the north-west horizon herald the coming monsoon and, as if to mark it, several trees such as *Syzygium* and *Vitex* put out succulent fruits.

The seasonal cycle described here in secular terms is also perceived by the Gidjingarli in a religious context. The winds are the manifestations of the great creation ancestors. The driving force for human actions is to carry out the crucial ceremonies at the right places and at the right times. During the stress periods of the full wet season, rules of what to eat, how and where to cook food and other prescribed behaviour are strictly adhered to, lest the Dreaming Serpent, asleep in the bends of the river, becomes angry and sends a devastating cyclone to destroy the human transgressors.

Stones and colours

The Gidjingarli have but two true colour terms, namely *gun-gungaltja* meaning light and also bright scarlet; and *gun-gungundja* meaning dark. All the world may be divided between these two terms.[29] They are important elements in classification systems, distinguishing species or life-cycle states of fish, different cloud states, fruits, and the races of man. In addition there are four pigments, namely red ochre, yellow ochre, black charcoal dust, and white pipe-clay. The terms for these may be used to describe other objects, using the device of saying that something has been 'painted' by the relevant pigment. Thus a bird may be described as being 'painted' with red ochre on its chest to distinguish it from another.

In geological terms, 'light' sand dunes are those associated with present-day coastal processes, whereas 'dark' ones are fossil, with well developed soil profiles. Features such as prior streams, or a fossil erosion gully that has cut across the grain of parallel chenier dunes, or a stone outcrop that seems out of place, are all keenly noted. Invariably the explanation is couched in mythical terms, namely, that the feature in question was formed by ancestral beings and that the essence of that creation still exists within it. The same is said of old archaeological sites, such as midden mounds, now located several kilometres inland and radio-carbon dated to 1500 years ago. We were told that these had not been formed by human actions but by the ancestral dingo, *Kula Kula,* as it travelled across the landscape, scraping up mounds of shell where it slept.

Stone tools used for artefacts are also divided on the basis of colour terminology. *Gun-garrema gun-gungaltja,* or light stone, includes sedimentary rocks such as cherts and quartzites and can be flaked for use as knives and spear tips. On the other hand, *gun-garrema gun-gungundja,* or dark stones, are volcanics, the basalts and granites made into axe heads and mortars by grinding. In some parts of Arnhem Land old men still know how to make stone blades at quarries of especially suitable stone. In 1981, at the celebrated quarry of Ngilipitji in eastern Arnhem Land, a small group of Rritharrngu men dug large spherical stones out of the ground and tested them for their suitability for flaking. Then, having prepared the cores, they struck a series of long blades and selected the flattest ones for further edge-trimming to make spear heads (see the film 'Spear in the Stone', directed by Kim McKenzie). In the recent past, before such tools were replaced by iron-tipped 'shovel-nosed' spears, Ngilipitji stone spears were the most prized and feared when used in duels and other combat. They formed a key element in a system of trade that helped to integrate the tribes of eastern Arnhem Land into a single cultural complex. The trade network was sanctioned through its relationship with rituals, including some of the most important secret ceremonies such as the Kunapipi.

At the quarry, the men spoke of the stone growing up in the ground. Only here at Ngilipitji did true 'killing stone' grow. The cross-sections of weathered rinds were compared to that of a kidney, with the best interior stone of pinky-grey silcrete referred to as *djukurr* or 'fat'. An esoteric oblique meaning of this word is power. It is this mystical power derived from supernatural sources integral to the site that gives the Ngilipitji stone blades their stupendous killing force. Once struck, man or beast is doomed.

PART II

Dualism and totemism

> What is that which is eternally and has no becoming, and again what is that which comes to be but is never? The one is comprehensible by thought with the aid of reason, ever changeless; the other opinable by opinion with the aid of reasonless sensation, becoming and perishing, never truly existent. Now all that comes to be must needs be brought into being by some cause . . . Of whatsoever thing then the Artificer, looking ever to the changeless and using that as his model, works out the design and function, all that is so accomplished must needs be fair: but if he looks to that which has come to be, using the created as his model, the work is not so fair.
>
> Plato, *Timaeus*

Aborigines make no distinction between religion, philosophy, and science. They have, however, developed a more or less unified and systematic ontology, which can be characterized as dualistic and totemistic. Let us briefly consider these two labels.

Like innumerable other peoples, Aborigines believe that reality comprises two coextensive domains. One is inhabited by living human beings, and knowledge of it is gained through the senses. The other is inhabited by gods, ghosts and demons. These extraordinary beings are normally invisible, though they may be fleetingly glimpsed or encountered. Entry into their domain may be achieved spontaneously through the act of dreaming; conversely, their entry into the domain of mortals may be contrived through the act of ritual. The Dreaming or Dreamtime, as this dimension of reality is often called,[30] is conceived as an ultimate reality, eternal and beyond explanation. The natural and social order in the domain of mortals is a dependent reality. Enlightenment consists in learning the nature of the dependence.

Totemism in its most elementary or vulgar form is a code that uses natural species as symbols to identify and distinguish social groups within a single frame of reference.[31] An example near at hand is the football competition, with its Bears, Tigers, Magpies, Eels and so on. Aboriginal totemism contains this codifying aspect, as well as the sentiments of solidarity that the symbols call forth and on which they crystallize. But it also encompasses a cosmological dimension that links the two domains of the dualistic universe and that cannot be satisfactorily explained as a mere epiphenomenon of social classification or collective consciousness. As conceived by Aborigines, totems are beings of great power who once roamed the earth performing wonderful deeds of creation and who now lie quiescent in focal points of the landscape. Before disappearing, they left behind in the care of men tokens of their being: carved stone or wood, songs, designs to draw on bodies, or bark, or on the ground, and so on. Undoubtedly such memorabilia function as badges of allegiance. But, at a deeper level, they serve as a palpable link with the departed dead, through whose hands and minds the works have passed from time immemorial, and with the transcendental world assumed to lie beyond the senses.

Land-owning corporations in traditional Aboriginal society are typically small groups of people related to each other by descent in the male line, though in some regions recruitment depends on place of birth or conception.[32] However constituted, each land-owning group is associated with a particular sacred site, or constellation of sites, and with the totemic powers from which such sites derive their significance. We may regard works of art symbolizing this link as being, in a sense, title deeds to land.[33]

Throughout Australia, Aborigines believed in the existence of a higher order of deities to whom they attributed a more general or fundamental cosmic significance than to the parochial totems. Usually conceived to be in human form, and sometimes referred to by expressions like 'Great Father' or 'Mother of All', high gods and goddesses were thought of as prime movers in the process of creation. In some cases they were supposed to have created the parochial totems, who in turn performed their own creative acts.[34] High gods (or sky gods, as earlier anthropologists called them) were commonly associated with thunder and lightning; and the rainbow (conceived as a serpent) was interpreted as a visible manifestation. Such deities were held in awe and reverence, not just by particular clans but by tribal groupings throughout broad regions. Male cults acted as guardians of ritual secrets associated with them; and from time to time initiates came together in large gatherings for the purpose of inducting young men into the mysteries.[35]

Conceptualization and control of nature

The above account is intended as no more than a rude sketch of the most salient features of the Aboriginal world-view as currently understood. Let us now focus on a particular aspect — the relationship between theory and practice or, more precisely, between conceptualization and control of nature.

About a century ago the British classicist Sir James Frazer began to develop his now well-known argument that humanity, in the course of its history, has attempted to influence the workings of nature in three distinctive ways through religion, magic, and science.[36] Religion assumes that the world is ruled by superhuman beings, human fortune depends on their beneficence, and divine goodwill may be produced by human acts of propitiation and conciliation. Magic is a spurious art based on misconceived notions of cause and effect. Like science, it seeks to act directly on nature without recourse to divine intervention. Science, however, achieves desired ends by developing techniques based on correct notions of cause and effect. So, to restore an ailing body to good health, the priest offers a sacrifice to the gods, the magician sucks out imaginary evil influences or objects, and the modern physician prescribes antibiotics.

Frazer believed that magic, religion, and science in that order represent successive stages in cultural evolution. Naturally, given prevailing assumptions, he also believed that the Australian Aborigines were exemplars *par excellence* of the first. In this regard his views were amply confirmed just before the turn of the century by the discoveries of Spencer and Gillen among the Arunta (or Aranda) of Central Australia. Spencer, professor of biology at Melbourne University, and Gillen, postmaster at Alice Springs, worked in close collaboration during and after the Horn Scientific Expedition to Central Australia in 1894; and in 1897 Frazer initiated a continuing scientific correspondence with them.[37] Frazer was particularly impressed by a class of ceremonies called *intichiuma,* in which human actors simulated the behaviour of various natural species and whose object was allegedly to cause such species to multiply. The performance was 'not associated in the native mind with the idea of appealing to the assistance of any supernatural being'.[38] Frazer, accordingly, had no doubt that *intichiuma* ceremonies were a type of magic designed to ensure a plentiful food supply. Yet there was a peculiar difficulty for this supposition: the tribe was divided into small groups, each responsible for a particular species, and every group was prevented by custom from eating more than token portions of its own totem species. Frazer concluded that the Arunta had devised a remarkable system of *co-operative* magic: 'Each group works its spells for the good of all the rest and benefits in its turn through the enchantments practised by others'.[39]

Neither in Spencer and Gillen's ethnography nor in Frazer's commentary are the nature and mode of operation of the spells and enchantments made particularly clear. Each totem group is associated with local sites containing spirits of future human members and also of the totem species (e.g. of kangaroos, and of human members of the Kangaroo totemic group); and the purpose of the *intichiuma* rites, according to Spencer and Gillen, is to 'drive out' these spirits in all directions so that, by entering living females, they become incarnate.[40] Songs appear to be the main instruments of propulsion, but here the data are disappointing. In a few instances song words are given without translation.[41] Elsewhere we are merely told that the chants 'invite' the totem species to be prolific.[42]

This lacuna in Spencer and Gillen's otherwise rich descriptions of *intichiuma* ceremonies (later generally referred to as 'increase ceremonies') was filled to overflowing with the publication of T.G.H. Strehlow's *Songs of Central Australia* in 1971. Strehlow spoke Aranda fluently (having spent his boyhood with native-speakers as the son of a missionary), and each song is presented in the vernacular with an English translation. His first example comprises eight verses believed to promote the increase of bandicoots.[43] Here are four of them:

Bandicoot Song

The bandicoots are rushing through the grass;
In and out of their nests they are rushing through the grass.

On the cracked swampflat they are brushing their fur;
The bandicoots are brushing their fur.

Crooking their little claws they are raking grass together;
With balled paws they are raking grass together.

They are snoring now —
Half-asleep they are snoring now.

Given their stated magical purpose (in the Frazerian sense), it is noteworthy that the Bandicoot verses are cast entirely in the descriptive mode (and therefore in the indicative mood). As Strehlow explains, 'the whole sequence is intended to reveal the daily life cycle of the totemic animal, from the moment it awakens to the moment when it falls asleep . . .'[44] He takes up the point later in a commentary on verses sung to create a plentiful supply of mulga sugar — a food enjoyed by the *ritjalitjala* bird as well as by humans.[45]

Mulga-sugar Song

The ritjalitjala *bird is speaking loquaciously.*
'Fragrance goes out from the bough, fragrance goes out for ever from
　　the bough;
Fragrance goes out from myself . . .
Fragrance goes out from the vessel . . .
Fragrance goes out from the sugar lump . . .
Fragrance goes out from the sweet cake . . .'
The hollow quivers, the hollow quivers and moves;
The vessel quivers, the vessel quivers and moves.

Strehlow notes that, in general, Aranda verses are not 'in the usual invocatory form commonly found in the corresponding spells uttered in other parts of the world'.[46] But, he says, we could easily give them such a form by changing the mood of the Aranda verbs in the song text from the indicative to the optative. The Mulga-sugar verses would then read as an invocation:[47]

'Let Ritjalitjala *speak incessantly!*
Let fragrance go out from the bough, let fragrance go out for ever from
　　the bough!
Let fragrance go out from thee . . .!
Let fragrance go out from thy vessel . . .!
Let fragrance go out from the sugar lump . . .!
Let fragrance go out from the sweet cake . . .!
Let the hollow quiver, let the hollow quiver and move;
Let the vessel quiver, let the vessel quiver and move!'

Strehlow performs a similar manipulation on a song from a grass-seed increase ceremony. Here is his initial rendering of the first three verses:[48]

Grass-seed Song

The ilbiritjilbira *bird is grinding the grain;*
Tall like a grass stalk he is grinding the grain.

The tall grass stalks are sprouting upwards;
Tall like a grass stalk he is grinding the grain.

The tall grass stalks are standing in full seed,
The tall grass stalks are sprouting upwards.

If we change the verbs from the indicative to the optative, the verses 'would have a very satisfactory sound both to our ears and to those of the natives'.[49] Thus:

'Let the ilbiritjilbira *bird grind the grain;*
Tall like a grass stalk let him grind the grain!
Let the tall grass stalks sprout upwards;
Tall like a grass stalk let him grind the grain!

Let the tall grass stalks stand in full seed;
Let the tall grass stalks sprout upwards!'

It is a moot point whether taking such liberties with the text is justified. Although there is no doubt that the songs are believed to be efficacious (e.g. the verses of the second song are said to produce an invisible flood of mulga sugar), the fact is that the words for the most part merely describe certain habits, aspects or associations of the desired species or foodstuff. Given that the Aranda language possesses the capacity to formulate wishes and exhortations, this in itself is a significant choice begging interpretation. It may help to throw some light on the problem if we make a comparison with some songs from a different part of Australia.

The people of central and eastern Arnhem Land, whose culture we have studied for some time, possess a number of song complexes called *manikay*. Each *manikay* is owned by a consortium of clans, usually belonging to several different tribes; and a *manikay* comprises verses about a distinctive set of totemic beings known generically as *wangarr*. *Manikay* songs are sung during complex mortuary rites, but they are also performed informally for enjoyment around the camp fire.

The song words in many instances do not occur in ordinary language, though sometimes there are recognizable relationships.[50]

The number of totemic beings contained in a *manikay* series may range from about twenty to forty. To take the one we know best, the Djambidj series comprises twenty-one subjects, six of which are bird totems (black bittern, silver-crowned friar bird, etc.), six are fish or marine animal totems (spangled grunter, porpoise, etc.), two are marsupial totems (brindled bandicoot, red-cheeked marsupial mouse), one is a reptile totem (king brown snake), and one is a plant totem (yam). From the repertoire, here are three examples to compare with the selections from Strehlow. Transcriptions and translations are by Margaret Clunies Ross.[51]

Bandicoot
Long snout and sharp teeth digs little holes,
long snout and sharp teeth digs little holes,
the long-eared One digs holes in His search for food,
scratches and digs small holes,
hops along digging the ground in search of food,
scratches and digs small holes;
long snout and sharp teeth digs little holes,
scratches and digs,
scratches and digs;
He hops along digging for food in the ground;
the long-eared One digs holes in His search for food,
hops along digging for food in the ground;
long snout and sharp teeth digs little holes,
scratches and digs,
long snout and sharp teeth digs little holes,
hops along digging in the ground for food.
The long-eared One digs holes in His search for food,
scratches with His sharp-nailed paws.
His long paws and yellow-furred chest scratching . . .
de-de-de de-de (refrain)
He hops along digging in the ground for food.

Wild Honey
A full Sugar Bag drips Honey, drips Honey,
From the mouth of dry Hollow Tree.
At Djubordaridja full, fermented Honey,
with food for the Bees' brood,
shows at the mouth of Hollow Tree.
Mature Wild Honey filled with Bees; food
overflows the dry trunk of Hollow Tree,
Treetrunk filled to overflowing.

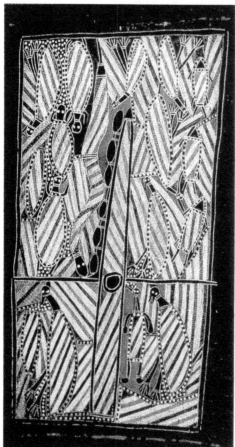

Bark painting — Wild Honey and Yam.
The central position is occupied by a
representation of the archetypal Hollow
Log containing Wild Honey (cross-
hatched) and Bee's Larvae (black circle).
Various other archetypes are
represented, including Yam (bottom
right segment, next to Sea Bird).
(Courtesy of the Australian Institute of
Aboriginal Studies)

Bark painting — Bandicoots. *Two*
bandicoots are represented in the top
right section. Archetypes of various
other species are included. (Courtesy of
the Australian Institute of Aboriginal
Studies)

Yam

True substance of the Wodbarridja plant,
with which the Spirits cram their dilly bags,
their dilly bags . . .
Fleshy stemmed Tuber
with full stems,
True substance of the Wodbarridja plant,
Dilly bags full to the brim,
with the Tuber Wodbarridja.
True substance of the Wodbarridja plant,
True flesh . . .
long Tuber,
the Tuber Wodbarridja.
Fleshy stemmed Tuber,
with bitter, hairy skin,
True substance of the Wodbarridja plant,
dar-dar-dar-dar dar-dar (refrain) . . .

The Arnhem Land songs quoted here, like their Aranda counterparts, are in the descriptive mode. One might regard them as epitomes of their totemic subjects, as evocations rather than invocations. While in general *manikay* songs are not sung with any specific intention of multiplying the totem species, Wild Honey verses may accompany certain ritual operations said to stimulate the production of this delectable foodstuff.[52] In certain other songs, features of gastronomic interest are singled out for attention (e.g. fatty chest and throat of turtle[53]). But it would seem that the magical efficacy attributed to songs depends less on their content than on their context (basically, whether or not they are performed with magical intent). Given that many of the *manikay* subjects form no important part of the diet, pragmatic concerns are unlikely to provide a dominant motive for either the composition of the songs or their regular performance. Even among the Aranda it is clear that ceremonies commemorating the totemic ancestors 'for their own sakes' are regarded as more important than those concerned with increasing the species.[54] There seems to have been a general tendency, no doubt better developed in rich environments than poor, for totems to figure in the Aboriginal mind more as objects of contemplation than sources of sustenance.

Totemism and the doctrine of forms

The contemplation of essences, so far as we know, received its most abstract and systematic formulation as a philosophical doctrine in Greece during the fifth century BC.[55] A crucial figure in this development was Pythagoras, a native of Ionia on the eastern fringes of Greek civilization where, in the sixth century BC, the decisive step towards science was made. Ionian thinkers (especially in the cities of Miletus and Ephesus on the west coast of what is now Turkey) turned away from dualistic explanations of nature enshrined in mythology and official religion and initiated the movement towards materialism and a systematic investigation of reality conceived as a single level of being. Significantly, this incipient science remains associated with the works of named individual thinkers, as contrasted with the anonymous doctrines of priestly orthodoxy.

Pythagoras left Ionia for political reasons and settled in Italy (c. 530 BC). On the way he spent some time in Crete, where he was inducted into a Dionysian cult. The Dionysian rites, so far as they can be reconstructed, are astonishingly similar to their Australian Aboriginal counterparts (i.e. initiation into the higher cults). The common motif is death and rebirth of the novice. Typically the symbolic destruction follows separation from the mother (either as a formal act at the beginning of the ritual, or as signified in a ratifying myth), and it may be represented as swallowing by the father (or a father-figure). In the Greek as well as the Australian rites, preliminary terrorization of the novice is achieved by the swinging of bullroarers, representing thunder or the voice of god. Rebirth may be represented as emergence from the womb of the Earth Mother.[56]

Orphism, which flourished at this time in the western regions of the Greek empire, was an attempt to purge the Dionysian cult of its cruder elements and give it a higher intellectual content. At Croton, in southern Italy, Pythagoras founded a fraternity devoted simultaneously to mathematics and religion. Just as herbs may be used to cleanse the blood, so purification of the soul may be achieved through contemplation of the principles of number, geometry, and harmonics, established by Pythagoras and his followers. The soul is to the body as mathematical laws are to the spatio-temporal objects they govern. Plutarch, expounding Pythagorean doctrine, says: 'The function of geometry is to draw us away from the sensible and perishable to the intelligible and the eternal'.[57]

Whereas his Ionian mentors had abolished the supernatural, Pythagoras thus sought to reconstruct the religious life on scientific principles. This involved the reintroduction of dualism, albeit of a much more sophisticated type than its predecessors. If we think of the mathematical concept of equality, for example, as compared with equal things, we are led to appreciate that true reality consists of principles — rational, perfect, and eternal — whereas the sensible world we take to be real is constituted by imperfection and ephemerality.

Although this doctrine undoubtedly took shape in Pythagorean circles, it achieved its fullest expression in the writings of Plato about a century after Pythagoras's death. According to Plato, the plurality of things that make up any natural kind (e.g. horses) is the spatio-temporal, imperfect, ephemeral manifestation of a perfect and eternal Form. The sensible world stands to the Forms as a shadow stands to a tree. True knowledge, therefore, is to be sought through a contemplation of essences, not by an investigation of appearances.

It might be argued that Aboriginal totemism, in its more refined and contemplative aspects, contains the lineaments of a doctrine of Forms. As a multiplicity of totems in *manikay* song poetry apparently serves no pragmatic objectives of a subsistence or social kind, except perhaps incidentally or tangentially, and as the *wangarr* beings are not objects of worship, obeisance, sacrifice, or invocation, it seems reasonable to assume that we are dealing for the most part with philosophy and art for their own sakes; or, to put it another way, with an attempt to represent or epitomize the structure of the cosmos. Regarded on that basis, *manikay* exhibit some striking features in common with Platonism. In both cases, particulars are seen as manifestations of an archetypal or ideal form; more importantly, the archetypes are credited with an independent existence. That is to say, they are not conceived merely as essential qualities residing in the members of a natural species or kind, but as separate entities inhabiting a transcendental dimension of reality. Forms and particulars coexist in their different realms, but whereas the latter are visible and ephemeral, the former are invisible and eternal.

An eminent historian of science has said: 'The Platonic point of view allured poets and metaphysicians, who fancied that it made divine knowledge possible; unfortunately, it made earthbound scientific knowledge impossible'.[58] While it would be unduly severe to blame Aboriginal dualism for the absence of a scientific tradition in pre-European Australia, there is no doubt that the indigenous religion systematically and pervasively turned men's minds towards the transcendental and directed a good deal of their energy into a celebration of the immutable. So profound was the conviction of dependence on the world beyond the senses that cultural products like songs, dances, sculptures and designs were themselves deemed to have emanated full-blown from transcendental sources. Yet totemic philosophy (despite its own subservient ideology) spawned representations of great subtlety and singular beauty. Although Australians of the future may not remember the indigenous inhabitants for their contributions to science, we can be sure that they will always contemplate with wonder and admiration the unique vision of nature enshrined in their art.[59]

Notes

1   Rhys Jones, 'The fifth continent: problems concerning the human colonization of Australia', *Annual Review of Anthropology*, 8 (1979), 445–66.

2   J. Allen, C. Gosden, N. Richardson, J.P. White, R. Jones and P. Gorecki, 'Pleistocene human colonisation of the Bismark archipelago, western Pacific', forthcoming (1987).

3   Rhys Jones,'East of Wallace's Line: issues and problems in the colonization of the Australian continent', in P. Mellars and C. Stringer (eds), *The Origins and Dispersal of Modern Humans: Behavioural and Biological Perspectives* (Cambridge, 1987, in press).

4   J. Golson, 'Australian Aboriginal food plants: some ecological and culture-historical implications', in D.J. Mulvaney and J. Golson (eds), *Aboriginal Man and Environment in Australia* (Canberra, 1971), pp. 196–238; Rhys Jones and Betty Meehan, 'Plant foods of the Gidjingarli: ethnographic and archaeological perspectives from northern Australia on tuber and seed exploitation', in D.R. Harris and G.C. Hillman, *Recent Advances in the Understanding of Plant Domestication and Early Agriculture* (London, 1987, in press).

5   J.M. Bowler, Rhys Jones, H. Allen and A.G. Thorne, 'Pleistocene human remains from Australia: a living site and human cremation from Lake Mungo, western New South Wales', *World Archaeology*, 2 (1970), 39–60.

6   Pawel P. Gorecki, 'L'homme et les glaciers en Nouvelle Guinée', *L'Anthropologie*, 90 (1986), 191–200.

7   C.E. Dortch and D. Merrilees,'Human occupation of Devil's Lair, Western Australia during the Pleistocene', *Archaeology and Physical Anthropology in Oceania*, 8 (1973), 89–115.

8   K. Kiernan, R. Jones and D. Ranson, 'New evidence from Fraser Cave for glacial age men in south-west Tasmania', *Nature*, 301 (1983), 28–32.

9   J. Glover, 'The geological sources of stone for artefacts in the Perth Basin and nearby areas', *Australian Aboriginal Studies*, 1 (1984), 17–25.

10 R.F. Fudali and R.J. Ford, 'Darwin Glass and Darwin Crater: a progress report', *Meteorites*, 14 (1979), 283–96.

11 R. Cosgrove and D. Ranson, pers. comm.

12 R. Luebbers, 'Ancient boomerangs discovered in South Australia', *Nature*, 253 (1975), 39.

13 Rhys Jones (ed), *Archaeological Research in Kakadu National Park*, Special Publication No. 13. Australian Wildlife Service, Canberra, 1985.

14 A. Watchman, 'Mineralogical analysis of silica skins covering rock art', in R. Jones (ed), ibid.

15 Rhys Jones and Betty Meehan, 'Anbarra concept of colour', in L.R. Hiatt (ed), *Australian Aboriginal Concepts*, (Canberra, 1978), pp. 20–39.

16 B. Meehan, *Shell Bed to Shell Midden* (Canberra, 1982).

17 B. Meehan, ibid., pp. 51, 54.

18 From B. Meehan, ibid., p. 51.

19 For example: R.L. Specht, 'An introduction to the ethnobotany of Arnhem Land', in R.L. Specht and C.P. Mountford (eds), *Records of the American-Australian Scientific Expedition to Arnhem Land, Vol. 3, Botany and Plant Ecology* (Melbourne, 1958), pp. 479–503; D.M. Smith and J.R. von Sturmer, 'The use of plants by the Aboriginal people in the Oenpelli region of Western Arnhem Land', unpublished paper held in the Australian Institute of Aboriginal Studies; N. Scarlett, N. White and J. Reed, '"Bush medicines": The Pharmacopoeia of the Yolngu of Arnhem Land', in J. Reed (ed), *Body, Land and Spirit: Health and Healing in Aboriginal Society* (St Lucia, 1983) pp. 154–91; P. Levitt, *Plants and People: Aboriginal Uses of Plants on Groote Eyland* (Canberra, 1981); J. Altman, 'The dietary utilisation of flora and fauna by contemporary hunter-gatherers at Momego Outstation, north-central Arnhem Land', *Australian Aboriginal Studies*, 1 (1984), 35–46; J. Russell-Smith, 'Studies in the jungle: people, fire and monsoon forest', in R. Jones (ed), op. cit. (n. 13), pp. 241–67.

20 J. Russell-Smith, op. cit. (n. 19).

21 B. Meehan, 'Nutrition and storage: some further notes on Aboriginal use of *Pandamus* in northern Australia', *Occasional Papers in Anthropology*, 10 (Brisbane, 1980).

22 J.C. Brand, C. Rae, J. McDonnell, A. Lee, V. Cherikoff and A.S. Truswell, 'The nutritional composition of Australian bush foods', *Food Technology in Australia*, 35 (1983), 293–8.

23 W. Beck, 'Technology, toxicity and subsistence: a study of Australian Aboriginal plant food processing'. PhD thesis, La Trobe University; Jones and Meehan, op. cit. (n. 4).

24 W. Beck, R. Fullager and N. White, 'Archaeology for ethnography: the Aboriginal use of cycad as an example', in R. Jones and B. Meehan (eds), *Is there Archaeology with Ethnography?* (Canberra, 1987), in press.

25 N. Scarlett et al., op. cit. (n. 19).

26 F. Péron, 'Observations sur l'anthropologie, ou l'histoire naturelle de l'homme, la nécessité de s'occuper de l'avancement de cette science, et l'importance de l'admission sur la flotte du Capitaine Baudin d'un ou de plusieurs naturalistes, specialement chargés des recherches à faire sur ce sujet', in J. Copans and J. Jamins, *Aux origines de l'anthropologie française* (Paris, 1978), pp. 177–85.

27 R. Jones, 'Hunters in the Australian coastal savanna', in D.R. Harris (ed), *Human Ecology in Savanna Environments* (London, 1980), pp. 107–46; Altman, op. cit. (n. 19).

28 R. Jones, 'Ordering the landscape', in I. Donaldson and T. Donaldson (eds), *Seeing the First Australians* (Sydney, 1985), pp. 181–209.

29 R. Jones and B. Meehan, op. cit. (n. 15).

30 The expression 'Dreamtime' derives from F.J. Gillen's translation of an Aranda term connoting the ancestral past: 'The natives explain that their ancestors in the distant past *(ulchurringa)* which really means in the dream-times ...' (Notes on some manners and customs of the Aborigines of the McDonnell Ranges belonging to the Arunta tribe), in B. Spencer (ed.), *Horn Scientific Expedition to Central Australia* (Melbourne, 1896). In the glossary of *The Northern Tribes of Central Australia*, Spencer and Gillen translate *Alcheringa* as the 'name applied by the Arunta, Kaitish, and Unmatjera tribes to the far past, or dream times, in which their mythic ancestors lived. The word *alcheri* means dream' (1904, p. 745). Elkin (*The Australian Aborigines*, 1964, p. 210) is incorrect in asserting that Spencer did not realize that 'dream' was one of the meanings of *Altjira* until he re-visited Alice Springs in 1926. For further information on this point, see W.B. Spencer and F. Gillen, *The Arunta* (London, 1927), pp. 304–6, 589–96; G. Roheim, *The Eternal Ones of the Dream* (New York, 1945), p. 210; W. Stanner, 'The Dreaming', in W. Lessa and E. Vogt (eds), *Reader in Comparative Religion* (Evanston, 1958), p. 514; T.G.H. Strehlow, *Songs of Central Australia* (Sydney, 1971), pp. xx–xxxviii, 614–15; D.J. Mulvaney and J.H. Calaby, *So Much that is New* (Melbourne, 1985), p. 124; J. Morton, 'Sustaining desire: A structuralist interpretation of myth and male cult in Central Australia', PhD thesis, ANU, 1985.

31 There is an enormous literature on totemism. For a convenient summary, see C. Lévi-Strauss, *Totemism* (London, 1962).

32 For a discussion of traditional land tenure, see L. Hiatt, 'Traditional land tenure and contemporary land claims', in L. Hiatt (ed), *Aboriginal Landowners*, Oceania Monograph No. 27, University of Sydney, 1984.

33 For a clear statement of this view, see N. Peterson, 'Totemism yesterday: sentiment and local organization among the Australian Aborigines', *Man*, 7 (1972), 12–32.

34 See, for example, Spencer and Gillen, op. cit. (n. 30), 1927, p. 356.

35 For a distinction between 'transcendental' and 'parochial' powers, see K. Maddock, *The Australian Aborigines*, 1982, Chap. 5.

36 The theory was foreshadowed in the first edition of *The Golden Bough* and developed subsequently. For a recent re-appraisal of Frazer's views on magic and religion, see P. Lawrence, 'Tylor and Frazer: the intellectualist tradition in social anthropology', in D. Austin (ed.), *Creating Culture* (London, forthcoming).

37 R. Marett, *Spencer's Scientific Correspondence* (Oxford, 1932). For a full account of the collaboration between Spencer and Gillen, and their relationship to Sir James Frazer, see D.J. Mulvaney and J. Calaby, op. cit., (n. 30).

38 W.B. Spencer and F. Gillen, *The Native Tribes of Central Australia* (London, 1899), p. 170.

39 J. Frazer, *Totemism and Exogamy*, Vol. 1 (London, 1910), p. 108.

40 Spencer and Gillen, op. cit. (n. 38), p. 206.

41 Ibid., pp. 191–2.

42 Ibid., pp. 172, 184, 186.

43 Strehlow, op. cit. (n. 30), pp. 132–4.

44 Ibid., p. 134.

45 Ibid., p. 290.

46 Ibid., p. 292.

47 Ibid., p. 292.

48 Ibid., p. 289.

49 Ibid., p. 293–4.

50 For a more detailed account of the relationship between *manikay* songs and social organization, see L. Hiatt, *Kinship and Conflict* (Canberra, 1965); I. Keen, 'One ceremony, one song: an economy of religious knowledge among the Yolngu of north-east Arnhem Land', PhD thesis, Australian National University, 1978; H. Morphy, *Journey to the Crocodile's Nest* (Canberra, 1984).

51 The first example is from 'Band descriptions', in M. Clunies Ross and S. Wild (eds), *Djambidj* (Canberra, 1982), p. 39; the second from 'Rom in Canberra', in S. Wild (ed), *Rom: An Aboriginal Ritual of Diplomacy* (Canberra, 1986), p. 54; the third from 'Band descriptions', op. cit., p. 29. See also M. Clunies Ross, 'The structure of Arnhem Land song-poetry', *Oceania*, 49 (1978), 128–56; idem, 'Two Aboriginal oral texts', in S. Knight and S. Mukherjee (eds), *Words and Worlds* (Sydney, 1983).

52 M. Clunies Ross and L. Hiatt, 'Sand sculptures at a Gidjingarli burial rite', in P. Ucko (ed), *Form in Indigenous Art* (Canberra, 1977), p. 142.

53 Clunies Ross, op. cit. (n. 51) 1982, p. 22.

54 Strehlow says: 'The ritual at animal increase ceremonies varies considerably from totem to totem. Rubbing of sacred rocks, blood-letting, and dramatic performance all occur, and these last are

sometimes of considerable elaboration, though they never rival, in point of variety, those acting performances which are purely dramatic in nature, that is, those commemorative ceremonies in which the totemic ancestors are introduced for their own sakes before the eyes of novices and initiates' (op. cit. (n. 30), p. 295).

55   Our brief notes on this development are based on W. Burkert, *Lore and Science in Ancient Pythagoreanism* (Cambridge, Mass., 1972); J. Burnet, *Early Greek Philosophy* (London, 1948); F. Cornford, 'Mystery religions and pre-Socratic philosophy', in J. Bury et al. (eds), *The Cambridge Ancient History*, vol. 4 (Cambridge, 1926); B. Farrington, *Greek Science: its Meaning for us* (Harmondsworth, 1961); J. Harrison, *Themis* (Cambridge, 1912); R. Mohr, *The Platonic Cosmology* (Leiden, 1985); D. Ross, *Plato's Theory of Ideas* (Oxford, 1953); G. Sarton, *A History of Science* (Cambridge, Mass., 1959).

56   The reference here to similarities between Dionysian and Aboriginal induction rites is in the nature of an aside. It is not being suggested that there is a connection between these rites and the doctrine of Forms.

57   Farrington, op.cit. (n. 55), p. 45.

58   Sarton, op.cit. (n. 55), p. 403. But cf. A. Koyré, 'Galileo and Plato', *J. Hist. Ideas,* 4 (1943), 400–28.

59   Although the two sections of this chapter were written individually (Part I by Rhys Jones and Part II by L.R. Hiatt), we share a long-standing association with the Gidjingarli people of Arnhem Land. Without their unfailing friendship and co-operation, the chapter could not have been written. We also owe a heartfelt debt of gratitude to Betty Meehan and Margaret Clunies Ross, whose own individual researches among the Gidjingarli are written large in the above pages.

2

Philosophical travellers at the ends of the earth: Baudin, Péron and the Tasmanians

Miranda J. Hughes

Any reference to a 'history of Australian science' generally evokes an image of the success stories of physics and medicine, the two major areas in which Australian scientists have gained international fame. At the other end of the spectrum is Aboriginal natural knowledge, which is immense but ignored, perhaps due to the difficulties of assimilating that perspective within the traditional framework of what it means to be 'scientific'. Between these two rather polarized fields lies much historical research devoted to other disciplines of science, technology and natural history. Yet surprisingly little has been done on the nexus of these two domains; that is, on the accounts of the meetings between the Aborigines and the European scientists, especially the first encounters in the early days of exploration of the continent. However, the encounter between Blacks and Whites tells us at least as much about the nature of European science, and especially anthropology, as it does about Aboriginal culture prior to European settlement. In particular, the encounter between the French scientific expedition captained by Nicolas Baudin and the Tasmanian Aborigines — the Diemenese — in the summer of 1802 illustrates well both the difficulties of achieving adequate field work in such situations and the perennial problem of cultural relativism.

The French voyages to Australia are relatively little known in our Anglo-orientated culture. The dominant influence of the British Empire in the history of Australia has led to a neglect of the contributions of other cultures to the Australian nation — not least those of the Aborigines. Most are relegated to a page or two in the standard histories of the continent and, due to this brevity, the accounts are often skewed. Baudin's voyage to the Southern Hemisphere between 1800 and 1804 is among the most maligned and misrepresented. A confluence of factors has led to its being denied an important place in the annals of Australian history, and to Baudin's being relegated to the role of tyrant captain, although recent histories have to some extent rectified this.[1] The voyage was, however, extremely successful in achieving its scientific aims: the official report claimed that more specimens were discovered and descriptions of new species recorded than in all previous voyages combined, including those of Cook.[2] Such success in the field of natural history suggests that the reason for the voyage being little known must be sought elsewhere.

Preparations, publications and philosophies

The most novel and significant feature of the research that was done was the anthropological work. Scientists and officers alike were aware of the importance of studying 'natural man' in his 'natural habitat' in a systematic scientific way, and instructions were issued on how this was best to be achieved.[3] Although a few natives were sighted in Western Australia, the major encounter occurred in Van Diemen's Land, and this resulted in numerous reports and illustrations of the Diemenese lifestyles by the ships' captains, naturalists, artists and seamen.[4]

This particular expedition provides, in both its planning and execution, a valuable source of materials for a history of science in Australia due to unique factors in the meetings between the Diemenese and the French. First, the reports furnish the major accounts we have of the Tasmanians prior to the English settlement of the island, and thus provide a useful and detailed account of their lifestyle. Secondly, there is the attempt to achieve a systematic anthropological study of the natives. No other voyage to the Southern Hemisphere had had the study of Man written into the official instructions as an explicit scientific aim, although most did include incidental descriptions of 'savages'[5] in their journals. Further, it can hardly be said of the first European settlers to New Holland that their interest in the Aborigines was mainly anthropological. Again, when Péron and Baudin present differing accounts of the same meeting, discrepancies in interpretations of actions and attributions of motives, customs and so on cannot be due to changes in tribal customs over time, a possibility that can figure in some debates about primitive societies (for example the Margaret Mead/Derek Freeman controversy about life in Western Samoa[6]).

Finally, this case provides historical antecedents for the recent trend of cultural relativism in the historical sociology of science. The French accounts of the Diemenese may not have achieved an unbiased cultural relativism, as, indeed, few works do even in modern anthropology; but we do find that the intention of evaluating each society according to its own internal standards rather than by the prevailing European customs is explicitly written into the anthropological instructions.[7]

The French were instructed to provide just ethnographic and anthropological details of the lives of the Aborigines. However, they did not, and could not, present a transparent rendering of Diemenese life, for observations are inextricably bound to theory. Descriptions of buildings, paintings, dances and actions inevitably require some theoretical assumptions, and even an apparently neutral description — such as that of a body — can become the means for establishing a hierarchy of humanity and human morals.

The concern of this chapter is *not* with whether the French reports achieve a neutral, objective description of Aboriginal life prior to White settlement. Rather, the interest in analysing them lies in explaining the disparities in tone and content; in examining their implicit biases and assumptions to establish the fallibility of the notion of independent, accurate observation; and in understanding them within the context of their broader social history.

The official four-volume account of the voyage was written by the naturalist, François Péron, and completed by the voyage's Lieutenant-Commander, Louis-Claude de Freycinet, after Péron's death in 1810.[8] As Freycinet's volumes provide mainly the scientific data of the voyage, it is to Péron's narrative that we must look in order to explain the subsequent reputation of the expedition. Péron's hostility towards his captain biases his narrative throughout. He constantly stresses his own achievements, and in fact presents the whole enterprise as something of a one-man show. An examination of Baudin's journal and other primary sources[9] reveals that the difficulties of the voyage were not caused by Baudin's bad seamanship, as Péron would have it, but rather, that a host of inopportune factors militated against its success. These included the slowness of the ships, the *Géographe* and especially the *Naturaliste,* unexpectedly variable winds and tides, bad provisions, and the inability to obtain good supplies at ports of call. Moreover, ten of the twenty-four scientists who embarked on the voyage had deserted the expedition by the time it left Timor. Some left because of illness, but most through disillusionment with life as explorer-scientists, which they had not envisaged as including eating food infested with maggots and drinking foul, stinking water while suffering *mal-de-mer.* Despite the excellent results the voyage produced, its full scientific potential was never able to be achieved.

Political factors also disrupted the voyage. Passports of diplomatic immunity did not prevent the English capturing the *Naturaliste* in the English Channel when the ship returned to a now warring Europe, and confiscating the entire collection. Only after much negotiation between the two nations, mediated by Sir Joseph Banks, were the scientific collections returned. Even then some pieces, inevitably, were missing. The political tension was heightened by the imprisonment of Matthew Flinders at Île-de-France (Mauritius), which raised doubts about the French claims that their voyage had been purely scientific in nature. However, the precise political motives of the voyage remain something of a mystery, and further discussion of them would lead us too far from the anthropological theme of this chapter; suffice to say, these factors meant the voyage would not attain a fame equal to Cook's, especially in the annals of English history.[10]

There are two major sets of documents on the anthropology of the Tasmanians from this voyage, one by the Captain, Nicolas Baudin, the other by the chronicler of the voyage, the naturalist, François Péron. While Baudin had had no formal training as a 'scientist', his knowledge of natural history was extensive, and many of the voyages he had commanded had also been voyages of discovery in which he had acquired many specimens, from camels to conches.[11] Lamarck referred to him as 'a most enlightened man',[12] while Jussieu told the Directory that 'of all the voyagers, he is the one whose achievements are the most meritorious'.[13] As an active and founding member of the *Société des Observateurs de l'Homme,* Baudin undoubtedly was familiar with its anthropological work. The quality of his previous work suggests that his observations during the Tasmanian expedition would also be of a high standard.

The naturalist, Péron, had been trained at the Muséum d'Histoire Naturelle and the École de Santé. He was admitted to the voyage on the strength of an article he submitted to the former, in which he argued that a naturalist specially assigned to study 'anthropology, or the Natural History of Man' ought to be included on this expedition, and that the person ought to be he.[14] The *savants* had already been chosen, but despite this, Péron was added to the voyage, although in the class of 'zoologist' and not the requested *anthropologiste*.[15]

Guided by the scientific spirit of the Enlightenment and the desire for an encyclopaedic knowledge of the world, the French government, in association with the Paris Académie des Sciences, had organized several scientific expeditions during the course of the eighteenth century to distant parts of the globe. In this respect, the voyages of La Pérouse and Bougainville are well known. To add to France's 'glory and honour',[16] Baudin in 1798 proposed to the Directory an ambitious world-wide expedition that would include exploration of the Americas as well as of *Terre Australes*. Though the financial situation of a country at war did not allow such an expensive tour to proceed, with Napoléon's support a modified plan for the Southern Hemisphere was accepted in 1799. As with previous voyages, much emphasis was placed on the preparation, collection and preservation of botanical and zoological specimens, and on the observation of geological, geographical and astronomical features.[17] As usual the task of preparing the scientific instructions was delegated to appropriate groups in Paris. In the case of the anthropological instructions, this was the Société des Observateurs de l'Homme.

The 'Observateurs de l'Homme'

Many of the assumptions about the nature of Man prevalent during the Enlightenment underlay the beliefs of the Consulate philosophers involved. However, the broad generalizations of the mid-eighteenth century were now being replaced by narrower treatises as different areas of inquiry became more formalized by the application of a scientific approach. This trend is exemplified by the formation of the Société des Observateurs de l'Homme, which stated as its aim 'the study of Man in all his settings' and which investigated wild-children, Chinese and Hellenistic cultures, the psychology of mind, and 'savage' tribes.[18] That a society *for* observers of Man could be formed at all indicates there was at least a limited consensus on what was to constitute anthropology. This is not to claim that anthropology had emerged as a clearly defined discipline with a well articulated methodology, nor even that the founders of the group had some concrete aim in establishing their Société; but it does reflect a fundamental shift in orientation that was taking place, which enabled the conception of a *science* of Man to be formulated and the creation of a more self-conscious and critical form of discourse to emerge.

The symbolism of the Société des Observateurs de l'Homme being established in the last three weeks of the 1700s — in December 1799 — is lost, for the Société had appeared, instead, in *frimaire, an* VIII of the Republican calendar that had been adopted to commemorate the victory of the Revolution and to emphasize the break from the Catholic Church. The main impetus for the Société's foundation came from the naturalist, Louis-François Jauffret, who was elected perpetual secretary at the inaugural meeting. The first president was the historian de Maymieux and the first vice-president, the linguist, Le Blond.[19] Within months, the membership had risen to over eighty and included such luminaries as Cuvier, Lamarck, Pinel and Hallé.[20] The Société met three times monthly. The topics with which it dealt reflected the broad range of disciplines represented. There were naturalists, linguists, philosophers, doctors, historians, economists, archaeologists and explorers.[21] The explorers were, for the most part, competent naturalists and included Levaillant, the explorer of Africa, and 'the famous Bougainville'.[22] Seven *Observateurs* were to become involved with the voyage of Baudin: Baudin himself, the botanists Riedlé and Maugé (who were Baudin's 'only friends'[23] on the voyage), the astronomers Bernier and Bissy, Baudin's second-in-command Hamelin, and the botanist Michaud who, having left the voyage at Île-de-France, travelled back to France via India.[24] Péron had had some contact with the Société, but was not a formal member.

Of the Société's eighty-three members, forty-three had been 'trained' in science or in medicine. For them, the acquisition of knowledge would proceed by the method of all good sciences: that is, through the 'gathering of many facts and observations' without the use of 'vain theories' or 'rash (French: *hasardées*) speculation'.[25] Concern for validity of approach is not surprising, given the professional and intellectual background of the members. They were aware of the specious nature of Cartesian systems and saw it as essential that the science of Man be based on the same objective, verifiable foundation as the physical sciences. Only then would the human sciences progress to a similar level. Various methodological approaches, pre-ordained by, and reflective of, the professions of the members, were utilized to achieve the aim: 'observer l'homme sous ses différents rapports physiques, intellectuels et moraux'.[26] The topics discussed embraced a diverse range of subjects: history, voyages, ideology, morality, language, education, culture, political science and jurisprudence were considered, together with anatomy, physiology, medicine and hygiene in the quest to establish a firm understanding of humanity.[27] The areas of special relevance here are 'the methodological classification of the races' and the attempt to understand human nature by an examination of its historical origins.

Explaining the origin of humanity and its spread upon the Earth was seen as the mutual aim of historians, Hellenists and explorers, their respective sources being archaeological remnants, the literature of ancient civilizations and observations of primitive tribes.[28] Inherent in this approach is an assumption of a significant identity between the societies of the past and those of present-day primitive cultures. This conflation of the actual origins of human society and discovered 'uncivilised' tribes is common throughout the history of anthropology.[29] Jauffret argues that despite the 'science of Antiquities' being severely hindered by its lack of access to the ancient world and its inhabitants, the problem was not insurmountable since the 'observations of navigators on the actual habitats of diverse regions ought to furnish precise light on the first epochs of the human species (French: *genre*)'.[30] Degérando expresses a similar point though far more poetically:

> The philosophical traveller, sailing to the ends of the earth, is, in fact, travelling in time: he is exploring the past; every step he makes is the passage of an age. Those unknown islands that he reaches are for him the cradle of human society.[31]

This reflexive aspect shows the dual nature of anthropology. The study of Man is as much a study of Man in general, of all humanity's origins and progress, as it is of Man in particular, of the primitive tribes. The French saw their own ancestry reflected in the natives of Van Diemen's Land, and their own present is reflected in their perceptions of the Diemenese.

The classification of Man and the study of 'savage tribes' utilized the approaches that came to be called physical anthropology, cultural ethnology and physiognomy. The *Observateurs* examined the physical differences among races, families and individuals, and investigated such features as build, hair type, cranial and skeletal structure and muscular strength.[32] The dominant medical theory of hygiene assumed an inseparable interaction of mind and body: hence a study of physiology necessarily involved the inclusion of the emotions and the will of Man.[33] Any physical anthropology thus involved comparative anthropology, which the Observateurs defined as 'the moral and cultural differences between the races'.[34]

Anthropological instructions

An extensive list of cultural features were suggested by Degérando in his essay of 1800 entitled *Considérations sur les diverses méthodes à suivre dans l'observation des peuples sauvages*,[35] one of the four commissioned by the Institut National to aid the philosophical travellers in their observations.[36]

Degérando's work must be classified as the first scientific treatise of anthropology, for it not only proposes which aspects of alien cultures are to be observed, but also includes a detailed analysis of the philosophical problems posed by studying the language and culture of an alien people, and a 'scientific' method for rectifying these difficulties by learning the language. This was to be achieved by generating a systematic grammar following the 'natural' acquisition of language. One began by understanding the gestures and signs of the natives, and then proceeded to concrete objects and their nouns and adjectives and, finally, to a comprehension of more complex and abstract ideas. Knowledge of the language of a tribe would lead to a proper understanding of the range and extent of its ideas.[37]

Degérando's *Considérations* were thorough in their comprehension of the crucial features of an anthropological understanding and of the faults that could befall such an enterprise. In practice, however, his scheme could not succeed for it relied too strictly on a mechanistic or rational scientific methodology. Knowledge of different lifestyles is not gleaned purely at this verbal level of rational communication: much may also be gained through *Verstehen* or hermeneutical understanding, that is, through assumptions of empathy and through intuition of common humanity. Much misinformation can result from verbal communication if there is either systematic misleading by the 'savages' or systematic misunderstanding on either part. Moreover, the proposed methods suggest complete comprehension as the aim, but this may be impossible ever to achieve and is definitely not manageable in the time allocated in this expedition for exploration on land. Péron and his fellow voyagers must receive praise for achieving as much as they did in a few weeks.

Unlike Degérando's essay, clearly the work of a lawyer and linguist, Georges Cuvier's 'Note instructive sur les recherches à faire relativement aux différences anatomiques des diverses races d'hommes'[38] reflects its author's interest in comparative anatomy. After giving a brief résumé of the works of Blumenbach, Camper and Daubenton on the anatomical, and especially the cranial, differences among races, Cuvier proposes some guidelines for philosophical voyagers to follow. He details methods of preserving skulls and skeletons and suggests graveyards and battle-scenes as potential sources of supply. He ambitiously wishes that living specimens — that is, human beings — could be brought back and, although 'insurmountable difficulties' might prevent this, these are in his view not so much moral as practical ones.[39]

Artists on the voyage are to forget 'the rules of proportion of art school'; in their place, Cuvier advocates the use of Camper's principles of anatomical representation. In strong contrast to Degérando's primarily cultural programme, Cuvier argues for the omission of all 'strange ornaments, bangles, pendants, and tattooes' from pictorial representation. Hair was to be portrayed simply and similarly for all members of a tribe so as to enable a clear assessment of the skull shape.[40] For Cuvier, the result was above all to be anatomically realistic; yet, curiously, it was not to be a strict 'photographic' representation of what the encountered savage was actually like. The basis of realism was solely biological.

In this respect, we can see that Cuvier heralds the approach of a scientific tradition much in favour later in the nineteenth century that reduced the various races of the globe and their complexities of lifestyle to physical characteristics. Cuvier's interest lies in the biological differentiation of races, which he views as having unalterable characteristics.

The works of the Observateurs later fell into obscurity. However, the revival of anthropology during the latter part of the nineteenth century saw, perhaps inevitably, the replication of many of the themes of the Observateurs' programme.[41] While the strict rote learning of language was not advocated, all systems of investigation required some access to a tribe's lifestyle through understanding of their language. The irony here is that very few anthropologists themselves became fluent in the languages of the tribes they studied. In practice, it was easier to find a native who could interpret through a 'pigdin' language. Instead of overcoming the problem of European bias, we now have a second bias introduced, that of the natives trying to understand what it is that the whites require of them.[42]

At the epistemological level of analysis, both modern anthropology and that of Degérando and the Observateurs suffer from the fundamental problem of justifying their systems of observations and methods. All attempts invariably appeal to basic assumptions about commonality in human experience, which can only be 'proved' through further assumptions. One cannot escape from these initial premises. Degérando was possibly the first to realize this: his 'fourth fault' is that explorers 'make the savage reason as we do, when the savage does not himself explain to them his reasoning'.[43]

This inherent difficulty is not to be confused with a claim that, at a practical level, anthropological research cannot be pursued; nor that one cannot list criteria that will separate the more methodologically comprehensive endeavours from an arbitrary collation of data.

The voyage to the ends of the earth

The expedition to the Southern Seas departed a day late, on 19 October 1800, from Le Havre. New Holland was sighted seven months later, on 27 May 1801, but the onset of winter resulted in the expedition's returning to Timor after careful exploration of the coast of Western Australia, rather than following the original plan of immediately travelling to Van Diemen's Land, and thence to Port Jackson. During their brief landing on West Australian shores, they met a few natives, but the encounters were brief and reticence on the natives' part also hindered the anthropological investigations of the French. It was not until Van Diemen's Land was reached, almost eight months later, that any extended encounter with the natives occurred. Even then, only thirteen meetings ensued. Baudin then attempted to explore the southern coastline of Australia, but illness — mainly scurvy — among the crew drove him to seek refuge at Port Jackson. Here, the expedition rested for five months, enjoying such delights as the new European settlement could offer before sailing to Kangaroo Island (off the coast of South Australia), King George's Sound, Timor and Île-de-France, where Baudin died, apparently from tuberculosis. As the major anthropological work was done on the Tasmanian Aborigines, it is on these meetings that we shall focus.[44]

The encounter between the French and the Diemenese parodies the archetypal anthropological endeavour. The reports of the voyage are not edited or depersonalized, and from them emerges a comic picture of the Aborigines trying to understand the French while the French endeavoured to advance the Science of Man. The French busily noted physical and cultural features of the Aborigines according, more or less, to their instructions: meanwhile, the Diemenese, in a far more intimate fashion, noted the physical features of the French. The Aborigines were as inquisitive about their visitors as the French were about them. Indeed, the French presented a far more unnatural group of humans than their own tribes. Perplexed by the apparent absence of white females, the blacks examined the most likely candidates, the young beardless sailors aged about fifteen, for breasts and other gender features, often making 'scrupulous inspections of the entire body' while 'uttering great bursts of laughter'.[45] Moreover, it seems that few if any of the whites took up the sexual offers of the women, who unambiguously displayed their intentions. The French seemed keen only to engage the men in arm-wrestling and other tests of physical prowess.

The first meeting with the Tasmanians occurred at d'Entrecasteaux Channel on 24 *nivôse an* X (14 January 1802) when some French officers — Baudin in one longboat, Péron in the other — rowed ashore. Baudin's account includes such observations as 'no harmful design', 'mutual trust immediately established', 'came up without the slightest distrust', and 'behaved in a very friendly way towards us'.[46] Péron met a group of four natives who made

TERRE DE DIÉMEN.

BARA-OUROU.

TERRE DE DIÉMEN.

NAVIGATION.

'gestures of surprise and wonder'. The first native's 'features were not the least austere or fierce. His eyes were bright and spiritual, and his countenence expressed at the same time benevolence and surprise'. The second, apparently in his fifties, possessed 'open and frank [features, and] in spite of some unmistakeable signs of uneasiness and fright, we could readily perceive his artlessness and good nature'. Both natives inspected the whites with 'astonishment and satisfaction', and the younger native's examination of the longboat provided to Péron 'the most striking example we ever had of attention and reason among savage people'. The two women were both kind and benevolent, while the younger one had 'an expressiveness and something spiritual about her eyes which surprised us and which since then we have not encountered again in any other woman of this nation'.[47]

Although Baudin and Péron concur on the general affability of the natives, important differences can be perceived in their accounts. Baudin presents qualified statements that are not presumptive of the Diemenese character, stating, instead, their overt actions and apparent intentions. Péron uses the nascent 'science' of physiognomy to gain insights into the character of the Diemenese and to attribute motivations for their actions. Physiognomy, however, is predicated on a European interpretation of facial expression, and Péron had no theoretical reasons to support an assumption of universality of this throughout humankind.

Péron's anthropological observations

Péron's final report, *Maria Island: Anthropological Observations,* written soon after leaving Van Diemen's Land, no longer characterizes the Diemenese as a noble race of nature, but rather as degenerate creatures low on the Great Chain of Being. The physiognomy is now thus described:

> In general, in all individuals ... their look always has something sinister and savage in it, and I strongly believe that basically their character corresponds with the expression on their features.[48]

Several pages later in the report, Péron again expands on this theme, and this 'objective description' is peppered with his own personal interpretation of underlying character:

> And how, amidst all the sensations which agitate them by turns, to distinguish those which belong more particularly to their natural disposition? In admiration and surprise, they appear stupid, insane in the expression of their joy, they are austere and tactiturn when suspicious and frightened; savage when confident and threatening, they are fierce in anger and its expression. But perhaps each of these states is normal for them, whence it follows that the characteristics of their physiognomy are extremely difficult to determine. When one wishes to study these with exactness, it is of particular interest to consider the influence of all the peculiar circumstances which modify their mental state and the nature of their feelings with regard to the observer.[49]

Clearly, Péron's interpretation of Diemenese character radically changed during the six-week sojourn in Van Diemen's Land, and this contrast in initial and final descriptions can be used to provide a framework for understanding the manner in which the encounters are portrayed. The initial observations were primarily guided by preconceptions of what type of human being would inhabit the region. The latter ones, although apparently based on and legitimated by a scientific method — the anthropology of the Observateurs — are merely a different set of opinions, which are no more valid than those based on the concept of the 'noble savage'. Analysis of the mode of description will also highlight biases that are implicit in the underlying theoretical framework. The physiognomy used by Péron provides a good example.

The two passages quoted above clearly presuppose that the purported character, namely, the sinister and savage nature of the Aborigines, is displayed in their facial features. However, the science of physiognomy can never rise above the inherent theory-laden nature of its observations. Like all physiognomists, Péron is aware of some of the difficulties, but seems oblivious to many faults in the methodology. Such errors include the assumption, firstly, of a definite meaning for certain features; secondly, that these meanings reflect on innate characteristics and not the transient moods; and thirdly, that although these unique circumstances will affect the Aborigines' response, Péron's interpretation will remain unaffected by his own exposure to strange settings.

The science of physiognomy, it appears, is a science of *Man:* women and children are excluded from it because they are viewed as incapable of deviousness in their expression and, hence, not susceptible to misinterpretation. Péron establishes friendly relations with a pretty young Aboriginal girl named Ouré-Ouré, who adorns herself with face paint and wears some feathers. While ignoring a similar amount of male decoration, the ear-ringed Péron[50] concludes that 'a fondness for ornament and sentiment of coquetry prevails in the hearts of the whole sex'.[51] We have here a striking example of the reflexive nature of anthropology, and of the way in which the general prejudices of a society stratify its members. Reflecting the mores of European society, the male bias denigrates the intelligence of women to the level of children, and both are considered to be somewhat feeble-minded and 'less affected by . . . the improvement of society'.[52]

Péron's *Observations sur l'anthropologie* makes specific reference to the 'robust majesty of natural man'[53] and, following the eighteenth-century advocates of the popular myth of naturalism,[54] suggests that investigations into how the savages maintain their 'nearly unalterable health and nearly miraculous longevity' will enable an improvement to be made in the pitiful state of 'our social establishments'.[55] The dynamometer[56] — an instrument to measure the muscular strength of the arms and legs — was used to test the hypothesis that 'natural' Man possessed a natural strength far above the capacity of physically degenerate civilized Man. However, the instrument's use also demonstrates the infiltration of bias and unquestioned assumptions into a methodology. Furthermore, the results obtained with it are believed to be endowed with more objectivity than mere observations alone could achieve. Péron's measurements resulted in a rank ordering by strength of English, French, Timorese, Aborigines and Diemenese. (The Diemenese refused to test the strength of their loins.)

Yet Péron's results did not, as he appears to believe, prove some objective truth about the strength of savages and that of Frenchmen, or even of the

English. Like many scientists, Péron has committed the ontological crime of reification, of believing that his numbers reflect on some true relation between savage and civilized strength. He fails to see as counter-evidence other observations, such as the Aboriginal display of marked endurance by running quickly for long distances, and easily outpacing the 'stronger' Frenchmen.[57] He does not understand that the dynamometer measures only a specific set of muscles in a specific way and, more importantly, in a way that advantages the French. He writes in *Maria Island* that 'the small number of trials I was able to make with the dynamometer cannot give an exact idea of their physical strength, at least that of the wrist'. The results he obtained were consistently less than with the English and French by a difference averaging twenty myriagrammes.[58] For Péron, the explanation of the 'physical cause of disproportion between the strength of their constitution and their effective strength' lies not in the limitations of the instrument, but in their 'habitual way of life'.[59] Péron's 'Expériences sur la force physique des peuples sauvages de la Terre de Diémen, de la Nouvelle Hollande et des habitans de Timor'[60] was presented to the Institut National, and a report prepared by La Billardière and Coulomb.[61] Its acceptance illustrates that Péron's inappropriate use of his results reflected contemporary understanding of research methodology.

Apart from drawing erroneous conclusions from the results and ignoring the glaring faults inherent in the design of experiment, the polemical use to which Péron puts his results is to show native resentment, bias and prejudice. Péron argues that, on registering a weaker reading than the French, the Diemenese first became agitated and then resentful and fearful that 'I had some secret and perfidious design in regard to them'.[62] He uses this to support his claim that the natives possess a 'general mistrust' and an inability to understand the Frenchmen's intentions and actions:

> What is of no importance to us is exceedingly important to them. The most innocent act in our eyes is often an act of open hostility in theirs. The use I wanted to make of the dynamometer furnishes a prime example of this . . .[63]

Degérando's *Considerations . . . on Savage Peoples* warned philosophical travellers about the difficulties of language acquisition and the understanding of cultures.[64] Despite this, Péron's account is permeated by biases and unfounded interpretations, albeit somewhat inescapably. Part of the existing vocabulary of an Aboriginal language that Péron used came from the vocabulary of Hunter.[65] He used some of these words on the various tribes he encountered in *Terre Australes* but found that he was often not understood. Péron blamed inaccuracies in the vocabulary[66] rather than considering alternative explanations — for example, that the words were written for English pronunciation, or that different dialects or even different languages could exist. This is indicative, more generally, of an assumption of homogeneity of race, where any one individual or small group is taken, despite Degérando's warning against such beliefs,[67] to provide a sufficient example of the whole race.

The vocabulary that Péron did produce is rather disappointing. All that remains of the work of the *anthropologiste* is a total of seventy-five words that Freycinet had 'come across' while going through Péron's papers. The words are nouns and verbs that could be defined ostensively and include 'the cheeks', 'spouse', 'to warm oneself', 'oyster', 'to fart', 'to piss', 'to put wood on the fire' and 'yes, it's all right'.[68] Degérando's proposals for language acquisition were bound to fail. Yet despite being equipped with an unworkable methodology, Péron ought to have been able to assemble a vocabulary comparable to those assembled by his predecessors. His results are not systematically organized and there does not appear to be any attempt, either then or during the years after his return, to analyse, with the aid of previous vocabularies, the nature of Aboriginal language.

A second aspect of language acquisition is the relative ability of the natives and the French to learn each other's languages. It is observed by Péron that the Aborigines were quick to pick up French words and pronunciation while the French had rather more difficulty with the Tasmanian language. The amount of laughter that the French efforts gave the Diemenese led Péron to concede in the draft of *Maria Island*[69] that the Diemenese had 'much intelligence'; yet in the final published version their linguistic skill is noted without comment. In particular, it was not interpreted as providing evidence for native intelligence or ability. Ultimately, nothing the natives could do would alter the French conviction of their low intelligence. In the absence of tangible constructions, clothes and other artefacts and trappings of civilization, the Eurocentric observer found it difficult to conceive the possibility that there might exist a wealth of culture and a highly organized system of kinship, religion and socialization and, behind that, an intelligence equal to the white man's but directed towards different ends.[70] This opinion was reinforced by the prevailing theory of socio-cultural evolution that argued that society had the same defined stages that an individual man has, with the same developmental and conceptual limits at each level. Consequently, the Tasmanian males were not considered to possess, or be capable of, a mental age equivalent to that of the French males, but only to that of male French children.[71]

The Diemenese: Péron 'interviews' and Baudin 'remarks'

The Aboriginal practice of burning the land, the explanation of which still engenders much heated controversy,[72] sparked comment from both Péron and Baudin. Although the incidence of fire is often noted by Péron, it receives little attempt at explanation and, at most, a poetical lamentation:

> Thus were destroyed these ancient and venerable forests which the sycthe of time had respected through the course of so many centuries, only to fall a sacrifice to the destructive instinct of their ferocious inhabitants.[73]

Baudin is also disturbed by this native practice, but he suggests reasons to explain such apparently destructive behaviour. After sighting a group of natives, one of whom was 'setting fire to everything as they went along', he states: 'this I believe is customary when they want to stave off, or begin a war amongst themselves'.[74] For the large fire on which Péron had commented, Baudin conjectures:

> It is possible not just for the pleasure of destroying that they set fire to their forests in this way, and it is reasonable to suppose that there is some useful

purpose behind it. The fire may perhaps have been started merely to burn off the grass and other undergrowth, which makes walking painful and difficult, or it may have been for hunting some quadrupeds sighted in the area.[75]

The observations on the native practice of burning the land exemplify the constant contrast in the writing of the two men, between observations laden with rhetoric on the one hand, and reasoned hypotheses to explain actions on the other. Prima facie, this would not be surprising, given that one writer is not a scientist but a sea-captain and the other is a scientist with a professed interest in anthropology. However, as has been shown, it is Baudin who presents the more 'scientific' of the two accounts, while Péron's account differs little, methodologically, from the literature of popular travellers' tales.

A comparison of the two anthropological reports of Baudin and Péron will enable this distinction to emerge, for the problems of encounter and research are approached from quite different perspectives. Indeed, the titles alone indicate this divergence: Péron writes of his 'Interview with the natives of [Maria] Island and description of a tomb found on the northern shore of East Bay',[76] Baudin on 'Particular Remarks concerning our sojourn in D'Entrecasteaux Channel',[77] or 'The natives we found at the Channel and their conduct towards us'.[78]

Both works were written soon after the expedition's departure from the south-west coast of Van Diemen's Land. They thus present a clearer account of initial perceptions that have not been modified by time than does Péron's *Voyage to the Southern Hemisphere*, which was composed some years after returning to France. Consequently, they may enable a more accurate portrait of the Tasmanians to be painted. *Why* these texts were written, however, provides some explanation for the different forms that they take. Baudin writes to inform the Ministre de la Marine et des Colonies and the Muséum d'Histoire Naturelle[79] of the progress of the expedition, and he includes amid other navigational and scientific details an extensive account of the autochthonous lifestyle. Péron, in contrast, writes to placate his captain,[80] to show evidence of industrious research during the Tasmanian sojourn. It begins with a mixture of excuses to Baudin as to 'why up to now have I not occupied myself in more specialized work?'[81] and a general discussion of the near impossibility, both in principle and in practice, of doing anthropology adequately. Had Péron argued solely for the inappropriateness of the methods, the report would have presented some interesting insights for the history of ideas. As it stands, the initial outburst reads as a truant's plea for leniency. After this, however, Péron, in his idiosyncratic style, imparts a variety of information about the Diemenese, which, although entertaining, is hardly a work of anthropology. Péron emulates the style of science by having a résumé at the side of the page to present the main point, which appealed to contemporary methodology in form but does not guarantee content. Since he is writing for his captain,[82] perhaps such literary or research devices are unnecessary.

The descriptions of the encounters in Péron's work are superficially adequate, but not perceptive, and many of the details of Tasmanian life that Baudin records are omitted. Chances for good anthropological field work are ignored in favour of philosophical flights of fancy concerning the nature of Man and his general relationship to the Universe. Two of Baudin's somewhat cynical remarks suffice to show the rashness of Péron's conclusions. Baudin describes the variety of foods the Diemenese eat, and adds:

> One of our scientists claims that they use a kind of fern whose root is starchy; he even added that he had absolute proof that they live on grass in times of scarcity. As he finds the evidence for his theory in some dried excrement, one can judge how far his hypothesis is well founded.[83]

While Péron could be praised for this imaginative approach to problem-solving, his solution here did not fall within the accepted scientific methodology of the epoch. From the information that Péron possessed, it was *unreasonable* to conclude that the natives subsisted on grass: the idea transgressed the canon of not using 'vain theories or rash speculation'. Baudin rightly criticizes it as an ill-founded hypothesis.

The second remark, also from this report, relates to the topic of native religion. According to Baudin:

> Some thought they had noticed that the natives often looked at the sun, and it has not been difficult to infer from this that the sun must be their divinity. Personally I saw nothing like this and believe them to be as lacking in ideas on this subject as they are in industry for their own preservation.[84]

It is clear that, in making assumptions about the absence of Aboriginal religion based on a lack of artefacts, Baudin himself has committed the fourth fault of Degérando: 'So they often pronounce excessively severe judgements on a society accused of ... atheism'.[85] One reads with a certain amount of incredulity that the Diemenese are 'lacking ... in industry for their own preservation', the mere existence of the Aborigines itself demonstrating their capacity for survival.

Péron's general approach is in strong contrast to Baudin's, who presents a rational description of events accompanied by suggested explanations of customs and incidents. Importantly, for unpleasant encounters Baudin always tries to give *les naturels* the benefit of the doubt by postulating that some unknown law has been transgressed. At a descriptive level, Baudin states specifically that the Tasmanians do not eat fish although they consume large quantities of crustasceans, whereas Péron fails to specify this. Nor, unlike Baudin, does 'our observer of Mankind'[86] include such details as the wearing of leaves of seaweed on the feet when they are sore; or the methods by which weapons are made; or how the Aborigines would use the knives and other French tools but, in the end, preferred their own. Baudin also notes that although the Diemenese were subject to yaws and ulcerated legs, it appeared that smallpox and syphillis were unknown to the islanders.[87] Yet Péron, who had suggested prior to departure the importance of gaining such knowledge of diseases and their local treatments,[88] fails to include these or any other observations on the medical aspects of Diemenese life.

The emphasis of interpretation is markedly different between the two authors, especially in relation to the assumed motives of the savages. This again illustrates the differences between reasoned and rash interpretation. Péron concludes that the savages possess no social conventions — that is, no

French or European conventions. Despite participating in, and describing, customs, he failed to recognize them as such, and criticized the Tasmanians for their 'uninformed state of social intercourse'[89] and for being 'almost entirely strangers to every principle of social order'.[90] The natives were 'treacherous and insincere'.[91] Physiognomy revealed that they were 'fierce and ferocious in their menaces, they appear at once suspicious, restless and perfidious'.[92] Baudin does note that the glance of the Diemenese is often restless: yet even here, he adds, 'This is perhaps the result of the distrust that men so different from themselves must have aroused in them.'[93]

Anthropological descriptions

The pervasiveness of 'observer bias', at both the cultural and personal level, is amply demonstrated in Baudin's and Péron's description of Diemenese physique. Moreover, it shows once again the contrast between Péron and Baudin. After some account of events, and of the native attitudes to weapons, fishing, European food and gifts, Baudin provides the following description:

> The natives of Bruny Island, like those in the other parts of the Channel, are in general not very tall, and quite awkwardly built. Almost all of them have slim legs, with slender thighs, a large stomach, a wide mouth, thick lips and prominent cheek bones; their nose is slightly flattened, though not as much as that of the Africans. Their wide prominent forehead makes their eyes seem small at first, which does not help their appearance.
>
> Their look, without being savage, is not pleasant, although it is alert and lively. The colour of their skin is light brown, but it can sometimes appear almost black according to whether they have smeared themselves with powdered charcoal or with red earth which they use to render themselves attractive or frightening. They all have very even teeth which are large and white and we saw no one who had lost any, but several had teeth which were decaying. Some of our scientists will not fail to say they found they had filed their teeth. This is not true! They have woolly hair but it is not very thick, and it is curly and short. The elegant ones among them draw it out with grease and red earth, with the result that they seem to be wearing a wig.
>
> Scarification is one of their customs, and I do not believe any of us was able to understand the means they use to cause the marks on their bodies to stand out like bumps on the surface of the skin. The places most affected are the face, the upper part of the arms, the back of the shoulders and the lower part of the stomach.
>
> The women are neither pretty nor graceful, or at least those we saw appeared so to me. They seem to be alone responsible for the upkeep of the household, and they have to provide for the whole family by fishing, an activity at which they do not lack skill.[94]

The passage clearly illustrates both attention to detail and a European perspective in the judgement of effects. Moreover, it shows Baudin's reluctance to arrive at conclusions that went beyond the data. Péron fails to present his information as concisely, the details being embedded in narrative:

> While they were thus amusing themselves in looking us over so particularly, I was myself occupied in considering them with profound attention. The majority of them were young men from fifteen to thirty-five years of age. One, older than the rest, seemed to me to be not less than fifty to fifty-five years of age, and he was the

only one wearing a kangaroo skin over his shoulders; the others were perfectly naked. In general, they were all fairly tall; among the oldest there was one who was thinner and more slender, and who was not less than five feet six inches tall. All the others varied between five feet two inches and five feet four inches in height. Only one among them had his hair in curls and reddened with ochre. This was a tall man twenty to twenty-two years old, a very fine fellow who was more handsome than all the others, although even he was disfigured by the general defect of his people that is to say associated with a well developed head, large and fleshy shoulders, a fine chest and very muscular buttocks, all his extremities were thin and feeble, especially the legs, and the belly was a little too large. Some other time I will have occasion to go into detail about the organic constitution of these people, and I will then point out the physical causes of the defects, but I will restrict myself now to some observations on their physiognomy.

In all, it expressed liveliness. The passions unite there vigorously and succeed one another rapidly. Changeable like their affections, all the lineaments of their faces change and become modified one after the other. When menacing, their faces show fierceness, when suspicious, apprehension and perfidy, and when laughing, a mad gaiety, in the oldest ones the look is sad and gloomy. In general, in all individuals and whenever they are being watched, their look always has something sinister and savage in it, and I strongly believe that basically their character corresponds with the expression on their features.[95]

It is apparent that, despite these two extracts being of similar length, they are not of similar quality. We seem to gain a clearer description of the Diemenese by reading Baudin. With Péron, as well as a clearly biased description, we get an interesting insight into the dynamics of contact between cultures: for example, the laughter is not 'gay', but has 'a mad gaiety', the phrase implicitly reinforcing the notion of 'savagery'.

Péron's anthropology can be criticized at two levels, the first being that he fails to reach the available scientific standard of his own time. This is the more serious problem because it was rectifiable. In Baudin's work, we can find an explicxt application of a belief in objective observation and analysis, and an apparently conscious utilization of the accepted methodologies of the late eighteenth century. In comparison, Péron's work is, *inter alia,* too subjective, lax and too prone to unfounded interpretations. The second criticism of Péron's anthropology is common to all interpretative work, both in the first collection of data and in the subsequent analyses. This problem belongs to the very nature of the enterprise for, in a strictly positivistic sense — that is, in the sense of being objectively true — no one interpretation is better than another. In some ways, these criticisms of Péron are reflexive on this current chapter: all interpretations are culturally bound and limited by the social facts in which the analyses find themselves. It may be argued that the writing here reflects more on the society of twentieth-century Australia than it does on the nature of the Tasmanian Aborigines, just as Baudin and Péron reveal more about early nineteenth-century French society than they do that of the Diemenese. It must be strongly emphasized that this is *not* to assert that 'anything goes': rather it is to argue that criteria other than truth must be used to evaluate texts. Such criteria include completeness, comprehensiveness, elegance and coherence (both logical and literary). Judgements thus become ones of value rather than 'Truth'.

Conclusion

The emphasis of this chapter has been on the encounter between the French and the Diemenese, and the resultant anthropological work. Consequently, little has been written concerning the other scientific aspects of the voyage, where the achievements were substantial. Any history of Australian biology, geology or oceanography would be incomplete without reference to Baudin's expedition. Péron wrote about a dozen memoirs on diverse topics such as the temperature of sea-water; a 'general history of jelly-fish'; and a 'Memoir on some zoological facts applicable to a theory of the globe'.[96] Leschenault published a work on the vegetation of New Holland and Van Diemen's Land.[97] Despite looting by the British and Prussians during the turbulence at the end of the Napoleonic era, many specimens survived in Paris, including black swans, wallabies and numerous plants and trees. There is also a vast quantity of archival material that has yet to be thoroughly examined.

The puzzle, then, is why this expedition has remained so little known. It has been demonstrated, at least for the case of anthropology, that the results deserve to be much more widely known, and it can be argued that the same is true for the other areas of science that were studied. The main explanation appears to lie in the change in attitude that transpired in France during the travellers' absence. Baudin's expedition departed in the sunset of the Enlightenment and returned to the dawn of the nineteenth century, where benevolence to the 'noble savage' had been displaced by a comparative anatomy that concentrated solely on the physical aspects of Man. A second reason is the untimely death of Baudin before the expedition returned to France. His *Journal* was not published until 1974. The official but biased account of Péron appeared in 1807 and, in English translation, in 1809.

As Péron's work was one of the few accounts of these Aborigines published during the early nineteenth century, one must ask if its portrayal of the Diemenese as ignoble savages influenced English policy. Baudin's attitude shows that alternative approaches were available: he attempted to empathize with the native viewpoint and he believed that the natives should not be arrogantly dispossessed of their land. Although Baudin described the Diemenese as 'essentially primitive natural men and, though not completely lacking in intelligence, they are certainly at the lowest level of civilization that one can imagine',[98] he neither equates (at least not explicitly) lowest with inferior, nor sees this as giving civilized nations the right to exploit or eradicate. Here, surely, lies one important episode for the history of Australia, and of European imperialism. It is in stark contrast that we read Sir James Barnard, eighty-eight years later, in an address to the 1890 AAAS congress:

> It has become an axiom that, following the law of evolution and survival of the fittest, the inferior races of Mankind must give place to the highest type of man, and that this law is adequate to account for the gradual decline in the numbers of the aboriginal inhabitants of a country before the march of civilization.[99]

Notes

1 Many earlier historians consulted only Péron's account of the voyages, and even such a standard reference work as Bernard Smith's *European Vision and the South Pacific, 1768–1850* (Oxford, 1960) asserts that 'Baudin was a poor navigator, paid little attention to the health of his men, and was quite out of sympathy with the work of the scientists' (p.147). More recent histories include N.J.B. Plomley, *The Baudin Expedition and the Tasmanian Aborigines, 1802* (Hobart, 1983); C. Cornell, *Nicholas Baudin's Australian Expedition, 1800–1804* (Adelaide, 1965); J. Copans and J. Jamin, 'Présentation: de la filiation déviée à l'oubli des origines', in *Aux origines de l'anthropologie française: les mémoires de la Société des Observateurs de l'Homme en l'an VIII* (Paris, 1978); J. Brosse, *Great Voyages of Exploration: The Golden Age of Discovery in the Pacific*, trans. Stanley Hochman (Sydney, 1983); Frank Horner, *The French Reconnaissance: Baudin in Australia, 1801–1803* (Melbourne, 1987).

2 The total number of specimens obtained for all disciplines exceeded 100 000, including 2542 new zoological species. Paris: *Archives Nationales*, AJ[15]569.

3 Originally, four papers were commissioned from the Société des Observateurs de l'Homme by the Institut National, which was responsible for co-ordinating the scientific work of the voyage. Papers prepared by Pinel and Hallé appear to be lost: those by Degérando and Cuvier, entitled respectively *Considérations sur les diverses méthodes à suivre dans l'observation des peuples sauvages* and *Note instructive sur les recherches à faire relativement aux différences anatomiques des diverses races d'homme*, are reprinted in Copans and Jamin, op.cit. (n.1).

4 The best collection — in either French or English — is in Plomley, op.cit. (n.1).

5 'Savage' is used only as a convenient label: it is not intended to convey that the encountered tribes were 'savage' in any pejorative sense of the word. Labillardière's *Relation du voyage à la recherche de Lapérouse* (Paris, 1796) presents much information on the Tasmanians, but it lacked the *systematic* background of Baudin's expedition.

6 See M. Mead, *Coming of Age in Samoa* (London, 1963) (orig. 1928); D. Freeman, *Margaret Mead and Samoa: the Making and Unmaking of an Anthropological Myth* (Canberra, 1983); and *Canberra Anthropology*, 6 (1), 6 (2), special editions: 'Fact and Context in Ethnography: the Samoa Controversy'.

7 Joseph-Marie Degérando, *Considérations sur les diverse méthodes à suivre dans l'observation des peuples sauvages* (Paris, 1800), reprinted in Copans and Jamin, op.cit. (n.2), pp.127-69. See especially 'Quatrième défaut', p.135 (English translation in F.T.C. Moore, *The Observation of Savage Peoples* (Berkeley, 1969), p.67). Future references to Degérando's work are to this edition.

8 François Auguste Péron and Louis Claude Desaulses de Freycinet, *Voyage de découvertes aux terres australes exécuté par ordre de Sa Majesté l'Empereur et Roi, sur les corvettes le Géographe, le Naturaliste, et la goëlette le Casuarina, pendant les années 1800, 1801, 1802, 1803 et 1804* (Paris, 1807-1816). Future references are to the English translation, *A Voyage to the Southern Hemisphere* (1809, reprinted Melbourne, 1975).

9 Nicolas Baudin, *Journal of Post Captain Nicolas Baudin: 1800–1803*, trans. Christine Cornell (Adelaide, 1974); Paris, *Archives Nationales*, AJ[15]565-76; Marine BB[4]995-7; Marine 5JJ 24-57. (Journals of the expedition), *Muséum d'Histoire Naturelle*, 2082.

10 When the voyage was first announced, it was stated that Baudin had been appointed to an expedition that would 'emulate the voyage of Captain Cook' (*Rapport: Ministre de la Marine*, 16 *thermidor an VI* (4th August 1798); Paris, *Archives Nationales*, Marine BB[4]995). Baudin's death at Timor enabled Péron to claim openly that he had had a secret mission as a government spy to evaluate Australia's potential for settlement. Governor King was sufficently disturbed to claim King Island as British under the very noses of the French (in such haste as to raise the flag upside down) and to establish (though only briefly) a small settlement at what is now Sorrento in Victoria (Baudin, op.cit. (n.9), pp.439-47).

Though not privy to the discussions between Baudin and Flinders when they met at Encounter Bay, Péron emphasizes the political nature of their dialogue (Péron, op.cit. (n.8), p.279). Baudin's *Journal* acknowledges that exchanges of information occurred and refers to the accuracy of Flinders' maps (Baudin, op.cit. (n.9), p.380). See also E. Scott, *Terre Napoléon* (London, 1910); J. Dunmore, *French Explorers in the Pacific* (Oxford, 1969); Horner, op. cit. (n.1).

11 See Paris, *Archives Nationales*, Marine BB[4] 995.

12 Paris, *Archives Nationales*, Marine BB[4] 995.

13 Moore, op.cit. (n.7), 'Introduction', p.9.

14 Péron, *Observations sur l'anthropologie, ou l'histoire naturelle de l'homme, la nécessité de s'occuper de l'avancement de cette science, et l'importance de l'admission sur la flotte du Captaine Baudin d'un ou plusieurs naturalistes, spécialement chargés des recherches a faire sur ce sujet* Paris, An VIII. The *Procès-verbaux* of the Institut National for 1 *thermidor, an VIII* (20 July 1800) state 'L'École de Médecine, ayant étendu dans sa dernière séance des *Observations sur l'anthropologie. . .*, communique à l'Institut ses idées à cet égard et lui recommende le C. Péron, de l'Allier, auteur des observations citées.' The *Observations . . .* are reprinted in Copans and Jamin, op.cit. (n.1), pp.177-85.

Péron's teachers included Lamarck, Lacepède and E. Geoffrey Saint Hilaire. For biographical details see L. Audiat, *François Péron: sa vie, ses voyages et ses ouvrages* (Moulins, 1855). J. P. F. Deleuze, 'Notice historique sur M. Péron', *Annales du Muséum d'Histoire Naturelle*, 17 (1811); Maurice Girard, *François Péron, naturaliste, voyager aux terre australes: Sa vie, appréciation de ses travaux* (Paris et Moulins, 1857); and C. Wallace, *The Lost Australia of François Péron* (London, 1984).

15 Paris, *Archives Nationales*, Marine BB⁴ 995. Ibid., AJ¹⁵ 569, f.367.

16 Ibid., AJ¹⁵569, f.282.

17 Paris, *Archives Nationales*, Marine BB⁴ 994 & AJ¹⁵ 569 contain the documents relating to the genesis and execution of the voyage. The Institut National organized the issuing of *Instructions* for the different sciences. Some were merely a re-issuing of those to La Pérouse, which can be found at the beginning of J.F.G. de la Pérouse, *A Voyage around the World performed in the years 1785–1788 by the Boussole and Astrolabe* (1799; facsimile ed Amsterdam, 1968).

18 Louis-François Jauffret, 'Introduction aux mémoires de la Société des Observateurs de l'Homme', in Copans and Jamin, op.cit. (n.1), pp.71-85.

19 M. Bouteiller, 'La Société des Observateurs de l'Homme, ancêtre de la Société d'anthropologie de Paris', *Bulletin et mémoires de la Société Anthropologie de Paris*, 7 (1956), 449.

20 Bouteiller, ibid., and Copans and Jamin, op.cit. (n.1), list just over sixty members; Reboul, op.cit. (n.3), lists eighty-three. As no membership lists survive and as all subsequent articles appear to derive from Reboul's list, it is not clear where the discrepancy arises. It is not possible to distinguish passive from active members.

21 See Reboul, op.cit. (n.3), p.34.

22 Reboul, op.cit. (n.3), p.127.

23 Baudin, op.cit. (n.9), p.340.

24 Reboul, op.cit. (n.3); Paris, *Archives Nationales*, Marine BB⁴ 995.

25 Jauffret, op.cit. (n.18), p.73.

26 Ibid.

27 Ibid., pp.77-8.

28 Ibid.

29 For example, Locke states: 'America is still a pattern of the first ages in Asia and Europe'; Lafitau (1724) writes: 'the customs of the savages have helped me to understand the ancient authors more easily and to explain several things in their work', while Montesquieu uses the same analogy in the first chapter of *L'esprit des lois.* Degérando claims that newly discovered tribes 'recreate for us the state of our own ancestors and the earliest history of the World' (Degérando, op.cit. (n.7), p.63).

30 Jauffret, op.cit. (n.18), p.77.

31 Degérando, op.cit. (n.7), p.63.

32 Jauffret, op.cit. (n.18), pp.73-7.

33 Hygiene was an elaborate version of eighteenth-century environmentalism, which argued that the complexity of the environment (in its broadest sense, including features such as diet and elasticity of air) determined the degree of civilization available to its inhabitants. Those who followed the doctrine of hygiene amongst the *Observateurs* were Cabanis, Hallé, Sue and Pinel. For further explanation of the doctrines of hygiene, see G. Gusdorf, *La conscience révolutionnaire: les idéologues* (Paris, 1978); F. Picavet, *Les idéologues* (Paris, 1891); and M. Staum, *Cabanis: Enlightenment and Medical Philosophy in the French Revolution* (Princeton, N.J., 1980).

34 Jauffret, op.cit. (n.18), pp.76-7.

35 Degérando, op.cit. (n.7).

36 See note 3.

37 Degérando, op.cit. (n.7), pp.73-6.

38 Reprinted in Copans and Jamin, op.cit. (n.1), pp.171-6.

39 Ibid., p.175.

40 Ibid., p.174-5.

41 For example, the Société d'Anthropologie was established in Paris in 1859, and its founding president, Paul Broca, paid homage to the Société des Observateurs de l'Homme as the 'noble ancestor of the Société d'anthropologie'. (Broca, *Histoire du progrès des études anthropologiques depuis la fondation de la Société, séance solennelle du 8 juillet 1869* (Paris, 1870).
    Referring to the early twentieth century, Stanley Diamond in *Anthropology: Ancestors and Heirs* (The Hague, 1980, p.13), writes: 'In the US, e.g. Franz Boas, established academic anthropology as the integrated study of Man, with each subdiscipline (linguistics, cultural anthropology, pre-historic archeology, human biology) corresponding to a critical aspect of human behaviour.'

42 The problem has become compounded because the native must evaluate what the anthropologist-explorer is asking. For example, does he want the name of all trees of this type, of this tree alone, what the name for branch is, or the purpose for which it is used? Not even the great anthropologists of the late nineteenth century — Baldwin Spencer, Evans-Pritchard, Boas, Malinowski, Radcliffe-Brown — were fluent in the language of their studied tribes.

43 Degérando, op.cit. (n.7), p.67. This is not stating that 'savages' cannot *in principle* reason as Parisians; rather, Degérando is arguing that we cannot *assume* that they reason like us unless we are provided with evidence that this is the case.

44 For an account of the voyage see Baudin, op.cit. (n.9); Péron, op.cit. (n.14); Plomley, op.cit. (n.1). For secondary accounts see Scott, op.cit. (n.10); Dunmore, op.cit. (n.10); Brosse, op.cit. (n.1); Horner, op. cit. (n.1).

45 F.A. Péron, 'Maria Island; Anthropological Observations' in Plomley, op.cit. (n.1), p.91.

46 Baudin, op.cit. (n.9), pp.304-6.

47 Péron, op.cit. (n.1), p.20.

48 Péron, op.cit. (n.45), p.86.

49 Péron, op.cit. (n.45), p.90.

50 Péron, op.cit. (n.8), p.221.

51 Ibid., p.178.

52 Ibid., p.179. The point here is that a different set of standards is being used to assess male and female personalities, as well as in establishing 'savagery'. That this characterization was common practice, both in literature and in scientific endeavours, only emphasizes that stereotypes were deeply engrained.

53 Péron, op.cit. (n.14), p.183.

54 The popularity of naturalism in the eighteenth century is reflected in the works of Enlightenment writers such as Diderot and Rousseau. The lifestyles of both ancient civilization, and the 'state of nature' such as Tahitian life was envisaged to be, were advocated as the solution to the degeneration of Man that had resulted from contemporary civilization.

55 Péron, op.cit. (n.14), p.183.

56 The portable dynamometer with which the expedition was provided had been developed by Régnier in 1798. Buffon had earlier requested such an instrument to aid 'objective' research on the natives.

57 For example, when the natives appeared to have carried off a young male sailor to determine his sex — the lack of females puzzled them — and the French gave chase, the Diemenese ran 'with such speed that they escaped us in a few moments and disappeared from sight'. Péron reports his questioning of Massé, 'a very swift runner': 'What then has become of that nimbleness you brag about all the time? "Good heavens, sir," he replied, "I run like a man and those bastards there run like deer." ' (Péron, op.cit. (n.45), pp.94-5.)

58 Ibid., p.93. A myriagramme is 10 000 grams. 'Of the seven individuals whom I have tested up to the present, three have not exceeded 48 kilograms, three others have reached 52-56 and only one made the needle rise to 61. All of them were nevertheless full grown men and the majority had a constitution which was extremely robust and strong, in appearance at any rate' (Ibid., p.93). Péron seems unable to reconcile the disparity between observations and the results obtained with the dynamometer. He realizes that the natives are strong and agile, but he fails to incorporate the information. (Cf. Péron, 'Experiment on the Strength of Savage People', in Plomley, op.cit. (n.1), p.149.)

59 Péron, op.cit. (n.45), p.93.

60 'M. Péron lit un mémoire sur la force physique des sauvages de la terre de Diemen, de la Nouvelle Hollande et de l'Ile de Timor. MM Coulomb et La Billardière sont nommés Commisaires' (Paris, Acadèmie des Sciences, *Procès-verbaux*, 21 prairial, an 13 (10 June 1805)).

61 La Billardière and Coulomb, *séance* du 3 *thermidor*, an 13 (22 July 1805) (*Procès-verbaux*, Académie des Sciences).

62 Péron, op.cit. (n.45), p.89.

63 Ibid., p.90.

64 Degérando, op.cit. (n.7), especially the eight 'Faults in the observations made up to the present' (pp.64-70).

65 Le Havre, Muséum d'Histoire Naturelle, MSS. 07121,07122. These are a copy, in Péron's hand, of some of the vocabulary of Hunter consisting of three sheets, recto-verso, in double column, in Aboriginal-English and not translated into French. The vocabulary can be found in John Hunter, *An Historical Journal of the Transactions at Port Jackson and Norfolk Island . . .* (1793, facsimile edn, Adelaide, 1968), pp.407-11. The voyage also possessed the vocabulary of La Billardière (Baudin, op.cit. (n.9), p.593); cf. 'In the course of the day we collected a large number of words that do not appear in La Billardière's vocabulary' (ibid., p.347).

66 Péron, op.cit. (n.8), p.218.

67 'It is no doubt unnecessary to warn observers that they must not restrict themselves to establishing these enquiries in the case of a single individual, but that it is necessary to repeat them in a large number of cases, and to compare the results gained', Degérando, op.cit. (n.7), p.87.

68 These are listed in Plomley, op.cit. (n.1), pp.73-4.

69 See F.A. Péron, *Ile Marie: observations anthropologiques,* draft, MSS 10840, Muséum d'Histoire Naturelle, Le Havre.

70 For debate about the degree of civilization attributable to the Tasmanian Aborigines, see, for example, R.V.S. Wright (ed.), *Stone Tools as Cultural Markers: Change Evolution and Complexity* (Canberra,

1977); L. Ryan, *The Aboriginal Tasmanians* (Brisbane, 1981) and L. Robson, *A History of Tasmania, Volume I, Van Diemen's Land from the earliest times to 1855* (Melbourne, 1983).

71   There are two inter-relating theories that operate here. Both rely on the idea of a progression from a simple state to a complex one, and both simplify the complexities of human nature. It is assumed that there are parallel psychological developments between societies and individuals. With respect to communicating with the natives, Degérando writes: 'It is clear that to establish an initial intercourse with them, we need to go back to signs which are closest to nature; with them, as with children, we must begin with the language of action' (Degérando, op.cit. (n.7), p.71). He further suggests that the explorers should use the general methods of teachers of deaf-mutes, 'For the deaf-mute is also a Savage, and Nature is the only interpreter to translate for him the lessons of his masters' (ibid, p.72). The four stages of society can be characterized as: hunter/gatherer; pastoral; agricultural; commercial. For an interesting discussion of these issues, see R.L. Meek, *Social Science and the Ignoble Savage* (Cambridge, 1976).

72   See, for example, A.M. Gill et al. (eds), *Fire and the Australian Biota* (Canberra, 1981); Rhys Jones, 'Fire-Stick Farming', *Australian Natural History,* 16 (1969), 224–8; and D.R. Horton, 'The burning question: Aborigines, fire and the Australian ecosystem', *Mankind,* 13 (3) (April, 1982), pp.237–51.

73   Péron, op.cit. (n.8), p.187.

74   Baudin, op.cit. (n.9), p.307.

75   Ibid., p.332.

76   This is the subtitle of Péron's *Maria Island* op.cit. (n.45).

77   Baudin, op.cit. (n.9), pp.344–50.

78   Baudin, 'Rapport au Ministre de la Marine et des Colonies'; Paris, *Archives Nationales,* BB[4] 995, reprinted in Copans and Jamin, op.cit. (n.1), pp.205–17. English translation in Plomley, op.cit. (n.1), pp.104–9.

79   Ibid., and Muséum d'Histoire Naturelle, MSS 2082. These reports are virtually identical.

80   'According to his report, the gardener (Guichenot) had collected more than one hundred and fifty different species of plant during his stay ashore and had sixty-eight pots of growing ones. This was work and not wit. I trust that Citizens Péron and Leschenault will have composed sixty pages of writing which, for a different reason, will be all wit and no work.' (Baudin, op.cit. (n.9), p.490).

81   Péron, op.cit. (n.45), p.83.

82   There is no evidence to suggest that Péron intended to publish this report. However, much of it was eventually included in the *Voyage.*

83   Baudin, op.cit. (n.78), p.107.

84   Ibid., p.108.

85   Degérando, op.cit. (n.7), p.67.

86   Baudin, op.cit. (n.9), p.490.

87   Ibid., pp.344–50.

88   Péron, op.cit. (n.14), especially pp.180–1.

89   Péron, op.cit. (n.8), p.175.

90   Ibid., p.210.

91   Ibid., p.186.

92   Ibid., p.217.

93   Baudin, op.cit. (n.9), p.344.

94   Baudin, op.cit. (n.76), pp.105–6.

95   Péron, op.cit. (n.14), pp.85–6. In the *Voyage,* Péron remarks that they are of 'completely average height'.

96   Wallace lists ten publications that appeared in the *Journal de Physique* and the *Annales du Muséum d'Histoire Naturelle* (Wallace, op.cit. (n.14), Bibliography). Some of these were presented to the Institut National; see *Procés-verbaux, an* XII (1804) to 1807.

97   J.B. Leschenault de la Tour, *Végétation de la Nouvelle Hollande et de la Terre de Diemen* (Paris, 1824).

98   Baudin, op.cit. (n.76), p.107. See, also, Baudin in F.M. Bladen (ed.) *Historical Records of New South Wales,* Vol.V, (Sydney, 1896), pp.830–3.

99   Sir James Barnard, Address to the Anthropology Section, *Report of the Meeting of the Australasian Association for the Advancement of Science,* 2 (Melbourne, 1890), p.597.

100   I would like to thank Rod Home, Homer Le Grand, Rhys Jones, Tom Perry, Fritz Rehbock and Roy MacLeod for their comments on an earlier version of this paper. I am grateful to the Archives de France, the Académie des Sciences, the Muséums d'Histoire Naturelle in Paris and Le Havre for permission to quote from their archives.

3

Cultural transmission: Science and society to 1850

Sybil Jack

The history of British colonization of Australia is inextricably bound up with the development of scientific knowledge in Europe. The pursuit of science contributed to the exploration and colonization of Australia and the colonization of Australia to the restructuring of scientific disciplines in Europe.[1] However, the precise nature and extent of that contribution remains problematic, and its relationship to the cultural transmission of science to the colony and the development of an independently transmissible tradition of science in the colony often tenuous.[2]

Interactions between science and the 'new' continent, however, are many and various. This chapter does not set out to cast new light on the scientific activity undertaken in Australia during the period discussed here or its contribution to the development of scientific thought and understanding in Europe, although undoubtedly this was considerable and its importance deserves a separate study.[3] The chapter will concentrate on a single issue, the growth of scientific consciousness and scientific activity in the colony by those who were, voluntarily or involuntarily, to make the colony their permanent home. It will try to do this in the context of the broader progress of scientific enquiry in Europe. Inevitably, the cultural transmission of scientific ideas from Europe shades into the wider problem of the cultural transmission of other ideas, but these will be handled only tangentially in so far as they relate to the transmission of science to the society that was establishing itself in the colony.

The free flow of ideas and ideologies to the Australian colonies was affected by a range of factors. The development of those ideas when they reached the colonies was then affected by a different if overlapping range of factors. Political structures and imperatives, religious concepts, economic needs and technological constraints, geographical problems and social expectations all played a part in creating the matrix within which the colonial scientific tradition was established.

Initially the most important factors were the political considerations, which influenced the government both in Britain and the colonies. Particularly in the first thirty to forty years of settlement, when the penal function of the settlements and their penal institutions were a primary preoccupation, the government understandably viewed most potential developments as deserving encouragement or prohibition in direct ratio to the probability of their stirring up disturbances. Studies of science in Britain show that science was from time to time in this period seen as potentially subversive. It seems likely that at such times the promotion of science in the colonies would have been discouraged and obstructed.

Paradoxically, from the time of Cook, science had been a major reason for European interest in the continent. Discoveries made in Australia in turn excited, stimulated and baffled those working in the areas of botany, zoology and geology in Europe. The impact, on the development of systems of classification, of the specimens, sketches and descriptions of the flora and fauna that were carried back to Europe from Australia was considerable. Australia became the object of much European curiosity[4] and her natural history in particular the subject of learned investigations.[5] Cook's voyages stimulated several further voyages of discovery, especially by the French, which produced their own quota of works. Phillip's founding fleet included a number of officials who hoped to profit by garnering material for publication.[6] Australian information was also rapidly included in the latest and most up-to-date geography books, together with the new astronomical discoveries of Dr Herschel and others.[7]

With government encouragement, therefore, Sir Joseph Banks sent out, at his own expense, the botanist and collector George Caley and at public expense Robert Brown, Allan Cunningham and Ferdinand Bauer.[8] On the whole, the more fully qualified they were, the less time they actually spent in the colony. The results from the activities of such observers, temporarily settled in Australia, were not primarily for the consumption of the public on the spot. Banks made his material readily available to appropriate scholars in Europe and his generosity in this respect was much admired, but its direct fruits sometimes took time to return to the land that had generated them. Flinders' voyages around Australia may have 'marked an epoch in the history of [botany]'[9] but in the colony at the time their importance was purely practical. Scientists in Europe were fascinated by the marsupials and especially the monotremes, and learned papers on their reproductive methods appeared in prestigious journals such as the *Philosophical Transactions* within six years of the foundation of the colony.[10] The views of Aborigines or colonials on these issues, however, were paid scant attention. Debate about their taxonomic place was conducted in Europe.

Despite the attention focused on Australia, we cannot assume that the growth of a scientific tradition in Australia would be the inevitable product of the passage of scientists through Australia and their study of Australian phenomena. Important in themselves, and intensely exciting to the scientists, they were not necessarily immediately relevant to a struggling penal settlement. After the initial burst of enthusiasm, moreover, the visits of scientists waxed and waned in response to the current trends in European science itself.

In short, the voyages of Captain Cook and the abiding interest of that key English patron, Sir Joseph Banks, in the country he had so fleetingly visited as a young man should not lead us to suppose that what was learnt *about* the country flowed immediately into the development of scientific knowledge *in* the country. Nicholson, in addressing the Sydney Mechanics School of Arts in 1839, might have tried to stimulate Australian interest in science by pointing to the geological discoveries in Australia that had been the means of establishing 'the important theories of the most eminent European philosophers',[11] but he was admitting that this was not a native achievement.

Creation of a scientific tradition

The creation of a scientific tradition in Australia during the first thirty years of European settlement did not have a high priority. Phillip and Paterson and the other men of upper ranks who had an interest in science communicated with and through Banks on matters of scientific interest,[12] but they were mainly birds of passage, destined after a longer or shorter term to depart and, in the first thirty years at least, limited in their transmission of their knowledge by the imperatives of a penal settlement.

While the small and often transient government circle may have been well informed and up-to-date on matters of European philosophical debate on scientific issues, what was known in the sub-culture of the ruled is a matter for speculation. Despite all that has been written by Australian historians about the convict era, we know far too little about the intellectual culture of the bulk of the convicts and their guards.[13] This has nothing to do with whether the convicts are perceived as moral outlaws or as potentially responsible householders and citizens needing only a helping hand to be rehabilitated. The problem relates to what they knew on matters scientific when they arrived, and what they subsequently learned. The currency lads and lasses may well have embraced respectable and acceptable *mores*; but what did they learn from their parents or at school about science, in what way did they learn, how were they taught and what stimulus to enquiry and investigation did they receive?[14]

One may perhaps assume certain general cultural norms that would have had some bearing on their possible scientific cast of mind among those born elsewhere, who were for the first forty years or more always in the majority. Diverse in other ways though the arrivals may have been, at the most basic level they and their masters presumably shared some cultural expectations so deep-rooted as to be unconscious: what constituted food (grain, type of animal flesh), appropriate clothing and housing, behaviour relating to physical needs, male/female roles, social conduct, even forms of punishment. It is noticeable in the forms of even the simplest houses they built, in their struggle to produce enough cloth for 'basic decency', in their tendency when not in barracks or prison conditions to live in households whose norm was the longstanding English four or five. At the same time, the rulers expressed and imposed their notions of social situation and hierarchy from the first in their town plans and public building arrangements.

These overriding European expectations, which established the objectives of the colony from the start and determined the colonists' frustration with the incalcitrant land, were constantly reinforced in the first period of settlement by the arrival of new migrants. If the community that arrived in 1788, about 1200 strong — about the size of a large village — had remained out of touch with the motherland, without reinforcements (although it was hardly culturally or even demographically viable) it might have begun to establish a distinct native tradition. But this was not the situation. For the first thirty years, migrants, for the most part involuntary, continued to arrive — at the rate of a thousand or so a year at first, more as the 'twenties and 'thirties progressed. As a result, the native-born were always a minority, though sometimes a sizeable minority, among the adults. Whether accurately or not, as late as the 1840s writers would routinely claim that the majority of their readers would not have been born in the colony, and would mostly have arrived within the past ten years.[15] Newcomers tended to assume their own superiority[16] and more up-to-date skills.[17] The motherland's traditions were therefore constantly reinforced by newcomers. This can only have strengthened the general unwillingness to perceive that the Aborigines' culture might contribute something useful to the new European culture they were establishing in Australia and the failure to accept the idea that there was a need to adapt themselves to the land. Instead, there was constant grumbling at the unco-operative, nay perverse, nature of the elements. Arriving in a place where their accustomed habits and practices were inappropriate, the colonists nevertheless resolutely sought to impose their own outlook and attitudes on an environment unsuitable for many of their practices.

Sources of scientific information

Although research has shown that a surprising number of the arrivals could read, possibly more than the average in the community from which they came, books had to be imported, which could be done only by the well-to-do.[18] In this period, once the residents had arrived, they were dependent on correspondents to send them necessary and recent works. Sending out parcels of books on consignment did not become a regular practice until the 1820s and booksellers did not appear until the 1830s, when they eked out their existence by selling other goods.[19] The better-off literate and articulate complained of problems in obtaining books, so one can only assume that, whether or not they could read, the convicts cannot have had much access to books and most of their knowledge must have depended on oral transmission.

This was the part of the community that by force of circumstances could expect to stay and make their life in the colony. The degree to which the unfamiliar surroundings of a world turned upside down afflicted them with *anomie* must remain uncertain, for they were historically voiceless and essentially passive, despite the outbursts of brawling and drunkenness so often referred to in correspondence with Britain. Even the aversion to the land expressed by repentant runaways cannot be taken at face value, for some at least lived with the Aborigines for considerable periods.[20] The majority of the guards were little different from the convicts, many of them being themselves in the colony as some form of military punishment for dereliction of duty.[21] The great majority were men who could on the balance of numbers have little expectation of transmitting their knowledge to their own children. How far they transmitted knowledge to the children of others is hard to tell.

From the founding of the colony to the conclusion of the Napoleonic wars, the only education available in the colony was of an elementary and practical kind.[22] The need for some schooling had, of course, been recognized from the start, particularly if the native-born generations were to be freed from the allegedly pernicious influence of convict parents. Nonetheless, the priority for school building was often less than that of housing the assistant surgeons,[23] and the provision of teachers whose educational and moral character was deemed appropriate by the government was often impossible. The early, recoverable history of education in Australia is the history not of teaching but of the political struggle between different religious denominations to dominate the syllabus and licensing of schools. The influence of religious infighting on the structure and curriculum of schooling in Australia at this period was enormous. From the start, the different denominations, well aware of the significance of control over education, seem frequently to have been more concerned to establish their own predominance in moral discipline than to share the burden of general instruction.[24] The teachers may have been no worse than they were in Britain but most taught little beyond the rudiments of reading and writing.

The absence of a native press, beyond the printing of the *Gazette* and some government papers, until the 1820s, also meant that the imported books, often virtually irrelevant to the colony, conveyed many other images to the learners as well as the art of literacy. Training for the colonial born was complicated from the start by the multiplicity of other cultural functions that education had to provide if the end products were to be adequately socialized for the role in life envisaged for them.

If science provided the specific ideology through which the hegemony of the upper classes was maintained, the convicts or their offspring stood in little danger of absorbing its tenets for, so far as records show, they did not encounter them. Throughout the time when, in Europe, scientific knowledge was changing more rapidly than ever before; when in every sizeable provincial town in Britain, societies were being established to pursue scientific knowledge; when the Royal Society in London was being challenged by the Royal Institution and a range of small but vigorous specialist societies; when, in short, science was emerging from natural philosophy and natural history to become increasingly professional, seeking to establish itself as the main validating form of knowledge; when (to follow some historians at least) science was a moving part of the restructuring of society that accompanied the Industrial Revolution, Australia remained, was indeed consciously kept, a backwater.[25]

For the first thirty years of the colony's history, then, the possibility of establishing a local scientific tradition was remote. Certainly there were educated men in the government circle and among the convicts. Most of the latter were rapidly elevated to places of responsibility by a government desperate for skills. There were some educated men, too, who came free, but until the 1820s many of these had little scientific interest.

The few men with any scientific experience who were permitted to come to the colony in its early days[26] came, from all the evidence, like John William Lewin, the naturalist and artist, because, disadvantaged by limited education in a competitive Europe, they hoped to use the fruits of colonial hardship to leapfrog to prominence at home.[27] As it turned out, Lewin neither realized his ambition to return nor prospered in the colony. He was before his time in seeking to profit from specialized skills in Australia. Although he notched up an Australian first by engraving his own plates in Australia, and his books on Australian insects and birds were well received in England, the list of local subscribers was short. A generation was to pass before another migrant produced plates locally, and yet a further generation before local book production, a critical factor if a native tradition were to establish itself, started to become established.

The young colony really had no place for a long-term resident scientist *per se*. Establishing a scientific tradition needed more than the mere presence of settlers with scientific background. George Suttor illustrates this point. He was a man with a working knowledge of horticulture and botany, who came out with a fair supply of books, knew all the botanists who visited the colony, and corresponded with Banks, who had promoted his migration. He was unsuccessful in his initial attempt to establish viticulture but managed to establish an orangery. His concerns, however, were with establishing his family first and with scientific matters only second.

Suttor's problems in educating his family indicate clearly why intellectuals might be unwilling to migrate to a country without an established cultural tradition and without good schools. 'We did what we could to teach their young minds how to shoot and to follow in a right course by reading the scriptures . . . [but] . . . early too our boys were obliged to tend the ewes and cows . . .' Eventually one son, Edwin, after attending King's School was sent to university in England, but this produced other problems. Pioneering did not lend itself easily to culture.[28]

The attitude and role of the governing class

Survival and making money were the critical issues for the settlers in the early period. Interest in the botany and zoology of the continent extended only so far as practical utility was involved. Concern for the exotic *per se* was stimulated only in as much as individuals became aware that unusual specimens could be sold for money to the collectors.[29] These on-the-spot informants no more needed to interest themselves in the 'real' significance of the material than did the primitives who were the guinea-pigs in anthropological studies. When interest in new species, or in further specimens of already identified species, waned in Europe, so too did the activities of the local collectors. Their interest might have been stimulated had any of the material justified Banks' confidence that among the mass of new species, something useful would be found:

> It is impossible to conceive that such a body of land, as large as all Europe, does not produce vast rivers capable of being navigated into the heart of the interior; or that, if properly investigated, such a country, situated in a most fruitful climate, should not produce some native raw material of importance to such a manufacturing country as England is.[30]

In practice, Banks' claim to parliament that 'it was not to be doubted that a Tract of Land such as New Holland, which was larger than the whole of Europe, would furnish matters of advantageous return'[31] long remained unjustified. Although there was early optimism, local flax did not prove itself a useful commodity, nor local bark a better way of tanning leather.[32] The economic justification of the settlement eventually came as a result of the spread of imported domestic animals and other species.

With so few among the settlers equipped intellectually, socially or economically to promote an interest in science, control over the transmission of an intellectually coherent culture and development of a native tradition, a scientific culture that would be vigorous enough to withstand deterioration, remained in the hands of the powerful. The majority of these had behaviour patterns geared to the political and social approval of the distant community from which they were an only temporarily divided part and whose cultural expectations they therefore sought to uphold. The ideas and attitudes of those who sent out the colonists and who thereafter for many years regulated and attempted to vet the flow of migrants were therefore bound to influence the development of an indigenous cultural tradition.[33] The attitude to science of the governing classes in both Britain and Australia was a critical element in the speed and direction with which science developed in the colonies.

The society that spawned the early colonists saw science as a hobby for the wealthy amateur, and as something that could be a useful tool in religious argument. Many of the most ardent students of natural history were ministers of religion. As yet there seemed no inherent conflict between science and religion. Most accepted Paley's claim in *Natural Theology,* the work that was a second Bible to British protestant studies for half a century, that the study of Nature was a path to God and that the argument from design was the most secure proof of the existence of the Almighty. This fitted well with the long-standing Baconian tradition of empiricism that infused the approach of the gentleman-scholar, for Bacon supposed that to investigate the workings of Nature was to study the works of God.[34]

Science might be a prop to religion, but to say this was not to admit that such a study should be lightly opened up to the untrained or illiterate mind. The study of natural philosophy and natural history might overheat such unprepared vessels. Only the upper classes, whose minds were tempered by the discipline of the classics, could safely embark on such studies. These were, in short, the prerogative, though not necessarily the sole preserve, of gentlemen because they and they alone had the true perspective. Its possession reinforced their prestige, even if it were not, as Berman suggests, a vital part of their hegemonic apparatus.[35] To many eighteenth-century minds, imbued with Baconian empiricism, science and technology were part of a continuous spectrum,[36] which could not be broken down into separable discrete parts.[37] They would have had no difficulty, therefore, in accepting that it was science that made possible the establishment of the colony and the creation of an Empire.

Science also provided them (and this may well apply equally to both convict and governor) with a self-justificatory argument for the moral fitness of their behaviour, especially vis-à-vis Aborigines who were commonly represented as having minds as deformed as their bodies. Science, in short, was embodied in the moral scheme; discoveries had a moral value, which would increase human happiness.[38] The Enlightenment had extended the notion of science to moral issues and, in this way, science indirectly contributed to the form that the early colony took, as the debate over the proper, scientific treatment of criminals developed.[39]

In other circumstances, the governors might have vigorously promoted scientific aspirations in the colony as the type of study that men should seek to undertake. Unfortunately, as most recent studies of science in Britain at the turn of the century indicate, the British government, made more nervous than usual by the outbreak of revolution in France, was coming to see a potentially subversive element in the opening up of science to a wider society. Science was becoming a sensitive issue unless its objectives were purely utilitarian. Provincials, whose status was due to their own endeavours, were interesting themselves in science and forming associations in imitation of the venerable Royal Society of London to pursue these interests. Many of them were at the same time non-conformist in religion and radical in their political views.[40] Whether the new associations were indeed an attempt to seize an ideology and remake it to the advantage of a different class,[41] to make the fostering of science among the settlers a government priority would clearly be an unnecessary complication in a penal colony.

The initial representatives of the government were in any case few enough and, furthermore, rarely drawn from that section of the upper class that had had leisure to develop a widely based interest in either natural philosophy or natural history. Those who came as rulers in the first thirty or forty years of the colonies generally had some experience of science or its philosophy, but as long-serving army or naval officers had skills that were in general more practical than theoretical. Their science manifested itself in the applied arts of surveying, navigating, drafting and describing. This was what institutions such as the Royal Observatory, the Board of Ordnance, the Board of Longitude and the Admiralty wanted.[42] They were not concerned with the leisured development of theory.

In Britain another group often involved in science were the doctors and apothecaries. Medical men as part of their training generally had courses in experimental philosophy, including studies such as chemistry, with its current terminology of salts, earths and waters, as well as courses on magnetism and electricity, given by university readers. Virtually all the early professional men who came out to the colony, however, were attached to the military forces. The earliest medical men were also military men, and military surgeons were rarely the best trained. The medical men who came with the first fleet and who might have taken an interest in science were practical in their orientation like the rest of the military. Recent patriotism has encouraged an over-optimistic interpretation of their early, desperate improvisations. They and the first settlers certainly made a great effort to utilize the local flora and fauna in lieu of familiar drugs, but they did so by crude analogy, which was all too often deceptive. The medical remedies tried so optimistically when nothing better was to hand, such as the use of the native currant or acid berry as an anti-scorbutic, were rapidly discarded by medical practitioners as ineffective once the pressure of the early, precarious months had gone.[43]

Finally, the clergy in Britain were often in the forefront of scientific endeavour. They usually imbibed some philosophy in their university training and were likely to have been influenced by the popularizing of science through the innumerable encyclopaedias and other works for the general audience.[44] This would mean they had imbibed the broad sweep of contemporary ideas, even if they were not highly scientific themselves. Unfortunately, most of those who came to the colonies, especially in the Church of England, men like Samuel Marsden, lacked a university degree, having been ordained for missionary work outside England only. Marsden, in fact, shows what could be done by a vigorous man even without formal training, but his orientation was strictly utilitarian.[45]

All the professionals who came to Australia depended on government favour in a variety of ways to establish themselves. The nascent colony, therefore, had neither the sort of upper class that normally devoted itself to science, nor a class with the resources and leisure to see in the pursuit of science a means of legitimizing their status. They were men who might have imbibed the broad sweep of contemporary ideas, but they were not necessarily highly scientific themselves. It was all very well for the superior convict like Michael Massey Robinson to refer to science in almost every ode he wrote to Albion's advancement, and even rhetorically to speak of:

> A British Chief! Who on Australia's shore,
> First cherished Arts, and bade young Science soar.[46]

In reality, science, weighed down with a convict ball and chain, struggled to crawl. The transference of an existing scientific tradition would therefore be difficult, its maintenance precarious and the development of an indigenous and perhaps alternative local culture probably impossible.

The scientific activities of the first settlers were thus strictly limited. At the outset men like John White, William Paterson and George Bass collected material but did not attempt any taxonomy. Even a supposedly careful and intelligent observer like David Collins, from whose work Britons obtained many of the better early observations of the 'new' birds and animals, did not seek to establish any systematic classification. In the spirit of the Enlightenment (and Sir Joseph Banks) he sought for the useful, as indeed did many mere mechanics who wanted substitutes for unavailable familiar artifacts. Collins in his writing presented himself as a man of the Enlightenment and, although his own sexual conduct may not bear scrutiny, deplored the morals and behaviour of the lower classes, which he principally attributed to drink. He viewed them across a great gulf and never suggested that science might guide them to a better appreciation of their position. The natives, too, he describes with equal parts of care and incredulity, apparently quite insensitive to the wider injustice of their dispossession since they live 'in that state of nature which must have been common to all men *previous* to their uniting in society'[47] (a convenient political assumption). He routinely condemned the petty thefts and injuries done them by the convicts to which he attributes all the sporadic displays of untrustworthy hostility and resistance; but he did not take the Aborigines' culture seriously. They were thus classified, to the detriment of the development of an indigenous understanding of the continent, as lower than the lower white classes whom poor moral fibre and general ignorance made incapable of scientific knowledge. Despite Governor Phillip's early sensitivity to Aboriginal culture,[48] Collins' view would prevail.

If a scientific tradition had to wait on the leisure of such men, then its development was going to be slow. Outside those in government positions who were appointed from Britain, the number of people who had the leisure to live and behave like the British ruling class was always small, and before 1820, minuscule. It is not necessary to adopt Hirst's semantic argument that there was effectively no ruling class[49] to believe that free settlers were in their way as dependent as the convicts on the government until 1831 and unlikely to be a major source of support for science. They were neither outstandingly well-born[50] nor, despite popular rhetoric, particularly wealthy.[51] Until the gold rushes, Australia was not a place where professional men made vast fortunes from their businesses. Even landowners were only moderately wealthy by British standards. They could not finance large-scale private interests in natural sciences or local investigations. Only the government could do that.

The imperial government and the fostering of science

Everything depended on the government and therefore on the expressed expectations and wishes of the British government officials and administrators who dispatched to a land they themselves had never seen, migrants, temporary and permanent, bond and free. They directed the overall policy. Without their active promotion the development of scientific training and enquiry would be unlikely to be supported by the local governors and might be stifled. Their expectations, of course, did not remain constant. Their attitudes shifted with changes in government and with the swings in the political winds. An erratic pattern emerged, which, after an inevitable time-lag, was reflected in a similar swing at the colonial end. Attitudes towards the sciences in Britain, their significance, role and cultivation were eventually carried to the colony within the context of the separately shifting pattern of attitudes towards crime and the appropriate means of handling the criminal. The varying strength of feeling between punishment and rehabilitation was itself an offshoot of scientific thought, the result of medical theories, especially on physiology and madness.[52] This determined what the British government permitted in its convict dump. The opening up of scientific knowledge was handled very differently in times when ideas of the potential regeneration of the criminal were in the ascendant compared with times when the grim notion, expressed most gloomily by Arnold of Rugby, that crime was visited even upon the children's children, prevailed.[53] At all times, however, down to the ending of convict transportation to the eastern states, it was believed that the potential moral corruption in a convict-based society required the application of highly specific educational considerations.

For the first thirty years, therefore, the government of the colony, which contained the vast majority of those able to diffuse the scientific knowledge of the mother country, did not encourage its pursuit even among the handful of respectable free settlers. Although the government in England was willing to encourage the scientist Robert Towson, author of the *Philosophy of Mineralogy,* to go as a settler, to the extent of allowing him £100 for the purchase of books and a laboratory, Bligh, as governor, took another view and when Towson arrived with these invaluable resources in July 1807 insisted that Towson consult the governor on its use. Whether Bligh 'saw it as his object to promote Towson's endeavours' as Finney suggests, or sought to restrict his activities to areas deemed appropriate by the government, Towson subsequently abandoned scientific pursuits and blamed Bligh for this. His correspondence abounds in complaints about his treatment. His potential influence as the most eminent scholar in the colony was limited to his personal friends, and his library was largely unexploited.[54]

When paranoia could lead the government in London to apply the seditious meetings act to scientific gatherings,[55] it is understandable that ordinary settlers in Australia were not to be encouraged. When the government sought to send out 'scientists', they did so for practical ends. A mineralogist, A.W.H. Humphrey, was sent when there was a prospect of coal, iron and other commercial minerals, not primarily for a scientific geological survey. Humphrey, however, like so many of the experts willing to accept so far-flung an appointment, proved hardly a success in either role, being once again more concerned with his personal problems.[56] Time and again, men who came with scientific ambitions turned on arrival to more commercial interests. The lesson is clear: science at this period did not pay, commerce kept body and soul together.

Reliance on Europe for a lead in all matters scientific was thus early a necessity. Imitation promised cheap and reliable returns. For the occasional aficionado, work as a correspondent for someone powerful, or for some prestigious institution in Europe, was the best way to that subtle necessity, the recognition of one's peers. Respectability and authority went together. Scientific hypotheses presented by unknowns were only going to be taken seriously if they were consonant with prevailing orthodoxies and only then if they were capable of being subjected to some form of existing verification. A few men regularly corresponded with England on matters of scientific interest. To argue as A.G. Serle does, however, that the main preoccupation of the best-educated settlers was with matters of scientific interest seems a great exaggeration of the actual activities of most settlers in the first thirty to forty years of the colony.[57] European philosophers often could not find correspondents at all in Van Diemen's Land, and even the governor regretted the colonists' indifference to the native botany of the region. Even the best educated men were primarily concerned with survival in an unfamiliar environment and were reluctant to spend time even to seek out the useful.[58]

Promoters of colonial science

The 1820s saw a gradual shift of position. For this, a number of concurrent factors combined, the most important of which was probably once again the attitude of the British government. After the Bigge report, free immigration was to be encouraged, so the penal aspects of the colony needed to be played down (though this, in practice, came in only slowly and erratically with much backtracking).[59] Science, moreover, was adopting in Britain a form that seemed less threatening to the government. The choice of governors contributed to this relaxation. Sir Thomas Brisbane was a keen amateur of science and brought Christian Carl Ludwig Rümker to the colony where he worked initially as Brisbane's private astronomer, and later as the first government astronomer, justifying his role by finding new comets in 1827. Rümker's career provides an interesting light on the problems that scientists had in getting on smoothly with government, even with so enlightened an enthusiast as Brisbane. Rümker's quarrels with Brisbane and later with Sir James Smith illustrate that science and government did not necessarily mix well.[60]

Brisbane's interest provided a temporary and artificial stimulus to the development of scientific institutions in the colony. If patronised by the government, the wealthier classes, dependent on that government, were prepared to participate in intellectual pursuits priced suitably beyond the means of the artisan. An attendance at lectures, as gratifying in size as it was misleading, was temporarily achieved in both NSW and Van Diemen's Land. The philosophical societies that were founded, however, were not only pretentious but premature, and they rapidly ceased to meet, succumbing, as much as anything, to the factionalism that was the first fruit of the still distant prospect of true self-government.[61]

Nevertheless, when Alexander Macleay arrived in the colony in 1826, there was a handful of individuals who could form the nucleus of a group of aficionados.[62] The 1820s thus saw the first stirrings of ambition for scientific development in the colonies and the first, largely abortive attempts of the residents to provide for themselves. If the beginnings of a scientific tradition are to be seen in local publications, then the 1820s in Australia saw two short-lived and over ambitious publications.[63] All these ventures must be seen in the context of local politics and it is probably significant that those involved were mainly those known, politically, as 'exclusivists' and conservatives.

Again, practical considerations, in the shape of the Agricultural Society of New South Wales, proved more durable than the scientific, even though the society was largely dominated by the same people who had been members of the Philosophical Society.[64] Men were easily diverted from science to technology; Busby, the new mineralogist, for example, was directed away from geology to the construction of a new water supply for Sydney. The government by this time was patronizing practical projects. The establishment of the botanic gardens, recommended by Bigge and actively promoted by the royal family and the British government as part of a wider vision of botanic gardens around the globe,[65] had the advantage of promoting simultaneously the scientific function of botany and the despatch of seeds and live plants to other parts of the globe. It also served the practical, economic purpose of introducing and naturalizing desirable foreign plants. The olive, lemon, lime, banana, date palm, passion fruit and others were all introduced in the late 1820s and early 1830s. Once the plants were established, layers and cuttings could be supplied to deserving settlers. In the 1820s this potential centre for scientific work was focused primarily on the practical. In the 1830s and 1840s it became a valuable focus for colonial scientific activity and competition for the post of superintendent became correspondingly fierce.[66] It was Brisbane who personally established observatories at Parramatta and in Van Diemen's Land. These were officially taken over by the government, which saw that they had practical as well as scientific advantages since the observatories, which were also concerned with meteorology and the weather, were clearly crucial to settlers. There were also associated magnetic surveys and various activities of use to navigation and surveying that made these practical and desirable.[67]

On the whole, it is hard to agree that the foundations of an Australian scientific tradition were laid in the 1820s, even though the dissatisfactions voiced by contemporaries show that the colonists were beginning to be aware of an unsatisfied need. Society was still too thinly structured to provide the patronage needed for scientific training and research, or even the audience that is as necessary to scientists as to artists. Although professionals came to Australia in increasing numbers in the 1820s and 1830s they rarely had the leisure or facilities to promote scientific endeavours. The ordinary working medical practitioner made no systematic attempt to analyze the potential medical properties of the local plants and probably lacked the equipment to do so. Despite the deterioration suffered in transit, drugs were safer imported, and so, incidentally, under government control. Local plants did not, therefore, replace ginger, senna or rhubarb, ipecacuanha, sal volatile, opium and nux vomica.

The total number of educated males in the upper ranks in NSW was probably no more than 300 or 400 and another 200 or so at most in Van Diemen's Land. It was thus unlikely that a native-born youth with brief schooling and the handicap of poverty might find a route to membership of a scientific community. Even a boy like Robert Lawrence had problems because he had no books and no-one with whom to talk.[68] The British myth was that scientific originality would allow many individuals, even those originating outside the Establishment, nonetheless to achieve entry.[69] This ignored the degree to which the looser structures of British society facilitated access to basic skills. Even in the 1820s, in Australia, such basic training was barely available. Even to keep abreast of the changes and the raging controversies that were occupying the minds of scientists in many fields in the first quarter of the nineteenth century required time and money. It increasingly also depended on imported books. What was perhaps stabilizing in the 1820s was the presence of a tiny group whose scholarship, essentially passive and dependent as it was, was nonetheless beginning to be more appreciated and encouraged by the government. In helping them to keep up to date, Alexander Macleay's library must have been at least equally important to them as Macleay himself.[70] Nevertheless it must have seemed a losing battle for the residents and one more form of colonial dependence for the favoured few. Men who would have barely assimilated Lavoisier had to learn afresh and belatedly of the theories of Dalton, the laws of chemical combination, constant proportions, equivalents and multiple proportions. The next shipload of arrivals would in prospect be more informed and have a better understanding, even possibly some of the convicts among them, like Bland.

The influence of free settlement

Nevertheless, free settlement brought with it the idea that the colony was part of the motherland, removed from it by geography but not by history, another form of province whose dwellers were a particular sort of provincial. Replicating the motherland implied replicating its social structure and its intellectual ideas, scientific and other. Here the nascent upper class left much to be desired. Even if the emancipists were excluded, the pushing, upwardly mobile group drawn from the fringes of respectable society in Britain was hardly the stuff from which administrators were conventionally drawn. Colonial administrators worried that the dominance of such people in the social life of the colonies made it increasingly impossible to maintain the sort of behaviour patterns they regarded as appropriate to the upper classes. Society became increasingly measured solely in terms of wealth and little, if at all, in terms of refined taste and intellectual capacity. This was not an atmosphere in which science was likely to flourish unless its utilitarian side was so plain as to override all argument. The majority of the successful had abandoned the elegant conventions whereby science was a part of the gentleman-scholar's claim to government, in favour of more mundane matters. Circumstances also rarely permitted their wives the leisure to take a genteel interest in botany or ornithology.[71] The rules for establishing status in the colony had been irrevocably simplified and had become more blatantly economic.[72] The government came to believe that for these classes education, including perhaps the absorption of scientific knowledge into the culture,[73] was becoming imperative.[74]

Before the mid-1820s the government's educational priorities had been for practical training, for carpenters, shoemakers, tailors and such like.[75] In many respects its interests continued to be purely practical as the complaints of the geologists appointed in the various colonies testify. Science was irrelevant. Nonetheless, by the mid-1820s a need for more scientific education, at least for some of the better off, was beginning to be accepted. Archdeacon Scott's report on schools in 1826 used the familiar rhetoric that played on the 'alarming apprehension of open resistance if not of open rebellion to the government'[76] to promote a plan for a boarding school in which the curriculum would include the rudiments of natural history (mineral, vegetable, animal), experimental lectures on the chemical elements, astronomy, higher mathematics and navigation.[77] Scott's plans were more forward looking than could easily be achieved, and significantly included the idea of exhibitions at Oxford and Cambridge so that colonists could drink at the fountainhead of learning.[78] The longstanding and legal responsibility of the Anglican church for education was in the Australian context something of a handicap in realizing any plans, since the church's resources were limited. The educational establishments of the early 1830s, the King's School or Lang's Australian Academy, were therefore more restricted in their scientific training, and maintained in strict imitation of the homeland. Imported textbooks and music must have created complex, European-dominated impressions even in boys with some elementary sense of their own Australian-ness. Nevertheless, it was a considerable step forward.[79]

By the later 1820s, attitudes in Britain towards the propagation of science and the advancement of education for the mechanical classes had shifted again. The government was prepared to listen to arguments made by men like Birkbeck and Brougham that understanding would better fit the mechanic for his trade. In this way, the movement in Britain to build mechanics institutes received its initial impetus at middle-class hands, hands that perhaps sought to use the institutions to maintain social order.[80]

It was with similar expectations in mind that the colonial administration, under the Whig governor Bourke, in 1833 promoted the opening of the Sydney Mechanics School of Arts even in a colony 'so peculiarly constituted as NSW'.[81] The movement was fostered by the presbyterian Scottish clergy, especially J.D. Lang. R. Carmichael's opening address might have been described as plagiarism, so closely did it follow the expressed hopes of Birkbeck's English followers.[82] Since Bourke was willing to patronize the undertaking, it was reasonably well received. The first meeting was attended by two hundred citizens. By 1837 it had received enough scientific equipment to make demonstrations in lectures possible. By 1839 it had six hundred members.[83] Van Diemen's Land followed suit. In 1839 Melbourne's ambitions prompted its local government representatives and a few tradesmen to open their own institution, again with the object of 'the diffusion of scientific, literary and other useful knowledge among its members and the community generally and particularly among the young as well as the operative classes'.[84]

Science had become fashionable. The Sydney Institute's management remained fairly exclusive; Bland, for example, was cold shouldered because he was an emancipist even though as a physician and inventor with a deep interest in phrenology, he was a *rara avis*.[85] The lecturers were enthusiastic, showing not only a sense of mission but, at least in some subjects such as geology, a commendable grasp of the latest European theories.[86] They were largely assisted in this by the importation of appropriate books not only into private libraries but also into more modest lending libraries of the schools of arts[87] and into the Australian subscription library. Thus the Sydney Mechanics School of Arts library catalogue for 1842 had thirteen works on chemistry, twenty on mathematics, astronomy and education, thirty-nine on mechanical arts and sciences, ten on medicine, twenty-four on natural history and thirty-four on natural philosophy including texts by such authors as Faraday. Most were of a popular or textbook nature but, for those who wanted to learn, the basics were beginning to appear.

The enthusiasm for science, however, was largely confined to the management, who did not persuade the general membership to share this view. In Australia, as in Britain, the mechanic was not much inclined to be stimulated towards natural history, especially geology and botany. Like their British counterparts, the Australian mechanics, many of them after all recent migrants, sought political rather than scientific enlightenment as a solution to their problems.[88] They, like their British counterparts, sometimes preferred the older, informal and customary ways of doing things to the canon of scientific method, which also promoted bureaucracy, division of labour and professionalism.[89] Sometimes they simply preferred escapism.[90] Enthusiasm for promoting science in this way waned somewhat in the 1840s. In Melbourne, the founders of the institute turned instead to imitating the British exclusive scientific club, the Athenaeum. In Sydney the popular functions largely overtook the scientific although the idea of founding such schools remained popular in country towns throughout the century.

*   *   *

The difficulties of sustaining 'colonial science'

By the mid-1830s, therefore, a scientific tradition of sorts had been established, but it was still dependent and impoverished. It was, moreover, still a largely amateur effort at a time when in Europe the amateur was rapidly giving way to the professional in the vanguard of scientific work. In Britain, science was surging ahead under the guidance of a sizeable intellectual group drawn from a wide range of emerging disciplines with a strong base in the universities and museums.[91] The foundation of the British Association for the Advancement of Science, its claims legitimated by the moderates in the Church of England who devised its strategies, able to turn the 'natural order' of science into the existing economic and political order[92] and propagate its views by its annual pilgrimages to various provincial centres, had once more changed the rules of the game to Australia's disadvantage. It had both defined 'science' in ways that excluded or downgraded certain subjects and elevated the 'science' so defined to a major force in the kingdom, widely accepted as a necessary basis for future material progress. It was also seen as the basis for progress in imperialism but the rhetoric was not at this time turned into a reality. As Morrell and Thackray say: 'in practice there was little scientific trade between the Colonies and the United Kingdom'.[93]

Certainly there was some continuing correspondence and some continuing interest. The arrival of the Reverend W.B. Clarke and Frederick McCoy among others stimulated geological debate in the colony, which was ultimately to make its contribution to the solution of the problem at issue.[94] The colonists also made a determined effort to develop museums along the lines newly fashionable in Europe, but although Australia's resident experts might now be able to promote science, they were not yet really able to advance it. British scientists were becoming professionals;[95] Australian scientists could rarely find employment that did not require them to turn away from theory and towards practice. Just as the town of Sydney was becoming large enough perhaps to emulate the British provincial towns with their active local philosophical societies,[96] the utility of such general institutions was fading, swept away by profound changes in the perceived nature of science itself. Advances were increasingly made only by those whose lives could be devoted to the pursuit of such knowledge.[97]

Promoters of science in Australia faced another difficulty. Although they could borrow the BAAS arguments about the moral dimension of its study and especially its methodology,[98] in a society that depended primarily on pastoralism and agriculture, the arguments about science's contribution to progress rang a little hollow. The direct bearing of science on the exploitation of Australian resources had been limited. The methods used in Australian farming were, from a European point of view, backwards. The hoe and spade long held their own against the plough. In the 1820s, promoters of progress like Sir John Jamison, president of the Agricultural Society, might try to establish experimental gardens and test new varieties of plants, but little progress was made. His reports as president see hope only in more migration. Similarly, James Atkinson castigates the generally slovenly habits of farmers and believes that much could be utilized if men would only devote themselves to enquiry.[99] This was of course published in London to promote migration.

Pitifully little had been accomplished on the practical side, in research into native plants, by 1830. No systematic attempt had been made to learn from the Aborigines, who had used the plants as their sole resource for so many millenia.[100] When Western Australia and South Australia were opened up, moreover, the pattern of the earlier colonies was repeated in this respect. Europeans wanted specimens of the unknown local species and collectors were paid to send them, but even with the new museums that formed collections, or under a man like James Drummond, the systematic study or exploitation of native plants was undertaken slowly, if at all.[101]

In seeking to turn what appeared to them barren to a useful purpose, as Robinson somewhat prematurely celebrated in his *Odes,* promoters still sought the answers in Britain even when they avowedly knew from experience that practices and techniques tried and true in one country could not be effectively employed in the other. 'Nature' in Australia was still categorized as freakish and perverse, contradicting otherwise universally reliable scientific opinions and, more immediately aggravating, contradicting most of the accepted signs of agricultural fertility. As Martin put it in 1838:

> The appearance of light green meadows lured squatters into swamps where their sheep contracted rot; trees retained their leaves and shed their bark instead; the more frequent the trees the more sterile the soil. The birds did not sing, the swans were black and the eagles white. The bees were stingless, some mammals had pockets while others laid eggs. It was warmest in the hills, coolest in the valleys.[102]

Despite the continued failure of a variety of attempts to introduce known European crops, this was doggedly pursued as a hoped-for method of overcoming the problems of drought conditions. There was a naïve belief that once technical problems had been overcome, British grasses would enable heavier stocking rates and a restructuring of agricultural society more on the lines of desirable British norms with smaller holdings and denser settlement.[103] Recognition of a need instead to adapt local grasses and other crops was slow in coming, in part because it was easier to blame the ignorance and slovenly habits of the ex-convicts and small-scale free settlers for their refusal to co-operate. No-one stopped to consider whether their experience in attempting to eke a living from their holdings might not have indicated a greater wisdom. Certainly the settlers did not regard the advice they got as worthwhile.

Science might have seemed a source from which a solution might come, had any scientist turned his attention to agricultural chemistry. Before Justus Liebig took up its study in 1838, however, there were only Humphry Davy's lectures (1802–12) to turn to[104] and they had little value in Australian conditions. Not until the 1840s, therefore, could any potential use be envisaged, and even in the 1840s lectures on agriculture in the Sydney Mechanics School of Arts were rare.[105]

The 1840s, however, were to be a watershed in the history of culture in Australia. In 1840, 'scientific' ideas on the handling of criminals in Britain led to the abolition of transportation to New South Wales, although convicts continued to be sent to Van Diemen's Land and were later to be sent to

Western Australia. 'Science' here was once more in conflict with the ideas of many Australians, for the settlers were reluctant at the time to see the system end and brought quite cogent arguments to support its continuance.[106] With the convicts gone, however, there was immediately a movement for self-government, which was not satisfied by the 1842 changes and was accompanied by a politically promoted upsurge of 'Australianism'. 'Native' values were asserted against those of the immigrant from 'the old country'. Clearly, despite the continued influx of people, an overall and distinct Australian cultural pattern was being manufactured. Somewhere within this, a place for science would have to be defined. What developed was the conscious promotion of the idea of the practical man as easily the equal of the man of science, who was usually still an immigrant. A deep-seated anti-intellectual strain is noticeable in the patriotic writings, and this encompassed a great scepticism for the value of theory as against practice.

The implications of this for the government and legislature were profound. If the values of the mother country were to be maintained, it must surely be time to introduce the keystone of the educational arch to the country: the university. The proper respect for religion and science might thus be inculcated into the children of the wealthy. A university, moreover, would mean the employment of men who could devote their lives to scientific study and so put Australia on an equal footing with Europe where science was already irrevocably professionalized. In the 1840s the group that was to promote and defend science in Australia was thus being defined, in opposition to a tradition of anti-intellectualism. It borrowed its beliefs and arguments from the rising scientific *push* in England, particularly from the younger radicals.

It is clear that the nature of the community that settled in Australia, and its exclusion of the older, Aboriginal community from any influence on its culture, shaped the way in which science was absorbed, employed, modified and occasionally developed in the first fifty years of the colonies. The dominance of the European concept throughout can be seen in the constant reiteration of the idea that Australia was the land of contrarieties. Even in the 1850s colonists saw their flora and fauna as freakish:

> There beasts have mallards bills and legs,
> Have spurs like cocks, like hens lay eggs
> There quadrupeds go on two feet,
> And yet few quadrupeds so fleet:
> And birds, although they cannot fly
> In swiftness with the greyhound vie.[107]

At no point does the conceptual framework show any signs of freeing itself from the European norms. Perhaps this is hardly surprising, given the way in which the colony was governed for most of the period by a small group of men who expected to return to Britain. It is also hardly surprising since their very dominance of the continent was due in part at least to European science. Although historians have found it difficult to demonstrate the symbiotic link that nineteenth-century rhetors so confidently proclaimed between science and technology, and have, in fact, increasingly protested that technology and science run on dissimilar tracks and have only partial, complicated and varying interactions,[108] some connection there undoubtedly is; and inasmuch as science contributed to the arts of war, medicine and engineering, as late nineteenth-century scientists were to claim,[109] then to science has to go some at least of the credit or blame for the extension of European occupancy of the surface of the globe.[110] The European conquest of Africa was hastened by the scientific development of the technology of guns, steamships and medicine; it may well be that the earlier conquest of Australia was perhaps delayed by scientific backwardness.[111]

Notes

1 For an interesting discussion of the effects on one branch of study, see Paul L. Farber, *The Emergence of Ornithology as a Scientific Discipline, 1760–1850* (Dordrecht, 1982), Chapter VIII and also, generally, L. Brockway, *Science and Colonial Expansion: The Role of the British Royal Botanic Gardens* (New York, 1979).

2 B. Smith, *European Vision and the South Pacific* (Oxford, 1968) is an admirable introduction to the issues involved.

3 A. Mozley Moyal, *A Bright and Savage Land: Scientists in Colonial Australia* (Sydney, 1986), undertakes this task far more expertly than I could ever do. Her earlier work, *Scientists in Nineteenth Century Australia: a Documentary History* (Melbourne, 1976) is also a fundamental source for such work.

4 For the lists of works, both solid and ephemeral, that were produced see: J. A. Ferguson, *Bibliography of Australia*, Vol. 1, 1784–1830; Vol. 2, 1831–38; Vol. 3, 1839–1845; Vol. 4, 1846–1850 (Sydney, 1941).

5 For example, in an area that fascinated the fashionable in the eighteenth century but is now scientifically peripheral, there was Thomas Martyn, *The Universal Conchologist: Exhibiting the Figures of Every Known Shell Accurately Drawn and Painted After Nature with a New Systematic Arrangement by the Author*, 4 vols, (London, 1789).

6 For example, the surgeon John White published his *Journal of a Voyage to NSW with Sixty five Plates on Nondescript Animals, Birds, Lizards, Serpents etc.* (London, 1789).

7 For example, W. Guthrie, *A New System of Modern Geography: or A Geographical, Historical and Commercial Grammar* . . . (London, 5th edn, 1782); and M. Adams, *The New Royal System of Universal Geography* . . . (London, 1793).

8 C.M. Finney, *To sail beyond the Sunset: Natural History in Australia 1699–1829* (Adelaide, 1984), pp.88–9, 105–6, 112–3, 119–24.

9 So at least said J.H. Maiden in *Sir Joseph Banks, the Father of Australia* (Sydney, 1909), p.111. Even this type of claim may be exaggerated. Voyages to South America, Africa and East Asia were more frequent than voyages to Australia, cf. Farber, op. cit. (n.1), pp.32–5.

10 Finney op. cit., (n.8), p.63.

11 Quoted in G. Nadel, *Australia's Colonial Culture: Ideas, Men and Institutions in mid-Nineteenth Century Eastern Australia* (Cambridge Mass. 1957), p.123.

12 This correspondence is contained in the Banks Papers, Brabourne Collection, Vols. 3–7, 8 and 10, Mitchell Library (ML) MS FM 4/1797–1754; and Banks Papers Vols. 18 and 19, Botanical and Horticultural 1796–1818 ML MS c213, pp.1–22, 29–91, 95–97 and the King Papers Vol. 8, Further Papers ML MS A1980-2 pp.116-21.

13 J.B. Hirst's book, *Convict Society and its Enemies* (Sydney, 1983), despite its title, is mainly concerned with convictism and the economic base of the class struggle.

14 P. Robinson, *The Hatch and Brood of Time* (Melbourne, 1985), argues the case for the solid citizenry; R. Hughes, *The Fatal Shore* (London, 1987), the more lurid view.

15 Nadel, op. cit., (n.11), pp.30–34.

16 Cf. Hirst, op. cit., (n.13), p.195 who disputes this for NSW migrants in the 1830s saying that the migrant conceded the 'boast of superior strength and staying power' to the native-born and 'could not boast of their British birth because the convicts . . . were British born as well' op. cit., (n.13), p.195. The argument is somewhat tendentious.

17 It was on such grounds that writers like Maclehose attempted to persuade would-be migrants of their opportunities in the colony. J. Maclehose, *Picture of Sydney and Strangers Guide to NSW for 1839* (Facsimile edition J. Ferguson 1977), pp.16–20.

18 The *Sydney Gazette* told one or two stories about convict theft of medical books that they could not read, presumably to amuse its literate readers. The shortage of books in a new colony is shown in Western Australia in the 1830s with the problems James Drummond had to overcome, cf. R. Erickson, *The Drummonds of Hawthornden* (Osborne Park, 1969).

19 E. Webby, 'A Checklist of Early Australian Booksellers' and Auctioneers' Catalogues and Advertisements 1800-1849', *Bibliographical Society of Australia and New Zealand Bulletin, 3* (1978), 123-48; 4, 1979, 33-61, 95-150. I am grateful to Wallace Kirsop for letting me see his unpublished paper on 'Scientific Information in nineteenth century Australia and New Zealand: the role of libraries and the book trade'.

20 D. Collins, *An Account of the English Colony in New South Wales, with Remarks on the Disposition, Customs, Manners etc of the Native Inhabitants of that Country*, 2 vols, B.H. Fletcher, (ed.) (Sydney, 1975), pp.356-7.

21 T.G. Parsons, 'The social composition of the men of the N.S.W. Corps', *Journal of the Royal Australian Historical Society (JRAHS),*50 (4) (1964), 297-305.

22 J.F. Cleverley, *The First Generation, School and Society in Early Australia* (Sydney, 1971), pp.23-43.

23 Collins, op. cit., (n.20), Vol. 1. pp.455-7.

24 See generally, Cleverley, op. cit., (n.22), C. Turney, 'The History of Education in New South Wales 1788-1900' (PhD thesis University of Sydney, 1964); and C. Turney (ed.) *Pioneers of Australian Education*, Vol. 1 (Sydney, 1969).

25 Cf. M. Berman, *Social Change and Scientific Organisation, the Royal Institution 1799-1844* (London, 1978) and M. Boas Hall, *All Scientists Now: The Royal Society in the Nineteenth century* (New York, 1983), pp.6-7.

26 The governor had to provide a permit or letter of permission for any free settler.

27 See P. Mander-Jones, 'John William Lewin' *Australian Dictionary of Biography*, Vol. 2, pp.111-2 and Finney, op. cit. (n.8), ch 5.

28 G. Mackaness (ed), *Memoirs of George Suttor F.L.S. Banksian Collector 1774-1859* (Sydney, 1948), esp. pp.58-60.

29 Hirst, op. cit., (n.13), pp.3-4, shows the convicts of the First Fleet collecting specimens to be carried back by the officers of the navy.

30 *Historical Records of New South Wales* Vol. 3, 383.

31 *House of Commons Journals*, 19 (1779), p.311.

32 For early references to flax see James Matra's proposal 23 August 1783, *Historical Records of New South Wales*, Vol. 1 (2) 3 and Heads of a Plan, encl. Sir George Young's plan, 13 January 1785, *Historical Records of New South Wales*, Vol. 1 (2) 19; Phillip's instructions 25 April 1787, *Historical Records of New South Wales*, Vol. 1(2), 84-91 contains directions concerning the cultivation of flax and his early despatches discuss its possibilities: *Historical Records of Australia* Series 1, Vol. 1, 24, 45, 99, 101. In his despatch to Nepean of 17 November 1788 he suggested that the quality of the flax would be superior to the European variety if there were someone available locally to dress it properly; *Historical Records of Australia*, Series 1, Vol. 1, 104. King's instructions to settle Norfolk Island include the provision that he is to begin cultivating flax immediately; encl. 15 May 1788, *Historical Records of Australia*, Series 1, Vol. 1, 33. The view that flax and timber were central to the decision to colonize NSW has been put by both Geoffrey Blainey and Alan Frost. For a summary of the foundation debate see G. Martin (ed), *The Founding of Australia: the Argument about Australia's Origins* (Sydney, 1978).

33 For a general discussion of the impact of British ideas and attitudes on Australian culture see most recently F.G. Clarke, *The Land of Contrarieties: British Attitudes to the Australian Colonies 1828-1855* (Melbourne, 1977).

34 C.C. Gillispie, *The Edge of Objectivity: An Essay in the History of Scientific Ideas* (Princeton New Jersey, 1960), p.264.

35 M. Berman, ' "Hegemony" and the Amateur Tradition in British Science', *Journal of Social History*, VII (Winter 1975), pp.30-50.

36 J. Morrell and A. Thackray, in *Gentlemen of Science: Early Years of the BAAS* (Oxford, 1981) argue that the rejection of the applied arts by scientists was the political outcome of the scientists' struggle in the 1830s to establish their own distinct ground and status.

37 This view is, of course, maintained by A.R. Hall, in *The Scientific Revolution 1500-1800* (London, 1954), pp.364-5.

38 M.E. Hoare 'Cook the discoverer: an essay by Georg Forster 1787', *Records of the Australian Academy of Science*, 1 (4), 7-16.

39 For this, see M. Ignatieff, *A Just Measure of Pain: the Penitentiary in the Industrial Revolution 1750-1850* (London, 1978).

40 C.A. Russell, *Science and Social Change 1700-1900* (London, 1983), pp.96-7, 136-130.

41 M. Berman, op. cit., (n.25), pp.xvii-xxv.

42 M. Boas Hall, op. cit., (n.25), pp.10-14.

43 For the optimistic view, see L. Gilbert, 'The bush and the search for a staple in NSW 1788-1810', *Records of the Australian Academy of Science*, 1 (1966), 6-17; for the more hard-headed view see G. Haines, *The Grains and Three Pen'orth of Pharmacy: Pharmacy in NSW, 1788-1976* (Kilmore, 1976), pp.6-7.

44 Cf. F. Sherwood Taylor, 'Science teaching at the end of the eighteenth century', in A. Ferguson, *Natural Philosophy Through the Eighteenth Century and Allied Topics* (London, Philosophical Magazine, Commemoration Number, 1948).

45 A.T. Yarwood, *Samuel Marsden, The Great Survivor* (Melbourne, 1977).

46 M.M. Robinson, *Odes of Michael Massey Robinson, First Poet Laureate of Australia*, with an introduction by G. Mackaness, (Sydney, 1946), p.58.

47 Collins op. cit., (n.20), vol. i, p.452.

48 See for example, Phillip to Sydney, 15 May 1788, July 1788, *Historical Records of Australia*, Series 1, Vol. 1, 24-29, 48-50.

49 Hirst. op. cit., (n.13), pp.169-174.

50 Ibid, pp.150-1, where he speaks of the 'embarrassingly humble origins of John MacArthur'.

51 W.D. Rubinstein, 'The top wealth holders of NSW 1817-1939', *Australian Economic History Review*, 20 (1980).

52 See generally, M. Foucault, *Discipline and Punish, The Birth of the Prison* (Harmondsworth, 1979).

53 Clarke op. cit. (n.33), pp.30-1.

54 V.W.E. Goodin, 'Robert Towson', *Australian Dictionary of Biography*, Vol. 2, pp.537-538; Finney, op. cit. (n.8), pp.126, 135; T.G. Vallance, 'Origins of Australian geology', *Proceedings of the Linnean Society of New South Wales*, 100 (1975), 13-43.

55 I. Inkster, 'London science and the Seditious Meetings Act of 1817', *British Journal of the History of Science*, 12 (1979), 196-6.

56 G.H. Stancombe, 'Adolarius William Henry Humphrey', *Australian Dictionary of Biography*, Vol. 1, 565-6.

57 A.G. Serle, *From Deserts the Prophets Come: the Creative Spirit in Australia 1788-1972* (Melbourne, 1973).

58 M.E. Hoare, 'Dr John Henderson and the Van Diemen's Land Scientific Society', *Records of the Australian Academy of Science*, 1(3), 8-10.

59 J. Ritchie, *Punishment and Profit: the Reports of Commissioner John Bigge on the Colonies of NSW and Van Diemen's Land 1822-3: Their Origins, Nature and Significance* (Melbourne, 1970), pp.6-11, 23-24, 61, 92, 251.

60 G.F.T. Bergman, 'Christian Carl Ludwig Rümker (1788-1862)', *Journal of the Royal Australian Historical Society*, 46(5) (1960), 247-289.

61 M.E. Hoare, 'Some primary sources for the history of scientific societies in Australia'. *Records of the Australian Academy of Science*, 1 (4), (1969).

62 S.G. Foster, *Colonial Improver, Edward Deas-Thomson, 1800-79* (Melbourne, 1978), pp.23-6, 41-3.

63 E. Newland, 'Forgotten early Australian journals of science and their editors', *JRAHS*, 72(1) (1986), pp.3-18.

64 See, for example, J. Atkinson, *An Account of the State of Agriculture and Grazing in New South Wales*, with an introduction by B.H. Fletcher (Sydney, 1976). Dr Fletcher is also writing a history of the RAS and I am grateful to him for all his help on this matter.

65 L.H. Brockway, op. cit. (n.1).

66 L.A. Gilbert, *The Royal Botanic Gardens, Sydney: a History 1816-1985* (Melbourne, 1986), Chap. 4; W.W. Frogatt, 'The curators and botanists of the Botanic Gardens', *JRAHS*, 18(3) (1932), 101-33.

67 Moyal, op. cit., (n.3), pp.129-31; J. Gentilli, 'A history of meteorological and climatological studies in Australia', *University Studies in History (Western Australia)*, 5 (1967), 54.

68 M.E. Hoare, op. cit., (n.58), p.10.

69 Repeated in H.I. Sharlin, *The Convergent Century: the Unification of Science in the Nineteenth Century* (London, 1967), pp.194-6.

70 On 1-4 April 1845 John Blackman auctioned 4000 of Alexander Macleay's books on the occasion of his moving to the country. This, presumably the most expendable portion, included a large number of volumes of the *Philosophical Transactions of the Royal Society* down to 1838 and eighteen volumes of the *Annals of Philosophy*, Davy's *Six Discourses* and other scientific works.

71 Mrs Georgiana Molloy appears to have been an exception; see A. Hasluck, *Portrait with Background: a Life of Georgiana Molloy* (Melbourne, 1955).

72 Nadel, op. cit. (n.11), pp.34-5.

73 For this idea see Sharlin, op. cit. (n.69), pp.182-3.

74 Arthur to Bathurst, 21 April 1826, *Historical Records of Australia*, Series III, Vol. 5, 150.

75 B. Earnshaw, 'The convict apprentices 1820-38', *Push from the Bush*, 5 (1979), pp.82-95.

76 Scott to Arthur, 13 February 1826, Encl. No. 1, Arthur to Bathurst, *Historical Records of Australia*, Series III, Vol. 1 at p.160. His report on New South Wales Schools and Churches of 1st May 1826 is encl. Darling to Bathurst, *Historical Records of Australia*, series I, Vol. 12, 309-321.

77 *Historical Records of Australia*, Series III, Vol. 5 at p. 160 and in the New South Wales Report, *Historical Records of Australia*, Series I, Vol. 12, at p.319.

78 *Historical Records of Australia*, Series I, Vol. XVI (1922), 220.

79 Alan Atkinson, 'Some documents and data from the King's School, Parramatta', *Push from the Bush* 4 (1979), 56-75.

80 S. Shapin and B. Barnes, 'Science, nature and control: interpreting mechanics institutes', *Social Studies of Science*, 7 (1977), 31-74.

81 Nadel, op. cit., (n.11), p.111.

82 Ibid, pp.114-21.

83 D.I. McDonald, 'The diffusion of scientific and other useful knowledge', *JRAHS*, 54 (2), 176-93.

84 R.W.E. Wilmot, *The Melbourne Athenaeum: 1839-1939* (Melbourne, 1939), pp.7, 11, 13, 15, 18-9.

85 Nadel, op. cit., (n.11), p.119.

86 I owe this information to Associate Professor Elizabeth Webby whose article on an Australian Historical Library (Fairfax, Syme and Weldon) is to appear in 1987.

87 For libraries before 1850 see the comprehensive analysis in E.A. Webby, 'Literature and the reading public in Australia, 1800-50' (PhD thesis, University of Sydney, 1971).

88 Shapin and Barnes, op. cit. (n.80), pp.31-74.

89 H.E. Schwartz, 'The radical artisan: democratic theory and political participation in early industrial England' (PhD thesis, Stanford University, 1981), pp.5, 9-10, 15-6.

90 Nadel, op. cit. (n.11).

91 Morrell and Thackray, op. cit. (n.36), pp.21-29.

92 Ibid., p. 31.

93 Ibid., p. 493.

94 See Moyal, op. cit. (n.3).

95 W.H. Brock, 'Advancing science: the British Association and the professional practice of science' in R.M. MacLeod and P. Collins (eds), *The Parliament of Science* (London, 1981), pp.89-117.

96 For this model see A.W. Thackray, 'Natural knowledge in cultural context: the Manchester model', *American Historical Review*, 79 (1974), 672-709 and I. Inkster (ed), *Metropolis and Province: Science in British Culture, 1780-1850* (London, 1983).

97 Sharlin, op. cit. (n.69), p.27.

98 For a discussion of methodology as the image of science, see Richard Yeo, 'Scientific methods and the image of science 1831-91', in MacLeod and Collins, op. cit. (n.95), pp.65-88.

99 J. Atkinson, *An Account of the State of Agriculture and Grazing in New South Wales* (London, 1826).

100 D.J. and S.M.M. Carr (eds), *Plants and Man in Australia* (Australia, 1981), p.3.

101 Ibid.

102 *Australian Sketchbook*, quoted in J.M. Powell and M.Williams (eds), *Australian Space, Australian Time: Geographical Perspectives* (Melbourne, 1975), p.23.

103 J.M. Powell, 'Conservation and resource management in Australia 1788-1860', in Powell and Williams (eds), op. cit. (n.102), pp.26-7.

104 Sharlin, op. cit. (n.69), pp.126-7.

105 Nadel, op. cit. (n.11), p.123.

106 Clarke, op. cit. (n.33), p.9; Hirst op. cit. (n.13).

107 F.G. Clarke, op. cit. (n.33), p.169.

108 Cf. C.A. Russell, *Science and Social Change 1700-1900* (London, 1983).

109 Ibid., Chap. 13 for scientists sedulously fostering an image in which scientific research is at the least the necessary matrix for effective technological advance.

110 D.H. Hardwick, *The Tools of Empire: Technology and European Imperialism in the Nineteenth Century* (New York, 1981), pp.2-3.

111 See for example, H. Reynolds, *The Other Side of the Frontier: Aboriginal Resistance to the European Invasion of Australia* (Ringwood, 1982).

PART II:
Science in a colonial society

4

The long arm of London: Sir Roderick Murchison and imperial science in Australia

Robert A. Stafford

King of Siluria

Historians of science have given considerable momentum to the study of the relationship between science and imperialism in the past few years, although historians of empire have almost unanimously stood apart from this debate. Yet, since Robinson and Gallagher established the concept of an informal empire based on economic and political influence forming a penumbra around the formal British empire,[1] it has become increasingly clear that cultural imperialism—the export to the non-European periphery of metropolitan technologies, standards, institutions, modes of organization, and patterns of thought—constituted an important aspect of European expansion.[2] Science played a major role in the process of extending and maintaining British paramountcy overseas: like trade and religion, it frequently led rather than followed the flag.

In his seminal essay on the diffusion of Western science, Basalla provided a periodized scheme for understanding the development of independent scientific institutions and habits of mind in peripheral cultures.[3] Useful modifications of Basalla's three-phase model have since been offered by MacLeod and Inkster, both of which rely largely on the history of Australian science for their reappraisals.[4] But as Kohlstedt has remarked from the same contextual vantage point—and indeed MacLeod and Inkster themselves admit—the time may yet be premature for the application of elaborate developmental models and fine weighings of internal versus external influence as appropriate means for analysing the meaning of colonial or imperial science.[5] Despite a growing corpus of excellent research, we remain deficient in rudimentary analysis and even basic facts concerning large areas of the history of Australian science. More work at the coal face is certainly required before we can construct with tolerable accuracy an historical framework that will incorporate individuals, institutions, intellectual trends, and the multifarious external factors impinging on the scientific enterprise, while at the same time expressing the dynamics that link them together.

The career of Sir Roderick Murchison offers an excellent point of departure for such correlative studies. His efforts to strengthen the imperial equation from both the metropolitan and colonial poles illustrate the complex interplay between the forces of science and empire in Australia. Murchison's career as a geologist and a 'statesman of science' in the Euopean context has been intensively analysed during the past decade,[6] but his equally impressive achievements in the imperial sphere have only recently been examined.[7] Because Australia held primacy of place in the imperial vision of the most powerful metropolitan savant since Banks to champion the cause of the Antipodean colonies, it is appropriate that he be considered in any retrospective of Australian science.

Sir Roderick Impey Murchison, Bart, KCB (1792-1871), was the eldest son of a Highland landowner who had amassed a fortune as a surgeon in the service of the East India Company.[8] After receiving a military education he served briefly in the Peninsular War, and in 1824 took up the study of geology in London, rising rapidly in scientific circles through his wealth, zeal, and connections. He was elected to the Geological Society of London in 1825 and to the Royal Society the following year. In 1830 he helped found the Royal Geographical Society, and in 1831 the British Association for the Advancement of Science. Besides serving for many years as an officer of these societies, he was a trustee of the British Museum and a power at the Athenaeum Club. During the 1830s and 1840s Murchison's labours as a Palaeozoic stratigrapher bore fruit in his definition of the Silurian System, which remained the cornerstone of his scientific reputation, the Devonian System, in conjunction with the Cambridge geologist Adam Sedgwick, and the Permian System, following the completion of three surveys of European Russia.[9] With the publication in 1854 of *Siluria*, a massive tome that incorporated new Palaeozoic research by other geologists throughout the world and set forth his views on the occurrence of gold and coal, his celebrity as the premier practical British geologist of his day was assured.[10] Among Murchison's methodological strengths were his reliance on palaeontological evidence to date strata and thus correlate formations in widely separated regions, and the Humboldtian view of earth science expressed in his axiom 'the geologist is but the physical geographer of former periods'.[11] Murchison's main weaknesses as a scientist were a defensive pride in his own accomplishments and an instinctive opposition to the new theories such as Lyellian gradualism and Darwinian evolution that were transforming perspectives within his discipline.

The patriotic and politically conservative geologist had named his Silurian System after a British tribe that resisted the Roman invasion, and he considered his creation as a territorial 'empire' that extended wherever its characteristic fossils occurred. Murchison reigned jealously as the world authority on Silurian rocks, supporting geologists who extended the sway of his system and fighting doggedly against those who sought to limit his writ over geological time or across geographical space. The Palaeozoic research of the 'King of Siluria', as Murchison was proud to be styled, was instrumental in establishing the international prestige of British geology, and he looked upon the spread of his own stratigraphic nomenclature as a scientific corollary of Britain's imperial and commercial expansion.[12]

In 1851 Murchison shifted his career focus from European geology to overseas exploration as he began his second term as President of the Royal Geographical Society (RGS). He was to preside over the RGS for fourteen years during the following two decades, transforming it into Britain's quasi-official directorate of exploration and the most popular scientific society in London. From this commanding position Murchison masterminded a breathtaking series of expeditions that linked his name with such national heroes as Franklin, Livingstone and Speke. As the organizer of scientific reconnaissances within and beyond the borders of the formal empire, and the interpreter of their results to the government and the public, Murchison played a key role in the process of empire-building.

In 1855 the scope of his imperial activities expanded when he succeeded Sir Henry De La Beche as the second Director-General of the Geological Survey of Great Britain. Besides its field staff, this institution comprised laboratories, a Museum of Practical Geology, and the Royal School of Mines.[13] De La Beche had used the survey to perform a variety of tasks for the imperial government, so that Murchison's own appointment as Director-General presented an opportunity not only to campaign for further Silurian extensions but to deploy his scientific expertise in the service of the empire. By providing control over mineral reconnaissance abroad, the Survey post perfectly complemented his role at the RGS. Murchison manipulated these two institutions to investigate the resources of British colonies and territories further afield in the interests of extending scientific knowledge and promoting imperial economic development.

The Victorian scientists most successful at winning government funding for large-scale research projects clothed their disciplinary goals in the trappings of utility. None was more adept at this tactic than the consummate promoter Murchison, whose success sprang from his ability to identify the aspirations of natural scientists with the practical needs of imperial Britain. This chapter examines Murchison's efforts to elaborate the symbiotic relationship between science and empire as it developed in Australia, the continent that held his attention more firmly than any other overseas region because it had the greatest immediate potential for scientific development. Much of this interest was purely scientific. Metropolitan geologists sought correlations between the strata and fossils of the Antipodes and those found elsewhere in order to explain the region's unique physical features and apparently anomalous flora and fauna.[14] But such research also had economic implications, for the discovery of exploitable coal and ore deposits affected the pace of colonial development. Murchison watched the progress of Australia with proprietary concern, for there he believed the role of the scientist as a pioneer of economic advance was being enacted with miraculous success.

Prophet of the gold rush

During his first term as President of the Geological Society in 1832–33, Murchison had encouraged further Australian research,[15] but his personal involvement began in 1838 when he provided the emigrating clergyman-geologist William B. Clarke with introductions to his cousin John Murchison and two key members of the small scientific community of New South Wales, P.P. King and W.S. Macleay.[16] King had been entertained by Murchison in London during the 1820s and had helped John, a former army captain, to find his feet as a farmer and squatter during the colony's great era of pastoral expansion.[17] Macleay had contributed to *The Silurian System*, and after Clarke delivered to him the first copy of this work to arrive in Australia, he began sending Murchison specimens to aid the intercontinental expansion of his domain.[18] Murchison promoted the metropolitan reputations of King, Macleay and Clarke in return for their proselytizing of his geological views. In common with other London savants, he strove through control of publications and honours and the collaboration of emigrant scientists to maintain the authority of metropolitan science over the work of colonial practitioners. According to this division of labour, taxonomy, theory, and the ultimate arbitration of data were reserved for the élite commanding the scientific institutions of the imperial capital, while colonial researchers, like their provincial counterparts in the home islands, were relegated to the subordinate role of field agents and collectors.[19]

Another early Australian contact of Murchison's was Sir John Franklin, FRS, FGS, FRGS, Lieutenant-Governor of Van Diemen's Land from 1837 to 1843. Franklin had shared his first lessons in geology with Murchison, proposed him to the Raleigh Club, a precursor of the RGS, and often met him socially.[20] Franklin played a decisive role in creating a scientific community in Tasmania, while metropolitan scientists evinced interest in the island as a relict environment that might explain what were seen in European eyes as the anomalies of Antipodean biology.[21] In 1838 Franklin introduced to Murchison the Czech naturalist John Lhotsky, who had explored southern New South Wales and examined Tasmania's coal formations.[22] While contemptuously dismissing Lhotsky as 'a mad Polish *friend* of yours', Murchison judged from his specimens that Palaeozoic strata might prove common in Tasmania and in consequence pressured Franklin for fossils after returning from his second Russian tour in 1842. 'I now place great importance upon the extension of the palaeozoic classification which I have established & have myself pushed the comparisons from the remotest parts of Europe into Asia', he wrote. 'I must try to do the same through my allies in the distant colonies and as you are one of my earliest playfellows in geology, I count upon your aid.'[23] In reciprocation, Murchison took care to praise the Lieutenant-Governor's scientific endeavours to the Geological Society.[24]

Franklin enlisted the services of Paul de Strzelecki, a Polish geologist exploring south-eastern Australia and evaluating Tasmanian coal mines being opened by the Governor. Together with Joseph Beete Jukes, geologist on the HMS *Fly* marine surveying expedition,[25] and armed with a copy of Murchison's *Silurian System*, Strzelecki discovered an excellent series of Palaeozoic strata in New South Wales, though without Silurian fossils. Murchison probably received news of this find in 1843 through Sedgwick, Jukes' former professor and principal scientific correspondent.[26]In 1844, following Strzelecki's return to Britain with an introduction to the King of Siluria provided by Franklin, Murchison saw his rock and fossil specimens as well as the manuscript of the book for which he was awarded an RGS gold medal in 1845.[27] Several years later, Murchison was to be instrumental in securing Strzelecki's election to both the Royal Society and the RGS.[28] The recalled Franklin also presented Murchison with a collection of Tasmanian rock specimens in 1845.[29] By the 1840s Murchison had thus realized not only that Australian researchers could supply the metropolis with invaluable new data, but that science could enhance both its local position and larger imperial opportunities by serving colonial governments in the cause of economic development.

As the founder of the Silurian System, Murchison developed a special interest in the formation and distribution of gold. His theories about gold derived from hypotheses about the cooling of the earth, the influence of magnetic and volcanic forces on its crust, the timing of faunal extinctions, and the mechanics of mountain building, glaciation and erosion. Murchison was also influenced by the French geologist Élie de Beaumont's theory of orogenesis, which defined twelve systems of mountain chains throughout the globe, each formed by a catastrophic event. Chains created synchronously should accordingly display parallel compass directions.[30] Combining his own observations in the Urals and those of Alexander von Humboldt in South America and Siberia[31] with this concept, Murchison erroneously theorized that mountain chains formed of Silurian strata had been thrown up around the world during the Permian period and later infused with gold by the agency of intrusive quartz veins. Auriferous deposits, he maintained, would therefore occur in mountain ranges exhibiting these diagnostic features. He believed meridional chains were the result of the Permian orogeny, and the principal gold-producing ranges—the Californian Sierras, the Rockies, the Andes, Australia's Great Dividing Range, the Urals, the Siberian ranges—did happen to be north-south chains displaying Silurian strata metamorphosed by igneous intrusions.[32] By the 1860s Murchison had come to assume that *any* auriferous metamorphosed rocks could be dated as Silurian: gold discoveries, like the discernment of typical fossils, thus represented extensions of his system.[33]

In 1844, in his first presidential address to the RGS, Murchison inferred the existence of gold in Australia's Great Dividing Range by comparing its geology as set forth by Strzelecki with that of the auriferous Urals.[34] In 1846, having learned of the discovery of small quantities of gold on the western flank of the Australian range at Bathurst, New South Wales, and among the copper-rich meridional ranges of South Australia, Murchison sought validation for his initially vague forecast by recommending that unemployed Cornish tin miners emigrate to Australia and search for gold.[35] During the next two years he received samples of the metal from William Tipple Smith, a mineralogist and manager of an iron works in New South Wales who had been inspired by his predictions to prospect in the Bathurst region, as well as further specimens of gold and geological evidence suggesting its widespread occurrence in the metalliferous ridges behind Adelaide from John Phillips, a Cornish mining surveyor of that city.[36] Murchison then advised Grey, Secretary of State for the Colonies, to commission a geological survey of New South Wales in 1848. If the results proved encouraging, he hoped the government might either begin state mining with convict labour or clarify the laws on mineral claims to encourage private prospecting.[37] Though Clarke, Strzelecki, King and Jukes had also campaigned for such a survey, Grey did nothing, fearing, as he later admitted to Murchison, 'that the discovery of gold would be very embarrassing to a wool-growing colony'.[38] In consequence, the colonial authorities remained unresponsive until 1849 when South Australia's copper boom and local reports of gold convinced them to follow the geologists' advice.

. De La Beche, as head of the British Geological Survey, was requested by the Colonial Office to recommend a qualified appointee to conduct a mineralogical and geological survey of New South Wales. After Jukes and another member of the home Survey declined the post, he chose Samuel Stutchbury, an English museum curator who had considerable experience as a coal viewer and had visited the colony during the mid-1820s.[39] De La Beche stipulated that Stutchbury conduct his research in conformity with the methods of the British Survey and collect specimens of economic minerals for the Museum of Practical Geology. In return, the Director-General promised to help Stutchbury date strata and evaluate ores, and to send him scientific news that might 'lead to the development of the mineral wealth of the colony'.[40]

Emboldened by this successful emplacement, De La Beche hinted to the Colonial Office in 1850 that a geologist be appointed in Western Australia to survey its rich new-found deposits of copper and lead, but his suggestion fell on deaf ears.[41] The following year, increasing production and discoveries of metal ores as well as the example of Stutchbury's employment also prompted South Australia's Governor, Henry Young, to request a geological surveyor. The colony's Legislative Council, however, refused to meet the contract terms insisted on by De La Beche to secure a candidate with the first-class qualifications desired.[42] The Director-General then fell back on a recommendation that a civil engineer emigrating to South Australia— Benjamin Herschel Babbage, son of the eminent mathematician Charles Babbage—be appointed to conduct a temporary survey. Babbage held the post of Geological and Mineralogical Surveyor for just over a year, but his conflicting duties as Gold Commissioner and Government Assayer prevented the accomplishment of any systematic field research.[43]

In New South Wales, Stutchbury's efforts were resented by W.B. Clarke, who had long coveted an appointment as government geologist himself,[44] and frowned upon by the colonists as the impractical researches of a visiting naturalist. They were upstaged in 1851, however, by the decisive gold strike of Edward Hargraves. Murchison boasted for the rest of his life that his 'scientific' forecast had anticipated Hargraves' find by seven years, though the government surveyor James McBrien had actually found gold in 1823, Lhotsky had predicted the discovery of precious metals in 1833 and publicly displayed a few specimens he found the following year, Strzelecki had rediscovered gold in 1839, and in 1844 Clarke had reported a discovery made three years before. While McBrien and Lhotsky had been ignored, Strzelecki and Clarke had been enjoined to silence outside official circles by Governor Gipps, who feared the social disruption that a gold rush would inevitably cause in what was still demographically a convict colony.[45] Because of this informal censorship and his own unassailable reputation as a London savant, Murchison was able to blunt Strzelecki's long-suppressed claim by gentlemanly agreement,[46] stave off Clarke's bid for full metropolitan recognition, reduce Hargraves' claim, despite its handsome reward by the Legislature of New South Wales, to the luck of a mere pick-and-shovel prospector, and reserve for himself the lion's share of scientific credit for what was in reality a fortuitous prognostication. Even the Australian newspapers, as another claimant sarcastically noted, had 'commanded in ... thundering peals ... "Fall down and worship the unimpeachable science of Sir R. Murchison".'[47]

Murchison's struggle with Clarke involved theory as well as priority. Clarke, who like Murchison had borrowed most of his ideas concerning gold from Humboldt, had in an article published by the *Sydney Morning Herald* in 1847 independently predicted the discovery of significant quantities of gold in Australia's Dividing Range on the basis of the similarities between that chain and the Urals. Clarke's forecast was based on his own gold discovery of 1841, his knowledge of the geology of the Australian cordillera, a French abstract of Murchison's book describing the Urals, and information supplied by his brother, who as physician to a Persian royal prince had travelled widely in Russia.[48] He maintained the occurrence of gold to be much more varied than allowed by Murchison's 'golden constants', as the King of Siluria termed the geological conditions he believed to determine the distribution of the precious metal. At the same time, Clarke insisted on an extreme interpretation of one aspect of Humboldt's views that did not sit well with the generalist Murchison, who also rejected the fully developed version of Élie de Beaumont's theory of parallelism because of its rigid and abstract geometry. Clarke believed that auriferous mountain ranges occurred along not only parallel but equidistant meridians—the Great Dividing Range lying precisely ninety degrees from the Urals—and that since the gold deposits of the Russian range were found on its dry eastern slopes, those of their Antipodean counterparts should occur on the dry western flanks. The evidence suggests that Murchison even resorted to using his influence at the Geological Society to suppress the full exposition of Clarke's views on gold, counter them with an over-riding publication of his own, and convince the Society's President to repudiate publicly Clarke's priority claim.[49] This dispute, like Murchison's more celebrated controversy with Sedgwick regarding the Cambrian–Silurian stratigraphic boundary, reflects the King of Siluria's extreme jealousy in regard to scientific precedence. It also illustrates the tendency of the London savants to quell bids for equal status from colonial scientists as threats to the domination of the metropolis.[50]

As successive gold discoveries in the 1840s and 1850s excited grave fears of currency devaluation in financial and political circles, Prime Minister Sir Robert Peel, as well as the Master of the Mint and many others, sought Murchison's opinion on the capacity of the monetary system to absorb the vast quantities of treasure being added to the world supply.[51] 'Well may political economists and politicians now beg for knowledge at the hands of the physical geographer and geologist', he boasted, 'and learn from them the secret on which the public faith of empires may depend'.[52] Murchison was convinced that only superficial deposits of alluvial gold, which he believed to have been accumulated by cataclysmic floods redistributing the detritus of parent Silurian matrices broken down by the agency of gradual erosion, could be profitably mined. He believed that gold 'was the last formed of the metals',[53] and followed Humboldt's hypothesis that gold in its original molten state of subterranean emission had been concentrated near the surface by an ill-defined process of sublimation toward the atmosphere. Murchison therefore argued that, as the history of mining seemed to suggest, auriferous veins in solid rock decreased in richness in direct relation to their depth,[54] and that 'just in proportion to the time a country had been civilized, the extraction of the precious metal had diminished'.[55] An oversupply could never occur because the extent of Silurian strata capable of producing gold was finite, and the production of new mines would be absorbed by demographic and economic growth. Citing the examples of Victoria and New South Wales, he thus expressed his hope that alluvial mining in the colonies would serve 'the purposes of Providence in providing for a great augmenting population, and in converting wild tracts into flourishing hives of human industry'.[56]

Following the Australian gold rush, Murchison's numerous publications cemented his fame as the acknowledged expert on the world's gold resources.[57] The great London geographer James Wyld dedicated a pamphlet on the world-wide occurrence of gold to Murchison, and the staff of the Bank of England was also schooled in his theories.[58] At the same time, De La Beche offered special courses at the School of Mines on gold prospecting for Australian emigrants. These lectures, also printed for public sale, were illustrated by Murchison's maps and diagrams, and the speakers relied on his theories. Among these experts was J.B. Jukes, who had returned from the Antipodes in 1846 and secured a post on the home Survey.[59] As Murchison boasted in 1852, 'the *public men* think much more highly of me for having been the first who *worked out* mentally the Australian gold . . . and for dwelling on it in successive years until the diggers discovered it'.[60] To safeguard his reputation and the status thus won for his science, Murchison attempted to thwart charlatans posing as geologists from exploiting the public's gold mania and its growing faith in science as a font of truth.[61] He also refused the presidency of an Australian gold mining company, fearing his name would be used to mislead investors, but he admitted that 'if I did start such a thing I might *waddle out* of the Stock Exchange a much richer man than I am . . . If I had been 20 years younger I might in 1846 have backed up my prediction . . . & have secured *lands* which would now have been worth a colossal fortune'.[62]

Still, Murchison felt no compunction in recommending the mineralogist Friedrich Odernheimer to the Australian Agricultural Company to evaluate the Newcastle coal fields and the gold potential of the Peel River region. He later presented Odernheimer's results to the Geological Society, publicized them at the RGS, and wove them into his gold theories.[63] He also influenced appointments to the new mints and assay offices being established to process the diggers' yield,[64] while mining promoters sent him ore specimens to publicize Australia's wealth and validate their speculations.[65] The Australian prediction, by enhancing his scientific reputation with the lustre of a prophecy of colonial prosperity, went far toward securing Murchison's appointment as Director-General of the Geological Survey in 1855. The colonial geologists and mining engineers thereafter trained at the Royal School of Mines imbibed his ideas about gold, as did students of provincial universities who laboured 'to prepare themselves for the colonies and gold-hunting'[66] under the tutelage of men who owed their appointments to his patronage.

Murchison's stubborn adherence to his gold theory prevented him foreseeing—or, for a long time, even accepting evidence of—the impact of hardrock and deep-level mining on durations of yield once technological innovations opened up the subterranean reserves of Australia, America and South Africa. As the guiding hand of British overseas mineral reconnaissance, he consequently emphasized a search for fresh alluvial deposits rather than intensive development of existing mines. Wherever they penetrated, Murchison's geologists sought the high ground, for the 'gold-finder's' theories dictated that regional 'dorsal spines' would exhibit the oldest strata and thus evidence of any coal or metallic ores. Paradoxically, Murchison's misconceptions actually facilitated mineral exploration, for he was right for the wrong reasons. Gold *was* usually found in mountain chains, whatever their direction, and frequently in quartz veins intruded in Silurian rocks. And while the idea of yield decreasing with depth was largely fallacious, the costs of quartz extraction at mid-century justified Murchison's argument that alluvial gold was more profitable to mine. Placers were also the easiest deposits to locate, and they pointed the way to parent lodes that might be exploited later.

Similarly, Murchison's belief that catastrophic inundations were responsible for the detritus characteristic of gold-producing regions did not deter the discovery of gold actually deposited by glacial and fluvial action.[67] Murchison's 'golden constants'—alluvia derived from Silurian rocks occurring in meridional mountain chains and exhibiting evidence of metamorphism and igneous intrusions—therefore provided a rough guide to the conditions in which gold could be expected to occur. His prestige as a geological authority may have slightly retarded the development of quartz mining, but innovators unacquainted with his ideas rapidly proved him wrong, and younger geologists then jettisoned a set of assumptions based on the outmoded theory of catastrophism.

Following the Hargraves strike, W.B. Clarke was commissioned to reconnoitre the gold districts of New South Wales while Stutchbury continued his general survey. Both appointments terminated in 1853.[68] Clarke discovered a series of Silurian fossils that confirmed and extended the significance of others he had found in the 1840s, corroborating Strzelecki's lithological evidence of an important development of Murchison's system in the colony.[69] These fossils were sent to Sedgwick's Woodwardian Museum in Cambridge for description, but when results were not forthcoming, Murchison—by then head of the British Survey—commandeered them for classification by palaeontologists at his own Museum of Practical Geology. The Director-General's vision of the imperial role of his establishment was clearly set forth in his semi-jocular admonishment to Clarke: 'Had you sent these *colonial* products to these British *Head Quarters*, you should long ago have had their description!'[70] This descriptive work was not completed until 1877, having been turned over eventually to the Belgian, L.G. de Koninck, but enough was accomplished by Murchison's subordinates during his own lifetime to demonstrate the existence of a conformable succession of Silurian, Devonian and Carboniferous strata in Australia.[71]

In the mid-1850s, Murchison and Clarke thus found enough common

Palaeozoic ground to forge a fragile truce based on recognition of their respective metropolitan and colonial claims to priority for the gold prediction.[72] They then joined forces to attack Hargraves' discovery as unscientific—both geologists having been nettled by his sneers regarding the practicality of their discipline, and the impoverished clergyman having been particularly provoked by Hargraves' reward of £5000 for his find.[73] Murchison also sponsored Clarke's election to the RGS at this time to help mollify the truculent cleric, and subsequently advertised the value of his researches in several publications.[74] Clarke, in turn, helped the RGS in its efforts to promote Australian exploration.[75] Simultaneously, however, Murchison was buying insurance against further eruptions of what he considered treachery on the part of Clarke by arranging Hargraves' election to the Geological Society during a visit by the celebrated prospector to London to oversee the publication of his book on Australian gold. This ploy of extending limited metropolitan recognition won Hargraves over to Murchison's side in the dispute with Clarke, and Hargraves' pronouncements in later years were consequently more deferential to the King of Siluria. His central position in the imperial capital thus enabled Murchison to play off his colonial rivals against one another in order to maintain his own ascendancy.[76]

The Murchison-Clarke dispute flared periodically for years, exacerbated by each claimant's continuing assertions of priority and the meddling of some friends of Murchison who wished to curry his favour. The controversy was never fully resolved, for each opponent felt his scientific reputation to be fundamentally threatened. The issue represented to the honours-hungry Murchison a triumphant vindication of his prescience as a geological sage, which had raised his stock in political circles and thus furthered his career. But for Clarke it meant something even more vital—the maintenance of scientific credibility, which in a utilitarian colonial culture constituted his only hope for further government employment as a geological surveyor or the official reward, amounting to £3000, which he was finally granted in 1861.[77]

Promoter of exploration

Before his own appointment as Director-General gave him a direct role in Australian mineral reconnaissance, Murchison concentrated on promoting geographical exploration of the southern continent in order to win new scientific data, glory for the RGS, and prosperity for the empire. As early as 1837 Murchison's zealous advocacy at the Geographical Society had helped secure approval from the Colonial Office for George Grey's expedition to north-western Australia in search of a new site for colonization. In gratitude, Grey had named the Murchison River in his patron's honour. Furthermore, he made field observations suggesting a Palaeozoic sequence among the rocks of Western Australia.[78] Addressing the RGS as President in 1844, Murchison encouraged further penetration of the interior and emphasized the need to establish permanent settlements on the north coast. Linked to the other Australian colonies by overland routes and to Britain's Far Eastern possessions by steamship lines, such ports could facilitate commercial expansion in the East Indian Archipelago, and defence of the Antipodean colonies and regional trade routes from French or Dutch aggression. Murchison was in this case advocating the promotion to Crown Colony status of the military outpost of Port Essington on the Cobourg Peninsula,[79] which had been founded in 1838 through the influence of George Windsor Earl,[80] a commercial promoter who saw himself as another Raffles. Sir John Barrow of the Admiralty, who likewise hoped to create a 'second Singapore' on the north Australian coast for strategic reasons, had brought Earl's scheme before the Colonial Office with the help of the naval hydrographers, Francis Beaufort and John Washington, who co-ordinated these appeals through the RGS. Port Essington was nevertheless abandoned in 1849 after Barrow's retirement eliminated its most powerful supporter.[81] Murchison was thus expressing an institutional commitment to a colony the RGS had helped found as well as demonstrating himself to be Barrow's successor in the field of Australasian imperial strategy.

In 1845 Murchison used his presidential address to the RGS to praise Sturt's South Australian explorations and call for a hydrographic survey of the seas north of the continent to stimulate the establishment of steam navigation links with Britain's oriental possessions.[82] The Australian sections of his addresses of 1852 and 1853 were largely concerned with defending the priority of his gold prediction, but he also promoted exploration, geological research, emigration, extension of the settlement frontier, and colonization of the Gulf of Carpentaria. Displaying cotton and silk samples donated by Stuart Donaldson, FRGS, member of the Legislative Council of New South Wales, he argued that gold digging must soon give way to agriculture. In 1853 he also announced that the proposal of an Austrian, Ernest Haug, to complete the lost Leichhardt's explorations of the interior was under consideration by the RGS.[83]

Haug stressed that his contemplated expedition south-east from the Victoria River might discover fertile lands and navigable bays, accomplish valuable scientific observations, and—to catch Murchison's eye—survey the northern mineral empire, where plenty of the precious metal will be found'.[84] The RGS recommended to the Colonial Office a modification of this plan concentrating on the region between the Victoria and the Gulf of Carpentaria, and after Murchison emphasized that scientific results if not land sales would repay the government's outlay, Newcastle, the Secretary of State for the Colonies, provisionally agreed to the expedition.[85] Meanwhile a swindler who hoped to profit from inducing a new Australian gold rush advertised the mineral potential of the mountains of the far north,[86] and the colonial promoter Trelawny Saunders, FRGS, published a pamphlet urging the government to support Haug in order to stimulate settlement in the Gulf of Carpentaria.[87] The grateful Murchison promised Saunders that Haug would explore 'the very tracts you wish to colonize'.[88] In the next few months, however, because of the explorer's indiscretion in publicizing the commitment of the RGS to a project that had not yet received formal official approval, the President seized on the pretext of Haug's foreignness and began recasting the expedition as an entirely English affair.[89]

During 1854 Saunders, having decided to emigrate to the Gulf of Carpentaria in anticipation of an appointment in the settlement he hoped the expedition would found, urged as its commander the naval surveyor John Lort Stokes, FRGS, who had charted much of the north coast in the early 1840s, officially testified at that time to the utility of maintaining Port Essington, and recently urged Newcastle to found a convict colony in the Gulf.[90] The RGS supported Saunders' proposal,[91] but Augustus Gregory, Assistant Surveyor-General of Western Australia, was chosen as leader. When Sir George Grey replaced Newcastle at the Colonial Office, Murchison renewed his pressure to 'get the N. Australian Ex. afloat'.[92] He also designated as the expedition's geologist James Wilson, a British civil engineer, who in the course of more than a decade's work in South Australia had familiarized himself with the structure of that colony's mineral districts. More significantly, Wilson also possessed two years experience in the Californian gold fields and held orthodox Silurian views on the metal's occurrence.[93] Because of his desires to extend the scope of his system and prediction, as well as to ensure practical results from the venture, Murchison clung to the hope that the Victoria River region would prove auriferous. With the advice of Sturt, Edward Eyre and Herman Merivale, FRGS, Under-Secretary of State for the Colonies, Murchison and Washington worked out the organization of the expedition.[94] Washington, Assistant Hydrographer of the Admiralty and former RGS Secretary, maintained that it offered an opportunity 'to press the immediate occupation of Cape York—the French, as you know, are at Tahiti and New Caledonia and what are *we* doing?'[95]

By 1855 an appropriation had been secured, a steamer arranged, and officers appointed. The scientific staff included Wilson, a zoologist, Ferdinand von Mueller of the Melbourne Botanic Gardens, and the artist Thomas Baines. Murchison maintained pressure on Palmerston's cabinet, via the British Association, to go through with the obligation incurred by Aberdeen[96] while suggesting to Sir William Denison, FRS, FGS, FRGS, the Governor of New South Wales, that a penal colony be founded if the Victoria River region proved fertile.[97] Once the expedition got under way, the reading of progress reports at RGS meetings permitted Murchison to highlight its imperial significance as tending 'to the establishment of a colony in North Australia which would materially strengthen the powerful position we already held in the East' as well as providing a terminus for a land route to the southern colonies.[98] The presentation of a memoir by Stokes on the strategic value of a northern convict settlement again elicited Murchison's support for this solution.[99]

Gregory's expedition soon confirmed the promise of the Victoria River region and northern Queensland.[100] Wilson's final reports were forwarded to Murchison via the Colonial Office and publicized through the RGS.[101] His consolidated geological and geographical account, containing descriptions of Silurian rocks, the newly named Murchison Mountains, and exploitable soils and pastures, was refereed by Murchison for the RGS in 1858. Its reading provoked Sir George Everest, the Society's Vice-President and the former Surveyor-General of India, to suggest the establishment of a penal colony for Indian mutineers on the Victoria River.[102] Murchison had argued similarly in his presidential address of 1857 that it remained of 'incalculable national importance' to found a penal settlement on the Victoria, and that since France had 'acquired a "point d'appui" on the eastern flank of our largest Australian colony' by annexing New Caledonia 'which our Cook discovered and named', Britain must establish naval stations on the north coast to defend her colonies and trade.[103] In 1858 he echoed Everest in stating that this 'high political object' might best be secured by the transportation of rebellious sepoys.[104] Gregory also advocated a penal colony as the most expedient method of exploiting his discoveries, completing the occupation of the continent, and establishing an entrepot to dominate regional trade.[105]

Though a settlement was not established on the north coast for some years due to the continued routing of steamships via Singapore, the North Australian Expedition reveals Murchison's tenacity in pursuing projects he believed to be of imperial importance. The opportunity to obtain new scientific data was subordinated to the larger goals of occupying the north coast, opening a new field for commerce and emigration, and establishing strategic bases for naval deployment and extended steam navigation. Murchison provided the continuity that saw the mission to conclusion through personnel shifts in the cabinet, the RGS, and the expedition itself—an impressive accomplishment in the midst of the Crimean War.

During the period 1859–61, Murchison also took part in a move to settle the north-western coast. Grey's explorations in the 1830s had suggested the region's fertility, and in 1858 Francis Gregory revealed vast tracts of excellent grazing north of the Murchison River. Gregory, well known to Murchison for his earlier geological researches in Western Australia,[106] wished to explore the coast of this promising region, but Governor Kennedy refused to sanction another expedition. To circumvent Kennedy, representatives of the expansive settler interest approached the RGS for support with the Colonial Office, since the discovery of resources would attract immigrants and capital to the colony. To arouse Murchison's interest, it was pointed out that since the voyages of Dampier, and possibly earlier, the region had been known as 'Provincia Aurifera'.[107] After the RGS council considered this as well as a similar proposal during 1859 from Dr Thomas Embling of Melbourne, a strenuous advocate of north-coast exploration in the interests of Victoria,[108] Murchison recommended Gregory's project to the Colonial Office.

In 1860 Gregory, several settlers, and John Septimus Roe, the Surveyor-General of Western Australia, arrived in England to promote the expedition through the RGS.[109] Murchison, now Vice-President of the society, wrote again to the Colonial Office as well as to the new President, Lord Ashburton.[110] Ashburton, believing that the RGS should 'bring to bear on Australian Exploration the united interests of the Cotton Lords & of the settlers', then tried to arrange a co-ordinated approach to the government involving the Manchester interests.[111] In light of the cotton famine resulting from the American Civil War and Grey's earlier reports that north-western Australia might become a great cotton-growing country, this was a wise tactic: it apparently failed, however, for the society was left to stress the possibilities for intensive cotton production using Indian labour. The RGS also maintained— and here the hand of Murchison was evident—that 'men of science have come to the conclusion that this district may be productive in gold and copper, as well as other minerals'.[112] Secretary of State for the Colonies Newcastle agreed to this proposal, promising £2000 if Western Australia put up a similar amount. Gregory then returned to Perth and, despite local fears that cotton cultivation would undermine the interests of established stockmen, rapidly raised sufficient funds to launch his expedition.[113] Murchison had meanwhile kept the Colonial Office up to the mark by direct exhortation as well as public pronouncements,[114] and had passed on to the Geological Society the results of Gregory's 1858 exploration. Received through the Survey, this report likewise stressed the mining potential and probable Palaeozoic stratigraphy of the far north west.[115]

In 1862 Murchison refereed Gregory's report for the RGS, and its reading prompted an encomium from the Perth faction on the capabilities of Western Australia.[116] The society's role in the expedition was recorded by Gregory's naming of rivers after presidents Ashburton and Earl Grey, Secretary Norton Shaw, and Chichester Fortescue, FRGS, the Colonial Under-Secretary who handled the negotiations and received the society's gold medal on Gregory's behalf.[117] The expedition had revealed extensive agricultural and grazing lands, but no valuable minerals. From the beginning, the RGS had made erroneous assumptions regarding the topography, climate, and resources of the region: now the society and the imperial government dropped the idea of a cotton colony because better opportunities beckoned in Africa and Fiji. When pastoralists began to arrive on the north-west coast in 1863, nothing was done to establish cotton plantations. In promotional terms, however, such mistakes were almost irrelevant. Since the examples of California and Victoria suggested that every wilderness must harbour useful products, the society's inveterate optimism constituted the best strategy for constructing the coalition of interest necessary to pressure the government into funding explorations.

Patron of colonial science

During this era Murchison maintained close ties with scientific Australian governors and officials, publicizing their efforts to stimulate exploration and development in exchange for news and dissemination of his own opinions. In 1857, for example, he sent Sir William Denison a set of the British Geological Survey's publications. When Denison then formally suggested that an imperial encyclopaedia of natural history would both benefit the colonial economies and reflect credit on Britain, the Colonial Office sought advice from the Royal Society and Murchison in his capacity as a geographer.[118] Murchison referred the proposal to the RGS Council while publicly extolling Denison's project as tending to 'unite by closer bonds all parts of our empire'.[119] Both societies reported favourably on the scheme, the RGS mentioning similar works in French Algeria and the Dutch East Indies.[120] The plan fell to the ground, however, when Gladstone's Treasury tried to limit the encyclopaedia to a series of unillustrated volumes, an expedient that defeated the purpose of furnishing field guides for the identification of useful colonial products.[121]

In 1860 Murchison formed a friendship with Sir Richard MacDonnell, FRGS, Governor of South Australia. MacDonnell provided Murchison with geological specimens, exploration news, and ongoing support for both his claim to have initiated the gold rush by scientific forecast and his theory of the metal's occurrence. He in turn supported MacDonnell's efforts between subsequent proconsular postings to establish a new colony as chairman of the North Australian Settlement Company.[122] The same year Sir George Bowen, FRGS, first Governor of Queensland, wrote to Murchison that while developing the colony's resources he hoped to 'add new conquests to Geography and to Science'. He begged the President 'to assist in making Queensland known in England'—a service repeatedly performed in future.[123] At the time of Bowen's appointment, Charles Sturt requested Murchison's help in obtaining the governorship of Queensland, as he did later that of South Australia. Throughout the 1860s Sturt acted as the RGS adviser on Australian exploration and was an agent for introducing Antipodean natural scientists into metropolitan circles.[124]

Murchison also maintained a link with Sir Thomas Mitchell, FGS, FRGS, who as Surveyor-General of New South Wales from 1828 to 1855 transmitted many scientific specimens to London and advocated colonization of the Gulf of Carpentaria.[125] Mitchell had provided Murchison with early indications of Silurian strata in Australia, and when he visited London in 1853 Murchison proudly hosted him at the scientific societies.[126] Stuart Donaldson, FRGS, an influential New South Wales politician, also kept the president informed on Australian affairs until his retirement in 1859, when he became active at the RGS.[127] As always, these activities demonstrated considerable overlap between Murchison's public and private capacities. In 1857, for example, he received a request through the Colonial Office from his old friend Charles Babbage that a type collection of British fossils and minerals be sent out to Babbage's son Benjamin in Adelaide for use in identifying rock formations discovered during the official geological explorations he conducted in the centre of South Australia between 1856 and 1858. In this instance the Director-General was unable to comply, having recently arranged for his Museum of Practical Geology to distribute twenty-two suites of duplicate fossils among provincial, colonial, and foreign museums, but he extended the courtesy of extolling young Babbage's labours at the Geographical Society.[128]

In 1854 Murchison helped secure the appointment of the palaeontologist Frederick McCoy to the chair of natural science at the new University of Melbourne. McCoy had become an embarrassment in Britain because of his zeal in providing fossil ammunition for Sedgwick's attack on the Silurian System. The professorship in Victoria, despite Murchison's encouragement to McCoy that 'a great deal may be done in that magnificent new colony & it is highly desirable to have a person of your real attainments there',[129] thus offered a handy means of transporting McCoy from the cockpit of scientific controversy to a colony where he might, as one Silurian supporter put it, 'do some good work and not stir up mischief between old friends'.[130] Murchison's recommendation also contributed to McCoy's success in winning a second appointment in 1856 as Palaeontologist to the Geological Survey of Victoria, a position he used to expand the Survey's fledgling collection into a comprehensive National Museum of natural history with functions parallel to those of the British survey's Museum of Practical Geology.[131] McCoy maintained his stratigraphic partisanship, however, by attempting to prove that Australia's chief gold deposits occurred not in Silurian rocks proper, but in the formations Sedgwick claimed as Cambrian.[132]

McCoy likewise embroiled himself in a thirty-year debate with W.B. Clarke over the age of Australian coal deposits. Clarke believed the coals to be Carboniferous, while McCoy maintained they were younger Jurassic formations. Since Carboniferous coals in the northern hemisphere were of much higher value as steam fuel than lignites, this stratigraphic battle, like the Murchison–Clarke dispute over the occurrence of gold, had direct implications for colonial development. The calorific quality of the abundant bituminous coal of New South Wales was by this time firmly established: still, public perceptions of the value of this resource could be significantly influenced by the pronouncements of scientific experts. This was especially so from the perspective of the imperial nerve-centre in London, where Eurocentric notions of the inferiority of Mesozoic coals prevailed and the views of geologists such as Murchison received more respect than did those of colonial colleagues within their own adopted societies. Thus an error in dating the coal formations could, at a stroke, either depress the apparent capacity of the Australian colonies to undertake industrialization or, conversely, grossly oversell their potential to prospective investors and immigrants. But Clarke's assessment that most of the deposits of New South Wales were at least Palaeozoic, if not specifically Carboniferous, gradually prevailed, and Murchison advertised this valuable fuel source as 'worthy of imperial notice'.[133] Ever anxious to promote the advancement of Britain's Antipodean possessions, he likewise gave forth his authoritative opinion that 'similar coal measures will probably be found in the coterminous colony of Queensland'.[134]

The Geological Survey of Victoria had been founded in 1852 by A.R.C. Selwyn, an appointee from the British Survey. He had explored the gold fields with one assistant until given an expanded staff and a direct link with McCoy's National Museum in 1856 and a chemical laboratory in 1864. The Victorian Survey rapidly achieved world-wide recognition for its precise cartographic and descriptive work, and its geologists went on to senior survey and academic posts throughout the Antipodes. Because of bureaucratic conflicts about the function of geological surveying and the overlapping duties of geologists and mining surveyors, Selwyn resigned his post in 1869 to take over the Canadian Survey, and the Victorian Survey was temporarily disbanded.[135] De La Beche and Murchison exchanged publications with Selwyn and, in the initial phase of his work, co-operated in the identification of Victorian minerals.[136] In 1853, for example, analysis by the Museum of Practical Geology of a heavy 'black sand' thrown aside in the gold diggings alerted Governor La Trobe to the wastage of an extremely rich source of tin ore, which soon became the basis for a subsidiary mining industry.[137] This relationship was particularly close during the governorship of Murchison's friend Sir Henry Barkly, FRGS, a keen amateur scientist. Besides their co-operation on geological issues, Barkly also supported Murchison's drive for a north coast colony, and Murchison reciprocated by securing Barkly's election to the Royal Society.[138]

Murchison had been impressed with Selwyn's geology before he left Britain, but once he began examining Australian gold deposits the Director-General was forced to acknowledge his skills anew, for grave discrepancies were revealed in the King of Siluria's theories. In the early stages of the Victorian rush, field data seemed to support Murchison's views,[139] but the accumulating evidence soon indicated that the mining of veinous gold from hard quartz rock was not only feasible, but would rapidly displace alluvial digging as the likeliest proposition on the gold fields.[140] The strained congruence between metropolitan theory and colonial experience was finally shattered in 1857 when Barkly sent the Director-General two reports on the progress and future prospects of the Victorian mining industry. These had been written by McCoy as chairman of the colony's Mining Commission, a short-lived forerunner of the Board of Science established by the governor in the same year. Selwyn was also a member of this commission and its findings were based on his research, but his name did not appear on the reports. Nor, as was soon to be revealed, did he agree with the chairman's conclusions. McCoy asserted in his second report that the Murchisonian theory of profitability decreasing with depth was being proven correct by Victoria's falling gold yields. In consequence, he discouraged both the private investment required to inaugurate the large-scale mining and quartz crushing operations that might exploit the veinous gold locked in the subterranean formations known as reefs, as well as the proposed railways to the gold field settlements that he believed could prove nothing more than evanescent boomtowns.[141]

McCoy's report fell like a bombshell among the miners of Victoria. In a vitriolic editorial, the *Bendigo Advertiser* attacked McCoy and Murchison in an attempt to restore shaken public confidence in the future of the gold-mining industry.[142] Barkly, caught between his respect for Murchison on the one hand and his desire to ensure that the economic viability of his colony was not unfairly prejudiced on the other, also argued in forwarding McCoy's reports to London that Victoria's deep-level mines might well prove permanently remunerative.[143] This was rank heresy in Murchison's view, and though he assumed Selwyn's agreement with McCoy[144] while drafting an official refutation of Barkly's position, it soon became clear that the governor had found a powerful ally in his chief geologist. In a report prepared at Barkly's request in 1858, Selwyn offered irrefutable evidence that, at least in Victoria, gold-bearing veins occurred plentifully in Cambrian as well as Silurian rocks, and that the exploitation of buried reef and placer deposits could prolong productive gold mining in the colony almost indefinitely. While Selwyn's report was couched in somewhat deferential language, Barkly's covering despatch overtly gloated at the blow thus dealt to what he termed the 'received geological doctrines' of Murchison. As the governor pointedly observed, 'the question is indeed not one of authorities but of facts'. The drop in yield from deep-level reef mining, he contended, would be so insignificantly gradual due to the vast extent of the deposits that large amounts of capital could be invested in such operations with perfect safety.

The future of Victoria was actually at stake in this esoteric debate mediated by the Colonial Office, for the second report of the Mining Commission had caused a faltering of confidence in quartz crushing companies until Selwyn publicly repudiated McCoy's position on the non-profitability of hardrock mining and, effectively, Murchison's theory.[145] This declaration of scientific independence from the authority of the metropolis could probably only have occurred at that time in Victoria, where the prestige of Selwyn and Barkly and the Director-General's desire to promote a colony whose progress he believed unparalleled 'in the annals of mankind' combined to persuade him to modify his views.[146] His subsequent publications, as Barkly had predicted, were accordingly revised to reflect Selwyn's findings.[147] And although Murchison sustained his old opinions on gold's formation, occurrence and rapid exhaustion as an exploitable metal, continuing to seek their confirmation in Australia through correspondence with Governor Denison of New South Wales,[148] his enforced shift of position on deep-level mining reopened the debate his pronouncements had stifled during the very decade when the great rushes were transforming the situation. By 1860, refereeing an article for the *Quarterly Review*, Murchison was compelled to admit that the publication of his own views would 'seriously impugn the character of the *Quarterly* in the estimate of scientific readers'.[149]

The difficult question of assessing the potential of Victoria's gold reefs could not be quickly resolved, however, for the practical knowledge of the miners contradicted the inadequate contemporary corpus of geological theory. Inevitably, then, the debate reopened in 1860, when Murchison responded to an article in the *Bendigo Advertiser* by a local geologist named C.A. Zachariae that castigated Selwyn as well as Murchison and McCoy for misinterpreting Victoria's gold deposits on the basis of a false comparison with those of the Urals. Zachariae argued that the veinous gold contained in the reefs would prove uniformly rich to a depth of over a mile and therefore remain 'an everlasting source of prosperity and wealth to this colony'.[150] Unable to restrain himself, Murchison thundered forth a characteristic reply, which defended his Ural work and his old views on gold, ridiculed Zachariae's rival theories as pure fantasy, derided his opponent's foreignness, and accused him of offering false hopes to the public of Victoria.[151] Zachariae's counter-rebuttal saw print five days later, but it merely reiterated his contempt for the cowardice of McCoy and Selwyn in being unwilling 'to overthrow a long established doctrine' and called on the Director-General to continue the debate.[152]

Murchison declined to do so, and was soon informed by Clarke, Selwyn and Barkly that even his initial fulmination had been unnecessary since Australian geologists put no stock in Zachariae's opinions.[153] But still the issue would not die. Miners wrote to the *Bendigo Advertiser* in support of Zachariae's position, former diggers offered further evidence in metropolitan periodicals, and Barkly informed Murchison of fresh examples demonstrating that the richness of the gold reefs continued undiminished to a depth of several hundred feet.[154] Geologists were to remain divided for years on the vexatious question of the Victorian reefs, and supporters and opponents of Murchison's doctrines continued their battle in the pages of Britain's scientific and mining journals.[155] It is possible that the wide publicity accorded Murchison's authoritative opinions may have temporarily deterred investment in the Victorian gold-mining industry during a crucial transition in its development, but there is no reliable evidence beyond the rhetorical debate examined here to support such a conclusion.[156] The influential views of Selwyn coupled with the indisputable profits of the miners rapidly carried the day with wary capitalists, and as the mountain of gold won from the reefs continued to grow due to technological innovations that made progressively deeper shafts viable, Murchison's theory was slowly crushed under the weight of colonial evidence.

Despite these complications Murchison loyally continued to support the Victorian Survey. In 1861, for example, as the Victorian Legislature contemplated a drastic realignment of the Survey's priorities under the urging of the Secretary of Mines, Robert Brough Smyth, he campaigned at the British Association for the maintenance of Selwyn's budget for systematic mapping.[157] The following year, after awarding Selwyn a medal as jury chairman for his mineral display at London's International Exhibition, the Director-General secured him a three-month extension of his year's home leave in order to study Britain's mining districts and also arranged to supply the Victorian Survey with promising graduates of the Royal School of Mines as new staff might be required.[158] Murchison had no influence, however, on the political decisions taken in Victoria that resulted in the abolition of Selwyn's Survey in 1869.

In 1870 Murchison was approached by Sir Redmond Barry for advice on founding a school of mines at Ballarat. As has been discussed, by the late 1850s deep-level gold and lead mining were demanding a reorientation of extractive technologies. In 1866 Murchison had similarly been requested by the Chancellor of the University of Sydney, Sir Edward Deas Thomson, to recommend a professor of geology who would be capable of introducing 'the newest and best methods'.[159] The Director-General's choice in this instance was A.M. Thomson, first appointed as reader in geology and promoted to a chair in 1870, a year before his death. As Barry in Melbourne made clear, the issue of building institutions capable of rendering the colonies independent of the metropolis for their supply of scientific and technical expertise was also involved. In a concise statement of the relationship that he believed should obtain between a colonial mining academy and its metropolitan parent, Murchison advised a curriculum for Ballarat based on that of the Royal School of Mines, but modified to emphasize practical mathematics and surveying rather than a complete course of instruction in natural history.[160]

In 1858 Murchison also installed a geologist in Tasmania. As Lieutenant-Governor from 1847 to 1855, Denison, aided by analyses carried out at De La Beche's museum, had continued Franklin's efforts to stimulate scientific research regarding local resources and promote coal mining.[161] When the Victorian rush raised expectations of further gold discoveries, W.B. Clarke informed Denison that the metal was likely to occur in Tasmania. Denison attempted to suppress this news,[162] but Clarke published his prediction and informed Murchison in 1852 that discoveries were already reported. The same year Murchison received the first specimens of Tasmanian gold sent to England.[163] In 1855 Lieutenant-Governor Sir Henry Young commissioned Selwyn to examine the island's coal fields and, on the advice of Clarke, who reported on some of Tasmania's auriferous formations for Young's government but declined an official proposal that he survey the entire colony, submitted the results to Murchison. The new Director-General duly recommended that a geologist be appointed.[164] On this advice the Tasmanian Legislature agreed to fund a limited survey, and Murchison chose for the post

Charles Gould, the son of ornithologist John Gould and a graduate of the School of Mines employed by the British Survey.[165] Before Gould's departure in 1859, the Director-General also forwarded a set of his Survey's maps and publications, and a type collection of British fossils, to the Royal Society of Tasmania in order to facilitate the comparative identification of strata in the colony.[166] He used this appointment to advertise the imperial utility of his own Survey, boasting that most colonial geologists 'had either been brought up in the establishment which I direct, or recommended by my predecessor or self '.[167]

The Tasmanian government had by now offered a reward for the discovery of a paying gold field, and Gould began his survey according to Murchison's instructions where the first specimens had been found. His observations, however, confounded the Director-General's expectations and reinforced Selwyn's conclusions. In contrast to the situation in Victoria, alluvia derived from Silurian rocks were too scanty in Tasmania to promise much gold but, as in the younger colony, quartz veinstones had proved so auriferous that several companies were constructing costly crushing mills. This anticipated boom never materialized, however: nor did Gould find significant amounts of gold on a subsequent prospecting expedition to the interior mountains, and after producing a coal report he began to complain that economic activities were hindering his scientific researches.[168] Murchison publicized Gould's coal findings at the British Association, used them to promote his own Survey as a training academy for colonial geologists, and recommended their publication as likely to benefit navigational and industrial development in Australia. Drawing on Gould's mistaken initial assessment of a Carboniferous age and an almost limitless extent for Tasmania's coal, Murchison at first advertised these fuel resources as among the most valuable in the empire, but even after Gould correctly deduced a Permian or Triassic age and a restricted occurrence of high-grade seams he remained reluctant to mute his optimism.[169] Reciprocally, Gould named a peak and a river in the West Coast Range in his patron's honour and classified a huge area in this region as Silurian, but though he contributed substantially to knowledge of the island's geology he made no major mineral discoveries. In 1862, for example, he had the misfortune to pass within a few hundred metres of the Mount Lyell copper deposits without finding them. In 1869 his contract was terminated by a disgruntled legislature and the Tasmanian Survey lapsed until 1882.[170]

Throughout the 1860s Murchison continued to promote Australian exploration, mineral reconnaissance, emigration, intercolonial consolidation, and strengthened ties of trade, communication and defence with the mother country. In the field of exploration, he awarded MacDouall Stuart the gold medal of the RGS and arranged special meetings to maximize the impact of his discoveries.[171] The Adelaide pastoral and mining firm of Chambers and Finke earned Murchison's praise for their 'bold and riskful' employment of the explorer, which some considered mercenary, and their transmission to London of his journals and geological specimens.[172] The disastrous Burke and Wills expedition likewise won Murchison's support, and two of its members received RGS awards, one bestowed via Governor Barkly of Victoria, whose local Royal Society had sponsored them. The President also used the deplorable fate of Burke and Wills, as he had that of Leichhardt, to promote further exploration in northern Queensland.[173]

In the field of mineral reconnaissance, Murchison encouraged explorers to collect specimens and processed their data through London's scientific societies. Contemporaries believed his judgement influenced the attractiveness of colonies to immigrants and investors. In 1856, for example, a report by the Assistant Surveyor-General of Western Australia suggested from evidence of the Director-General's 'constants' that a gold field might exist on the Murchison River. Murchison, refereeing the paper for the RGS, allowed this statement to stand in hopes that it might stimulate an extension of his Silurian prediction.[174] But in 1864 Edward Hargraves, now employed under contract by the Western Australian government, contradicted this view, though he shielded Murchison from any appearance of error because by this time he owed the metropolitan geologist a debt of gratitude.[175] Hargraves' report was also refereed by Murchison for the RGS. At its presentation both he and the visiting Selwyn—the latter with decided reservations echoing his earlier confrontation with Murchisonian dogma—defended Hargraves' negative assessment of gold potential against protests levelled by one of the same Western Australian activists who had recently promoted Gregory's north-west expedition that this publicity constituted an unfair attack on the colony's capabilities.[176]

Other geological reports received included those of Richard Daintree, regarding Queensland, which were forwarded by Governor Bowen with accompanying remarks by W.B. Clarke.[177] Daintree, a ten-year veteran of the Victorian Survey who had interrupted his service with Selwyn for six months' study at the Royal School of Mines, had pursued amateur research in Queensland from 1864 to 1867, when his discovery of gold prompted his appointment, together with C.D.H. Aplin, another subordinate of Selwyn's, as a government geologist. Both men lost their posts by 1870, but their discoveries accelerated economic development of Queensland, the first Australian territory examined by indigenous scientists. Their example stimulated the permanent refounding of a local Survey in 1876, and their mineral collection later formed the nucleus of the colony's museum of natural history.[178] As Bowen remarked, Daintree's evidence of Queensland's gold deriving predominantly from Silurian and Devonian strata proved gratifying to Murchison as a further realization of his prediction. The Director-General enthusiastically recommended the publication of these findings by the RGS, but he died shortly before Daintree's extended observations and pioneering map of the colony appeared in the *Quarterly Journal* of the Geological Society.[179]

Murchison's promotion of Australian emigration even included advising a temperance society on founding a settlement,[180] but his pleas for colonization of the north coast met opposition from the orientalist John Crawfurd, FRGS, who had testified against the maintenance of Port Essington on climatological grounds and remained sceptical of the region's suitability for European settlement.[181] Improvements in communications between the southern continent and Britain by telegraph and steamship were also advocated, and the imperial loyalty of Australians received comment while the American Civil War made secession a fashionable topic. When Herman Merivale explained to the British Association that Britain must remain on good terms with her existing colonies as well as continue to found new ones, Murchison apprised Barkly: 'Merivale has done good service in smothering that grovelling and unworthy sentiment of a few *doctrinaires* as to the inutility of our colonies. I am furious when I read their cold and heartless reasoning, and I shall take good care to show at our evening affairs how warmly the Australian colonists support the mother country, and sympathize with it.'[182]

In 1856 Murchison also formed an association with the Australian politician, businessman and scholar, Sir Charles Nicholson, FGS, a vociferous advocate of colonial loyalty. He elected Nicholson to the RGS and the Athenaeum, and Nicholson reciprocated by hosting the President at London's first Australian reunion and defending the priority of his gold prediction.[183] Nicholson wished to expand Australian influence into neighbouring islands—a concern that had been stimulated by the French occupation of New Caledonia in 1853 and that meshed with Murchison's desire for north coast settlements. P.P. King and John Washington had railed to Murchison of Britain's failure to pre-empt this annexation,[184] and John Lort Stokes had urged him to press for northern settlements to circumvent the second Derby administration's distaste for colonial expansion and 'prevent the French taking the wind out of our sails there, as they most certainly have done in New Caledonia. We may rest assured that Napoleon IIId. has in his far seeing eye, the value & importance of the islands fronting the North coast of Australia.'[185] Murchison publicly deplored the French move in 1857 and in the following year, while whipping up anti-French sentiment at the RGS, he sounded the Colonial Office on Nicholson's behalf about establishing a port in north Australia, citing Denison and Stokes to emphasize that 'now that the French have occupied in force the New Caledonia of our Cook, the case seems to me to be one of absolute necessity'.[186] In 1859 his RGS address again raised the issue in phrases redolent of Nicholson, and two years later Queensland's Governor Bowen, who had himself proposed to the Colonial Office that a settlement be established in the Gulf of Carpentaria, followed the President's advice in founding a coaling station near the tip of Cape York.[187] When Nicholson returned to England in 1862 Murchison installed him on the RGS Council, engineered his presidency of the Geographical Section of the British Association in 1866, and continued to help him promote Australian expansion and defence.[188]

Heir to Banks' scientific *imperium*

Murchison's promotion of geographical and geological exploration as concomitant instruments of development was most consistently demonstrated in Australia. Exploring expeditions roughed out the basic features of unknown territories, providing data for capability assessments preliminary to occupation and development facilitated by more systematic surveys. Murchison regarded Australia as an exclusively British preserve wherein a long-term experiment in the scientific colonization of an alien environment might be played out in splendid isolation, untainted by the meddlings of rival powers or competing cultural influences. For this reason he jealously guarded the inviolability of Australia's empty coastline, remained suspicious of foreign intentions in neighbouring islands envisioned as a hinterland for future expansion, and advocated improvements in communications that would knit the colonies into a cohesive whole and bind them permanently to the mother country. Because of his own part in initiating development of its vast potential, Murchison remained fascinated with the southern continent. In this respect, as in his role as an unofficial representative of Australian interests in London, he followed directly in the tradition of Banks and Barrow.[189] The coincidence of his own rise to sustained authority in scientific and imperial circles, the saturation of the official mind with confidence in science as an engine of progress, and the stimulus given mineral exploration by the gold rushes in New South Wales and Victoria gave Murchison unparalleled opportunities for action in the Antipodes. His diverse influences on the exploration, settlement, mineral development, and establishment of scientific institutions in Australia were consequently greater than in any other part of the empire.

Vallance has demonstrated that Australian geology remained dominated by European experience and concepts until the mid-1870s.[190] While Eurocentrism certainly distorted the progress of colonial science, it was nevertheless inevitable that colonial geologists, necessarily trained in Europe during the era before local educational institutions could provide specialized curricula, would initially rely on European theories, models, and techniques for guidance in classifying the structure of new environments that they confronted with no local data base. Modifications in scientific theory and practice eventually followed as field evidence failed to agree with the imported interpretive framework, and as the inhibiting influences were overcome, an indigenous geology developed.[191] Viewed against this paradigm of scientific growth, the activities of the King of Siluria appear simultaneously as both an impediment and an enhancement. Despite his rigid adherence to a European stratigraphy in many respects inappropriate to Australian geology, his shortcomings as an expert on the occurrence of gold, and his reliance on catastrophist explanations for geological change, Murchison's attempts to extend his Palaeozoic systems, his theorizations about the southern continent, and his installation and ongoing aid of Australian geologists directly stimulated colonial research as well as metropolitan interest concerning Australia's structure.

Colonial governments controlled the funding and therefore the scope and immediate goals of their respective surveys. But at the same time, Murchison's dominance of metropolitan public opinion regarding geological matters and his powerful influence on the views of geologists themselves constituted an indirect system of control that, in conjunction with his official patronage of survey and academic appointments, worked to perpetuate the subordination of colonial science to the imperial capital at both the individual and institutional levels. His creation of a shadow sovereignty of patronage and obligation within the service of the empire exemplifies how the lines of imperial scientific authority paralleled political authority throughout much of the nineteenth century. While imperial science as practised by Murchison consisted largely of the organization and projection of scientific activities from Britain rather than an overt imposition of metropolitan authority over colonial practitioners and data, elements of scientific autocracy nevertheless leavened his administrative omnipotence. He never attempted, for example, actually to dictate the ages of Australian strata from London. But his ubiquity as a metropolitan mediator of colonial research and the lengthy disputes he carried on with antipodean geologists whose work he perceived to threaten his own reputation demonstrate the influence he exercised on the disciplinary development of geology in Australia.

Murchison's disputes in the imperial sphere, like his better known controversies with British geologists, illustrate the role of social and political factors in shaping the creations of scientific knowledge.[192] Though Murchison attempted to stifle Clarke's rival theories and claims—and in fact succeeded in the case of Zachariae—it is also true that Clarke as well as Selwyn and Gould successfully challenged his authority, forcing him at least to negotiate intellectual compromises. Having once admitted the validity of their findings, Murchison brought them to the attention of metropolitan scientists, so that truth eventually emerged from this trans-hemispheric dialectic. Viewing the territorial expansion of his scientific domain as analogous to that of the British empire, Murchison may well have perceived that the partial devolution of authority represented by the resulting consensus had a political parallel in Westminister's granting of responsible government to the colonies of white settlement. Ultimately, hegemony was best consolidated by adaptation and redefinition.[193] The fact that the re-importation into Britain of data and hypotheses generated by Australian geologists whom Murchison appointed or helped to maintain in their posts significantly affected theoretical debates within the discipline of geology demonstrates that imperial science was a bilateral arrangement benefiting both contracting parties. This long remained an unequal exchange, however, and colonial scientists often experienced difficulty in winning metropolitan acceptance. In the field of geology, Murchison, the patron best placed to effect this integration, paradoxically remained its greatest obstacle because of his own vested interests.

Still, Murchison's activities in fostering scientific institutions and in reconnoitering and publicizing Australia's resources contributed to the cultural unification and economic development of the colonies. The benefits of science were diffused from the geological surveys he helped found by the establishment of museums and laboratories, the training given local assistants, the encouragement offered independent researchers, and interaction with educational institutions.[194] The work of the surveys helped to guide and regulate efficient mineral exploitation, just as the topographical and geological maps they produced aided the classification and sale of land and the routing of railways and telegraphs. The maps resulting from the exploring expeditions that Murchison organized similarly encouraged the inland expansion of the settlement frontier.

As RGS president, Murchison also made available to forward-looking Australian colonists an important metropolitan venue for promoting immigration and investment. The North-west Australian Expedition demonstrates how special interests in the colonies, by casting their projects as vehicles for the simultaneous achievement of imperial and scientific goals, could mobilize his support for expansionist aims thwarted by local administrations and economic interests anxious to maintain the status quo. As both a geologist and a geographical promoter, Murchison performed an entire spectrum of services for Australia. Organizing systematic research into the physical make-up and resources of the several colonies and publicizing their potential for development were two of his more significant accomplishments. On balance, it is clear that his influence helped to accelerate rather than retard Australian scientific and economic development.

Murchison's activities defy neat classification according to the various models that have been proposed to periodize the development of colonial science.[195] Quintessentially mid-Victorian, Murchison also represents an unusually durable anachronism whose accrued power and influence, operating through the centralized but loosely-knit structure of the imperial civil service, allowed him to function in a 'Banksian' style until nearly the final quarter of the nineteenth century.[196] In this respect, combined with his overt espousal of imperialism during the Palmerstonian era when British hegemony was still largely preserved by informal means, he is atypical, perhaps unique, among Victorian scientists. Certainly in his enthusiasm for a Greater Britain, his creation of multiple institutional power bases, and his maintenance of a tentacular network of political and social connections, he far surpassed his only rivals for authority in the field of imperial science, the anatomist Sir Richard Owen and the botanists Sir William and Sir Joseph Hooker.

Yet Murchison is too important a figure to be ignored in any attempt to categorize imperial science, and his incongruence with existing models suggests that we have far to go in perfecting such schemes. Similarly, his ambivalent attitude as a metropolitan co-ordinator toward science as practised in the colonies and the record of his accomplishments in this field conflate the terms colonial science and imperial science that some scholars have striven to differentiate and define.[197] The revolt against the autocracy of Murchison first attempted by Clarke and fought to a victorious conclusion by Selwyn in 1858 also offers a minor corrective to the assertion that the years 1861–63 marked the first overt clash between Australian scientists aspiring to equal recognition from their British overlords.[198]

While imperial concepts, colonial data, and colonial career opportunities affected the development of several of the natural sciences in Victorian Britain, these disciplines in turn provided the empire's rulers with new tools of administration and development and new methods for conceptualizing the world. As a disciple of two pre-eminently territorial sciences, it was logical that Murchison adopt an extensive rather than intensive strategy toward their development. Britain's possession of an empire, the unknown resources of which required inventorying, made this approach all the more rational, for the unexplored colonies offered not only a ready-made justification for demands that government subsidize an expansion of the public science sector, but a gigantic laboratory in which the then-dominant taxonomic sciences could accomplish heroic feats of classification, reduction, and containment. More than merely a scientist exploiting the opportunities inherent in empire, however, Murchison was an active imperialist in his own right, bending his science, organizational skills, and promotional talents for some forty years to the strengthening of Britain's position overseas. Throughout this period

Australia remained the focus of his attention, as the number of natural features bearing his name testifies. And as an admirer once remarked to Murchison: 'The monuments of Alexander's victories have for ages been swept away. Yours will be coeval with the everlasting rocks.'[199]

Notes

The following abbreviations are used in the notes below:
ADB   Australian Dictionary of Biography. BL(AM), MuP   British Library, London (Additional Manuscripts), Murchison Papers. CUL, SeP   Cambridge University Library, Sedgwick Papers. EUL, MuP   Edinburgh University Library, Murchison Papers. GSL/M/(a–z)   Geological Society of London, Murchison Papers (incoming correspondence). GSL/M/J(no.)   loc. cit., (fair-copy journals). IGS   Institute of Geological Sciences, London, British Geological Survey Archives. JRGS   Journal of the Royal Geographical Society. ML   Mitchell Library, State Library of New South Wales, Sydney. ML, ClP   loc. cit., Clarke Papers. PRGS   Proceedings of the Royal Geographical Society. QJGS   Quarterly Journal of the Geological Society of London. RGS   Royal Geographical Society Archive, London. RGS, GC   loc. cit., General Correspondence. RGS, LB   loc. cit., Letter Books. RGS, MuP   loc. cit., Murchison Papers. RGS, NAE   loc. cit., North Australia Expedition Papers. RS   Royal Society, London.

 1   LR. Robinson and J. Gallagher, 'The imperialism of free trade', *Economic History Review*, 2nd ser., 6(1953), 1–15.

 2   L.H. Brockway, *Science and Colonial Expansion: The Role of the British Royal Botanical Gardens* (London, 1979); R. Dionne and R. MacLeod, 'Science and policy in British India, 1858–1914: perspectives on a persisting belief', *Proceedings of the Sixth European Conference on Modern South Asian Studies; Colloques Internationaux du C.N.R.S. No. 582* (Paris, 1979); D.R. Headrick, *The Tools of Imperialism: Technology and European Imperialism in the Nineteenth Century* (Oxford, 1981); E. Said, *Orientalism* (New York, 1978); R.A. Stafford, 'Geological surveys, mineral discoveries, and British expansion, 1835–71', *Journal of Imperial and Commonwealth History*, 12 (1984), 5–32; M. Worboys, 'Science and British colonial imperialism, 1895–1940' (D.Phil. thesis, University of Sussex, 1980).

 3   G. Basalla, 'The spread of Western science', *Science*, 156 (1967), 611–21.

 4   R. MacLeod, 'On visiting the "moving metropolis": reflections on the architecture of imperial science', *Historical Records of Australian Science*, 5(3)(1982), 1–16; I. Inkster, 'Scientific enterprise and the colonial "model": observations on Australian experience in historical context', *Social Studies of Science*, 15(1985), 677–704.

 5   S.G. Kohlstedt, 'Australian museums of natural history: Public priorities and scientific initiatives in the 19th century', *Historical Records of Australian Science*, 5(4)(1983), 1–29.

 6   J.A. Secord, *Controversy in Victorian Geology: the Cambrian–Silurian Dispute* (Princeton, 1986); J. Morrell and A. Thackray, *Gentlemen of Science: Early Years of the British Association for the Advancement of Science* (Oxford, 1981).

 7   J.A. Secord, 'King of Siluria: Roderick Murchison and the imperial theme in nineteenth-century British geology', *Victorian Studies*, 25 (1982), 413–42; Stafford, op. cit. (n.2); R.A. Stafford, 'The role of Sir Roderick Murchison in promoting the geographical and geological exploration of the British Empire and its sphere of influence, 1855–1871', (D.Phil. thesis, University of Oxford, 1985), forthcoming from Cambridge University Press.

 8   A. Geikie, *The Life of Sir Roderick Murchison*, 2 vols (London, 1875).

 9   R. Murchison, *The Silurian System*; 2 vols (London, 1839); M.J.S. Rudwick, *The Great Devonian Controversy: the Shaping of Scientific Knowledge Among Gentlemanly Specialists* (Chicago, 1985); R. Murchison, E. de Verneuil, and A. von Keyserling, *The Geology of Russia in Europe and the Ural Mountains*, 2 vols (I, London, 1845; II, Paris, 1845).

10   R. Murchison, *Siluria* (London, 1854). Unless otherwise noted, the 4th edition of 1867 is used for citations in this essay.

11 R. Murchison, 'Address to the RGS', *JRGS*, 35(1865), cviii–clxxxvii, at cxciii.

12 Secord, op. cit. (n.7).

13 P.J. McCartney, *Henry De La Beche: Observations On An Observer* (Cardiff, 1977).

14 T.G. Vallance, 'Presidential Address: Origins of Australian geology', *Proc. Linnean Society of N.S.W.*, 100 (1975), 13–43.

15 R. Murchison, 'Address to the Geological Society', *Proc. Geol. Soc. of London*, 1(1832), 362–86, at 365–7.

16 Murchison to Clarke, 17 Dec. 1838, ML, ClP, Mss. 139/43, ff.665–8; see also E. Grainger, *The Remarkable Reverend Clarke: The Life and Times of the Father of Australian Geology* (Melbourne, 1982), 63–4, 77. This book, while offering the most comprehensive biography of Clarke to date, nevertheless remains inferior in many respects to James Jervis' earlier sketch, 'Rev. W.B. Clarke, "The father of Australian geology" ', *Royal Australian Historical Society Journal and Proc.*, 30(1944), 345–458, in which see 410–15 for Clarke's relations with Murchison.

17 King to Murchison, 24 Jan. 1834, GSL/M/K3/1.

18 Macleay to Murchison, 3 Ap. 1841, BL(AM)46127, MuP, ff.160–1; R. Murchison, 'Anniversary Address of the President', *Proc. Geol. Soc. of London*, 3(1842), 637–87, at 645.

19 For Murchison's most blatant attempt to stifle an Australian competitor, see E. Newland, 'Sir Roderick Murchison and Australia: a case study of British influence on Australian geological science', (M.A. thesis, University of New South Wales, 1983), which despite some inaccuracies offers a valuable analysis of Murchison's dispute with Clarke. (See also below.) For the efforts of other British scientists in this regard, see A. Moyal, 'Sir Richard Owen and his influence on Australian zoological and palaeontological science', *Records of the Australian Academy of Science*, 3(2) (1975), 41–56; K.G. Dugan, 'The zoological exploration of the Australian region and its impact on biological theory', in N. Reingold and M. Rothenberg (eds), *Scientific Colonialism: A Cross-Cultural Comparison* (Washington, 1987), pp. 79–100. For an early attempt to institutionalize the dependency of provincial and colonial scientists on the London elite, see M.J.S. Rudwick, 'The foundation of the Geological Society of London: its scheme for co-operative research and its struggle for independence', *Br. Journ. History of Science*, 1(1963), 325–55.

20 GSL/M/J6, 1824–6, ff.7–8; Raleigh Club Minute Book, entry for 11 May 1829, RGS.

21 M. Hoare, 'All things are queer and opposite—scientific societies in Tasmania in the 1840s', *Isis*, 60(1969): 198–209; K. Fitzpatrick, *Sir John Franklin in Tasmania 1837–1843* (Melbourne, 1949), 192–205.

22 *ADB*, 2, 114–5; Vladislav Kruta, *et. al.*, *Dr John Lhotsky, the Turbulent Australian Writer, Naturalist and Explorer* (Melbourne, 1977).

23 Murchison to Franklin, 12 Feb. 1842, Scott Polar Research Institute, Cambridge, Lefroy Bequest, Ms. 248/222.

24 Murchison, op. cit. (n.18), p.646.

25 Joseph Beete Jukes, *Narrative of the Voyage of H.M.S. Fly*, 2 vols (London, 1847).

26 Franklin-Strzelecki correspondence, ML, CY Reel 1759, MLA 1604; Jukes to Sedgwick, 28 Nov. 1842, CUL, SeP, Add.Ms.7652.I.F.Portfolio II.90.

27 Franklin to Murchison, Nov. 1842, Scott Polar Research Institute, Cambridge, Lefroy Bequest, Ms. 248/225/3, quoted in H.M.E. Heney, *In A Dark Glass: The Story of Paul Edmund Strzelecki* (London and Sydney, 1961), p. 132—a book containing numerous inaccuracies; R. Murchison, 'Address to the R.G.S.', *JRGS*, 22(1852), lxii–cxxvi, at lxxxii.

28 In 1853—see G. Rawson, *The Count: A Life of Sir Paul Edmund Strzelecki, K.C.M.G., Explorer and Scientist* (London and Melbourne, 1953), p.168.

29 Franklin to Murchison, 30 Jan. 1845, GSL/M/F17/1.

30 J.B. Élie de Beaumont, 'Researches on some of the revolutions which have taken place on the surface of the globe', *Philosophical Magazine*, 10(1831), 241–64; for discussion see Mott T. Greene, *Geology in the Nineteenth Century: Changing Views of a Changing World* (Ithaca, N.Y. and London, 1982), pp.69–121; for Murchison's views, see his 'On Russia and the Ural Mountains', discourse at the Royal Institution reported in *Athenaeum Journal*, no.920(15 June 1845), 591–2.

31 While Murchison constantly reiterated his debt to Humboldt (e.g., R. Murchison, 'A few observations on the Ural Mountains, to accompany a new map of the southern portion of that chain', *JRGS*, 13(1843), 269–78, at 271; op. cit. (n.10), pp.458–9); others felt he had plagiarized the Prussian—see Sedgwick to Clarke, 11 March 1853, quoted in Jervis, op. cit. (n.16), pp.369–70.

32 Murchison, op. cit. (n.10), pp.448–75.

33 GSL/M/J7, 1826–38, ff.112–3.

34 R. Murchison, 'Address to the R.G.S.', *JRGS*, 14(1844), xlv–cxxxviii, at xcix–c.

35 R. Murchison, 'A brief review of the classification of the sedimentary rocks of Cornwall', *Trans. Roy. Geol. Soc. of Cornwall*, 6(1846), 317–26.

36 Smith to Murchison, 27 Feb. 1852, EUL, MuP, Gen.523/6; R. Murchison, 'On the distribution of gold ore in the crust and on the surface of the earth', *Report of the Br. Assoc., . . . 1849, Trans. of Sections*, (1850), 60–3, at 63; see also the tendentious but useful L.R. Silver, *A Fool's Gold? William Tipple*

Smith's Challenge to the Hargraves Myth (Milton, Qld., 1986). Cf. A. Mozley, 'The foundations of the Geological Survey of New South Wales', *Proc. Royal Society of N.S.W.*, 98 (1965), 91–100, at 100, which mistakenly identifies Thomas Chapman, the gold-finding shepherd of Victoria, rather than Phillips as Murchison's second donor.

37 Murchison to Grey, 5 Nov. 1848, enclos. in 'Further papers relative to the recent discovery of gold in Australia', *Parl. Papers*, 1852–3, LXIV(1684.), 465–700, at 521.

38 Murchison, op. cit. (n.27), xxxiii.

39 Mozley, op. cit. (n.36); *ADB*, 6, 216–7.

40 De La Beche to Hawes, 4 Aug. 1849, IGS, GSM 1/5, ff.188–91.

41 De La Beche to Hawes, 12 Aug. 1850, IGS, GSM 1/6, ff.17–18.

42 Hawes to De La Beche, with enclos., 5 July 1851, IGS, GSM 1/6, ff.98–9; reply, 16 July 1851, ff.104–6.

43 B. O'Neil, *In Search of Mineral Wealth: The South Australian Geological Survey and Department of Mines to 1964* (Adelaide, 1982), 25–31; *ADB*, 3, 65–6.

44 Clarke to Sedgwick, 2 Feb. 1843, CUL, SeP, Add.Ms.7652.I.F.II.91.

45 T.G. Vallance, 'John Lhotsky and geology', in Kruta, op. cit. (n.22), pp.41–56: Heney, op. cit. (n.27), pp.187–97; Grainger, op. cit. (n.16), pp.163–4. Cf. Geoffrey Blainey, 'Gold and governors', *Historical Studies Aust. and N.Z.*, 9(1961), 337–350; *ADB*, 3, 420–2; Mozley, op.cit. (n.36), p.91. Blainey and Mozley reject the traditional historical interpretation based on the testimony and actual behaviour of Clarke and Strzelecki—the former having observed public silence about his find until Gipps' death in 1847 and the latter, except for brief mentions of specimens in an obscure British parliamentary report and a Sydney newspaper article, until 1856. (See note following.) They contend that Gipps did not 'gag' the two discoverers and prevent an earlier gold rush from fear of a breakdown in social order, but rather that he remained insufficiently impressed by their small specimens to officially promote the exploitation of a resource whose potential to generate revenue for the colonial exchequer he would not, given adequate evidence, have been able to ignore.

46 Strzelecki to Murchison, 17 Jan. 1854, EUL, MuP, Gen.523/6; P.E. de Strzelecki, *Gold and Silver* (London, 1856).

47 S. Davison, *The Discovery and Geognosy of Gold Deposits in Australia* (London, 2nd edn., 1861; 1st publd. 1860), p.ii.

48 W.B. Clarke, 'Geology: comparison of Russia and Australia', the *Sydney Morning Herald*, 28 Sept. 1847; Newland, op. cit. (n.19), especially pp.94–105. For Clarke's brother, see Clarke to Murchison, 5 Nov. 1852, GSL/M/C13/a+b; Grainger, op.cit. (n.16), p.55.

49 W.B. Clarke, 'On the discovery of gold in Australia', *QJGS*, 8(1852): 131–4; Clarke, *Researches in the Southern Gold Fields of New South Wales* (Sydney, 1860); R. Murchison, 'On the anticipation of the discovery of gold in Australia, with a general view of the conditions under which that metal is distributed', *QJGS*, 8(1852), 134–6; W. Hopkins, 'Anniversary Address of the President', *QJGS*, 8(1852), xxi–lxxx, at liv–lvii; Clarke to King, nid.[1853], ML, King Papers, A3599, pp.327–34; King to Murchison, 24 Nov. 1853, EUL, MuP, Gen.523/6; Murchison to Clarke, June 1852, 1 Jan. 1853, ML, ClP, Mss.139/43, 677–81, 683a–86; Newland, op.cit. (n.19); pp.96, 102.

50 Newland, op. cit. (n.19); cf. Grainger, op. cit. (n.16), pp.164–6, 192, who underestimates the significance of this dispute. For Murchison's defence of his prediction, see correspondence in EUL, MuP, Gen.523/6; ML, ClP, Mss.139/43, 'Further Papers', op. cit. (n.37), 522; Murchison, op.cit. (n.49).

51 Murchison to Peel, 27 June 1844, BL(AM)40547, Peel Papers, ff.284–5; Herschel to Murchison, Jan. 1851, 10 Jan. 1852, EUL, MuP, Gen.523/6.

52 Murchison, op. cit. (n.34), lxix.

53 Murchison, op.cit. (n.10), p.473.

54 Ibid., 448–75.

55 Murchison, op.cit. (n.36), p.61.

56 R. Murchison, 'Address to the R.G.S.', *JRGS*, 23(1853), lxii–cxxxviii, at cxxvii.

57 E.g. R. Murchison, op. cit. (n.27), pp. lxxxii–lxxxvi; op. cit. (n.10), pp. 448–75.

58 J. Wyld, *Notes on the Distribution of Gold Throughout the World* (London, 3rd edn: 1853; 1st publd. 1852); Brayley to Murchison, 3 Feb. 1852, EUL, MuP, Gen.523/6.

59 Newspaper notices of 'Lectures on gold', 26 June 1852, IGS, GSM 1/57, f.5; Jukes to Murchison, 1 July 1852, EUL, MuP, Gen.523/6; J.B. Jukes et al., *Lectures on Gold for the Instruction of Emigrants About to Proceed to Australia, delivered at the Museum of Practical Geology* (London, 1852).

60 Murchison to Whewell, 19 Feb. 1852, quoted in Geikie, op.cit. (n.8), ii, p.131.

61 For Australian mountebank John Calvert, see his *The Gold Rocks of Great Britain and Ireland, and a General Outline of the Gold Regions of the World, with a Treatise on the Geology of Gold* (London, 1853); Clarke to Sedgwick, 13 Sept. 1853, CUL, SeP, Add.Ms.7652.II.M.22a; Murchison to Clarke, 16 Nov. 1853, ML, ClP, Mss. 139/43, pp.687a–694; Clarke to Murchison, 23 Jan. 1855, EUL, MuP, Gen.523/4/6; Hargraves to Murchison, 3 and 6 Feb. 1855, EUL, MuP, Gen.523/4/23, /24; Murchison, (1854 edn), op.cit. (n.10), p.436.

62 Murchison to Sopwith, 24 March 1852, American Philosophical Society Library, Philadelphia,

Murchison Papers, B:M93p; cf. Newland, op.cit. (n.19), p.105, who makes the erroneous and unsubstantiated claim that Murchison sat on the boards of several gold mining companies.

63 Odernheimer to Murchison, 22 June 1853, GSL/M/O1/1; Friedrich Odernheimer, 'On the geology of part of the Peel River district in Australia', *QJGS*, 11(1855): 399-402; R. Murchison, 'Address to the R.G.S.' *JRGS*, 27(1857): xciv-cxcviii, at clxxv-clxxvi; Murchison, op.cit. (n.10), p.463.

64 Herschel to Murchison, 2 Jan. 1853, EUL, MuP, Gen.523/4/43.

65 E.g., Terry to Murchison, 21 May 1853, RGS, GC.

66 Nicol to Murchison, 22 Nov. 1852, GSL/M/N8/6a+b

67 Murchison, op.cit. (n.10), pp.454-7.

68 Mozley, op.cit. (n.36); Grainger, op.cit. (n.16), pp.163-204.

69 Clarke to Murchison, 7 July 1851, EUL, MuP, Gen.523/6; 23 Jan. 1855, Gen.523/4/6; Jervis, op.cit. (n.16), pp.385-6, 412-14.

70 Murchison to Clarke, 12 Feb. 1856, ML, ClP, Mss. 139/43, pp.699a-700.

71 Jervis, op.cit. (n.16), p.423; L. De Koninck, 'Descriptions of the palaeozoic fossils of New South Wales (Australia)', *Memoirs of the Geological Survey of N.S.W., Palaeontology*, 6(1898), 1-298.

72 Murchison to Clarke, 1 Jan., 16 Nov. 1853, ML, ClP, Mss. 139/43, pp.683a-694; replies, 18 Nov. 1854, EUL, MuP, Gen.523/4/8; 23 Jan. 1855, Gen.523/4/6.

73 Grainger, op.cit. (n.16), pp.171-7.

74 Clarke to Murchison, 30 May 1855, EUL, MuP, Gen.523/4/7; Murchison, op.cit. (n.56), pp.cxxv-cxxvi; op.cit. (n.10), p.18.

75 Jervis, op.cit. (n.16), p.450. Before the gold dispute flared, Clarke had in fact provided the Society with news of Australian explorations via Murchison—see Jackson to Clarke, 13 Feb. 1841, RGS, LB, v.1841-44.

76 Hargraves to Murchison, 17 Jan. 1855, EUL, MuP, Gen.523/4/22; *QJGS*, 11(1855): 497; E. Hargraves, *Australia and its Gold Fields* (London, 1855); also see notes 175, 176.

77 Grainger, op.cit. (n.16), pp.215-21; Newland, op.cit.(n.19), pp.100-5.

78 Murchison, op.cit. (n.11), p.cxli; Murchison, op.cit. (n.18), p.645.

79 Murchison, op.cit. (n.34), pp.xcvii-ciii.

80 Earl later emerged as another rival for priority regarding the Australian gold prediction. He based his claim on a memoir submitted to the Geographical Society in 1845 in which he argued that the mountain ranges of Australia and the islands of the East Indian Archipelago were a continuation of the richly mineralized ranges of Asia. Murchison opposed Earl's postulation at its presentation on mineralogical and palaeontological grounds, and, probably feeling that his own prediction of 1844 had been plagiarized, countered Earl's bid for precedence following the gold rush of 1851. See G.W. Earl, 'On the physical structure and arrangement of the islands of the Eastern Archipelago', *JRGS*, 15(1845), 358-65; *Athenaeum Journal*, no.920 (15 June 1845), 590; G. W. Earl, *A Correspondence Relating to the Subject of Gold in Australia* (London, 1853); Murchison, op.cit. (n.10), (1854 edn), pp. 450-1.

81 D. Howard, 'The English activities on the north coast of Australia in the first half of the nineteenth century', *Proc. Royal Geographical Society of Australasia, S. Australian Branch*, 33(1931-2), 21-194; G.S. Graham, *Great Britain in the Indian Ocean: A Study of Maritime Enterprise, 1810-1850* (Oxford, 1967), pp.402-43; C. Lloyd, *Mr Barrow of the Admiralty* (London, 1970), pp.160-4.

82 R. Murchison, 'Address to the R.G.S.', *JRGS*, 15(1845), xli-cxi, at lvii-lxi.

83 Murchison, op.cit. (n.27), pp.lxxxii-lxxxvi; op.cit. (n.56), pp.cxxiv-cxxx.

84 Haug to Murchison, 23 Ap. 1853, RGS, NAE, 2/7.

85 Shaw to Newcastle, 24 Aug. 1853, RGS, LB, v.1850-9, ff.91-5; Newcastle to Ellesmere, 23 Dec. 1853, RGS, GC.

86 J. Calvert, 'Mineralogy of Australia: some observations on the interior of the Australian continent', *The Mining Journal*, 23(1853), 580-1. See also note 61.

87 T. Saunders, *The Asiatic Mediterranean* (London, privately printed, 1853).

88 Murchison to Saunders, 6 Oct. 1853, Dixon Library, State Library of NSW, Ms.Q421, Gregory Papers, ff.86-8.

89 Murchison to Shaw, n.d. and 6 Oct. 1853, RGS, MuP; Murchison to Haug, 6 Oct. 1853, RGS, LB, v.1850-9, ff.95-6.

90 Howard, op.cit. (n.81), pp.139-40; J.L. Stokes, 'On steam communication between England, Australia, and the Cape of Good Hope', *JRGS*, 26(1856), 183-8; *PRGS*, 1(1855-6), 79-82.

91 Saunders to RGS, 2 Jan. 1854, RGS, NAE, 5/7; Shaw to Merivale, 14 Jan. 1854, Dixon Library, State Library of NSW, Ms.Q421, ff.17-18.

92 Murchison to Shaw, 14 Aug. 1854, RGS, MuP.

93 Wilson to Shaw, 23 Nov. 1853, RGS, NAE, 5/1; Murchison, op.cit. (n.10), pp.462-3; J. Wilson, 'On the gold regions of California', *QJGS*, 10(1854), 308-21.

94 Wilson to Murchison, with enclos., 16 Sept. 1854, RGS, NAE, 5/1; Murchison to Sturt, 22 Dec. 1854, Rhodes House, Oxford, Sturt Papers, MssAustral.S.5, ff.143-4; Sturt to Murchison, 24 Feb. [1855], GSL/M/S85/2.

95 Washington to Sturt, 8 Oct. 1854, Rhodes House, Oxford, Sturt Papers, MssAustral.S.5, no.44, ff.122-3.

96 Council Minutes, v.1841-57, entry for 27 June 1855, Bodleian Library, Oxford, British Association Papers.

97 Denison to Murchison, 21 May 1855 and 16 Aug. 1856, quoted in W. Denison, *Varieties of a Vice-Regal Life*, 2 vols (London, 1870), i, pp.309-10, 362-3.

98 A.C. Gregory, 'Progress of the North Australian Expedition', *PRGS*, 1(1855-6), 32-5.

99 Stokes, op.cit. (n.90).

100 A.C. Gregory, 'Journal of the North Australian Exploring Expedition', *JRGS*, 28(1858), 1-137; *PRGS*, 1(1856-7), 490-501.

101 Murchison to Elliot, 13 Nov. 1857, IGS, GSM 1/7, f.500.

102 Murchison to Shaw, 19 Ap. 1858, RGS, Journ.MssAust., referee's comments; J. Wilson, 'Notes on the physical geography of Northwest Australia', *JRGS*, 28(1858), 137-53; *PRGS*, 2(1857-8), 210-17.

103 Murchison, op.cit. (n.63), pp.clxxvi-clxxx.

104 R. Murchison, 'Address to the R.G.S.', *JRGS*, 28(1858), cxxxiii-ccxviii, at cxcv-cxcviii.

105 'Correspondence respecting the western boundary of Queensland', *Queensland Parl. Papers*, 1861, 1009-17.

106 J.W. Gregory and F.T. Gregory, 'Remarks to accompany a geological map of Western Australia', *QJGS*, 4(1848), 142.

107 DuCane to Shaw, 23 June 1858 and n.d. (received 14 June 1859), RGS, GC; this section also relies on information kindly supplied by Sister Mary Albertus, O.P., of 'Santa Sabina', Lalor Street, Scarborough, WA.

108 Embling to Murchison, 8 Jan. 1858, RGS, GC; M.E. Hoare, 'Learned societies in Australia: the foundation years in Victoria, 1850-1860', *Records of the Australian Academy of Science*, 1(1967), 7-29, at 22.

109 F. Gregory to Shaw, 24 Jan. 1860, ML, F. Gregory Papers, Mss.A306, pp.19-22; W. Burgess, L. Burgess, and Sanford to Murchison, 30 May 1860, pp.23-4; W. Burgess to Murchison, 31 May 1860, RGS, GC.

110 Murchison to Shaw, 23 July 1860, RGS, MuP; W. Burgess to Murchison, 18 July 1860, GSL/M/B69/1.

111 Ashburton to Shaw, two of 8 Aug. 1860, RGS, Ashburton Collection.

112 Shaw to Fortescue, 24 Aug. 1860, RGS, LB, v.1859-79, ff.23-7.

113 Shaw to Treasury, 24 Jan., 20 March 1861, RGS, LB, v.1859-79.

114 Murchison to Fortescue, 8 June 1861, RGS, LB, v.1859-79; R. Murchison, 'Address to the R.G.S.', *JRGS*, 31(1861), cxi-clxxxvi, at clxxv; *Report of the Br. Assoc., ... 1861, Trans. of Sections* (1862), 197.

115 F.T. Gregory, 'On the geology of a part of Western Australia', *QJGS*, 17(1861), 475-83.

116 Murchison to Shaw, 12 May 1862, RGS, Journ.Mss.Aust., referee's comments; F.T. Gregory, 'Expedition to the north-west coast of Australia', *JRGS*, 32(1862), 372-429; *PRGS*, 6(1861-2), 54-9.

117 *JRGS*, 33(1863), cx-cxi.

118 Elliot to Royal Society, 19 Ap. 1859, RS, MC.6.22; Lytton to Murchison, 19 Ap. 1859, RGS, Colonial Office Corresp.; Denison to Murchison, 27 May 1859, GSL/M/D45/1.

119 Murchison to Elliot, 9 May 1859, RGS, MuP; R. Murchison, 'Address to the R.G.S.', *JRGS*, 29(1859), cii-ccxxiv, at ccxix.

120 Council Memo., n.d. 1859, RGS, Colonial Office Corresp.

121 Elliot to Sharpey, 5 Nov. 1859, RS, MC.6.44; Denison, op.cit. (n.97), i, pp. 455, 479-80.

122 MacDonnell to Murchison, 26 Oct. 1860, BL(AM)46127, MuP, ff.142-3; 24 Dec. 1863, GSL/M/M26/1; R.G. MacDonnell, *Australia: What It Is, and What It May Be* (Dublin, 1864), pp.17, 59; R. Murchison, 'Address to the R.G.S.', *JRGS*, 34(1864), cix-cxciii, at cxlix-cl.

123 Bowen to Murchison, 30 Ap. 1860, BL(AM)46125, MuP, ff.197-8; Murchison, op.cit. (n.114), p.clxxvi; also see notes 177 to 179.

124 Sturt to Shaw, 24 May 1858 and 23 Aug. 1860, RGS, Sturt Collection; Sturt to Murchison, 9 Feb. 1861, BL(AM)46128, MuP, ff.161-2.

125 T.L. Mitchell, *Journal of an Expedition Into the Interior of Tropical Australia* (London, 1848), pp.vi-vii, 427-30.

126 T.L. Mitchell, *Three Expeditions Into the Interior of Eastern Australia*, 2 vols (London, 2nd edn, 1839; 1st publd. 1838), ii, pp.353-9; Murchison, op.cit. (n.18), p.645; Murchison to Mitchell, 19 Jan. 1853, ML, Mitchell Papers, A294, pp.133-5. These rocks have since been reclassified as Devonian.

127 Murchison, op.cit. (n.119), p.ccxvi.

128 See n.43; Merivale to Murchison, with enclos., 13 March 1857, IGS, GSM 1/7, ff.376-7; Murchison to the Committee of the Privy Council on Education, 29 July 1857, GSM 1/7, f.434; Murchison to C. Babbage, 24 March 1857, BL(AM)37197, Babbage Papers, ff.178-9; Murchison, op.cit. (n.119), p.ccxiii.

129 Murchison to McCoy, 29 May 1854, ML, McCoy Papers, CY Reel 499, frames 587-8.

130 Nicol to Murchison, 9 July 1854, GSL/M/N5/6; Secord, op.cit. (n.6), pp.271-2; T.G. Vallance, 'Pioneers and leaders—a record of Australian palaeontology in the nineteenth century', *Alcheringa*, 2(1978), 243-50, at 246.

131 Murchison to Sedgwick, 1 Aug. 1854, CUL, SeP, Add.Ms.7652.D. notebook V.68; Kohlstedt, op.cit. (n.5).

132 McCoy to Sedgwick, 25 June 1857, Add.Ms.7652.II.J.50. In a sense, both Murchison and McCoy were correct, for the chief gold deposits of Victoria occur in rocks of the Ordovician System—the stratigraphic classification established by Charles Lapworth in 1879 as a compromise between the rival Cambrian and Silurian systems.

133 R. Murchison, *Annual Report of the Director-General of the Geological Survey of the United Kingdom, 1866* (London, 1867); T.G. Vallance, 'The fuss about coal: troubled relations between palaeobotany and geology ', in D.J. and S.G.M. Carr, eds, *Plants and Man in Australia* (London, 1981), pp.136-76. The Palaeozoic coals of New South Wales, Queensland, and Tasmania are now known to be Permian: seams of varying Mesozoic age also occur in these localities as well as elsewhere in Australia.

134 Murchison to Egerton, 10 Oct. 1866, IGS, GSM 1/8, ff.419-21. Such coal was found in the immensely productive Bowen Basin and elsewhere.

135 R.K. Johns, ed., *History and Role of Government Geological Surveys in Australia* (Adelaide, 1976), pp.17-19; M.E. Hoare, '"The half-mad bureaucrat" Robert Brough Smyth (1830-1889)', *Records of the Australian Academy of Science,* 2(4) (1970-3), 25-40; T.A. Darragh, 'The Geological Survey of Victoria under Alfred Selwyn, 1852-1868', *Historical Records of Australian Science,* 7(1) (1987), 1-25.

136 Sally Kohlstedt ably discusses the general relationship between the British Survey and its Australian counterparts (op. cit., n.5), but she overemphasizes the authority exercised by colonial governments and the Colonial Office itself in the appointment of geologists—a process which, with local funding once assured, was in fact controlled by the Director-General of the British Survey.

137 Elliot to De La Beche, 6 Sept. 1853, IGS, GSM 1/6, ff.377-8; Reeks to Elliot, 20 Sept. 1853, ff.278-9.

138 Barkly to Lytton, 27 Jan. 1859, ML, A2346, p.6385; Murchison to Herschel, 3 Nov. 1863, RS. Herschel Papers, HS.12.430.

139 Compare, e.g., G.H. Wathen, 'On the gold fields of Victoria or Port Philip [sic]', *QJGS*, 9(1853), 74-9; and Murchison, op.cit. (n.56), p.cxxvi.

140 E.g., J. Calvert, 'Australian quartz veins—No.1, No.2, No.3', *The Mining Journal*, 24(1854), 741, 773, 798.

141 'Second progress report of the commissioners appointed to enquire into the mining resources of the colony', *Victorian Parl. Papers*, 1856-7, IV, 1463-74. For the Board of Science, see Hoare, op.cit. (n.135).

142 'Facts and theory in quartz mining', the *Bendigo Advertiser*, 14 Dec. 1857, p.2, col.4; see also Newland, op.cit. (n.19), p.133.

143 Barkly to Labouchere, 31 Oct. 1857, printed in *Despatches From the Governor of Victoria, Sir Henry Barkly to Secretary of State Henry Labouchere,* 1857-8, II, 1145-6.

144 As recently as 1854 Selwyn had followed Murchison's poverty-with-depth theory regarding gold veins in a memoir published by the Geological Society of London. But even then Selwyn had implied that Victoria's vast alluvial deposits alone were capable of sustained yield. See A.R.C. Selwyn, 'On the geology and mineralogy of Mount Alexander, and the adjacent country, lying between the rivers Loddon and Campaspe', *QJGS*, 10(1854), 299-303.

145 Murchison to Merivale, 27 March 1858; Barkly to Stanley, 12 July 1858; Selwyn to the Private Secretary, 13 July 1858—all printed in 'Geological Survey', *Victorian Parl. Papers*, 1859-60, I, 1209-16. See also Murchison to Clarke, 15 Ap. 1858, ML, ClP, Ms.139/43, pp.707-10. Besides the report he wrote for Barkly, Selwyn also published a second article through the Geological Society of London, which asserted that Victoria's alluvial gold deposits alone were of such immense extent that they could be profitably worked until an 'indefinitely remote' period—see A.R.C. Selwyn, 'On the geology of the gold-fields of Victoria', *QJGS*, 14(1858), 533-8. For the actual stratigraphy of the Victorian gold, see note 132. The dispute between the members of the Mining Commission destroyed the confidence of the mining community in the organization's work and brought about its dissolution in late 1857 after less than two years of service. Its successor, the Board of Science, had a similarly short lifespan between 1858 and 1860.

146 Murchison, op.cit. (n.63), p.clxxv.

147 E.g., Murchison, op.cit. (n.10), pp.464-7.

148 Denison to Murchison, 27 May 1859, GSL/M/D45/1; same to same, April 1860, quoted in Denison, op.cit. (n.97), i, pp.479-80.

149 Murchison to Murray, 8 Jan. 1860, John Murray Archive, London.

150 C.A. Zachariae, 'The Australian gold formation', the *Bendigo Advertiser*, 23 Nov. 1859, p.3, col.3; 24 Nov., p.2, col.4; 26 Nov., p.2, cols.5-6; 28 Nov., p.2, col.6, p.3, col.1; 29 Nov., p.2, col.4; 30 Nov., p.2, cols.4-5; see also Newland, op.cit. (n.19), pp.138-75, for this conflict.

151 Murchison to Hutchinson, 27 Feb. 1860, printed as 'The geology of the goldfields' in the *Bendigo Advertiser*, 11 May 1860, p.2, cols.5-6; see also editorial of same title, p.2, col.3.

152 C.A. Zachariae, 'A reply to Sir Roderick Murchison's letter on the Australian gold formation', the *Bendigo Advertiser*, 15 May 1860, p.2, cols.4-5.

153 Murchison to Clarke, 31 Oct. 1860, ML, ClP, Mss.139/43, pp.711-14; Barkly to Murchison, 17 May 1860, GSL/M/B3/1.

154 "A quartz miner", 'Sir Roderick Murchison's letter on the goldfields'—a letter, the *Bendigo Advertiser*, 15 May 1860, p.3, cols.1-2; "Murchison's ghost", 'Murchison's theory'—a letter, 16 May 1860, p.3, col.3; W.J. Morgan, 'Note on gold-drifts at Ballaarat [sic]', *The Geologist*, 3(1860), 153-4; Barkly to Murchison, 17 May 1860, GSL/M/B3/1.

155 E.g., F. McCoy, 'Deep gold diggings of Melbourne', *The Geologist*, 6(1863), 36; J.A. Phillips, 'Gold mining and the gold discoveries made since 1851', *Society of Arts Journal*, 10, no.495(1862), 419-29.

156 Newland makes this claim in 'Sir Roderick Murchison and Australia', op.cit. (n.19), pp.123, 131, 140-1, 179, 181, citing neither any pronouncements by investment advisors or admissions by dissuaded investors that could directly link Murchison's views to a faltering of financial confidence, nor even any statistics of company failures or unsubscribed share issues to demonstrate that a significant slackening of investment occurred. Instead, Newland adduces a single article and several unspecified editorials backing Murchison's position printed in the British *Mining Journal* which, while certainly the most influential organ of the industry, likewise served as a forum for contending opinions and consensus formation. Newland seems to have taken the opinions of Barkly, Selwyn, and the editor of the *Bendigo Advertiser*, as well as representations made to Barkly by the proprietors of several quartz crushing companies which were mentioned in the Governor's despatch to the Colonial Office (note 145), that such a falling off of investment *might* take place, as proof that it actually did.

157 R. Murchison, 'Address of the President of the Geology Section', *Report of the Br. Assoc., ... 1861, Trans. of Sections* (1862), 95-108; Hoare, op.cit. (n.135), p.30.

158 R. Murchison, *Annual Report of the Director-General of the Geological Survey of the United Kingdom, 1862* (1863); Murchison to Darling, 23 Oct. 1863, IGS, GSM 1/8, ff.275-6; Darragh, op.cit. (n.135).

159 Thomson to Murchison, 20 Ap. 1866, BL(AM)46128, MuP, ff.195-6.

160 Barry to Murchison, 4 Jan. 1870, BL(AM)46125, MuP, ff.117-20; reply, 15 March 1870, IGS, GSM 1/9, ff.179-80.

161 Denison to De La Beche, 1 March 1849, IGS, GSM 1/14, f.34; De La Beche to Hawes, 27 July 1850, GSM 1/6, ff.11-13.

162 Denison to Thomson, 16 June 1851, ML, Denison Papers, A1531^{-2}, pp.599-600.

163 Clarke, op.cit. (n.49, first citation); Clarke to Murchison, 18 Feb. 1852; Jackson to Murchison, 2 July 1852 — both in EUL, MuP, Gen.523/6.

164 Merivale to Murchison, 22 Dec. 1855, IGS, GSM 1/7, f.268; reply, 7 Jan. 1856, ff.266-8; Clarke to Murchison, 20 June 1856, GSL/M/C13/2.

165 Merivale to Murchison, 6 Nov. 1856, IGS, GSM 1/7, ff.354-5; reply, 14 Dec. 1858, ff.577-8; Margaret Reeks, *Register of the Associates and Old Students of the Royal School of Mines, and History of the Royal School of Mines* (London, 1920), pt.2, 72; ADB, 4, 277-8.

166 Murchison to Fortescue, 30 July 1857, IGS, GSM 1/7, ff.433-4.

167 Murchison, op.cit. (n.119), p.ccxviii.

168 Young to Murchison, 11 Dec. 1859, GSL/M/Y1/1; Gould to Clarke, 16 Sept. 1860, ML, ClP, Mss.139/38, pp.301-8; same to same, 9 Dec. 1863, quoted in A. Moyal, ed., *Scientists in Nineteenth-Century Australia: A Documentary History* (Melbourne, 1976), pp.149-50; Gould to Murchison, 20 June 1861, IGS, GSM 1/15, f.160.

169 C. Gould, 'Results of the Geological Survey of Tasmania', *Report of the Br. Assoc., ... 1861, Trans. of Sections*, (1862), 112-13; Murchison, op. cit. (n.157); *Annual Report of the Director-General of the Geological Survey of the United Kingdom, 1867* (1868).

170 G. Blainey, *The Peaks of Lyell* (Melbourne, 4th edn., 1978; 1st publd. 1954), pp.5-11.

171 JRGS, v.31, civ-cx; MacDouall Stuart, 'Journal of an expedition across the centre of Australia', *JRGS*, 31(1861), 65-145; PRGS, 5(1860-1), 55-60, 104-6.

172 R. Murchison, 'Address to the R.G.S.', *JRGS*, 33(1863), cxiii-cxcii, at cxxiv; Chambers and Finke to Murchison, 10 May, 26 Nov. 1861, BL(AM)46125, MuP, ff.377-80; *ADB*, 3, 377-8.

173 JRGS, 32(1862), c-cii; Murchison, op.cit. (n.172), pp.clxiii-clxvii.

174 Murchison to Shaw, 19 March 1856, RGS, Journ.Mss.Aust., referee's comments; R. Austin, 'Report of an expedition to explore the interior of Western Australia', *JRGS*, 26(1856), 235-74; PRGS, 1(1855-6), 30-1.

175 See notes 73 to 76.

176 Murchison to Markham, 29 Jan. 1864, RGS, Journ.Mss.Aust., referee's comments; E. Hargraves, 'Report on the non-auriferous character of Western Australia', *PRGS*, 8(1863-4), 32-4.

177 W.B. Clarke, 'The auriferous and other metalliferous districts of northern Queensland', *PRGS*, 12(1867-8), 138-44.

178 *ADB*, 4: 1-2; Johns, op.cit. (n.135), pp.46-8; Kohlstedt, op.cit. (n.5), pp.5-7.

179 Murchison to Markham, 16 Jan 1868, RGS, Journ.Mss.Aust., referee's comments; Clarke, op.cit. (n.177); R. Daintree, 'Notes on the geology of the colony of Queensland', *QJGS*, 28(1872), 271-360.

180 MacMaster to Murchison, 22 Aug. 1861, BL(AM)46127, MuP, ff.166-7.

181 Crawfurd to Murchison, 17 June 1863, BL(AM)46125, MuP, ff.487-8; Howard, op.cit. (n.81), pp.137-8.

182 H. Merivale, 'On the utility of colonization', *Report of the Br. Assoc., . . . 1862, Trans. of Sections* (1863), 161-2; Geikie, op.cit. (n.8), ii, p.294.

183 Murchison to Lytton, 2 May 1859, Hertfordshire Record Office, Hertford, Lytton Papers; Murchison, op.cit. (n.10), pp.461-2.

184 King to Murchison, 28 Aug. 1852, EUL, MuP, Gen.523/6; see above, pp.79-81.

185 Stokes to Murchison, n.d. [1858], BL(AM)46128, MuP, ff.144-5.

186 See notes 103, 104; *PRGS*, 3(1858-9), 91; Murchison to Lytton, 29 Dec. 1858, Herts. Record Office, Hertford, Lytton Papers.

187 Murchison, op.cit. (n.119), pp.ccxx-ccxxi; op.cit. (n.122), p.cl; op. cit. (n.105); G.F. Bowen, *Thirty Years of Colonial Government*, 2 vols (London, 1889), i, pp.214-5.

188 Murchison to Galton, 25 Feb. 1866, University College, London, Galton Collection, 190; Murchison, op.cit. (n.122), p.cli.

189 G.R. Crone, *Modern Geographers* (London, 1970), p.11; Lloyd, op.cit. (n.81), pp.160-4.

190 Vallance, op.cit. (n.14).

191 For discussion of the paradoxes and ambiguities involved in assessing the impact of Eurocentrism on colonial science, see G. Seddon, 'Eurocentrism and Australian science: some examples', *Search*, 12 (1981), 446-50.

192 Secord, op.cit. (n.6); Rudwick, op.cit. (n.9); for a more general appraisal of this phenomenon in Victorian science, see Morrell and Thackray, op.cit. (n.6).

193 For the connection between Australian desires for political and scientific equality, see Moyal, op.cit. (n.168), p.172; MacLeod, op.cit. (n.4).

194 Kohlstedt, op.cit. (n.5).

195 They can be interpreted as Phase I holdovers in the midst of Phase II of Basalla's model as well as of Inkster's modified version of the same scheme; in MacLeod's model they would occupy the 'Colonial' time-frame while exhibiting many of the characteristics ascribed to an earlier 'Metropolitan' period— see Basalla, op.cit. (n.3); Inkster, op.cit. (n.4); MacLeod, op.cit. (n.4).

196 Newland reaches the same conclusion—see op.cit. (n.19), pp.39, 180.

197 MacLeod, op.cit. (n.4).

198 Moyal, op.cit. (n.168), pp.125, 172. This period witnessed rebellions by two other Victorians—von Mueller and McCoy—against, respectively, Kew Gardens for commissioning a British botanist to write *Flora Australiensis*, and the British Museum for attempting to buy and ship to England both of the Cranbourne meteorites. Like Clarke's effort, these were less successful bids for equality than Selwyn's.

199 Everett to Murchison, 15 Aug. 1854, GSL/M/E11/1.

This research was supported in part by a post-doctoral research fellowship awarded by the University of Melbourne. I wish to thank Professor Rod Home and the participants in the History and Philosophy of Science seminar series at the University of Melbourne for comments and criticisms on earlier drafts of this essay. Permission to quote from manuscripts has been granted by the Syndics of Cambridge University Library, Edinburgh University Library, the Council of the Geological Society of London, the Royal Geographical Society, and the Royal Society.

5

Support for the scientific enterprise, 1850–1900

Ian Inkster and Jan Todd

During the nineteenth century Australia demonstrated many of the features of economic and social development associated with areas of recent settlement.[1] Such areas were relatively large territories (South Africa, Canada, New Zealand) populated by European settlers, the economic growth of which was linked to staple export products — fisheries, gold, wool, wheat and meat — rather than to large-scale developments in manufacturing industry.[2] The Australian case illustrates how such territories were also characterized by the presence of groups of European intellectuals devoted to the establishment of a scientific culture that increasingly addressed itself to the economic demands of staple exploitation and the social requirements of cultural settlement and security. Until well into the nineteenth century such scientific intellectuals were peripheral thrice over — in their position at the far edge of European culture, in their often part-time position as scientific workers, and in their role as human capital for the exploitation of staple products of economic significance to the European and Atlantic economies.[3] The scientific and technical culture that emerged in such a region was unlikely just to mimic that of established or newly industrializing nations in Europe, or that of their highly populated, traditionally settled colonial acquisitions such as India.[4] No purposeful indigenous culture resisted the encroachment of new ideas, artifacts, instruments or techniques for, as in Canada or South Africa (or, before them, the territories that became the United States of America), established cultures were sufficiently disabled by force (and eventually disestablished by time) to reduce the frictions that resulted from political, military and cultural clashes in more densely populated regions. In areas of recent settlement, the history of science became the history of the transplantation and modification of Western science in the hands of European cultural activists, in a context relatively unaffected by the requirements of a manufacturing sector. Here, the function of science would be expanded and altered but not revolutionized.

As is shown below, the public agendas of Australian scientific societies, wherever they arose, were of a type. They emphasized the development of natural resources, locally specific knowledge, the collection of native artifacts and specimens, and applications of European science in such immediately practical pursuits as agriculture, mining and mineralogy, together with the public services required for efficient drainage, water supply, transport and public health. Firstly, for pre-eminently utilitarian reasons, natural history loomed far larger than in the European science of the period, and science resembled more the European science of the preceding century. Secondly, the imperial science of Britain forged deliberate instruments of cultural suasion, direction and reward. Australian investigators sent their specimens or discoveries to the herbarium at Kew, and Kew sent its botanists back. Botany was, indeed, an eminently practical pursuit, yielding precise indications of soil quality, carrying capacity and mineral content. Most of the expeditions of any official nature included a botanist. So, specifically *colonial* pressures added to the local economic imperatives of a region of recent settlement to hammer out a utilitarian and initially subservient scientific programme. Thirdly, natural history provided psychological as well as material and social rewards to intellectuals settled in such regions. As Fleming states: 'For the colonial investigator himself, natural history was the ideal refuge from the more perilous enterprise of embarking upon theoretical constructions by which he would be pitched into naked competition with the best scholars of all countries'.[5] These several forces converged to create the utilitarian and localized profile of Australian science in the nineteenth century. Though they were never sufficient to exclude enquiry of a less sheltered type, they were strong enough to ensure that the scientific community in Australia did not take on the intellectual and cultural responsibilities of the European scientific enterprise until the end of the period considered in this chapter. Reconnaissance (looking at nature) gave way to acclimatization and specific applications (the manipulation of nature) before the ethos of utility was seriously disturbed by the cultural requirements of *Australians*.

Significant elements in the support for the scientific enterprise may be classified as *basal, infrastructural* and *superstructural. Basal* may be defined as support for or constraint on science that arose from the natural, local imperatives of Australia as an area of recent settlement and a British colonial outpost. This includes such factors as population size, rate of growth and occupational distribution, the level of urbanism (rising in our period), effective external and internal distance (falling due to transport and communication improvement), and the nature of staple exploitation in different periods (variable). Together these dictated the problems of survival and settlement.

Australian nature was first described by those who dominated the agendas of the *superstructural* institutions. During utilitarian exploring expeditions scientists found egg-laying mammals, flightless and songless birds, trees that retained their leaves but shed their bark, white eagles, warm hills and cool valleys and the inexplicable and annoying flooding of river systems. Resource management required explanation as well as observation.[6] It is hardly surprising that Barron Field, writing in Sydney in 1819, commented that 'botanists are many, and good cheap', nor that Sir Thomas Mitchell, writing in 1839 of his three expeditions into the interior, claimed that 'every settler is under the necessity of becoming a geologist'.[7] Here was the American artisan homesteader in Australian guise, for in both continents curiosity had to give way to informed pragmatism. The difference lay in Australian time and space. Victoria and South Australia were generally settled only as late as the 1830s. Railway expansion may well have quickened and increased traffic generally, but the decline of the road-building programme may have reduced the number of communication linkages in and between urban systems. Information flows were constrained by a relative sparseness of newspapers, postage and telegraph services, all of which improved vastly during our period. If intellectual peripheralization was three-fold, as suggested above, intellectual isolation was two-fold in a continent of less than half a million settlers in 1851. As we shall see, the gold rush and population boom of the 1850s modified such constraints considerably. The population of Australia rose to over one million, that of Victoria from 77 000 to 540 000; and as a result the culture of science was able to surmount at least some of the problems imposed by size and distance.

The colonial connection, however, remained perfectly intact as vice-regal patronage of the superstructure gave way to bureaucratic participation and support.[8] Formal linkages between colonial land and emigration commissioners and the Secretary of State in Britain, and the governors, legislative assemblies, colonial secretaries, commissioners and surveyors-general in Australia, spilled over into the scientific enterprise. Together with deputy surveyors, university professors, postmasters-general and many individuals employed in the colonial administration, such powerful bureaucrats and politicians dominated cultural cliques and popular societies alike. All areas of recent settlement shared the same stamps of authority, but colonies were obedient to a distant 'centre' also. Local imperatives and opportunities for employment meant that 'government from a distance' extended far into science itself. Many of Australia's scientific intellectuals — surveyors, engineers, geologists — were employed in government departments. Scientific experts employed in other areas were called on for official service. Systematic colonization, from Wakefield to the Colonial Land and Emigration Commissioners, required information that was truly informed and not merely observed.[9]

Support for science filtered not only into overtly scientific associations, such as universities, surveys and observatories, but also into activities often regarded as ancillary[10] to the major organizations of science and technology. In Australia, science frequently came to government through the side door. As is suggested below, what we term *infrastructural* institutions provided a great deal of support for scientific workers, both because they represented an audience and because they were frequent and large-scale recipients of government finance. During most of the period under review here, none of the Australian colonies were populous enough to support the development of strong, sustained scientific communities, and the cost of communication delayed the advent of a federal scientific organization until 1888 and beyond. Given the problems of distance and a lack of manufacturing industry — and therefore the slightness of applied research fostered by individual economic interests and market forces — scientific discussion, research and teaching could only emerge if the general participation in science and its applications was continuously raised. The developing *superstructure* of science very much depended on the good offices of government and the interests of a wide audience.[11]

* * *

Base trends

Base trends represent those movements in the economy and social structure that had demonstrable impacts on the development of scientific enterprise. The most spectacular effects were in the boom decade of the 1850s and the depression and recovery of the 1890s, but it can be argued that variations at the basal level continuously conditioned fluctuations in the size, location and nature of Australian scientific culture.

During the 1850s the colony of Victoria provided one-sixth of British woollen imports and one-third of the world's supply of gold. Melbourne, with its population reaching 125 000, was thus an urban outpost of the European, especially the British, economy. A.G. Serle has suggested that massive migration and urbanization created, in essence, 'a new, large self-governing county, automatically recreating British institutions and re-forming familiar clubs and societies'.[12] All the institutional indicators show that the dramatically changing base affected the scientific enterprise considerably. This was the decade of the Geological Survey, the Museum of Economic Geology, the Victorian Institute for the Advancement of Science and the Victorian Science Board, the active membership of all of which was dominated by government servants. Science entered a phase of civic profesionalism. The most pressing public project arising from the demands of the new urbanism was the building of the Yan Yean reservoir, to supply not only drinking water but also irrigation and hydraulic power services. At a time when the Surveyor-General, Andrew Clarke, was a lecturer to both the Victorian Institute and the Philosophical Society, a committee of the latter was established to evaluate the project. Robert Brough Smyth, later Secretary of Mines and a member of the Board of Science, utilized the scientific work of Charles Lyell and Alexander von Humboldt to investigate the contentious issues of micro-climatological influences. The significance of such civic and public scientific work is summarized by Powell:

> Methods of catchment analysis and water quality control were quickly improved and engineers were persuaded to look more closely into geology and hydrology, weather forecasting and climate interpretation also received attention; scientists and technicians generally combined more readily to provide an essential community service and the links with the elected government were strengthened.[13]

Against such a background the Victorian government appointed a number of expert scientific bureaucrats. Men such as A.R.C. Selwyn, G.B. von Neumayer and Ferdinand von Mueller became the leaders of the scientific enterprise.

Again, the major indicators suggest that in the economic boom of the 1880s the number of people engaged in scientific activities rose, and that the following depression saw a fall. Associations subject to government financing and bureaucratic involvement were bound to be highly sensitive to such large-scale basal movements. The narrowing of the economic gap between Victoria and New South Wales led to a shift in the scientific enterprise towards the latter region. Similarly, the expansion of wheat and prosperity in South Australia had an immediate impact on that region's scientific enterprise —

from the formation of the University of Adelaide to the creation of some seventy-five country technical institutions.

Of still greater interest is the stimulus to science created during the recovery years of the later 1890s and beyond. Attempts to diversify the economy led to a search for scientific and technical aid in doing so, to an increase in innovative activity (see below) and increasingly explicit and selective government aid to science and technology. With manufacturing industry stagnating until the early years of the twentieth century, the initial recovery was based on mining and agricultural diversification.[14]

Developments in the scientific infrastructure

Between the years 1839 and 1892, some 400 mechanics institutes for scientific and technical discussion were formed in the colony of Victoria alone.[15] Whatever their value as formal educational institutions, these and similar formations were of particular importance in providing an infrastructure within which the science of the philosophical and 'royal' societies could grow. Infrastructure science involved far greater numbers of people and boasted a longer and more continuous tradition than did the science of the 'superstructure'. Furthermore, such ancillary associations served as a training and recruiting ground for membership of the scientific community, provided an audience for Australia's *savants* and represented a solid platform for the rhetoric of those who proclaimed science publicly.[16] The infrastructure took science out of the small, private arena of the philosophical society and promoted a level of discussion that the emergent universities could not generate until the twentieth century.

In New South Wales alone, nearly 18 000 individuals attended 269 government-aided institutions in 1897; by 1909 these figures had risen to over 40 000 persons and 394 associations, with access to some 565 000 library volumes.[17] The formation of such associations perceptibly accelerated during the 1880s. Of 132 government-aided institutions active in New South Wales in 1888 (the year in which the Australasian Association for the Advancement of Science was founded), 64 per cent had been founded post-1875 and 48 per cent post-1880.[18] This group of institutions represented a relatively heavy investment. The thirteen largest institutions in New South Wales in 1888 had cost £66 000 to build, held 51 000 volumes and claimed many thousands of library readers per annum. The largest, the Sydney Mechanics School of Arts, with a membership that peaked at 5000 in 1891, received government subsidies totalling £23 000 between 1880 and 1900. No association of the nineteenth century scientific community 'proper' could boast such a membership and financial status.

From the first, the scientific agenda of the mechanics institutes was wide-ranging, but there was a continuous emphasis on localism and utility. In 1828, the Hobart Institution planned a museum of 'models, machines, minerals and natural history', lectures on science and the trades, experimental workshops and a laboratory.[19] While both general cultural and specific 'social control' arguments were often aired by proponents of institutions, such views expressed the standard rhetorical baggage of the nineteenth century emigré.[20]

The actual membership, provisions and interests of the infrastructure reflected a wide social and intellectual appeal, a search for the natural peculiarities and utilities of Australia and a localism represented in frequent lectures and discussions on agricultural chemistry, geology, mineralogy and mineral resources, and exploration.[21] At a time when philosophical societies were non-existent or lacked vitality, the buoyancy of even small infrastructural associations was evident enough.[22] By the end of the 1850s the 400 or so members of the Hobart Mechanics Institute were exchanging 10 000 volumes per annum; by the 1880s, 500 subscribers at Launceston circulated between them 20 000 volumes per year.[23]

The more popular culture of science was imbued with a spirit of egalitarianism and utility not so easily discerned in the intellectual cliques of the Australian superstructural associations or the cultures of European nations at this time. This may be illustrated with reference to developments in Victoria, South Australia and New South Wales.

During 1858–59, when he was in receipt of £12 000 of government subsidy, Ferdinand von Mueller estimated that 200 000 visits had been made to the Melbourne Botanic Gardens during the year. In any one month some 5000 visitors utilized the collections in Melbourne's Public Library.[24] The Museum, at the University of Melbourne, recorded 32 000 visits in the same year. By 1860 the 120 active mechanics institutes of Victoria had received some £47 000 of government aid, at a time when the University of Melbourne was obtaining £9000 annually from general revenue.[25] The highly successful, well attended Industrial and Technological Museum of Victoria, with its public lecture courses on chemistry, mineralogy, mathematics and telegraphy, epitomized the success of the infrastructure by the 1870s.[26] It becomes fairly obvious that infrastructural institutions benefited enormously from the demographic, urban and bureaucratic boom in Victoria during the 1850s.[27] Among the 400 000 occupied population of the colony in 1857 were large numbers employed in the government bureaucracy (4000), the learned professions (2000), scientific, literary or educational pursuits (3000), and those of independent or mercantile wealth (14 000). While this may have represented a threshold population for some learned scientific discussion, the infrastructural associations and institutions also gathered in a portion of the 25 000 individuals who described themselves as 'mechanics and artizans' and perhaps a smaller fraction of the 62 000 miners and 75 000 individuals engaged in pastoral and domestic services.[28] Melbourne was a boom town, and the Trades Hall and Literary Institution served as a centre for intellectual and educational activities as well as 'a labour mart'.[29]

The emergence of a scientific community in mid-nineteenth century Victoria depended on the prior and contemporaneous development of a complex infrastructure for scientific discussion and reportage. Gerard Krefft, later curator of the Australian Museum, first 'wrought as a miner' at Forrest Creek, Fryers Creek, Ballarat, Avoca and Maryborough before discovering Gould's *Sketches of Australian Mammals* in Melbourne's Public Library. The Melbourne Museum then engaged Krefft to participate in the expedition to the Lower Murray and Darling rivers. On Krefft's return in late 1857, Professor Frederick McCoy engaged him as assistant at the museum.[30] In

January 1858 a meeting of the Philosophical Institute of Victoria held at the mechanics institute resolved to marshal public support 'in carrying out the design of the institute to fit out a Victorian expedition for the exploration of the interior'.[31] At that time the secretary of the Philosophical Society, John Macadam, MD, was a regular lecturer on chemistry and physics at the mechanics institute, as well as natural science teacher at Scotch College. Macadam gained a further audience for science in his regular public lectures delivered from his own residence.[32] Such venues and activities nicely capture the inter-relationships between different levels of the emergent scientific enterprise. The 700 members of the mechanics institute boasted (and financed) the lecture courses of university professors. The scientific *savant* C.H. Le Souef became manager and promoter of the Victorian Industrial Society, which mounted trials of reaping machinery and the public exhibition of colonial manufactures, for which it received a government grant.[33]

From this popular background science was financed. In early 1859 plans were underway for the foundation of a Victorian zoological society (government-aided with £3000) as well as for the establishment of an agricultural museum, laboratory and experimental farm, initially financed through a government grant of £3700.[34] By 1859–60 the Chief Secretary's Department was spending approximately £43 000 annually on the financing of Victorian scientific enterprise. The provision of colonial funds on such a scale served to bolster and integrate infrastructural and 'learned' associations, with relatively generous sums directed to the encouragement of agriculture, botany, economic geology, exploration, indigenous commercial plants, mining and other subjects 'connected with the development of the resources of Victoria'.[35] Such official expenditure excluded the expenses of the colony's university, the professors at which were delivering 500 lectures per annum on mathematics and scientific subjects.[36] Within this rich, if close, environment it is hardly surprising to find that the Victorian Philosophical Institute considered itself as at least 'in the vestibule of the temple of science', nor that it should stress the provision of intellectual opportunities 'curbed and confined elsewhere'.[37] As Redmond Barry, acting Chief Justice of the Supreme Court, emphasized, 'the wants of this country demand, and must exact, scientific innovation'.[38]

The history of scientific enterprise in the much smaller colony of South Australia illustrates how intellectual forums emerged from provisions at the infrastructural level. Between 1856 and 1884 the scientific community of South Australia was organized within the government-financed South Australia Institute (SAI). The SAI was formed as a result of the deliberations of a select committee of the Legislative Council, sitting from September 1854.[39] The Act of 1855–56 provided for the financing of a library, museum, classes, lecture courses and exhibitions, under the management of a board of governors. The SAI incorporated the activities of several scientific and technical forums — the Adelaide Mechanics Institute, the Adelaide Philosophical Society, the South Australian School of Arts, the Royal Society of South Australia, a large number of country infrastructural institutions (see below), the Philomathic Society (from 1863) — and in turn spawned a major library and museum, newly-founded country institutes, a school of design, a natural history and mineralogical museum, a museum of economic botany, an experimental laboratory, a technological and patent museum, a series of technological and trade exhibitions, and regular public lectures on scientific and technical subjects. Clearly, the SAI represented the centralization of scientific culture within an institutional complex controlled and financed by colonial bureaucrats, legislators and intellectuals.

Such organizational features encouraged the growth of a disproportionately large audience for scientific and technical discussions when compared with Victoria. Membership was divided into subscribers to the SAI *per se* and members of the several affiliated country institutions. SAI membership settled around 700–900 between 1860 and 1884. But many others were involved in its activities, from the 200 or so members of J.G.O. Pepper's classes at the school of design to the 100 000 visitors to the library and museum during 1884–85.[40] The membership of the affiliated institutions peaked at 5500 in 1884 and 14 500 in 1909. Success at this level owed much to the official backing. The affiliated institutions alone (numbering 40 by 1865, 81 in 1876, 125 in 1885, 192 in 1909) were receiving annual grants of over £5000 by the early 1880s. The SAI itself obtained large grants for specific purposes, including £36 000 for the construction of a new building following a considerable parliamentary debate and a parliamentary commission between 1874 and 1879.[41]

The smaller scientific community operated within this wider context. Scientific experts were engaged as lecturers or in various salaried positions: J.W. Haake as museum director; J.C. Cloud as mineralogist; A. Zietz as *préparateur*.[42] Other men associated with the work of the SAI (Professor Ralph Tate, James Stirling, Clement Wragge) all became active in the later Royal Society of South Australia. The SAI acted as a venue for their lectures, exhibitions and meetings, and for the activities of several scientific groupings. Savants were attracted by the library of the SAI which regularly received learned journals and official government surveys and reports. In nine months of 1884 the library acquisitions included the transactions and records of thirty-seven scientific and learned associations, of which eighteen were foreign.[43]

The Philosophical Society, which rarely achieved a membership of more than eighty, was the most overtly learned and expert of the intellectual forums to benefit from large and regular attendances of other SAI members.[44] The 129 meetings of the Philosophical Society to 1866 were addressed by fifty different authors, seventeen of whom delivered eighty-seven papers.[45] Although encompassing a general scientific agenda, the Philosophical Society established itself as practical and specialist. Lectures and discussions during the 1860s and 1870s concerned such subjects as the construction of iron tramways, commercial gums and other resources, the chemical analysis of South Australian soils, iron manufacture, viniculture, the adoption of a decimal system, wool growing, drainage, meteorology, rust in wheat, meat preservation, the cultivation of silk, irrigation, copper smelting, explorations and colonial geology and land tenure systems — hardly the normal fare of a nieteenth-century scientific society.[46] A large proportion of members were employed as scientific or technical experts in government, a smaller number were medical men. A significant number had gained fellowships in British learned societies. With this membership and the links with the SAI, it is hardly surprising that discussion focused on matters of immediate concern to the colonial government. One of the first papers was that of B. Herschel Babbage on the theme of 'the adaptation of machinery to calculating and printing mathematical tables'. This was followed by a more locally useful lecture by W.H. Light on the 'geology of the Bendigo Gold Field', a result of his two-year stay on the field.[47] Although the Philosophical Society's efforts to promote an inland exploration in 1855 foundered on the rocks of economy, its members constantly aimed to 'encourage every legitimate enterprise that has for its object the further development of our natural resources'.[48] Robert Kay introduced members to the utility of Nasmyth's direct-action steam hammer, B.H. Babbage addressed the SAI on the vital statistics of the colony, and frequent discussions took place on the usefulness and potential exploitation of Australian woods, plant and wildlife. The society that received public support adopted public responsibilities.

The voluntary cultural associations comprising the infrastructure provided a basis for the intervention of governments in the last twenty years of the century. In 1878 the New South Wales government voted £7000 for the establishment of a technical college in the Sydney Mechanics School of Arts (SMSA). Under the presidency of H.C. Russell, 1000 students were enrolled at the outset; by 1890, 6580 students were attending the Sydney Technical College and its provincial affiliates.[49] By the mid-1890s, 200 separate classes produced over 2500 students presenting for examinations conducted by the City and Guilds of London Institute. Meanwhile scientists such as Archibald Liversidge had joined with politicians such as Edward Coombes in establishing the New South Wales Technological Museum, which had received a foundation grant of £5000 in 1879. Again, this highly public institution reflected the historical role of the scientific infrastructure: 'to assist in every possible way the development of the natural resources of the colony . . . and publish all scientific and useful information which may aid the producer, manufacturer and consumer in their choice'.[50]

Members of the scientific community were active in the later years. The Natural History Association of New South Wales, an adjunct of the SMSA, had as its president Dr George Bennett, and committee members in 1888 included Dr G.P. Ramsay and C.S. Wilkinson.[51] Several of the scientists active at the infrastructural level were to become of great importance in the early years of the Australasian Association for the Advancement of Science — Russell, Liversidge, Bennett, and Wilkinson joined forces with G.H. Reid, Sir Henry Parkes and W.H. Keyte.[52] Lecture courses sponsored by the government's Board of Technical Education from the later 1880s utilized the services of leading scientists and attracted attendances of 250 and more.[53] The college employed the agricultural chemist, Angus Mackay, and B. Dunstan, later assistant government geologist in Queensland, and claimed the pedagogical services of E.H. Rennie (later professor of chemistry at Adelaide and a leader of AAAS in the 1890s) and W.A. Dixon, who had previously taught in a private capacity from his chemical laboratory in Hunter Street.[54] The chemical laboratory begun by Norman Selfe, consulting engineer, within the SMSA during the 1860s was revamped and transferred in 1889 to the

Department of Public Instruction, under which it developed a capacity for over 100 students. Even the physics teaching of the Sydney High School was associated with the technical college during the 1890s.[55]

The scientific infrastructure was clearly performing a series of fundamental supportive roles by the later decades of the nineteenth century. Having harboured emergent scientific communities, the institutions of the infrastructure gave way to government control and policy. Much of the tentative 'science policy' of the various Australian colonies was, in fact, devoted to promoting ancillary scientific services. Similarly, much of the 'applied' and local orientation of Australian science was first established within this level of cultural activity. Finally, the existence of a scientific audience provided an immediate environment for the scientific programmes and careers of members of the scientific community as well as for the public pronouncements of scientific devotees.

The scientific superstructure

Heroes of science in nineteenth-century Australia may have been few, foot-soldiers more numerous, but all appear here as the scientific *superstructure*, a term reserved for the individuals, activities and institutions more conventionally described in the history of science — those most clearly involved in the 'doing' of science.

A scientific *community* could not be readily discerned in Australia in the first half of the nineteenth century.[56] However, beginning in the 1850s, amid the dramatic changes wrought by self-government and gold, scientific communities in each colony gradually became visible through various institutions designed either to support science or to draw on its resources by marshalling scientific knowledge and skills.

At the beginning of the period under review, the fledgling scientific communities could be located most readily within the small number of voluntary associations established in the 1850s for the purpose of fostering science in an environment of little intellectual vigour. In each colony the first generation of scientific societies, termed 'philosophical', laid the groundwork for the second generation, the 'royal' societies.[57] For some time these societies constituted the nucleus and symbol of the emerging scientific communities.

From the 1850s other sources of succour for an infant scientific community began to appear. The colonial separation embodied in self-government implied a new self-perception on the part of the colonies, one of permanence and self-sufficiency. Universities were opened in Sydney (1852) and Melbourne (1855), each with two chairs of science. A branch of the Royal Mint, which began operations in Sydney in 1855, required staff with chemical and analytical skills. Museums, botanical gardens, geological surveys and observatories were established to reconnoitre the natural life, resources and phenomena that could be harnessed to colonial development.

Many of the scientific personnel required for these institutions were recruited from overseas.[58] Others had been attracted to the colonies by gold and its attendant opportunities, only to be sorely disappointed.[59] Either way, a welcome boost was given to the number of active scientists in the colonies when positions that provided remuneration for scientific work were created. Within each of these institutions, the scientists were initially few in number, sometimes a single individual, and often limited by pragmatic brief.[60] Under these conditions scientists sought out kindred spirits and stimulation within the fellowship of the local scientific society.

These early societies were under the leadership of influential men. In New South Wales, Governor Denison, newly arrived from Tasmania in 1855, pressed the leaders of the ailing Australian Society to reorganize under the title the 'Philosophical Society of New South Wales'.[61] The new society began with twelve other prestigious councillors including the Colonial Secretary, the provost of Sydney University, the Deputy Surveyor General, three university professors and the Deputy Master of the Mint.[62] Vice-Regal leadership was undoubtedly effective in attracting an impressive roll-call of membership from the more educated strata of society — doctors, engineers, lawyers, businessmen and civil service administrators.[63] Nor was this kind of membership restricted to New South Wales. Even in a small colony like South Australia, the governor was frequently in the chair at meetings of the Adelaide Philosophical Society and the bishop, the Chief Justice, the judges, the Surveyor-General, the Postmaster-General, the leading educationists and the newspaper editors were active members.[64]

The typical agenda of Australian science was expressed through these scientific societies for some time, though it did not always accord exactly with the objects set out by their founders.[65] In the 1850s and 1860s public issues associated with colonial development frequently predominated, manifested in discussions on water supply and drainage, transport and communications, and public health. Though often having a technical component, these papers could hardly be described as original scientific contributions.

Governor Denison led the proceedings of the new Philosophical Society in Sydney with a paper on 'Development of the railway system in England, with suggestions as to its application to the colony of N.S.W.'. This was followed by two papers from Edward Deas Thomson, the Colonial Secretary, on 'Steam communication with England'.[66] In Victoria, a vigorous debate was maintained on the merits and evils of the Yan Yean water supply and resulted in representations to the government.[67] The South Australian Society considered matters such as 'National education', 'Mesmerism' and 'The probabilities of gold in South Australia'.[68]

Scientific subjects contemplated by the early societies were usually in the fields of astronomy, meteorology, botany, zoology or geology. While much of this fare was lacklustre and without originality, there were some significant contributions in each of the colonies.

Victoria was at the time blessed with a greater number of scientists than the other colonies. There the government botanist, Dr Ferdinand von Mueller, was prolific in describing the results of his labours as he plunged enthusiastically into a survey of the plants of the colony. The artist and naturalist, Ludwig Becker, described by Lady Denison as 'one of those universal geniuses who can do anything', revealed his eclectic talents on subjects as diverse as the platypus, Donati's comet and geological speculations. Georg Balthasar von Neumayer, meteorologist and director of the Victorian magnetic survey, authored sophisticated contributions including papers on the application of Dove's 'law of veering' to Australian winds. In 1859, Paul Howard MacGillivray, a medical practitioner and surgeon, ventured the first of his many important illustrated descriptions of Australian polyzoa (commonly known as 'sea-mosses').

Elsewhere, the South Australian clergyman–geologist, the Reverend J.E. Tenison Woods, began in 1865 his descriptions of South Australian Tertiary fossils. In Queensland, the first Chief Justice of the Colony, James Cockle, displayed his mathematical competence and Dr Joseph Bancroft his talents as a field naturalist and critical experimental biologist.[69] The New South Wales society was enriched by the contributions of the astronomer John Tebbutt, who operated from his own observatory at Windsor, and Gerard Krefft, by now curator at the Australian Museum, whose studies ranged from Australian extinct and living mammals to birds, snakes, fish and whales.

By the 1870s it was clear that the programme that had unfolded in these societies was one largely committed to the collection, description and classification of Australian natural history, phenomena and resources, combined with the discussion of practical matters involved in colonial development. This reflected the mood of the times, which had little patience with abstract theorizing. It also reflected the nature of the leadership and membership of the societies and the way they saw science in relation to their world.

Contributors were a mixture of full-time practising scientists and part-time, so-called 'amateurs'. The latter were often men of public affairs, conscious of the imperatives of development. For others, the collection or description of natural history specimens could be a rewarding hobby in a largely uncharted territory. Of the professional scientists, most were in positions created by the state. Therefore, their overall brief was a product of colonial needs as perceived by and pressed upon governments.[70]

The university professors provided little counterbalance to this pragmatic empirical emphasis. Not only was their role at the universities defined and clearly perceived as being confined to teaching, they were often in close relationship with the state and its priorities. In Victoria, Frederick McCoy, professor of natural science, was also director of the National Museum and all his spare time was given to the identification and classification of specimens.[71] In Sydney, Professor John Smith, an expert on water and other analysis, gave evidence at various parliamentary select committees and in 1867–69 chaired a royal commission into the water supply of Sydney and suburbs.[72]

A further consideration for contributors was what would be acceptable to an audience that was largely non-scientific. In 1870, in the Royal Society of New SouthWales, 60 per cent of the membership of 127 had no scientific background or involvement. Of the remaining 40 per cent, only 11 (8.7 per cent) were practising scientists (excluding engineers), though others, from the nature of their work or training, had a natural predisposition to some area of science.[73]

During the 1870s, most of the societies underwent some kind of self-assessment, re-evaluation and rejuvenation. The arrival of an energetic professor at the local university could give particular stimulus to reorganization. From the arrival in 1876 of Ralph Tate, professor of natural science at the new university, the whole status of the Adelaide society was changed, new rules were drawn up, and publication was put on a sound and permanent basis by the foundation of a regular journal. Scientific contributions came in abundantly for the first time and soon exchanges were arranged with societies around the world. Archibald Liversidge, professor of geology and mineralogy at the University of Sydney, had a similar effect on the Royal Society of New South Wales and also ensured an international distribution for its journal. This provided a potential audience of vast dimensions for contributors, and also, through exchange, secured for the society valuable scientific works from overseas.

In the 1880s business was booming in the colonial royal societies. Memberships reached new heights in all colonies (see the table below).[74] By this time the former preoccupation with pragmatic issues of development was receding and a more consistent offering of original scientific papers was being maintained. Yet contributors were still being urged to keep to collecting facts related to local phenomena.[75]

Membership of the royal societies and antecedents 1855–1910

	1855	1860	1865	1870	1875	1880	1885	1890	1895	1900	1905	1910
NSW	153	154	108	127	176	457	494	461	420	374	326	313
Vic.	230	302	118	117	115	193	236	262	166	136	156	154
	(1857)											
SA	39	47	64	60	85	85	128	79	86	74	77	70
QLD		29		30			146	143	120		75	
		(1862)		(1869)			(1886)					
Tas	330	207	–	140	165	168	–	196	–	111	118	153

Even though memberships were rising, papers still came from a small minority, more often, now, professional scientists reporting on investigations made during the course of their paid employment. They were still, however, being received by a largely non-scientific audience, though this was gradually changing (see the next table).

Scientific background of members of the Royal Society of NSW

Year	Total membership	Proportion with no scientific background[76] (%)
1870	127	60
1880	457	59
1890	461	47
1900	374	35
1920	362	20

As membership peaked in the late 1880s, the royal societies had become established scientific institutions. They had provided a nucleus around which young scientific communities could grow and mature. They had attracted a broad-based audience for science and had established communication links within and beyond Australia. They had also done much to shape the scientific agenda of the colonies.[77] Change, however, was on the way. The structure of the scientific community became more complex as other foci of scientific activity appeared — other scientific and technical societies, centres of research in universities, more government departments with scientific programmes.

The Linnean Society of New South Wales, founded in 1874, had established itself as the first really viable alternative to a local royal society. Specializing in natural history and priding itself on a superior quality of contributions, by 1885 it had 176 members. The 1880s saw the beginning of other more specialized societies,[78] then in 1888 came the first meeting of the Australasian Association for the Advancement of Science.[79] Modelled on the British Association and reflecting the wider movement towards federation, this helped to fuse the dispersed and isolated scientific communities of the various colonies into a national community. It aspired, as well, to provide more systematic direction to scientific enquiry.[80]

In the universities, towards the end of the century several extremely capable new professors, who were determined to integrate fundamental research with their teaching role, were recruited. David Orme Masson, who took up the chair of chemistry at the University of Melbourne in 1886, argued that no university should be merely 'a second-hand Science Shop'.[81] He was supported by his colleague W. Baldwin Spencer, biology professor from 1887, and physicist Thomas R. Lyle who arrived in 1889 as professor of natural philosophy to complete a brilliant trio of scientific professors dedicated to attracting new graduates into their research programmes. In this they were aided by the further inducement of the new degree of doctor of science, first instigated at Melbourne University in 1887, and the 1851 Exhibition scholarships, established in 1891 to allow proven creative colonial researchers to refine their skills in Britain.[82]

In Adelaide, Professors Tate and Lamb had established an early precedent of research on their own account before the days of research students.[83] Lamb's successor, William Bragg, became an ingenious experimenter and in the early 1900s began his seminal investigations of radioactive substances that were finally to lead him to a Nobel Prize. As collaborators Bragg recruited Honours students (two of whom were awarded 1851 Exhibition scholarships on the basis of their contributions to the programme), and colleagues J.P.V. Madsen, lecturer in electrical engineering, and E.H. Rennie and W.T. Cooke from the Chemistry Department.[84] Rennie, the first professor of chemistry at Adelaide (1885–1927), himself founded an influential school of organic chemistry, which retained its research prominence well into the twentieth century.[85]

In Sydney, Professor Richard Threlfall, previously demonstrator at the Cavendish Laboratory under J.J. Thomson, built up a competent and well equipped physics department and, aided by his research students, carried out electrochemical research until he returned to England in 1898. The geology professor T.W. Edgeworth David soon earned world-wide repute built largely on his work on a remote coral island in the Pacific.[86] William Haswell, Challis Professor of Biology from 1889 and a former pupil of Huxley, represented the new wave of researchers who, under the influence of Darwin, were to transform biological studies in Australia.[87] Another was J.T. Wilson, professor of anatomy from 1890, who quickly earned a reputation from several papers on the anatomy and homologues of marsupials. He also attracted the young British physiologists Charles Martin and J.P. Hill to Sydney and inspired the original researches of Grafton Elliot Smith, a graduate of the department. This talented group pursued fundamental studies on the morphology, neurology and evolutionary physiology of Australian marsupials and monotremes.[88]

By the 1890s, then, in chemistry, physics, biology, geology, physiology and anatomy, thriving new centres of research in the universities were embarking on exciting scientific programmes. The nature of those programmes was now determined more often by the theoretical questions of European science than by the need to survey local resources. Darwin had provided inspiration in the natural sciences. In the case of the chemist, Masson, the theories of solution developed by Ostwald, Arrhenius and van't Hoff had captured his imagination since their publication in the 1880s and were to provide continuing stimulation, leading to the presentation of his own gaseous theory of solution in 1891. A continuation of his Bristol and Edinburgh work provided research topics for students.[89]

Further diversity in the scientific community came as more government departments felt the need for some kind of scientific research. The observatories in New South Wales, Victoria and South Australia had all been presided over by men who became both organizational and intellectual leaders of their respective scientific communities through the medium of their royal society.[90] As government scientists found themselves with more internal colleagues and more government outlets for publication of their work they were less likely to concentrate their energies on the older associations.

The geological surveys, which began uncertainly in the 1850s, had secure roots by the end of the century. In New South Wales, the new geological survey formed in 1875 had six geologists on its staff by 1878, engaged in surveying and mapping the goldfields and other mineral deposits and in building up collections of minerals, rocks and fossils. During the 1880s, detailed reports were published, and the periodical *Records of the Geological Survey of New South Wales* began publication in 1889. By 1890 there were ten professional staff in the survey. In Victoria, four professional staff in 1890 had swelled to 16 by 1900.[91]

Most colonies also established departments of agriculture in the late part of the century to help with the diversification of the rural economy. Adopting a philosophy of educating the farmer in the application of scientific principles, these departments evolved scientific programmes designed to generate knowledge of local conditions and problems in a form that could be readily assimilated by the man on the land. Professional staff employed included chemists, plant pathologists, entomologists, bacteriologists, biologists, dairy and other experts — almost a microcosm of a scientific community. Though often weighed down with a multitude of routine tasks, several of these scientific officers produced work of significant value, most of which was published in the departments' own journals and bulletins. Daniel McAlpine laboured for twenty-five years as vegetable pathologist to the Department of Agriculture of Victoria, unravelling the mysteries of rusts and smuts in wheat and the diseases in citrus fruits and other plants. His prolific departmental publications made him a world authority, and served as standard texts for many universities and colleges overseas.[92]

By the end of the century the museums in each colony offered rich collections for world scientific reference as well as public display. Curators were no mere organizers of exhibits. In Queensland, for instance, Charles Walter de Vis maintained a flow of articles on birds, fish, reptiles, batrachians, marsupials and his own palaeontological explorations. He began the *Annals of the Queensland Museum* and drew attention to the ethnological and biological products of New Guinea.[93]

In the wake of the natural history museums came the technological museums, directed at elucidating the economic value of Australian natural products.[94] It was at the Sydney Technical Museum (established in 1880) that the team of Richard Baker and Henry George Smith carved out a niche at the forefront of the new field of phytochemistry. Spurred on by the curator, J.H. Maiden,[95] they were generally orientated towards commercial applications but their theoretical creativity was demonstrated in their *Research on the Eucalypts, especially in Regard to Their Essential Oils* (Sydney, 1902). Though their theories were challenged by many Australian botanists their work was often praised overseas.[96]

Operating now within a more diversified institutional framework, scientists at the end of the century were still usually members of their local royal society, but were less committed to it. Their best efforts were often directed elsewhere. The national stage provided by the new AAAS, for instance, was, for many, more prestigious and attractive. It was here that Orme Masson first presented his new gaseous theory of solution in his presidential address of 1891.[97] Furthermore, in a specializing world, specialist societies promised a more stimulating environment, and Masson went on to help found the Society of Chemical Industry of Victoria, and the Australian Chemical Institute.[98]

To AAAS would go the task of compensating for this tendency towards specialization and fragmentation. It was hoped that its national character would promote research and higher-level discussions and thrust 'scientific workers out of their isolation' into 'public regard' by bringing a 'stronger impulse and more systematic direction' to Australia's scientific enterprise.[99]

The support for AAAS embraced a familiar range of forces. To scientific practitioners and public scientists were added the influence of the infrastructural institutions, of the scientific and technical bureaucracy, and of the booming economic conditions of 1888. Perhaps for the first time Australia could boast a coherent scientific movement gathering together to celebrate an intellectual 'grand fate'.[100] The editor of the *Sydney Morning Herald* applauded the association as a gathering directed to economic improvement. The newspaper was not far wrong. The early council included diverse interests, from Professor H.M. Andrew representing Melbourne University's Science Club to A.O. Sachse, MSc, consulting engineer, member of the Royal

Geographical Society of Australia and technical entrepreneur of the
Melbourne-based Ice Supply Co. Ltd.[101] The majority of papers were devoted
to natural history, engineering and civil science. Major contributions were
drawn from surveys, technological museums, universities and observatories as
well as from government departments. In general format, the new association
wore the clothes of earlier times.

Economy, government and scientific enterprise: Case studies

Technology and economic growth

Before 1890 and after 1910, much of the growth of the Australian economy was
derived from exploitation of natural resources and new lands, population
growth, staple expansion and the inflow of British capital. The period from the
economic recovery of the mid-1890s to around 1910 witnessed a significant
qualitative change in the nature of Australian economic development.
Although manufacturing efficiency remained relatively stagnant, the
economy as a whole benefited from efficiency gains induced by technological
and institutional change over a fairly wide front.

Recovery entailed a more intensive use of land, a replacement of
dependency on the wool staple by wheat, dairying, meat, sugar and fruit
exports, improvements in shipping sector technologies (from steel to marine
engineering and refrigeration), the development of new strains of crops (e.g.
Farrer's work on wheat), fallowing and fertilizers. Such manufacturing growth
as occurred was related to the servicing of such change — in the provision of
agricultural machinery, food processing, refrigeration equipment and urban
services. In this period the fairly *ad hoc* applied research programme
associated with acclimatization or more efficient exploitation of resources
gave way to specific applications of systematic knowledge in areas such as gold
production and mineralogy. Efficiency increases arising from such
applications peaked around 1910, and thereafter further economic growth was
associated with large capital inflows, railway development, irrigation works
and further expansion of settlement. Thus, the interim period is of special
interest in demonstrating new relationships between the scientific enterprise,
technological change and government policy. If technical change in
manufacturing was dampened by the ease with which labour could transfer
from construction and domestic service, the structural changes that occurred
in other sectors (gold recovery versus gold mining, meat and butter versus
wool, the growth of urban services) were all associated with technological
change, much of which originated within Australia.

Of over 38 000 patents for new inventions applied for in the four major
colonies between 1848 and 1899, 34 000 (90 per cent) were lodged between
1880 and 1899, 68 per cent between 1890 and 1899.[102] Victoria had provided
some 44 per cent of all patentees, New South Wales 26 per cent. But in the five
years 1900-1904 (involving a further 13 600 applications), 30 per cent of
patents originated in New South Wales, 33 per cent in Victoria. By the end of
the period 1904-1918, during which 35 000 patent applications were made
under new Commonwealth regulations of which 70 per cent were lodged by
Australian residents), New South Wales had firmly overtaken Victoria as a

source of Australian inventiveness. This trend correlates well with the nature of economic and demographic change and the movement of the scientific enterprise towards New South Wales and away from Victoria's earlier dominance. New inventions were readily available, and the Australian propensity to patent was higher than in such nations as Britain, USA or Germany at this time. Further, patented technology ranged widely: at the turn of the century over 50 per cent of applications were in the manufacturing and metal-processing field. There can be little doubt that a combination of Australian inventiveness and foreign transfers was sufficient to supply the economy with specific information needed for further increases in productivity.

Acclimatization

From the later 1850s the subject of plant and animal acclimatization served as a practical and immediate focus for the emergent scientific community of Australia. Although the export of Australasian plants to England was a well established tradition by mid-century,[103] only during the 1850s did Australians attempt to reverse the procedure through a purposeful programme of regulated acclimatization of foreign flora and fauna into the several colonies. However faulty were many of the assumptions on which it was based, acclimatization became a significant element in the process whereby Australian scientific effort was localized, made more immediate and practical and subject to official enquiry, interest and support.

In the 1850s in Europe, acclimatization was seen as part of the development of an imperial, over-arching scientific enterprise. The first such society was the Société Impériale d'Acclimatation, founded in 1854, which in the words of Isidore Geoffrey Saint Hilaire was 'composed of agriculturists, naturalists, landowners, all the scientific men, not only of France, but of every civilised country'.[104] Following this model, the Society for the Acclimatization of Animals was founded in London in mid-1860 with the aim of introducing foreign animals to Britain and her colonies. It was at an early meeting of this society that Edward Wilson, erstwhile radical of Melbourne's *Argus* newspaper, gained the idea of translating the programme into Victoria.[105] Before this, Wilson had been active in the Philosophical Institute of Victoria,[106] and on his return to Melbourne in 1861 he founded the Acclimatization Society of Victoria and was subsequently influential in the formation of similar societies in New South Wales, Tasmania, South Australia and New Zealand.[107] The provision of large establishment grants from government (eg. £5500 in Melbourne, following the award of £10 000 to its precursor, the Royal Park Zoological Society, in 1857), together with continued annual assistance, lent the movement a stability that the contemporaneous 'philosophical' societies did not yet share; by 1873 the nearly 500 members of the Victorian society alone outnumbered the total membership of the 'philosophical' societies in New South Wales, Victoria, South Australia and Queensland.[108]

Australian science of the 1860s and beyond was closely associated with the acclimatization movement. The early phase witnessed the serious involvement of such scientific figures as W.L. Martin, A.R.C. Selwyn,

A.A.C. Le Souef, Ferdinand von Mueller, Frederick McCoy and Dr Thomas Black (Victoria), G.W. Francis and R.M. Schomburg (SA) and George Bennett, W.B. Clarke and W.J. Stephens (NSW). In the long term, several of the more authoritative scientific works of von Mueller, Bennett and J.H. Maiden were the result of acclimatization research or sponsorship.[109] Holding the firm belief that 'having paved the way by opening up foreign countries by conquest or colonisation, the grand scheme of distribution should be elaborated scientifically, systematically and exhaustively', the scientific and commercial acclimatizers directed themselves to research on a variety of potential staples — flax, silk, cotton, game and fisheries were all grist to the mill of applied Australian science, and were subjects of trial and experimentation from the earliest years of acclimatization societies to the initial discussions of the Australasian Association for the Advancement of Science.[110] The acclimatizers disowned the several disasters or costly mistakes (e.g. a variety of commercial plants and grasses, the rabbit, cashmere goat, Asiatic mongoose, fox, squirrel, sparrow and cattle egret); for example, the botanist G.M. Thomson claimed that 'it is to the Government, acting on the instigation of large landed proprietors that we owe the introduction of stoats, weasels and ferrets'.[111] On the other hand, several failures and environmental nuisances resulted from over-zealous acclimatization promoted by the societies, particularly in the case of the hare, alpaca, several varieties of deer and the unprofitable ostrich and angora goat.[112] Plant acclimatization was of greater practical effect, although even here the considerable and lengthy involvement of society members, governments and commercial interests in all aspects of sericulture did not result in the establishment of a staple product line, even though Charles Brady, one of the most successful and creative of Australian silk experimentalists, was appointed by the New South Wales government as state organizer of the industry.[113]

Although much of the costly experimental work in acclimatization was in fact absorbed by private activity, the Australian scientific community promoted the subject publicly throughout the period under consideration.[114] As part of an international movement, Australian acclimatization demanded regular contact with researchers and collectors throughout Britain, Europe, Asia and America, and in turn represented an issue of Australian, rather than merely individual colonial, concern.[115] Furthermore, in so far as the early widespread acclimatization programme encouraged the growth of a more formal economic and commercial botany in the 1890s and beyond, it represented a proving ground for government involvement in experimental work.[116] In the absence of large-scale manufacturing in Australia, natural history took the role played by mechanical engineering or industrial chemistry in large, industrial economies in the nineteenth century.

The cyanide process

Against a background of declining gold production and technical frustrations, the cyanide process of gold extraction was developed in Britain through a programme of laboratory research designed to reveal a solvent that could recover gold from refractory ores. The resulting patent of 1887, by J.S. Macarthur and R.W. and W. Forrest, enshrined the discovery that potassium cyanide solutions could selectively dissolve and extract gold (and silver) from complex physical and chemical combinations in natural ores.[117]

Despite the depressed state of Australian gold-mining at the time, this new low-cost chemical process was slow to exert its attractions in Australia.[118] Scattered and remote mining districts, economic depression, and the lack of scientific skills and sympathies all played their part.[119] Cementing anti-'theorist' sentiment was the allegedly excessive patent royalty of 10–20 per cent and the associated conditions demanded by the patentees for access to this product of science.[120]

By 1897, however, the reality of economic rewards, proved elsewhere,[121] was dispelling prejudice. In that year Charters Towers in Queensland was reported to be 'cyanide mad'.[122] By 1900, two million bullion ounces of gold had been produced by the cyanide process in Australia, second only to the Transvaal.[123]

Implicit in the extensive adoption of the cyanide process was an incorporation of the trappings and culture of science into the mining sector. The process imposed certain minimum scientific requirements on the user. In practice, the chemically active potassium cyanide could be lost in many side reactions. Sensitivity to the chemical processes at work, and the facilities to control them, were thus required. As a contemporary text[124] on the process instructed, a 'commodious and well-equipped laboratory forms one of the most important and necessary parts of the whole plant'. Furthermore, a hierarchy of skills was required in laboratory staff — a metallurgist to tailor the operations to different classes and grades of ore, a trained assayer to measure extraction efficiency, and numerate laboratory assistants to prepare reagents and carry out routine support work.

For substantial modification of the process to local ores, people capable of technical industrial research had to be called in. Western Australia, where extremely rich but often intractable ores threw up both the technical challenge and the incentive to meet it, was a prime example of this. In the atmosphere of experimentation that resulted, the 'rule of thumb'[125] men were out of their depth and the infiltration of the trained technical expert was confirmed. It took a Dr Ludwig Dhiel, for instance, to confront the slow extraction rate of gold from telluride ore and to perfect the addition of bromo-cyanide as an accelerator.[126]

In this environment, graduates from Australian schools of mines were soon snapped up. However, need for technical expertise was met largely by an influx of metallurgists and mining experts from other countries and this served to highlight the very meagre provision of training in Australia and to create demands for improvements in the technical infrastructure servicing the mining industry.[127] As a most immediate response to the demand, all states had respected schools of mines by 1903.[128]

If the economic rewards of the cyanide process inspired support for science in the private sector of mining, this was encouraged by the facilitating actions of governments alerted to its employment and production potential.[129] The South Australian government chose to build its own cyanide customs plant. Victoria and New South Wales preferred experimental plants where the Government Metallurgist could test ore samples and advise miners on the suitability of cyanide extraction. Western Australia followed South Australia, erecting government customs works in areas where the industry and cost structure were likely to prohibit private cyanide extraction.

The private sector could also appropriate the benefits of research within government. The chemical investigations of South Australia's government analyst, G.A. Goyder, made a success of the Mt Torrens plant, the demonstration effect of which was felt within South Australia and beyond. Likewise, the freely communicated scientific detail of Goyder's work was quickly soaked up. Disseminated through the staff, students and reports of the new South Australian School of Mines, and the meetings and journals of the Royal Society of South Australia and the newly formed Australasian Institute of Mining Engineers, his work reached the vanguard of the Australian mining community, who were quick to acknowledge its worth.[130]

Government advocacy achieved its highest and most controversial profile in the cyanide patent disputes that raged in Australia from 1895 to 1901, with the purpose of securing the process on terms agreeable to Australian users. Prodded by an alliance of local nationalists, innovators and miners, and of British gold-mining companies,[131] governments intervened on the side of the industry. In New South Wales, where the Attorney General and the Minister for Mines put the case as formal objectors in the court, the patent was ruled invalid. In Victoria, the government provided both verbal and financial support for the formal objectors, and eventually purchased the patent rights for £20 000. Similar events occurred in New Zealand. In Western Australia, where patent validity could not be challenged, the government finally passed legislation specifically to prevent the extension of the patent.[132]

Most states were therefore able to enter the twentieth century with the burden of royalty payments lifted. By 1903, when Australian gold production peaked, cyanide extraction was contributing approximately 20 per cent.[133] Perhaps of even more lasting significance, the *Australian Mining Standard* in 1901 could comment that 'success no longer depends upon rich returns, but upon the steady average from the low-grade ores. The metallurgist, the geologist, the chemist, and the engineer are the men to whom the great expansion of this industrial field is due'.[134]

Microbiology

In the 1880s and 1890s in Australia, various forces coalesced to bring the new science of microbiology into rapid prominence, generating demand for its application in disparate quarters, and hastening the evolution of infrastructural support.

The public health crisis in the cities of Melbourne and Sydney in the 1880s was the most dramatic and impelling of these forces. A succession of epidemics focused attention on the germ theory and demanded some practical response. In New South Wales the panic created by the 1881 smallpox epidemic brought forth the Board of Health to devise and implement measures of control. Composed principally of leading medical men who were also active in the world of science, this board became a major instrument for the incorporation of bacteriological knowledge into state responses.[135] As the author of detailed reports and the draftsman of consequent legislation, Dr John Ashburton Thomson, chief medical inspector to the board from 1885, was a key figure in this process.[136] The Public Health Act of 1896 consolidated advances already made and created the need for a continuing bacteriological service to support the board's new obligations. The microbiological laboratory was built in 1897 and on 1 January 1898, Dr Frank Tidswell took up duty as microbiologist to the Board of Health at a salary of £800 per annum.[137]

In Melbourne, the Royal Commission on the Sanitary State of Melbourne (1888-90), chaired by Harry Allen, professor of anatomy and pathology at Melbourne University, brought the matter of bacterial action to public attention. This led to the appointment, in 1892, of Thomas Cherry M.D. as demonstrator in pathology at the university. He had responsibility for incorporating bacteriological instruction into the medical curriculum.[138] By 1897 Cherry's laboratory was also operating as an outpost of the Victorian Board of Health, performing bacteriological diagnoses of diphtheria, typhoid and tuberculosis, quarterly examination of the Melbourne water supply and other bacteriological services as required, all for the sum of £150 p.a.[139] As this service expanded, so did the funding from government, allowing the establishment of bacteriology as a separate department from 1900, under Cherry's direction. From its new building, the bacteriological laboratory was soon servicing a wide range of government and other institutions.[140]

Also in the 1880s, the pastoral industry, especially in New South Wales, experienced two major problems that highlighted the new insights of bacteriology. In 1885, the bacterial disease anthrax was identified by the government veterinarian as the dreaded killer decimating flocks in several areas.[141] This discovery initiated investigations by the Stock Department of the anthrax vaccine developed by Louis Pasteur in 1881.[142] Meanwhile the pastoral industry generally was reeling from the effects of the explosion in the rabbit population. In desperation the New South Wales government, in August 1887, offered a reward of £25 000 for the successful demonstration of a method to cope with the problem. Pasteur responded with the proposal that an infectious disease specific to rabbits could eliminate the pest. By April 1888 his representatives had arrived in Australia to demonstrate the method.[143]

Though the rabbit experiments were unsuccessful, Pasteur's agents convinced a board appointed by the government of the efficacy of anthrax vaccine.[144] Many affected pastoralists were soon clamouring for the vaccination of their flocks. Though several difficulties with Pasteur over the rabbit experiments intervened, in June 1890 Pasteur's nephew, Adrien Loir, began the production of anthrax vaccine on Rodd Island in Sydney Harbour.[145] Unfortunately, the vaccine was not always successful.[146]

The problems encountered with Pasteur's vaccine under Australian conditions[147] motivated John Alexander Gunn to begin his own experiments in a laboratory built on the property he managed.[148] At first successful in preparing a vaccine according to Pasteur's method, he went on, in partnership with John McGarvie Smith from 1895, to supply a vaccine with vastly improved keeping qualities, and then a single-dose vaccine. By 1898 the success of the McGarvie Smith and Gunn vaccine had almost eliminated the Pasteur product from the market.[149] By the early 1900s anthrax had been virtually brought under control in New South Wales. This very tangible product of bacteriological science had a great demonstration effect on the pastoral industry, the representatives of which thereafter made consistent approaches to government to initiate research on animal diseases and the production of a range of vaccines.

The effect of the 1890s depression on the pastoral industry confirmed government views that diversification into other agricultural activities was a high priority. The dairy industry became one of the special targets for development and in 1895 the New South Wales government decided to appoint a dairy expert as the chief agent of the process. A suitable recruit, who would also have bacteriological expertise, was sought from Britain;[150] M.A. O'Callaghan began duties in January 1897. A year later he was insisting that 'the technique of Australian dairying requires to be more guided by bacteriological research than the dairy work of any other country with which Australia competes'.[151] The sanitary control of dairies for public health reasons had laid the groundwork for the application of bacteriological knowledge to the industry.[152] Now bacteriology was also to inform product improvement and development. With this as his brief, O'Callaghan addressed people at all levels: back-blocks farmers, dairy inspectors, dairy students at Hawkesbury Agricultural College, and the AAAS. He also carried out investigations in his government laboratory and was soon supplying the industry with starter cultures.[153]

The emerging scope of microbiological science had even broader implications for policies of diversification. For example, the speciality of soil bacteriology developed from the 1890s with discoveries of the nitrogen-fixing properties of nodule-forming bacteria in the roots of leguminous plants. The role of a whole range of micro-organisms in plant as well as animal disease was also being revealed. In response, the NSW Government Bureau of Microbiology was established in 1890 with Dr Frank Tidswell as director. With a programme combining routine examinations and applied research,[154] the bureau developed a pool of microbiological expertise which, as well as serving a wide range of industries, was to provide a source of recruitment for other institutions.[155]

The various developments already described, which brought forth a supportive infrastructure for bacteriology and, more broadly, microbiology, were paralleled by activities in the superstructure of science. Sanitary science, for instance, became incorporated in the programmes of the Royal Society of New South Wales and AAAS. William Macleay was an early exponent of the value of bacteriology[156] and made the laboratory at Linnean Hall available for bacteriological research by Oscar Katz during the second half of the 1880s.[157] With this patronage, Katz quickly established himself as a bacteriological expert, being appointed as such to the Rabbit Commission in New South Wales and the Sanitary Commission in Melbourne. In 1891 Macleay had bequeathed £12 000 to found a chair of bacteriology at Sydney University, though on rejection of his conditions by the senate the money passed to the Linnean Society of NSW to establish a fellowship. The first beneficiary was Robert Greig-Smith whose diverse investigations regularly brought bacteriology to readers of the Linnean Society's journal. Greig-Smith went on to make important original contributions in the field of soil bacteriology.[158]

By 1910 microbiological science was being diffused and applied across a broad front. It was being taught in a range of institutions to students of medicine, sanitation and public health, pharmacy, dentistry, agriculture, veterinary science and dairy science. An industry that produced vaccines, yeasts for fermentation industries, starter cultures for the dairy industry, cultures of nodule bacteria for agriculture and other products of microbiological science was emerging. Government departments and laboratories were extending the benefits of the science in various directions. Some original research was also being carried out.

Conclusions

The history of the scientific enterprise in our period was closely related to changes in the more general economic, spatial and demographic characteristics of Australia. The early concentration on staple exploitation led to an emphasis on a widely-based 'natural history' programme within the scientific enterprise as a whole. Natural history was suited to both the economic needs of Britain and her colonies and to the psychological requirements of the scientific workers concerned. But although some focus was provided in the emergence of the 'acclimatization' movement after mid-century, it was not until the last two decades of the nineteenth century that Australian science began clearly to respond to local imperatives rather than to 'imperial' demands.

Until then, Australian scientific development was intermittent, heroic, dependent and pragmatic. Australian urban communities spawned scientific endeavour in the context of a fragile demand for manufacturing technologies, a small population and slim colonial budgets. This was in stark contrast to the earlier and larger area of recent settlement, North America. There, urban scientific culture and invention built on a very large rate of population growth (from 8.5 million in 1815 to 17 million in 1840 and 63 million in 1890) and a sturdy machine culture. Once improvements in transport had lifted the level of effective demand for manufactured industrial goods, this all combined to produce a fast rate of economic growth, a cumulation of technological skills and a large, state-supported scientific enterprise.[159] In Australia, the sparsity of the scientific audience, the weakness of the market for manufactured goods and slender budgets combined with colonial direction to minimize the extent of official support for science and to confuse the goals and purposes of the scientific community itself. Were intellectual forums to be labelled 'royal', 'philosophical' or 'acclimatization' societies?

Only when, from the 1880s, the demand for the applications of science became acute, did the character of scientific investigation alter. It was the onset of depression and the need for resource-based efficiency improvements, rather than an increase in population or industrial production, that eventually transformed the state of things scientific. The previous section of this chapter has shown how new techniques or processes won official support because of their economic potential. The subsequent emergence of a tentative 'science policy' exerted a significant impact on the scientific enterprise more generally. In this, the existence of a reasonably buoyant scientific infrastructure was of some importance, for this was often the first recipient of state activity. In the

1880s the recognition of science as a potential asset rather than a mere commodity was hastened with the appearance of a number of public scientists who devoted much of their energy to proclaiming the utility of scientific research and advocating its official imbursement.[160] With improvements in transport and communications between the Australian colonies, such elements converged to produce a new phase in the development of the Australian scientific enterprise.

Notes

1   J.W. McCarty, 'Australia as a region of recent settlement in the nineteenth century', *Australian Econ. History Review*, 13 (1973); B. Dyster, 'Argentine and Australian development compared', *Past and Present*, 84 (1979), 91-110; R. Pomfret, 'The staple theory as an approach to Canadian and Australian economic development', *Australian Econ. History Review*, 21 (1981), 133-46.

2   That is, in such areas, significant growth and structural change occurred in the absence of a discrete 'industrial revolution'. This had a systematic impact on the nature of scientific and technological development in such regions. See Chapter 8 of Ian Inkster, *Science and Technology in History* (London, forthcoming).

3   E.A. Shils, 'Towards a modern intellectual community', in J.S. Colemen (ed), *Education and Political Development* (Princeton, 1965), 498-518; idem., *The Intellectuals and the Powers and Other Essays* (Chicago, 1971), esp. Chap. 17; R. Macleod, 'On visiting the "moving metropolis": reflections on the architecture of imperial science', *Hist. Records of Australian Science*, 5 (3) (1982), 1-16.

4   This is the principal reason for rejecting the generality of the Basalla approach: I. Inkster, 'Scientific enterprise and the colonial "model": observations on the Australian experience in historical context', *Social Studies of Science*, 15 (1985), 677-704; G. Basalla, 'The spread of western science', *Science*, 156 (1967), 611-22.

5   D. Fleming, 'Science in Australia, Canada and the United States: some comparative remarks', *Proc. Tenth International Congress of the History of Science, 1962*, Vol. 1 (1964), 179-96 (quote p.181).

6   J.M. Powell, 'Conservation and resource management in Australia 1788-1860', in J.M. Powell and M. Williams (eds), *Australian Space, Australian Time* (Melbourne, 1975) pp.18-60.

7   B. Field, *First Fruits of Australian Poetry*, 2nd edn (Sydney, 1823); idem., *Geographical Memoirs of New South Wales* (London, 1825); T.L. Mitchell, *Three Expeditions into the Interior of Eastern Australia* (London, 1839).

8   Inkster, op.cit. (no. 4), pp.680-3; many of the editorial comments and analyses in A. Mozley Moyal, *Scientists in Nineteenth Century Australia* (Melbourne, 1976); M.E. Hoare, 'Science and scientific associations in eastern Australia' (PhD thesis, ANU, 1974), esp. Chap. IV.

9   E.G. Wakefield, *A Letter from Sydney* (London, 1829); idem., *England and America* (London, 1833); R.C. Mills, *The Colonisation of Australia: The Wakefield Experiment in Empire Building* (London, 1915) (reprinted London, 1968); J. Philipp, 'Wakefieldian influence in New South Wales, 1830-1842', *Historical Studies*, 9 (1960), 178-80; the magnificent D.H. Pike, *Paradise of Dissent: South Australia 1829-1857* (Melbourne, 1957); and D.N. Jeans, 'The impress of central authority upon the landscape: southeastern Australia 1788-1850', in Powell and Williams, op.cit. (n. 6), 1-17.

10 For arguments as to the essential role of 'ancillary' institutions see Shils, op.cit. (n. 3, 1965), esp. p.498; Inkster, op.cit. (n. 4), 678-9, 690-93.

11 For further analysis along such lines see J. Todd, 'Colonial adoption: the case of Australia and the Sydney Mechanics' School of Arts', in I. Inkster (ed), *The Steam Intellect Societies: Essays on Culture, Education and Industry, 1820-1914* (Nottingham, 1985), pp.105-130, and I. Inkster, 'Science, public science and science policy in Australia circa 1880s-1916', Workshop on the History of Science in Australia, Australian Academy of Science, 1982 (Basser Library), esp. Appendix A and pp.1, 25-30.

12 A.G. Serle, *The Golden Age: A History of the Colony of Victoria 1851-1861* (Melbourne, 1963). p.381.

13 Powell, op.cit. (n. 6), p.52. For an excellent acccount of the growth of Victorian civic science in these years, see this source pp.41-56.

14 N.G. Butlin, *Investment in Australian Economic Development 1861-1900* (Cambridge, 1964). pp.201-10: idem., *Australian Domestic Product, Investment and Foreign Borrowing 1861-1938/39* (Cambridge 1962), pp.460-1; R.K. Wilson, *Australia's Resources and their Development* (Sydney, 1980), esp. Ch. 5 and pp.36-41; E.A. Boehm, *Prosperity and Depression in Australia 1887-1897* (Oxford, 1971), pp.120-1, 306; B. Fitzpatrick, *The British Empire in Australia: An Economic History, 1834-1939* (Melbourne, 1947, 1961, etc.)

15 A Wessan, 'Mechanics' institutes in Victoria', *Australian Journal of Adult Education*, 12 (1972), 3–11.

16 For the general role of public proclamation and advocacy in science see F.M. Turner, 'Public science in Britain, 1880-1919', *Isis*, 71 (1980), 589–608.

17 *N.S.W. Statistical Register*, various years.

18 Ibid. For further detailed indicators see J. Todd, op.cit. (n. 11).

19 *Rules and Orders of the Mechanics' Institution, Hobart Town, Van Diemen's Land* (Hobart, 1828).

20 For which see G. Nadel, *Australia's Colonial Culture* (Melbourne, 1957); D. Whitelock, *The Great Tradition: A History of Adult Education in Australia* (Brisbane, 1974); J. Martin, *Popular Amusements*, Society for the Promotion of Morality (Melbourne, 1870).

21 For example, *Annual Report of the V.D.L. Mechanics' Institute for 1838* (Hobart, 1839); 'Minute Books of VDL Mechanics Institute 1839-1867' [Mitchell Library, Sydney, MS A583 (1-3)], Vol. I, 1839-46, general meeting of 29 April 1839; Vol. II, 1846-53, general meeting of 30 Jan. 1851; Vol. III 1853-67, Meetings of 1 Feb. and 2 May 1856, 15 Dec. 1865.

22 M.E. Hoare, 'Doctor John Henderson and the Van Diemen's Land scientific society', *Records Aust. Acad. of Science*, 1 (3) (1968), 7–24.

23 *Annual Reports of the Launceston Mechanics' Institute, 40–42 (1882–1884)* (Launceston 1883-1885). At this time the colonial authorities were voting money to the Launceston Institute for the establishment of a School of Mines and a technological class, and lectures were delivered on mining of coal and geology. For the originally general aims of the institute see *Rules of the Launceston Mechanics' Institution*, 8 Oct. 1844 (Launceston, 1845).

24 *Votes and Proceedings of the Legislative Assembly, Victoria 1858–59*, I, 455-83; *Age* (Melbourne) 1 April 1858, 5.

25 'Proceedings of the Council of the University of Melbourne', ibid., 1859-60, III, 667-72.

26 *Victorian Year Book*, Vol. I (1873) (Melbourne, 1874), p. 46; vol. II (1874), p. 186.

27 For summaries of which see R. Cotter, 'The golden decade', in J. Griffin (ed.), *Essays in Economic History of Australia, 1788–1939* (Brisbane and Melbourne, 1967), pp.113-34; T.H. Irving, '1850-1870' in F. Crowley (ed.), *A New History of Australia* (Melbourne, 1974), pp.124-54.

28 'Victoria in 1857-60', Supplement to *The Herald*, 25 October 1860, pp. 8–9.

29 *Age*, 8 January 1858, p.5.

30 'MS Disposition of Gerard Krefft, Curator and Secretary of the Australian Museum' [Basser Library, Australian Academy of Science, Canberra MS 21, n.d., incomplete], p.1. (of 39 pp.).

31 *Age*, 1 January 1858, p.1.

32 Ibid., 27 January, 16 February and 6 April 1858.

33 Ibid., 12 and 21 January 1858.

34 Original correspondence between the Port Phillip Farmers Society and the Chief Secretary, Melbourne, indicated that funds were to be made available for an agricultural laboratory (£1000) as well as for an experimental farm (£3720).

35 *Votes and Proceedings 1859–60*, 955-1044.

36 Ibid., IV, 667.

37 R. Barry, *Inaugural Address to Members of the Victorian Institute, 21 Sept. 1854* (Melbourne, 1854).

38 Ibid., 7.

39 J.I. Hall, 'The sources and history of the SAI' [MS S.A. Archives, State Library of S.A., D.4953(T)]; 'General Committee of the Adelaide Library and Mechanics' Institute appointed at the General Quarterly Meeting of 15 Jan. 1853' [MS S.A. Archives, A61 (B1)]: *South Australian Register*, 25 November, 9 December 1853; *Adelaide Observer*, 14 January 1884, 2 February 1861. For the early relations between the foundation of South Australia and scientific-intellectual groupings see I. Inkster, op.cit. (n. 4), 682-3; D.H. Pike, *Paradise of Dissent* (Melbourne 1957); *South Australia*, 7 July 1838, 24 August, 4 and 18 September, 13 and 27 November 1839.

40 Figures derived from the *Annual Reports of the SAI, 1861–1884* [bound volume. S.A. Collection, State Library of S.A., Z027.05].

41 Large financing frequently led to debate and 'public science' rhetoric; see, for example, 'Petition for addition to SAI buildings', *S.A. Parliamentary Papers* (1871), no.85, House of Assembly, printed 6 Sept. 1871 [S.A. Collection, 2022, Adelaide Philosophical Society, (c) Pamphlets]; *South Australian Register* (25 Nov. 1853); much material in 'MS and Reports of Royal Society of S.A. 1876-1940' [State Archives, SRG, 10].

42 SAI, op.cit. (n. 40), various.

43 *Final Report of the SAI* (Adelaide, 1885), pp.10-12.

44 *Royal South Australia Almanack for 1855* (Adelaide, 1855?); 'MS. and Board Annual Report of Adelaide Philosophical Society, 1853-1876' [State Archives, Z per 506].

45 Calculated from material in ibid. (n. 44). On the early activists see also R.S. Rogers, 'A history of the society, particularly in its relation to other institutions in the state', *Trans. and Proc. Roy. Soc. of S.A.*,

46 (1922); C.T. Madigan, 'The past, present and future of the society, and its relation to the welfare and progress of the state', *Trans. and Proc. Roy. Soc. of S.A.*, 60 (1936); and the treatment of particular sciences in papers by Mawson, Prescott, Rogers and Dickinson in *Trans. Roy. Soc. of S.A., Centenary Volume* (Adelaide, 1954).

46 For the general tendency of Australian science towards natural history, acclimatization, exploration and application see elsewhere in this chapter as well as Inkster, op.cit. (n. 4).

47 *First Report of the Adelaide Philosophical Society* (Adelaide, 1884), 3; *Second Report of the Adelaide Philosophical Society* (Adelaide, 1855).

48 'Special Committee Report', *Third Ann. Report of Adelaide Philosophical Society* (Adelaide, 1856), pp. 2-4.

49 *The Australian Technical Journal of Science, Art and Technology* (22 Feb. 1897).

50 Ibid. (25 March 1897); *Sydney Morning Herald*, 3 September 1888.

51 *Sydney Mail*, 17 March 1888.

52 C.E. Wilkinson FGS was President of the RS of NSW; Reid, Parkes and Keyte were behind plans for a union of infrastructural institutions in NSW (e.g. *Sydney Morning Herald*, 28 Aug. 1888).

53 *Sydney Morning Herald*, 25 Aug. 1888.

54 A. Mackay, 'Science in agriculture: what it means', *Sydney Morning Herald*, 1 Sept. 1888; W.A. Dixon, *The Constitution of Acids* (Sydney, 1886); idem, *Technical Education, as at the SMSA Technical College* (Sydney, 1881); A. Mackay, *Helpful Chemistry for Agriculturalists* (Sydney and Melbourne, 1894); *Australian Technical Journal of Science, Arts and Technology* (25 March 1897).

55 W.A. Dixon, 'The Technical College chemistry laboratory', ibid., 30 June 1897.

56 For details of science before 1850 see M.E. Hoare, op. cit. (n.8).

57 In New South Wales, the Australian Society formed in 1850 gave way to the Philosophical Society of NSW (1855), then the Royal Society of NSW (1866). The Victorian Institute for the Advancement of Science and the Philosophical Society of Victoria, both formed in 1854, merged in 1855 into the Philosophical Institute of Victoria, which in 1859 became the Royal Society of Victoria. The Adelaide Philosophical Society, founded in 1853, took up the royal title in 1880. The Queensland Philosophical Society founded in 1859 became the Royal Society of Queensland in 1883. The Tasmanian Society of Natural History and the Royal Society of Van Diemen's Land for Horticulture, Botany and the Advancement of Science amalgamated under the royal title in 1849.

58 For example, A.R.C. Selwyn of the Geological Survey of Britain was appointed to head the Geological Survey of Victoria in 1852. The Reverend William Scott was chosen by the Astronomer Royal to fill the post of Astronomer at the new Sydney Observatory in 1856.

59 For example, Gerard Krefft; *Australian Dictionary of Biography*, Vol. 5, p.243.

60 The first geological surveys were born of the need for expert geological advice following the gold discoveries. For details on the conflict over the role of the geologists see R.K. Johns (ed), *History and Role of Government Geological Surveys in Australia* (Adelaide, 1976), and M.E. Hoare, 'The half-mad bureaucrat': Robert Brough Smyth (1830-1889)', *Records Australian Academy of Science*, 2(4) (1973), 25-40.

61 J.H. Maiden, *Journal and Proceedings of the Royal Society of N.S.W.*, (1918), 260.

62 Ibid., pp.260-1.

63 Ibid., pp.263-314.

64 C.T. Madigan, *Transactions of the Royal Society of South Australia*, 60 (1936), vi.

65 For stated objects see R.T.M. Pescott, *Proc. Roy. Soc. Vic.*, 73 (1961), 5; Maiden, op. cit. (n.61) p. 261; Madigan, op. cit. (n.64), p. vi; E.N. Marks, *Proc. Roy. Soc. Queensland*, 71 (1959), 18.

66 J. Smith, 'Anniversary Address', *Journal and Proceedings of the Royal Society of NSW*, (1881), 6; P.R. Shergold and I. Inkster, 'Civil engineering and the admiralty: Thomas Tredgold, Edward Deas Thomson and early steam navigation, 1827-28', *The Great Circle*, 4 (1982), 41-52.

67 M.E. Hoare, 'Learned societies in Australia: the foundation years in Victoria, 1850-1860', *Records of the Australian Academy of Science*, 1(2) (1967), 13-5; and above, Section II.

68 Madigan, op.cit. (n. 64), p.vi.

69 For example in contributions on Pituri and allied plants of the genus *Duboisia* and their physiological action. E.N. Marks, *Proc. Roy. Soc. Queensland*, 71 (1959), 22.

70 For example, the Government Astronomer in Victoria, R.L.J. Ellery, was appointed in 1858 to conduct the geodetic survey of the colony. This took a large part of his time until 1874 and had a large influence on the astronomical programme of the observatory, *Australian Dictionary of Biography*, Vol. 4, 135-46.

71 *Australian Dictionary of Biography*, Vol. 5, 134-6.

72 The chemical analysis for the commission itself provided the subject for a paper to the Royal Society of New South Wales in 1869.

73 The largest single group in the membership comprised merchants and other businessmen (bankers,

company directors etc.) at approx. 17 per cent. Another 7 per cent were legal men, 9 per cent were non-technical civil service administrators, and nearly 10 per cent were members of Parliament (both Houses). Those who had some scientific background included 16 (13 per cent) who were surveyors, draftsmen, engineers or architects, and another 20 were medical men, dentists or chemists and druggists. (Numerous sources including various issues of the *Sands Directory of Sydney*, the *Australian Dictionary of Biography*, and membership records of the Society.)

74 Figures are derived from annual reports of the societies. Membership lists usually exceeded the numbers of subscriptions actually paid. The figures of Table I include ordinary, corresponding and associate members, but not honorary members.

75 See, for example, *Journal and Proc. of Roy. Soc. of N.S.W.*, 11 (1877), 277; *Journal and Proc. of the Roy Soc. N.S.W.*, 12 (1878), 1-2; Madigan, op.cit. (n. 64), p.iv.

76 Those classified as having no scientific background include two groups: those whose occupation required no scientific training, knowledge or experience, and those for whom there is no evidence of any overt intellectual activity in science.

77 This occurred in other ways besides the overall direction and objects of the society, e.g. Presidential Addresses often identified areas and priorities for research, prizes were offered for essays on specified subjects, delegations and other representations were made to government on issues affecting science or having a scientific component.

78 For example, the Geological Society of Australasia, the Geographical Society of Australasia, the Victorian Institute of Engineers, the Naturalists' Society of NSW, the Meteorological Society of Australasia, the Field Naturalists' Club of Victoria. In the 1890s came, for instance, the Australasian Institute of Mining Engineers, the Anthropological Society of Australasia, astronomical societies and medical societies. D.F. Branagan has attempted to chart the foundation of the scientific societies in his 'Words, Actions, People: 150 Years of the Scientific Societies in Australia', *Journal and Proceedings of the Royal Society of NSW*, 104 (1972), 123-41.

79 Archibald Liversidge, the organizing force behind the formation of AAAS, found during the course of his preparations in 1886 that there were some 25 to 30 recognized scientific societies in the Australasian colonies embracing 2500 to 3000 members. *J. and Proc. Roy. Soc. of N.S.W.*, 20 (1886), 36-7.

80 M.E. Hoare, 'The intercolonial science movement in Australasia, 1870-1890', *Records of the Australian Academy of Science*, 3(2) (1975), 7-28.

81 G. Blainey, *A Centenary History of the University of Melbourne* (Melbourne, 1971), p.105.

82 R.W. Home, 'The beginnings of an Australian physics community', in N. Reingold and M. Rothenberg (eds), *Scientific Colonialism: A Cross-Cultural Comparison* (Washington, 1987), pp.1-30. For details on careers of 1851 Exhibitioners see I.W. Wark, '1851 Science Research Scholarship Awards to Australians', *Records of Aust. Acad. of Science* 3 (3/4) (1977), 47-52.

83 Tate in geology, botany and zoology (see A.R. Alderman, 'The development of geology in South Australia', *Records of Aust. Acad. of Science*, 1(2) (1967), 29-52; Lamb, in hydrodynamics (A. Mozley Moyal, *Scientists in Nineteenth Century Australia* (Melbourne, 1976), 226-7).

84 For further details on the life and work of Bragg see G.M. Caroe, *William Henry Bragg, 1862–1942: Man and Scientist* (Cambridge, 1978); R.W. Home, 'The beginnings of an Australian physics community', op.cit. (n. 82); R.W. Home, 'The problems of intellectual isolation in scientific life: W.H. Bragg and the Australian scientific community, 1886-1909', *Historical Records of Australian Science*, 6 (1) (1984), 19-30; R.W. Home, 'W.H. Bragg and J.P.V. Madsen: collaboration and correspondence, 1905-1911', *Historical Records of Australian Science*, 5 (2) (1981), 1-29.

85 A. Mozley Moyal, op.cit. (n. 83), p.222.

86 *Australian Dictionary of Biography*, Vol. 8, p.219.

87 In his Presidential Address to the Biology Section of AAAS, in 1891, Haswell described the new, and late, influence of theory on the field of biology, mainly due to Darwin's writings. *Report 3rd Mtg. AAAS*, (1891), 174-92.

88 Moyal, op.cit. (n. 83), p.223.

89 Joan Radford, *The Chemistry Department of the University of Melbourne* (Melbourne, 1978), pp.54-60.

90 In South Australia, Charles Todd, appointed Government Astronomer (and Superintendent of Electric Telegraphs) in 1855; in Victoria, R.L.J. Ellery, appointed 1854; in NSW, H.C. Russell, appointed 1870.

91 See R.K. Johns, op.cit. (n. 60), pp.35-6.

92 Daniel McAlpine Papers, La Trobe Manuscript Collection, State Library of Victoria, MS8729.

93 *Australian Dictionary of Biography*, Vol. 4, pp.63-4.

94 Melbourne 1870, Sydney, 1880, Adelaide 1893.

95 First curator of the Sydney Technical Museum, later director of the Botanical Gardens, Joseph Henry Maiden soon established a reputation through his own botanical work on Australian plants of economic value.

96 *Australian Dictionary of Biography*, Vol. 7, p.154.

97 Orme Masson, 'The gaseous theory of solution', Presidential Address, Section B, *Report 3rd Mtg. AAAS*, 1891.

98 Moyal, op.cit. (n. 83), p.222.

99 The *Australasian*, 1 Sept. 1888, pp.490-91; J. Steele Robertson, 'Natural science in Australia', *The Centennial Magazine*, 2, No. 7 (July 1889), 523-7.

100 *Sydney Mail*, 8 Sept., p.504. On the importance of a 'scientific movement' in mediating between the scientific community and society as a whole see I. Inkster, 'Aspects of the history of science and science culture in Britain, 1780-1850 and beyond' in I. Inkster and J. Morrell (eds), *Metropolis and Province: Science in British Culture 1780-1850* (London, 1983), pp.11-54, and the review of the book by James Paradis in *Victorian Studies* (Winter, 1986).

101 *The Australasian*, 18 Aug. 1888, p.352, *Sydney Morning Herald*, 5 Sept. 1888, p.5.

102 The following figures derive from a continuing study of the magnitudes and role of the international patent system prior to 1914. For the problems involved in patenting statistics as historical indicators see I. Inkster, 'The ambivalent role of patents in technology development', *Bulletin of Science, Technology and Society*, 2 (1982), 181-90, and for Australia, idem. (n. 11 above), 45-8 and Appendix A, pp.10-11.

103 See L.A. Gilbert, 'Botanical investigations of the eastern seaboard, 1788-1811' [BA Hons. Thesis, New England, 1962] I, 52-73, 171-5, Appendix A56-72.

104 The commercial and imperialist intent of the Zoological Society of London had been established during the 1820s, which included in its agenda 'the introduction, exhibition and acclimatization of subjects of the animal kingdom'. While the 1860 London group may have been a gastronomic laboratory for gentlemenly naturalists, the French Society boasted 2000 members, affiliated groupings and awarded medals of success. It is noteworthy that its early interests included sericulture.

105 *The Edinburgh Review*, 224, January 1860; The *Times*, 22 Sept., 1860; A.S. Wilson, *John Wilson and Edward Wilson*, private printing (London, 1884); much had been going on prior to Wilson's conversion to acclimatization, and in 1873 Thomas Blair claimed to be at least a joint founder of the Australasian movement; *Proceedings of the Zoological and Acclimatization Society of Victoria*, 2 (1873), 13.

106 Apart from the papers on acclimatization, Wilson joined in discussion and plans for an inland expedition, improvement in Melbourne's water supply and the establishment of an astronomical and magnetic observatory: *Transactions of the Philosophical Institute of Victoria*, Vols 2-5, Melbourne, 1857-1860.

107 *First Annual Report of the Acclimatisation Society of Victoria* (Melbourne 1862).

108 *Proceedings, op.cit.* (n. 105), membership lists pp.20-5.

109 See *Third Annual Report of the Acclimatisation Society of N.S.W.* (Sydney, 1864); G. Bennett, *Acclimatisation, Its Eminent Adaptation to Australia* (Sydney and Melbourne, 1862); F. von Mueller, *Select Extra-Tropical Plants, Recently Eligible for Industrial Culture or Naturalisation* (Sydney, 1881), originally a series of five contributions to the Victorian Acclimatisation Society between 1871 and 1878, published also in India, and enlarged for the Sydney edition; J.H. Maiden, *The Useful Native Plants of Australia and Tasmania* (Sydney, 1889); [William L. Morton, 1820-1898], *Yeoman and Australian Acclimatiser*, various issues 1861; V.M. Coppleson, 'Life and times of Dr George Bennett', *Bulletin of Post-Graduate Committee in Medicine, University of Sydney*, 9 (1955), 207-64; and articles on Bennett and Clarke, *Medical Journal of Australia*, 42 (1955), 273-8; *Journal of Royal Aust. Hist. Soc.*, 30 (1944), and debate between Maiden and Frederick Turner in *Australian Agriculturist*, February 1899.

110 E. Wilson, *Acclimatisation*, Royal Colonial Institute (London, 1875); W.H.D. Le Souef, 'Acclimatisation in Victoria', *Report of the Second Meeting of the Australasian Association for the Advancement of Science* (Melbourne, 1890), pp.476-82; G.M. Thomson, 'On some aspects of acclimatisation in New Zealand', ibid., *Third Meeting* (Wellington, 1891), pp. 194-213.

111 Thomson, op.cit. (n.110), p.213.

112 *Australian Encyclopaedia*, various, under headings of 'George Bennett', 'Acclimatisation', 'Mammals', 'John Harris', Zoology'.

113 *New South Wales, Silk Correspondence Relating to Cultivation, Presented to Both Houses of Parliament*, 1870, [Mitchell Library, Sydney 638.2/N]; S.F. Neill, *The Silk Worm, Its Education, Reproduction and Regeneration on M. Alred Roland's Open-Air System* (Melbourne, 1873); 'Parkes Correspondence', [Mitchell Library, ML A922, ML A879] 9, letters to Sir Henry Parkes, 27 July 1893, 12 Feb. 1894.

114 C.F.H. Jenkins, *The Noah's Ark Syndrome* (Sydney, 1977), pp.46-81; 'Letters and MS Book, 1862-85', Royal Park Zoological Society, Victoria', [Mitchell Library, MSS A345]; G. Winter, 'For the Advancement of Science: The Royal Society of Tasmania, 1843-1885', [BA Hons Thesis, University of Tasmania, Hobart, 1972]; G.P. Whitley, 'MS History of the Australian Museum 1827-1962', [Basser Library, nd., MS Box 22/1].

115 *First Annual Report of Acclimatisation Society of N.S.W.* (Sydney, 1861); T. Hutton, *Remarks on the Culture of Silk in India* (Bombay and London, 1869); letter of W. Monro, Home Dept India, E.C. Bayley, Secretary of Government in India in Calcutta, F.O Adams (Silk Supply Association), Charles Brady, A.B. Johnson, Home Dept in *Silk Correspondence*, op.cit. (n. 143); 'Deane Family Papers' [National Library of Australia, MS 610, relating mostly to Henry Deane 1847-1924, Sydney engineer and botanist], file 30 'Correspondence with Baron von Mueller 1886-89' [items 568-88].

116 For which see acclimatization/economic botany articles in *The Rural Australian*, December 1887, *Gardeners Chronicle*, 23 Jan. 1897, *Sydney Morning Herald*, 16 Jan. 1911, 14 Sept. 1912; 'Secretary of Advisory Council of Science and Industry Executive Committee (Melbourne), 24 May 1916 to Frederick Turner (botanist)', in 'Frederick Turner MS Collection', [3 Bound Volumes, National Library of Australia, MS 392], III, 1916; Frederick Turner, 'The Effect which Settlement in Australia has Produced Upon Indigeneous Vegetation', [Deane Family Papers NLA op.cit. (n. 115), file 12, 5 Aug.-13 Sept. 1892.]

117 J.S. MacArthur 'The MacArthur-Forrest process of gold extraction', *Journal of the Society of Chemical Industry*, 9 (1890), 267-70, and J.S. MacArthur, 'Gold extraction by cyanide: a retrospect', ibid., 24 (1905), 311-5.

118 One attraction was that it could be used to extract the residual gold in the piles of tailings remaining from past mining activities. Previously regarded as waste, these low-concentration tailings could now yield significant product for small cost and effort, boosting the productivity of the industry. The process could also be applied directly to freshly mined ore, even when a high concentration of the troublesome 'sulphides' was present.

119 For example, Report of Directors of Australian Gold Recovery Co. (Ltd), The *Times*, London, Dec. 23, 1893, p.11(d); ibid., March 19,1897, p.3(e); G. Blainey, *The Rush That Never Ended* (Melbourne, 1978), p.251: The *Australian Mining Standard*, June 12, 1889; ibid., May 13, 1893, p.253.

120 See e.g. *Queensland Parliamentary Debates*, Legislative Assembly, 27 September, 1895, p.1078. Users had to buy cyanide from the British patent holders, the Cassel Gold Extracting Co. (Ltd), and had to employ one of Cassel's own chemists. They were also committed to use of the process for the lifetime of the mine if possible, but at least for the whole duration of the patents and any extension of them. (Cassel Gold Extracting Company, *Board Minutes*, 14 March, 1889, Cheshire County Records Office, Chester, England).

121 Initially, this was in the Transvaal, where it was well established by 1891-92. In New Zealand it was established by 1893 and in America by 1895.

122 *Australian Mining Standard*, Dec. 7, 1897, p.2477.

123 Calculated from figures presented to the International Congress of Applied Chemistry at Berlin, June 1903, by George T. Beilby, published in *Queensland Government Mining Journal* (Nov. 14, 1903), p.589.

124 James Park, *The Cyanide Process of Gold Extraction* (London, 1900), Chapter 3.

125 *Australian Mining Standard*, Feb. 12, 1903, p.217.

126 Blainey, op.cit. (n. 119), p.201.

127 *Australian Mining Standard*, November 23, 1899, p.395; ibid., Aug. 2, 1900, p.104; ibid., 29 Oct. 1903, p.578.

128 Schools of mines were established at Charters Towers, Queensland in 1900, at Kalgoorlie in Western Australia in 1902 and in Zeehan, Tasmania in 1902. At the South Australian School of Mines, the number of students increased from 600 in 1895 to 1600 in 1900.

129 For example, see *South Australian Parliamentary Debates* (October 11, 1893) p.2076; *Australian Mining Standard, February 8, 1900, p.116; N.S.W. Parliamentary Debates*, 1896, Vol. 85, p.3703.

130 G.A. Goyder, 'Trials of gold extraction by cyanide of potassium with South Australian ores', *Annual Report of South Australian School of Mines*, (1893), 4-24 and 'The action of cyanide of potassium on gold and some other metals and minerals', *Transactions of the Australasian Institute of Mining Engineers*, 1 (1893), 84-7; G.A. Goyder, 'Reactions of double cyanide, bearing upon the cyanide-process for gold extraction', *Transactions and Proceedings of the Royal Society of South Australia.* (1895), 25-6.

131 For example, the NSW Cyanide Opposition Committee, formed in 1897.

132 An amendment confirming the validity of the patent had passed through unnoticed in Wester Australia in 1895 when the goldfields there were only just opening up.

133 This figure is a recent estimate by Jan Todd. It includes use of modifications of the original cyanid process.

134 *Australian Mining Standard*, 3 Jan., pp.1-2.

135 From 1883 the President of the Board was the Medical Adviser to the Government.

136 J. Cummins, *A History of Medical Administration in New South Wales, 1788-1983* (Sydney, 1979) p.80 and *passim* on Thompson.

137 *N.S.W. Civil Service List* (Sydney, 1898).

138 University of Melbourne Faculty of Medicine Minutes, 15 October, 29 October and 7 December 1891

University of Melbourne Archives. At Allen's suggestion Cherry had prepared himself for the position by travelling to Europe in 1891 to further his studies in bacteriology.

139 *University of Melbourne Calendar* (1899), p.394.

140 Ibid. (1901), p.428; ibid. (1904), p.331.

141 Appendix C of the Annual Report of the Stock and Brands Branch of the Dept of Mines for 1885, in *Journal of the Legislative Council of N.S.W.* (1885-86), Vol. 40, Pt 2, 649; L.W. Devlin, 'The advent of the vaccination of stock against anthrax in Australia', *Australian Veterinary Journal* (Aug. 1943), 102-11.

142 Annual Report of the Stock and Brands Branch for 1886, in *Journal of the Legislative Council of N.S.W.* (1887-88), vol. 43, Pt 2, p.1047.

143 'The rabbit pest (papers on proposals to effect destruction of rabbits by means of diseases)', *Journal of the Legislative Council of N.S.W.* (1887-88), Vol. 43, Pt. 4, pp.693-709.

144 Report of the NSW Anthrax Board, Sydney (1889).

145 Rodd Island in Sydney Harbour had been set aside and equipped for the experiments for the Rabbit Commission. The Government, despite many public fears about spreading microbes, now allowed the Pasteur Institute free use of the Island.

146 *Annual Report of the Stock and Brands Branch* for 1890, p.18.

147 The vaccine did not always survive the long distances in the extreme conditions of inland Australia. It was also quite costly and there were delays in obtaining it. The fact that two inoculations were required was a particular disadvantage in Australia where flocks grazed over vast areas and two musterings for inoculation were time consuming and costly.

148 R.H. Webster, *Bygoo and Beyond* (Sydney, 1956), p.86.

149 *Annual Report of the Stock and Brand Branch* (1895-1898).

150 Department of Agriculture Special File, *Appointment of Expert for Dairy Industry, 1896-1924*, N.S.W. State Archives, 6/883.

151 M.A. O'Callaghan, 'Dairy bacteriology', *Agricultural Gazette of N.S.W.* (April 1899), 291.

152 Through the Dairies Supervision Act of 1886, administered by the Board of Health.

153 Report of the Dairy Expert, in *Annual Report of the N.S.W. Dept. of Agriculture* (1901) (1902).

154 'Report of the Government Bureau of Microbiology for the year 1909' in *Joint Volumes of Papers presented to the Legislative Council and Legislative Assembly* (N.S.W.) (1910), Vol. III, p.3.

155 C.J. Cummins, op.cit. (n. 136), p.189.

156 Minutes of Linnean Society Council Meeting, February 20, 1884, *Linnean Society of N.S.W. Minute Books 1884-1886*.

157 *Linnean Society of N.S.W. Proceedings* (1886), p.907.

158 Robert Greig-Smith was Macleay Bacteriologist from 1898 to 1927. He had studied microbiology in Edinburgh, Bonn, Newcastle-on-Tyne and Copenhagen; F. Fenner, 'Trends in microbiological research in Australia: historical perspectives', *Aust. Journal of Science*, 31 (1969), 239-47.

159 J.W. Oliver, *History of American Technology* (New York, 1956); H.J. Habakkuk, *American and British Technology in the Nineteenth Century* (Cambridge, 1962); P.A. David, *Technical Choice, Innovation and Economic Growth* (Cambridge, 1975); A. Pred, 'Industrialisation, initial advantage and American metropolitan growth', *Geographical Review* 55 (1965) and I. Feller, 'The urban location of United States invention 1860-1914', *Explorations in Economic History*, 8 (1970-71).

160 See Turner, op.cit. (n. 25) and Inkster, op.cit. (n. 11) Sections 2:1 to 3:2.

6

Baron von Mueller: Protégé turned patron

A M Lucas

I have spoken of the gardens of Melbourne generally as contributing to the spacious dimensions of the town; but I must not omit to make special mention of the botanical gardens and their learned curator, Dr. Von Mueller. Dr. Von Mueller, who is also a baron, a fellow of half the learned societies in Europe, and a Commander of the Order of St Jago, has made these gardens a perfect paradise of science for those who are given to botany rather than beauty. I am told that the gardens·and the gardener, the botany and the baron, rank very high indeed in the estimation of those who have devoted themselves to the study of trees, and that Melbourne should consider herself rich in having such a man. But the gardens though spacious are not charming, and the lessons which they teach are out of the reach of ninety-nine in every hundred. The baron has sacrificed beauty to science, and the charms of flowers to the production of scarce shrubs, till the higher authorities interfered. When I was at Melbourne there had arisen a question whether there should not be some second, and alas, rival head gardener, so that the people might get some gratification for their money. The quarrel was running high when I was there. I can only hope that the flowers may carry the day against the shrubs. (Anthony Trollope)[1]

The flowers did win out. In July 1873 von Mueller was dismissed from the directorship of the Melbourne Botanic Gardens, removed from his house in the grounds, but left in his post of government botanist. In vain did he appeal to his correspondents in Europe for support. He was unable to use the influence of his patrons as he had done in the past. Von Mueller accumulated more honours, more medals and more recognition in the twenty-three years between his loss of control of the gardens and his death, and was instrumental in having memberships and honours bestowed on others. He had the attention of governors, the leaders of science in Europe, and of the settlers and amateurs in the colonies, through his publications, letters and speeches.

He was prolific in his publications, perhaps publishing more than any other scientist, with over 1500 items in his bibliography. Even allowing for the minor letters to newspapers and republished papers and translations, he produced more than the 995 items attributed to Sir George Caley by de Solla Price in *Little Science, Big Science*.[2] In addition to these formal publications he was a prolific letter writer, claiming to have been writing up to 6000 letters a year, to officials, friends, and to those whom he encouraged to collect for him.[3]

The benefits obtained and sought from correspondents, the obligations incurred by doing so, and the marks of recognition available to be bestowed and received were all part of a formal and informal system of encouragement and rewards that were operating in nineteenth-century science. Mueller provides a useful example to consider when examining this system in operation in the Australian colonies during the second half of the century. He was a protégé of some of the English botanists, and benefited from their indirect and direct patronage; he was recognized as a skilful botanist, especially by the English workers at the Royal Botanic Gardens, Kew, although there were some dissenting voices, notably over botanical nomenclature and a supposed tendency to publish names prolifically;[4] he was admitted to many of the senior scientific learned societies, and played a part in arranging the admission of other Australian members; he was active in fields other than botany, notably exploration and some zoology, so he was in correspondence with a wide range of organizations in Britain and around the world; and he had an active correspondence with many settlers, both in his official capacity and in privately encouraging botanical exploration by amateurs and school children.

Outline of Mueller's career

Born in Rostock on 30 June 1825, Ferdinand Jacob Heinrich Mueller began his formal scientific work in 1847 when he graduated PhD from Kiel University with a thesis on the embryology of *Capsella*[5] ('shepherd's purse') after having graduated as a pharmacist from Kiel earlier the same year. He did not, however, have a high opinion of Nolte, his professor, who had 'sunk into obscurity ever since he ascended (and that was before I was born) to his Professorship'[6] and claims to have been much influenced in coming to Australia by Ludwig Preiss, who had collected extensively in Western Australia from 1838 to 1842, and also by Darwin's *Voyage of the Beagle*.[7] Late in 1847, after the health of one of his two surviving sisters gave concern (his parents and elder sister had already died of consumption), he brought them both with him to Adelaide, where he worked as an assistant in a chemist's shop. He was befriended by Samuel Davenport, a connection he maintained throughout his life.[8] He bought land in the Bugle Ranges in the Adelaide Hills,[9] but he could not have devoted much time to farming as he made extensive journeys in the colony, into the Flinders Ranges and the Murray scrub. His first articles on these explorations were published in a German newspaper in Adelaide, and sent to German scientific journals.[10] It was not long, however, before he began to send articles to Britain,[11] seeing himself as a British subject, having been naturalized in South Australia in 1849.[12] In 1853 he became government botanist of Victoria, a position he was to hold until his death on 10 October 1896. He had some periods of leave of absence, notably for the North Australia Exploring Expedition led by Augustus Gregory from July 1855 to the end of 1856, and two short periods in Western Australia. After his return from the Gregory expedition he was also given charge of the Botanic Gardens, the position he lost in 1873.[13]

During his service in Victoria Mueller was active in scientific societies and other affairs, particularly those connected with the German community or exploration, and he became a member of many societies and organizations in

Victoria, other colonies, Britain and elsewhere.[14] His memberships were by no means merely nominal; he was active in the Victorian Philosophical Society, being its president on its transformation into the Royal Society of Victoria, remaining on its council until 1862 when he resigned to 'devote more time to [George] Bentham's work' on the *Flora Australiensis*.[15] He was an active member of the committee of the Victorian Branch of the Royal Geographical Society of Australasia, some committee meetings being held at his house.[16] At the time of his death he was still active, his last recorded letter to Kew being written on 26 September 1896.

Mueller wrote regularly to Kew, usually each mail when he was in Melbourne, and whenever possible when he was engaged in botanical exploration. Sometimes the letters were brief serving merely to keep the correspondence alive, or to introduce a visitor to London. In the letters we find news of his own family and that of the Hookers. There are reminiscences of Mueller's life in Europe; details of his hopes, disappointments and disasters; comments on, and disputes about, botanical matters; and comments on other botanists, in Australia and abroad.[17] Many of Mueller's early English-language publications came from his letters to Sir William Hooker, director of the gardens at Kew, which Hooker extracted for publication in his *Journal of Botany and Kew Garden Miscellany*. Hooker omitted private comments, and made minor editorial corrections, but the published versions are taken from the letters, the earliest of which do not appear to have been written with publication in mind. Although the letters were not as carefully written as his early articles prepared for publication, they are the foundation of the extensive bibliography.[18]

Mueller published widely, in the standard English journals, in the leading German, French and Russian journals, and extensively in Australia, including in his own *Fragmenta Phytographiae Australiae*, a Latin botanical work written in the classic style and published in fascicles from 1858. His habit of publishing in a multitude of journals infuriated some of the English botanists; Bentham urged him to limit the range of his publications and J.D. Hooker complained of his using 'an infinity of periodicals'[19] but he justified this practice as one designed to 'excite the interest of the different contributors to my collections, to continue to do so in future'.[20] He did recognize the difficulty of obtaining access to descriptions of plants that appeared in scattered publications, and early in his career he began to prepare indexes of his plants, extending the listing to the publication details of all Australian plants in his 1882 *Census*.[21]

Mueller saw the literary work of describing plants as additional to his official duties, and often alluded to the impossibility of being a pure botanist; for example, in 1888 he wrote to the director of Kew Gardens: 'Botany...can *here* be only subsidiary or collateral to the practical professions, and what is absolutely necessary for them puts such a science as botany quite back. The "struggle for existence" is so hard to professionalists'.[22] He claims to have spent much of his capital and an extensive proportion of his salary directly on his botanical work. In 1862, he reports, the Melbourne bookseller Bailliere had valued the botanical part of his private library at £1000, and in his final years he paid two juniors in the herbarium from his own pocket.[23] Some at least of his early spending on books and specimens appears to have been motivated by the desire to remove the argument that he was not in a position to prepare an Australian *Flora* because he had no access to the literature, for he reports in 1861 that he had 'secured for [my] private library almost all the important works which I require'.[24]

Other aspects of Mueller's work are treated only incidentally in later sections of this chapter. Details of his private life, and of his work not discussed here, can be found in the sources listed in note 13, but some of the 'facts', and especially the interpretations, in Kynaston's biography should be treated with caution. Here, we are concerned with the patronage he received, the way in which he returned obligations, and the way he promoted Australian biological exploration by the use he made of nominations for society membership and eponymization through plant names and geographic features.

Early patrons and the return of obligations

How did Mueller come to be given the post of government botanist in Victoria? He was relatively young, had no substantial body of published work to show for five years in Australia, and he had few English connections at that time.

At the time of Mueller's appointment in 1853 Sir William Jackson Hooker was director of the Royal Botanic Gardens at Kew, the centre for economic and taxonomic botany for the British colonies.[25] Hooker, from the time he held the chair of botany at Glasgow, 'encouraged botanical students, not only fitting them for a career but making sure with his influence in high quarters that he started them on the right road in a worthy post'.[26] Mueller was not one of 'Hooker's young men', but it is generally assumed that Hooker was influential in getting him his post as government botanist of Victoria. His biographers make this claim, Roach describing Hooker as 'bringing influence to bear'; Daley has him 'accrediting' Mueller and Kynaston describes him as giving approval.[27] None cites the evidence, and no evidence has been found in the Kew Archives correspondence between the Victorian Governor, Charles La Trobe, and Hooker, nor in La Trobe's dispatches preserved in the Colonial Office files in the English Public Record Office. We know, however, from a letter from La Trobe to R.C. Gunn, the Tasmanian botanist and correspondent of J.D. Hooker, that Mueller was in Melbourne and impressed La Trobe as the best botanist he had met in the colony.[28]

Despite the difficulties Mueller felt himself to be labouring under after his appointment, with minimal diagnostic resources, a lack of 'authentic' specimens with which to compare his own, and the absence of a powerful lens,[29] his early work satisfied La Trobe, who reported to the Colonial Secretary that Mueller's 'ability . . . is beyond all question'.[30]

Whether or not he was involved with Mueller's appointment, there is no doubt that it was W.J. Hooker who encouraged Mueller, publishing his letters as journal articles and praising him in print, both 'services' to Mueller that Mueller saw as necessary to support his precarious position in the Colonial Service after La Trobe had left. The supportive comments Hooker made would, Mueller hoped, encourage the 'Government to offer me the support which I since enjoyed'.[31] Hooker also appears to have supported Mueller for the position of botanist on the Gregory expedition, a position previously offered to the prolific collector and Hooker correspondent James Drummond of Western Australia, who declined the post because of his age.[32] It was to Hooker that Mueller turned for advice about the conditions of the post: he sensibly inquired about his rights to the descriptions of the plants and was assured that the relevant officials had been told that Mueller should receive a full set of the plants collected.[33]

Throughout his correspondence with W.J. Hooker, from 1853 until Hooker's death in 1865, Mueller treated him as a father figure, commenting in his letter of condolence to Joseph Hooker that '... I above most others, shall feel the loss keenly, because an almost paternal kindness and an affectionate solicitude characterized Sir William's numerous letters and all of his actions to me'.[34]

There is no hint in the correspondence that Mueller saw his relationship changing with time: W.J. Hooker was always the master, at whose feet Mueller almost worshipped.

Mueller showed his recognition of W.J. Hooker's help in a number of ways, for example by dedicating the Victorian Flora (*The plants indigenous to the Colony of Victoria,* [1862]) to him as 'a token of esteem and affection', and by arranging for Augustus Gregory to name geographic features after Hooker,[35] but he named very few plants for him, not least because, as author of so many names, 'there is no square mile of the globe where the vegetation does not speak of him'.[36]

Mueller's assistance for collectors such as W.H. Harvey is, at one level, an indirect return for his support by Hooker. Harvey, then curator of the herbarium of Trinity College, Dublin, visited Australia and met Mueller in Melbourne; Mueller presented him with algae collected at Sealers Cove and at Port Phillip Heads.[37] Mueller would help Hooker's friends whenever he could; for example he collected information for Darwin at the instigation of the Hookers.[38] He also employed people at Hooker's request, for example taking on a former Kew gardener in 1859.[39] But such favours are only a 'return' to Hooker at a superficial level; people did not have to be well connected to be helped by Mueller; many collectors were helped financially and in many other ways. Some, such as A.F. Oldfield, stayed in Mueller's house.[40]

Mueller's attitude to Joseph Hooker was always one of respect for his botanical work, and even after he succeeded his father as director of Kew Joseph was treated as an equal, not as a mentor. Joseph Hooker was only eight years older than Mueller, and shared the same birthdate, a coincidence that delighted Mueller.[41] Mueller wrote to Joseph Hooker as a friend and colleague, not as disciple to master. He kept track of his son, Brian, when he came to Australia, introducing him to people of influence who might find him a job, and passing on messages through others, such as John Forrest of Western Australia, with whom he was in correspondence.[42] Mueller praised Joseph's work of which he approved, for example the Tasmanian *Flora* and the *Genera Plantarum*, and criticized that which he did not. He was concerned about some of Hooker's nomenclature and identifications, and was particularly critical of 'omissions' of references to Mueller's own publications in the *Genera*

Plantarum.[43] Early in his career Mueller stated that he 'did not like to anticipate any work of [Joseph's] Tasmanian *Flora*', although this did not prevent him publishing Tasmanian plants, much to Joseph's chagrin.[44]

Of a friend it was possible to ask favours, ones that the friend could arrange or help to promote. It is symptomatic of the difference in the relationship between Mueller and the two Hookers that except for help in revising Mueller's early manuscripts, Sir William was rarely asked for direct help. Mueller did ask him for advice, for example on whether to join the Linnean Society, and for favours for others (which would often indirectly benefit him). He asked Sir William to act as a go-between in seeking permission to name *Josephinia eugeniae*, and requested him to publish an illustration of *Duttonia* to honour 'my friend, the South Australian Senator'.[45] But for help in making major professional advances he turned to the younger Hooker. It was to Joseph that Mueller wrote seeking nomination to the Royal Society of London, although he also asked that Hooker not 'campaign' on his behalf, at the same time pointing out that Mueller's election as FRS would also honour the newly formed Royal Society of Victoria.[46] Although not directly asked, W.J. Hooker did support Mueller's candidature for FRS, being the first signatory of those supporting Mueller 'from personal knowledge'. The others who signed in this category were J.D. Hooker, W.H. Harvey, Rob. Wight and Roderick Murchison. Seven others, including Charles Darwin, signed 'from general knowledge'.[47] Of these only Harvey appears to have met Mueller personally.

George Bentham, in contrast to Joseph Hooker, was both venerated and reviled; recognized as the 'foremost among plain phytographers of this age',[48] or rejected as having wrong-headed views, especially on the Darwinian evolution of plants and on the limits of genera and species. Despite Bentham's rebuttals, Mueller maintained that field experience was important in judging species limits, and that Bentham, being restricted to herbarium specimens, could not recognize the variability induced by environmental factors.[49] Yet Bentham's views were respected, and appealed to when Mueller wanted to be sure that one of his genera would stand, especially when the genus was intended to honour someone politically important to Mueller; he asked Bentham to act as 'chief judge' of the generic validity of *Barklya* and *Denisonia*, genera named for Colonial governors.[50]

The issue of the authorshop of *Flora Australiensis* colours the whole relationship between Mueller and Bentham. The account by Daley[51] should be consulted for the letters between Mueller, the Hookers and Bentham that led up to the decision to place the responsibility for the authorship with Bentham, with the title page acknowledging the 'assistance' of Mueller. But Daley's account does not explore the role of Henry Barkly, the governor of Victoria. W.J. Hooker wrote to Barkly a number of times on this issue, and Barkly reports many discussions with Mueller over the authorship. Barkly commented that although Mueller expressed willingness to

> cooperate with Mr Bentham in the Editorship, . . . it would be very painful to his feelings to resign the greater part of the honor & glory to be gained from such a publication to another, as must inevitably be the case. It is only human nature that it should be so. He has dedicated his existence to this particular task. He has

spent fourteen years in Australia, travelled 26 000 miles over it, collected six or seven thousand specimens of Plants, which he is working night & day to describe and determine, and he dreads his peculiar province being invaded, his views set aside, and himself reduced to a subordinate position — the Knoth of some mighty Humboldt.[52]

Barkly's judgement was correct. Mueller returned to these themes at times of depression, especially after 1873 when he lost *his* garden, the garden that was one of the reasons why Mueller did not accept the invitation of the Hookers to go to Kew to write up his Australian material on his return from the Gregory expedition.[53]

Despite the cloud between them, Bentham was considered a great botanist by Mueller, and it was he who proposed that the Linnean Society create a Bentham medal to honour his memory. Mueller was also responsible for Bentham receiving the CMG.[54] Bentham appears not to have directly honoured Mueller, or arranged for him to be honoured, yet his presence as simultaneous collaborator over the *Flora Australiensis* clearly spurred Mueller to systematic work on his specimens. The need to send off complete groups of plants to Kew provided Mueller with a pattern of descriptive activity. He feared sending off material undescribed, ostensibly in case it was lost at sea, for he had had losses before,[55] but perhaps so that Bentham had to acknowledge his authorship. When Bentham criticized his species he responded vigorously in his feelings of injustice; when Bentham praised his skills Mueller responded equally strongly, thanking Bentham profusely.[56]

Other botanists in Britain and elsewhere were important to Mueller as he arranged for groups of plants such as mosses, ferns, and fungi to be described, sometimes publishing these descriptions in Australia.[57] But these botanists were not active patrons, even though Mueller valued their influence. He saw the American Asa Gray, for example, as a potential supporter, and was disappointed when he did not get his help over the loss of the garden directorship, and feared that some of Gray's published comments, to the effect that Mueller was better off out of his garden, would aid his enemies.[58]

To Mueller, one of the major disappointments in his English correspondents was their failure to help him retain, or be reinstated in, the Botanic Gardens at Melbourne. He appealed to them all, asking Darwin, for example, to intercede with his neighbour Edward Wilson, one of the proprietors of the Melbourne *Argus,* which was campaigning against him. He complains that only M.T. Masters, editor of the horticultural journal, *The Gardeners Chronicle*, supported him. 'Sir William Hooker would have' written a letter of support, he complained to Joseph.[59] Particular distress was caused by the fact that those who supported Joseph Hooker in his fight to retain control of Kew Gardens did not lend their weight to help Mueller in his similar battle. Mueller was politically naive to think that botanists at 'home' would, or could, influence a fiercely independent colonial legislature, but he continued to resent the apparent desertion, alluding to it years after the event.[60]

Of the English botanists, William Hooker was clearly the influence that enabled Mueller to become established, arranging the publication of his papers that established his work. Bentham was a stimulus, but hardly a patron, while Joseph Hooker did much to gain him public honours, both scientific (FRS) and secular (KCMG in 1879) visibility.[61] In return, Mueller used whatever means were in his power to recognize their support in public ways: by dedications of his work; by publishing biographical notes;[62] by conferring their names on geographical features; and by arranging memberships or instigating memorials. Few more direct honours could be given to those who had so many already. J.D. Thissleton-Dyer, Joseph Hooker's son-in-law and successor as director at Kew, was an exception. Perhaps because he had comparatively few honours Mueller tried to get him an award from Halle, through his correspondent Alfred Kirchhoff, the professor of geography. This attempt, which does not appear to have been successful, was explicitly to 'show you by some continental honour my gratitude' for moving the motion to award Mueller the Royal Medal of the Royal Society in 1888.[63]

His lay patrons were a different matter. Here he could honour actual or potential supporters by naming a plant or a geographic feature for them. He always tried to so honour the governors of Victoria and some of the other Australian colonies. Perhaps he misunderstood their political role in an increasingly independent and autonomous representative democracy, but Mueller seemed to think that they could and would protect and help him.[64] Certainly Sir Henry Barkly promoted his cause, praising him publicly and defending him against 'those who think he ought ere this to have published a *popular* handbook of Australian Botany'.[65] But Barkly was particularly interested in science, and had a wide circle of scientific friends for whom he collected, or arranged to have collected, many natural history specimens.[66] Later Victorian governors were not so strongly inclined to science.

Mueller was accused of buying honours by honouring, or sending specimens to, those who were in a position to award distinctions: 'He thinks no trouble too great to be undergone in order to obtain the order of St Jingo, or the cross of the Spotted Donkey, from the Prince of Pumpernickle, or the King of the Cannibal Isles'.[67] There may be some truth in this charge, for he certainly wished to pay homage, to Queen Victoria, for example, and to Empress Eugénie. Did he expect to get rewarded for so doing? It is impossible to be sure. Perhaps when the King of Württemberg was honouring various men as representatives of a variety of fields he would think of Mueller as the representative for science if Mueller had first helped his son obtain specimens during Prince Paul's incognito visit to Melbourne in 1858.[68] But was the receipt of a barony the motive? It is not surprising that a person of his background was in active communication with centres of learning in Europe, especially as he felt himself 'scientifically exiled' in Australia, where he 'never met any Botanist [but Harvey] for more than a few hours'.[69] Even if he was a 'minnow among tritons',[70] he was for many years effectively alone as a taxonomist working in Australia; other botanic garden directors had typically sent their collections to Kew for identification or description there, and Walter Hill of Brisbane wished to continue to do so rather than supply Mueller.[71]

But Mueller certainly overtly and actively bought financial and other support for his projects from members of the public by rewarding them for their interest. It is this use of his power to offer rewards that is considered in the next two sections of this chapter.

'Immortalization' as a reward and incentive

As well as naming plants to honour European royalty and the 'great' names of science, Mueller also used the opportunity to confer nominal immortality on those who, by collecting for him, helped to advance botany. He was quite open in his use of this tool. For example, in writing to tell one of his correspondents, Mary Bate of Tilba Tilba, that the new material she had sent enabled him to complete the description of the new species named for her, he added that 'I hope this scientific acknowledgement will encourage you to continue your researches, as doubtless a whole host of rare plants and some new ones remain there yet to be discovered'.[72]

In addition, he used this power to stimulate promising young botanists whom he sought to encourage. An instructive example is the honour given to William Guilfoyle by Mueller in April 1873 when he created the genus *Guilfoylia* and dedicated it as follows:

> Dedicatio honorabit amicum Guilielmum Guilfoyle, hortulanum, qui a praedio suo ad flumen Tweed regiones circumjacentes botanices causa jam ample perlustravit et ipse fructum hujus arboris detexit et mecum communicavit.[73]

Elsewhere he described Guilfoyle as 'distinguished as a collector [who] evidenced great ardour',[74] and the distinction of naming a genus after him was evidence of his hopes for the young collector.

But when Guilfoyle was appointed to take Mueller's place in the Botanic Gardens in July 1873 Mueller's opinion rapidly changed. From being 'distinguished' he became a 'nurseryman [with] no claims to scientific knowledge whatever',[75] who got the job not on ability but because he was the 'nonscientific cousin of the wife of the Minister' responsible.[76] Instead of being a promising botanist, he was now an ignorant man, who 'unscrupulously praise[d] himself daily up on [Mueller's] expense'.[77]

However, what Mueller gave he could also take away: in his *Census* of 1882 he suppressed *Guilfoylia* as part of the genus *Cadellia*, without noting it as a synonym, despite his introductory comment that he had quoted abolished genera 'such as found their way actually into the literature of Australian plants'.[78] The reduction was not botanically inconsistent for, to Mueller, a genus, unlike a species, was an artificial entity, and the limits of genera were a matter of botanical judgement. By inclination Mueller was a 'lumper', tending to reduce the number of genera because of the ease of working with them in the field, so the reduction of genera was not unusual.[79]

There is at least one other occasion when he comments that extraneous factors should result in the removal of the honour given by a name: he wrote to Dyer that the 'name peroffskiania ... ought to be banished after the impiratricidal horrors of St Petersburgh'.[80]

Mueller honoured all sorts of people by naming plants after them. He asked Bentham to honour his one-time fiancée Euphemia Henderson, but more frequently he himself conferred the honour. A listing of those so honoured is not appropriate here, but it contains the now-famous, like [Sir] John Forrest, politicians and public servants of the day such as the Minister for Mines William McLellan and the Secretary of the Mines Department Major Couchman, as well as those whose names are now otherwise forgotten, like

Mueller's brother-in-law E.F.D. Wehl and enthusiastic collectors such as Jessie Hussey and Mary Bate.[81]

In addition to naming plants, Mueller used his geographical interests as an opportunity for immortalization. We have already seen that he recognized Bentham and Hooker in this way. He bestowed the names of many whom he wished to honour on places discovered by W. Ernest P. Giles in his first exploration of Central Australia. Mt Olga and Lake Amadeus were named in recognition of 'the royal patrons of science'; the Krichauff Range after his partner in land ownership in South Australia; the Petermann range after the German geographer whom Mueller kept informed about Australian exploration and whose maps Mueller supplied to expeditions.[82] But the most overt use of this power was to raise money for exploration. As well as himself financially supporting Giles, he also obtained money by promising to name geographic features after supporters. He actively solicited funds by promising that 'each subscriber shall be honoured by his name being given to some new geographic locality'. Features named for these donors are singled out by an asterisk in Giles' account of his journey.[83]

Perhaps because of the implicit contract Mueller was concerned that names stayed associated with the features mapped. In 1876 he wrote a short article intended for publication by the Royal Geographical Society in London. In this article he argued that just because minor errors of surveying occurred, later explorers should not rename mountains and other features. To do so was unjust not only to the original explorers, but also to those honoured by having their names bestowed. He wished to see rules of geographical nomenclature established, giving an absolute priority to the first author, as in botany and zoology. The article was not published; it was 'too utopian' commented Francis Galton, who refereed it.[84]

The geographical names manuscript is entirely consistent with Mueller's stand over botanical nomenclature. His concern for priority is a constant theme in his letters to Kew, and he published articles on the issue, for example on the name *Vahea*, which he maintained was valid because Lamarck had illustrated the plant under this name in 1792, although a description, as *Landolphia*, did not appear until 1806. Similarly, he continued to use *Candollea* for the plants that others called *Stylidium*.[85] He went so far as to say that he was not bound by the decisions of a botanical congress that he did not attend, and he resigned his membership in the Le Mans Academy, apparently as an expression of his rejection of de Candolle's views. In his letters to Kew and in his nomenclature publications the case for justice is made, both to the author of the original name, and, although rarely, to the person honoured. He believed in an *absolute* rule of priority.[86]

Although in arguing this principle Mueller was probably more concerned for his own claims as author of names, it is also true that if the rule did not apply he could not guarantee immortality of their names to his collectors and supporters. If it turned out that he had used a name that was already occupied, or for a form rather than a true species, then he could find another plant to fulfil his wish;[87] but if one of his otherwise valid names could be suppressed by a botanical congress, perhaps after his death, then the honour would be lost.

Names, of genera, species, rivers, mountains and other geographic features, were a currency in which collectors, donors, friends and potential colleagues could be rewarded and his own claims to fame advanced. Mueller was vigorous in his defence against possible debasement of that currency.

Recruiting agent or scientific arbiter? The use of society membership

One of the benefits of membership in a select academy is the right it gives to suggest others to whom the honour should also be given. Mueller belonged to many societies, of various degrees of esteem, and it is instructive to look at how he used the nominating power. Three London-based societies can be compared: the Royal Geographical Society, the Linnean Society, and the Royal Society. Mueller was a member of all three.

He was elected to the Royal Geographical Society on 21 January 1858, on the nomination, from 'personal knowledge', of Count Paul Strzelecki, who had refereed his early letters for publication in the Royal Geographic Society's Journal, and Thomas Baines, artist on the Gregory exploring expedition. Their nomination was supported by the signatures of Roderick Murchison and Francis Galton. Mueller had apparently not sought nomination, for he wrote to the secretary expressing pleasure at this 'unexpected distinction', being particularly pleased at the support of Strzelecki and Murchison.[88]

It is not clear who first suggested that Mueller become a member of the Linnean Society, but it was probably one of the Hookers. In one of his letters to W.J. Hooker, Mueller mentions that the secretary of the Linnean wanted to see him as a member, but later in the same letter he thanks both W.J. and J.D. Hooker for their support. Mueller appeared to be in two minds about accepting membership: he couldn't really afford to pay the life membership fee — 'unless this will ensure publication' — and as he couldn't attend meetings or use the museum and library it might appear a 'mere vanity' to belong. But if W.J. Hooker thought it sensible he would accept, and he asked Hooker to pay the entrance fee and the first year's subscription out of the funds Hooker held as his agent.[89] In the event, Mueller was elected on 20 January 1859, on the nomination of both Hookers, W.H. Harvey, J.J. Bennett, Thomas Bell, George Bentham, Robert Heward, John Hogg and Charles Darwin.[90]

Fellowship of the Royal Society was, as we saw above, a distinction that Mueller sought. The letters to the Hookers do not contain signs of feeling rejected or disappointed at failing on the first ballot; it was not unusual to miss out first time around, and presumably Mueller knew this.

In nomination for all three societies Mueller was supported by acknowledged authorities. Bentham, both Hookers, and Darwin all had reputations that survive to this day; they were not mere members. Even in the Royal Geographical Society, which appears to have had the broadest criteria for membership, Mueller was supported by leaders of the society, who were active in its affairs.

How then did Mueller use his power to nominate new members? Did he discriminate carefully between candidates whom he could have nominated? Full evidence is difficult to obtain, but an examination of the certificates of members of the Royal Geographical, the Linnean, and the Royal Societies, as well as the list of nominations for the Royal Society (which includes the details of both successful and unsuccessful candidates) adds to the picture that emerges from Mueller's surviving letters.

In the period from Mueller's nomination as FLS in 1859 to his death in 1896 at least 150 residents of Australia were nominated as members. Mueller was one of the nominators of 126 of them (84 per cent), and only 15 of the others had other Australian nominators without Mueller being involved. All nominators were outside Australia for 5 nominees and data are missing for 4 candidates. It is more difficult to obtain equivalent data for the Royal Geographical Society because of the way that certificates are filed, but there were at least 110 members resident in Australia in 1897, 33 (30 per cent) of whom had been nominated by Mueller. From his own election until his death, Mueller supported 10 of the 20 Royal Society candidates resident in Australia at the time of their candidature for a Fellowship, including 4 of the 9 successful candidates.

A number of interesting points emerge from an examination of these data.

The discrepancy in the proportions of members who had Mueller as one of their nominees between the RGS and the Linnean is quite large, but it is not surprising. Membership in the Linnean suggests an active, and at least partly skilled, interest in botany or zoology; that in the RGS an interest in geographical exploration, which did not require the same expertise to understand papers in the journal. Young men in, or going to, the colonies appear quite frequently to have been nominated from 'home'. Those nominated for the Linnean entirely from without Australia appear to have achieved this status by at least serious collecting for the professionals in Britain, if not by personal scientific work.

What is more surprising is the virtual monopoly Mueller had on nominating Australian members of the Linnean Society for many years. It was not until the group in the Australian Museum in Sydney became active in nominating their colleagues and collectors that a significant number of members was nominated without Mueller's involvement. Members he nominated did not themselves quickly begin to nominate. Perhaps this is partly because many of them were not fully engaged in science. For most, it was a hobby interest, even if a serious one, especially for medical practitioners such as P.H. MacGillivray, an active invertebrate zoologist, and priests such as J.E. Tenison Woods, who as well as collecting for Mueller was active in geology and a writer on geographic exploration. They would not be as likely to have people sending them specimens and data; they would not be as visibly concerned with science as the professionals such as Mueller and W.A. Haswell, the marine zoologist of the Australian Museum in Sydney.

Mueller was a clearly established official man of science, well, even notoriously, known to be a member of scientific societies, and he was the obvious person to ask if one wanted to add the initials FLS to one's name. And approach him some did. Rev. John Bufton explicitly asks to be 'rewarded' for collecting at Port Davey in Tasmania. Mueller annotated the letter 'Ans. 27.12.92 Proposed Bufton 4.1.93, French and Leuhmann seconded'.[91]

Not all Mueller's nominees to the Linnean had approached him. Some who had been successfully elected never actually joined, because of a failure to pay the necessary fees. One possible reason for this is that not all of them had been consulted by Mueller before he sent in the nomination. Certainly Leonard Rodway appears to fall into this category. He declined, as he did not 'esteem fellowship a sufficient distinction'.[92]

Mueller appears, therefore, sometimes to have proposed people as members specifically because he thought their abilities deserved recognition by an established scientific society, even if at other times he acceded to requests for 'rewards'. (We don't know whether he refused any such requests, and if so, on what grounds, so it is possible that even in the cases where he was approached he exercised some judgement.) But, whether the initiative was Mueller's or the nominee's, membership was a reward, a method of encouragement and a bestowal of approval.

The Royal Society was treated similarly. He appears to have suggested nomination on his initiative. He points out that Tenison Woods did not request nomination, and adds, perhaps significantly, that as Woods was 'no botanist' his proposal had no advantage to Mueller.[93] Frederick McCoy, professor of natural science at Melbourne, was also reported to have been nominated on Mueller's own initiative, although there is some evidence that McCoy at least suggested that Mueller campaign on his behalf.[94] Mueller appears to have been a little uncertain of the criteria to be employed in making nominations to the Royal Society. He was very cautious in his advocacy of James Ruddall, his friend and collaborator in studies of the reproductive biology of the echidna. He explicitly asked J.D. Hooker, who by then had been president of the Royal Society, not to allow Ruddall to be put forward unless he was sure of the support of council.[95] Ruddall does not appear in the Royal Society lists of nominees.

The power to confer distinctions, was, we have seen, one that was recognized by recipients and used by Mueller. The effectiveness of these rewards cannot be explored fully here. Were such distinctions useful in developing the careers of the recipients, by putting them in line for career positions and other distinctions perhaps? Did Bufton see the distinction of FLS as stepping stone to the FRS, a distinction he later sought from J.D. Hooker?[96] These questions need investigation by careful study of the careers of those so honoured, but in the case of many of the recipients of Mueller's benefice the distinctions would have little more than symbolic value to the recipients in their social relationships. Only a few were professional biologists. Of these, most were botanists, or had serious botanical interests. The most notable were Charles Moore and J.H. Maiden, directors of the Botanical Gardens in Sydney.[97] But both these men already had their positions before Mueller initiated the election, and in this sense it was not an act of patronage helping their careers. The majority of his nominees were not professional scientists, although many published scientific work. Examples include the Sydney railway engineer Henry Deane (elected June 1855) who published some botanical papers; the Hobart city engineer R.A. Bastow (December 1885) who had a well developed interest in mosses; and the Victorian Registrar General W.H. Archer (November 1874) who was encouraged by Mueller to use his skills with the microscope to see whether spores could be used as a diagnostic feature for ferns.[98]

It is tempting to view FLS membership nominations as a method of encouraging serious collectors to send their specimens to Mueller, a payment available to be used in addition to the honour conveyed by naming plants and the more mundane offer of useful seeds and small cash payments in exchange for specimens.[99] But this view of Mueller's nominations is too simplistic. He was acting with some discrimination, and many of the people whom he nominated FLS were not mainly botanical collectors. People such as MacGillivray were publishing zoological descriptions in their own right, others like J. Bracebridge Wilson of Geelong Grammar School had a serious interest in an animal group and only incidentally collected botanical specimens for Mueller, and some such as C.A. Topp, the secretary of both the Science Society of Ballarat and the Ballarat Branch of the Field Naturalists Club of Victoria were actively interested in natural history, without necessarily being interested in a special group of organisms. It appears, from a comparison of the people who had plants they had collected named after them and those who were put up for memberships, that Mueller reserved the membership offer for people who were more than collectors. He seems to have suggested only those with a serious interest and a developed knowledge in natural history, whereas he did not expect his botanical correspondents to have this degree of knowledge. He was, it seems, not acting as an undiscriminating recruiting agent for the Linnean Society.

In the case of the Royal Society he was more circumspect, and while many of his nominees were unsuccessful, he did not support people who had few claims to the distinction, although his letter of regret about Tenison Woods' failure suggests that he was as much concerned with length and breadth of scientific activity as with its quality: '[I] can't see why 30 years labours in various branches of science with extensive fieldwork count for nothing'.[100]

It is more difficult to judge the motivation for his RGS nominations. Some, such as Alexander Forrest, the brother and partner of John Forrest, were explorers; others, for example the Rev. W. Potter, who was secretary of its Antarctic Exploring Committee, were active members of the Australasian geographical societies, but the geographical interests of the majority of those Mueller nominated are not known. His effective patronage of exploration was much more direct. He was an active member of the exploration committee of the Royal Society of Victoria that organized the ill-fated Burke and Wills expedition; he was the main force behind the equally ill-fated Ladies Leichhardt Search Expedition, Duncan McIntyre having his leadership contract with Mueller; and Thomas Elder asked him to be heavily involved in the planning of the Elder expedition of the early 1890s.[101] In addition to these, we have seen that he was a direct patron of Giles' first and second expeditions. Direct interventions, his membership of the Committee of the Victorian Branch of the Royal Geographical Society of Australasia, and his presidency of the Geography Section of the Australasian Association for the Advancement of Science were more effective methods of promoting geography than nominations to the RGS.

*   *   *

Mueller and the reward systems of science

Most accounts of the reward structure in scientific communities concentrate on the system *within* science,[102] and ignore the very extensive relationships that were built up by the scientific naturalists during the period of biological exploration in the eighteenth and nineteenth centuries. Mueller, who was locked into the scientific part of the system, with his memberships and awards from the Royal Society down, also demonstrates the important role that a subsidiary system could play. His encouragement, of serious amateurs by recommendation of Linnean Society memberships, of collectors by naming plants after them and by ensuring that every extension of geographic range was noted under the name of the collector,[103] and of geographic exploration by appealing to the vanity of sponsors to whom he promised eponymization, is a reminder that when we examine the use of rewards in the advancement of scientific practice we also need to include those who receive rewards, but who are not themselves scientists.

This chapter is not a complete study of the sub-scientific layer of the nineteenth century reward system, even that part of it operated by Mueller. To look completely at Mueller's role would require a much more detailed and systematic analysis of his publications, especially the use of personal names for plants,[104] and of the herbarium sheets, with which are stored many of the letters from his correspondents. In addition, the relationship between the use of memberships in Australian scientific societies and those of Europe has not been examined here; some at least of those whom Mueller nominated for memberships in England were also active members of local societies and this activity formed part of the case for nomination as FLS (see C.A. Topp, above). Did Mueller and other professional scientists make distinctions between these in terms of offering memberships according to the prestige of the society?

But even the level of analysis used here, restricted to only three societies and the major groups of available correspondence in English, shows that the development of botany, geographic exploration (and, to a lesser extent zoology) was influenced by the assiduous cultivation of the common people, from the mounted policemen and bushmen who acted as Mueller's' guides in Western Australia, to settlers he met in his travels[105] as well as those of money and influence whom he has been accused of hounding for favours.[106]

Most of Mueller's protégés were not destined to become as well known in science as he himself became; he did not establish a school of followers; and he was not in a position to have students he could send into positions of influence in the way that Sir William Hooker did from his department at Glasgow. In studies of the formal scientific networks Mueller would be a receiver of patronage, not a source. Yet, as we have seen, he had his own network, feeding him specimens and geographic distribution data. His network was managed by a graduated system of rewards, with those who made more informed contributions being entered into the formal scientific recognition system. In the development of botanical exploration Mueller's network, and its associated reward system, may in the end have proved to be more important than the formal recognitions he himself received from the scientific and political leaders of the world.[107]

Notes

The following abbreviations for archives are used in these notes: BMO British Museum (Natural History), Library, Owen Correspondence. BMB British Museum (Natural History), Botany Library, Berkeley Correspondence. KM1 Royal Botanic Gardens, Kew: Library and Archives; Kew Correspondence: Australia: Mueller 1858-1870. KM2 loc.cit., 1871-1881. KM3 loc.cit., 1882-1890. KM4 loc.cit., 1891-1896. KAL Royal Botanic Gardens, Kew; Library and Archives; Directors' Correspondence; LXXIV, Australian Letters, 1851-58. KDC loc.cit., LXXV, Australian and Pacific letters, 1859-1865. KJH Royal Botanic Gardens, Kew: Library and Archives, Letters to J.D. Hooker, Vol.16. LIN Linnean Society of London, Archives. LINC loc.cit., Certificates of Fellows, Foreign Members and Associates. MEL National Herbarium of Victoria, Library, Mueller manuscripts. RGS Royal Geographical Society Archives. Folio numbers refer to the first folio of the source.

1   Anthony Trollope, *Australia and New Zealand* (London, 1873), pp.392-3.
2   See D.J. Carr and S.G.M. Carr, eds, *People and Plants in Australia* (Sydney, 1981), p.135 for a discussion of the claim that Mueller was the most prolific publishing scientist. See D.M. Churchill, T.B. Muir and D.M. Sinkora, 'The published works of Ferdinand J H Mueller (1825-1896)', *Muelleria*, 4 (1978), 1-120, with a supplement by the same authors in *Muelleria*, 5 (1982), 229-48.
3   See, for example, Mueller to Dyer, 30 April 1896, KM4, f.83; Mueller to Holmes, 20 June 1884, LIN, E.M. Holmes collection.
4   See, for example, the anonymous obituary in *The Gardeners Chronicle* (17 October, 1896), pp.464-6, but note that J.D. Hooker, the former Director of Kew Gardens, explicitly rebuts these comments in introducing the obituary notice by W.B. Spencer in the *Journal of Botany,* 35 (1897), 272-8.
5   Churchill et al., *Muelleria,* op.cit. (n.2), p.8.
6   Mueller to J.D. Hooker, 28 July 1866, KM1 f.223.
7   Mueller, letter to the Editor, The *West Australian* (8 May 1891), p.4d. (Preiss), and Mueller to Owen, 24 August 1861, BMO, Vol. 19, f.359, (Darwin).
8   See letters from Mueller to Davenport, South Australian Archives, PRG40/20.
9   N. Gemmell, 'Some notes on Ferdinand von Mueller and the early settlement of Bugle Ranges', *The South Australian Naturalist*, 49(4) (1975), 51-4.
10 See *Muelleria,* op. cit. (n.2), p.73.
11 His manuscript paper on 'The vegetation of the district surrounding Lake Torrens' was read at the Linnean Society on 21 December 1852. LIN.
12 *South Australian Register* (14 August 1849). He was also naturalized in Victoria, and applied for naturalization in Western Australia (Battye Library, Archives, CSO, 1877, 852, Mueller to the Governor of WA, 20 November 1877).
13 For details see C. Daley, *Baron Sir Ferdinand von Mueller,* 'reprinted from the Victorian Historical Magazine, Vol. X, Nos. 1, 2. May and December, 1924', Government Printer, Melbourne (n.d., ?1925); M. Willis, *By their Fruits: A Life of Ferdinand von Mueller, Botanist and Explorer* (Sydney, 1949); E. Kynaston, *A Man on Edge: a Life of Baron Sir Ferdinand von Mueller* (Ringwood, 1981).
14 For a list of memberships see a printed four-page sheet, 'Liber Baro Ferdinandus de Mueller'; copy enclosed with Mueller to Lord Kelvin, 23 April 1893, Royal Society, Miscellaneous Correspondence, Vol. 16, f.31.
15 Mueller to W.J. Hooker, 24 April 1862, KDC, f.149.
16 Mueller to Keltie, 3 December 1895, RGS.
17 For example, the death of Mueller's brother-in-law, Mueller to J.D. Hooker, 24 February, 1876, KM2, f.173; Brian Hooker, Mueller to J.D. Hooker, 20 October 1885, KM3, f.159; reminiscences about student days, Mueller to J.D. Hooker, 16 August 1879, KM2, f.241; dismissal from the Botanic garden, e.g. Mueller to Baker, 15 August 1881, unbound ms filed with KM4; hopes for preparing a definitive Australian flora, see letters published in C. Daley, 'The history of *Flora Australiensis',* published in parts in *Victorian Naturalist,* from 44(3) (1927), 63-9 to 44(10) (1928), 271-8.
18 The article in *Hooker's Journal of Botany and Kew Garden Miscellany,* 7, 1855, 233-42 is based on Mueller to W.J. Hooker, 16 December 1854, KDC, f.147.
19 See Bentham to Mueller, 12 February, 1863; (excerpt in Daley, op. cit. (n.17), p.130) and J.D. Hooker to Mueller (undated) (in Daley, p.273).
20 Mueller to W.J. Hooker, 11 October 1857, KAL, f.170.
21 Indexes to the *Fragmenta* and to the plants described in the *Proceedings of the Philosophical Institute of Victoria* were being prepared by 1859, Mueller to W.J. Hooker, 15 June 1859, KDC, f.114. For the origins of the *Census of Genera of Plants Hitherto Known as Indigenous to Australia* (Sydney 1882), see Mueller to J.D. Hooker, 7 October 1882, KM3, f.32.
22 Mueller to Dyer, 20 October 1888, KM3, f.259.
23 For book valuation, Mueller to W.J. Hooker, 26 August 1862, KDC, f.155; for employment of juniors, Mueller to Dyer, 8 October, 1895, KM4, f.79.
24 Mueller to W.J. Hooker, 24 March 1861, KAL, f.130.

25 See L.H. Brockway, *Science and Colonial Expansion: the Role of the British Royal Botanic Gardens* (New York, 1979).

26 M. Allen, *The Hookers of Kew* (London, 1967), p.244.

27 Daley, op. cit. (n.13), p.8; Kynaston, op. cit. (n.13), p.80; B.S. Roach, 'Ferdinand von Mueller', *South Australian Naturalist*, 2 (1922), 78.

28 See extract from La Trobe to Gunn, 8 October 1852, printed in Kynaston, op. cit (n.13), pp.79-80.

29 Mueller to W.J. Hooker, 18 October 1853, KAL, f.137.

30 La Trobe to Duke of Newcastle, 24 November 1853, Royal Botanic Gardens, Kew; Library and Archives; Miscellaneous Reports: Melbourne, Mueller 1853-96, f.23. (Extract of dispatch.)

31 Mueller to W.J. Hooker, 9 May 1855, KAL, f.151.

32 Daley, op. cit. (n.17), p.66, for support for Mueller; *Inquirer and Commercial News* (Perth, WA, 17 and 31 January 1855), for Drummond.

33 W.J. Hooker to Mueller, 4 January 1856, printed in Daley, op. cit. (n.17), p.67.

34 Mueller to J.D. Hooker, 24 October 1865, KM1, f.176.

35 For Victorian *Flora*, Mueller to W.J. Hooker, 20 February 1862, KDC, f.145; for Hooker Creek, Mueller to W.J. Hooker, 6 April 1857, KAL, f.159.

36 Mueller to J.D. Hooker, 25 January 1866, KM1, f.193.

37 See, for example, Mueller to W.J. Hooker, 14 July 1854, KAL, f.141.See also S.C. Ducker, 'Australian phycology: the German influence', in Carr and Carr, op.cit. (n.2), pp.116-38, at p.124 for extract of letter, Harvey to his sister-in-law Hannah Todhunter, 14 September 1854.

38 See Mueller to W.J. Hooker, 15 April 1858, KAL, f.177, and Mueller to J.D. Hooker, 15 October 1858 for references to early contacts. See Darwin to Mueller, 8 December 1858 (Cambridge University Library, DAR.181) for a request to Mueller.

39 Mueller to W.J. Hooker, 9 January 1859, KAL, f.174.

40 Mueller to Bentham, 5 February 1866, KM1, f.196.

41 Mueller to J.D. Hooker, Easter 1896, KJH, f.26.

42 On finding work for Brian, see, for example, Mueller to J.D. Hooker, 2 October 1885, KM3, f.159 and Mueller to Dyer, 20 October 1885, KM3, f.160. For messages via Forrest see Mueller to J.D. Hooker, 5 August 1894, KJH, f.14.

43 For examples of praise see Mueller to J.D. Hooker, 15 November 1867, KM1, f.288; Mueller to W.J. Hooker, 27 October 1854, KAL, f.145; Mueller to J.D. Hooker, 28 March 1893, KJH, f.3. For disagreements see Mueller to W.J. Hooker, 11 June 1857, KM2, f.4 (synonymous names); Mueller to J.D. Hooker, 30 October 1879, KM2, f.249 (errors in illustrations); Mueller to J.D. Hooker, 24 April 1880, KM2, f.268, Mueller to W.J. Hooker, 25 October 1864, KM1, un-numbered folio after f.154, and Mueller to J.D. Hooker, 16 March 1880, KM2, f.260 (supposed slights to Mueller's own work).

44 Mueller to W.J. Hooker, 9 May 1855, KAL, f.151; see also Mueller to J.D. Hooker, 15 October 1858, KM1, f.44, and Mueller to W.J. Hooker, 11 January 1857, KAL, f.155. For Hooker's reaction see J.D. Hooker to Mueller, 23 August 1858, printed in Daley, op.cit. (n.17), p.214.

45 For requests for assistance see Mueller to W.J. Hooker, 3 February 1853, KAL, f.185; on joining the Linnean Society, Mueller to W.J. Hooker, 17 May 1858, KAL, f.178; as a go-between regarding *Josephinia eugeniae*, Mueller to W.J. Hooker, 21 June 1857, KAL, f.164; and on illustrating *Duttonia*, Mueller to W.J. Hooker, 11 June 1857, KAL, f.163.

46 Seeking nomination, Mueller to J.D. Hooker, 15 October 1859, KM1, un-numbered folio after f.44; not to campaign, Mueller to J.D. Hooker, 16 January 1860, KM1, un-numbered folio after f.44.

47 Royal Society, Library and Archives, Certificate of Proposal, dated 12 January, 1860. Not elected, 7.6.60; elected, 6.6.61.

48 Mueller to J.D. Hooker, 7 October 1884, KM3, f.119.

49 For example, on Darwinian evolution, Mueller to Bentham, 9 January 1865, KM1, f.158; on limits of genera, Mueller to Bentham, Christmas 1866, KM1, f.246; on field variability, Mueller to [Bentham? (only salutation is 'Dear Sir')], 9 January 1865, KM1, f.158. For Bentham's rebuttal see Bentham to Mueller, 26 October 1862, in Daley, op.cit. (n.17), p.130.

50 Mueller to Bentham, 15 June 1858, KM1, f.44. See also Mueller to W.J. Hooker, 15 May 1858, KAL, f.179.

51 Daley, op.cit. (n.17).

52 Barkly to W.J. Hooker, 20 April 1861, KDC, f.11. See also Barkly letters to W.J. Hooker, from 14 June 1859 to 24 June 1861. KDC, ff.7-13.This series includes Barkly to Mueller, 24 July 1861.

53 Mueller to W.J. Hooker, 1 February 1857, KAL, f.157.

54 For the Bentham Medal see Mueller to J.D. Hooker, 6 November 1884, KM3, f.126; for the CMG, Public Record Office, CO 447, 30, dispatch 217 of Governor Bowen, 22 November 1878.

55 For example of fears see Mueller to Bentham, 26 August 1867, KM1, f.278, and Mueller to Bentham, 24 September 1864, KM1, f.38; on losses, of manuscripts, Mueller to W.J. Hooker, 18 October 1853, KAL, f.137, and of specimens from the Gregory expedition, Mueller to W.J. Hooker, 6 April 1857, KAL, f.159.

56 For example, on injustice, Mueller to Bentham, 29 March 1868, KM1, f.310; for praise, Mueller to Bentham, 23 October 1862, KM1, f.80.

57 W. Mitten described the mosses and Mueller arranged for papers to be published by the Linnean Society of NSW and the Royal Society of Victoria, see letters from Mueller to Mitten, Kew Correspondence, Letters to W. Mitten, 1848-1905, at ff.205-19. J.G. Baker did much with the ferns. publishing jointly with Mueller (e.g. 'Note on a collection of ferns from Queensland', *Journal of Botany*, 25 (1887). 162-3). See also Mueller to Baker, 15 August 1881, unbound mss. filed with KM4. The Reverend M.J. Berkeley received the majority of the fungi; for Mueller's letters to Berkeley, from 16 February 1857 to 1 January 1882, see BMB volume 9.

58 Mueller to Gray, 21 February 1882 (typescripts of F. Mueller correspondence with Asa Gray. chronologically arranged by D. Sinkora from photocopies of material from the Gray Herbarium. Harvard University. MEL).

59 Mueller to Darwin, 12 June 1874, Darwin papers, Cambridge University Library, DAR 171. For Masters see Mueller to Owen, 6 October 1874, BMO, f.378, and for W.J. Hooker's probable views see Mueller to J.D. Hooker, 7 July 1877, KM2, f.199.

60 For example, Mueller to Berkeley, 9 September 1873, BMB; Mueller to Bentham, 12 February 1877. KM2, f.191. For the Kew battle, see Roy Macleod 'The Ayrton incident: a commentary on the relations of science and government in England, 1870-1873', in A. Thackray and E. Mendelsohn (eds), *Science and Values* (New York, 1974), pp.45-78.

61 Mueller did not know whom to thank for recommending his knighthood. He assumed that Barkly and J.D. Hooker had played a prominent part (Mueller to J.D. Hooker, 27 May 1879, KM2, f.229). J.D. Hooker drew attention to Mueller's work in a circular announcing the completion of the series of Colonial Floras (19 August 1878), but the suggestion for Mueller's promotion to KCMG originated in the Colonial Office, in a minute on Mueller's own recommendation that Bentham be honoured, Public Record Office, op.cit. (n.54).

62 Mueller to J.D. Hooker, 23 January 1884, KM3, f.80 is the covering letter of some biographical notes written by Mueller and published in the Melbourne *Leader* to accompany a woodcut of Joseph Hooker.

63 Mueller to Dyer, 18 December 1888, KM3, f.267. He was still apologizing for the delay in the award in Mueller to Dyer, 20 August 1891, KM4, f.12.

64 For example, the monotypic genus *Dennisonia* was named for Sir William Denison in 1859 (*Journal of the Proceedings of the Linnean Society, Botany*, 3, 157-9); *Rhododendron lochae* was named for the governor of Victoria, Sir Henry Loch in 1887 (*Gardeners Chronicle*, ser. 3, 1:543-4). For examples of hoped-for vice-regal patronage see, re Sir James Bowen, Mueller to J.D. Hooker, 25 March 1873. KM2, f.81, and re Sir Henry Loch, Mueller to J.D. Hooker, 21 May 1884, KM3, f.91.

65 Barkly to W.J. Hooker, 24 April 1860, KDC, f.8.

66 The fragmentary scrapbook known as the Barkly Papers (Westminster City Libraries, Archives Department, Accession 618) contains evidence of Barkly's communication with Agassiz (fossil fishes),Galton (skulls), Rolliston (bushman skulls), Owen (South African fossils), as well as the Kew botanists. Barkly was elected FRS in 1864.

67 *Australasian*, 6 January 1872. For other examples see J.M. Powell, 'A baron under siege: von Mueller and the press in the 1870s,' *Victorian Historical Journal* 50(1) (1979), 18-35.

68 *Grevillea victoriae*, Transactions of the Philosophical Society of Victoria, 1 (1855), 107; *Josephinia eugeniae, Hooker's Journal of Botany and Kew Garden Miscellany*, 9(1857) 370-1. For Prince Paul's visit, see Mueller to Governor Barkly, 21 August 1858, in Public Record Office of Victoria; VPRS 1189, G58/7190, Governor Barkly to Chief Secretary, 21 August 1858. Mueller was created a baron, as a representative of science, on 20 December 1867.

69 Mueller to J.D. Hooker, 28 July 1866, KM1, f.223.

70 William Archer to W.J. Hooker, 22 May 1865, KDC, f.6.

71 W. Hill to W.J. Hooker, 17 February 1863, KDC. f.78.

72 Mueller to Mary Bate, 20 November 1881 (Photocopy at MEL).

73 *Fragmenta Phytographiae Australiae*, 8 (1873), 33.

74 Mueller to Kippist, 18 March 1869, LINC.

75 Mueller to J.D. Hooker, 15 July 1873, KM2, f.100.

76 Mueller to J.D. Hooker, 6 November 1873, KM2, f.125.

77 Mueller to J.D. Hooker, 'Queens Birthday' [24 May] 1880, KM2, f.273.

78 *Census*, op.cit. (n.21).

79 For an early statement of the artificiality of genera see Mueller to W.J. Hooker, 15 October 1853, KAL, f.137. On Mueller's reduction of species, see Barkly to W.J. Hooker, 14 June 1859, KDC, f.7, and Mueller to Bentham, 23 September 1868, KM1, f.347.

80 Mueller to Dyer, 5 September 1891, KM2, f.317.

81 For *N[ematolepis] euphemiae* see Mueller to Bentham, 20 February 1863, KM1, f.88; for *Eremophila forrestii, Fragmenta Phytographiae Australiae*, 7 (1869), 49; *Trematocaryon mclellani* and *Pleioclinis*

couchmanii are described in 'Reports of the mining surveyors and registrars', Victoria, *Parliamentary Papers, Votes and Proceedings of the Legislative Assembly, 1882–83*, 2, No.27, at pages 13 and 19 respectively; *Wehlia, Fragmenta Phytographiae Australiae*, 10,22. *Myoporum bateae, Proceedings of the Linnean Society of New South Wales*, 6 (1882), 92; for Jessie Hussey see D.N. Kraehenbuehl, 'Jessie Louisa Hussey', in Carr and Carr, op.cit. (n.2), pp.388-9.

82 Giles had originally wanted to name Mt Olga (Queen of Württemberg) and Lake Amadeus (King of Spain) after Mueller, but Mueller substituted the present names (B. Threadgill, *South Australian Land Exploration 1856–1880* (Adelaide, 1922), p.135). For the other names, see Giles, *Geographic Travels in Australia* (Melbourne, 1875), passim.

83 Mueller to Mackinnon, 7 July 1873 asking for £10 subscription to support Giles' journey. La Trobe Library, Manuscripts, Mackinnon, Box 12/6. For features named for subscribers, see Giles, *Australia Twice Traversed* (London, 1889), especially p.179 at fn.1.

84 'On the rules of priority of geographic names', November 1876 (RGS, Journal mss. 1877, Mueller).

85 There are a great many letters on this theme, although most are after 1880. For *Vahea/Landolphia* see *Chemist and Druggist of Australasia, Supplement*, 5 (1882), 39-40, and Mueller to Dyer, 31 December 1881, KM2, f.349; for *Stylidium/Candollea* see Mueller to Dyer, 10 September 1881, KM2, f.318. Mueller's last use of *Stylidium* for a new species was in his description of *S. trichopodum* in *Fragmenta Phytographiae Australiae*, 10 (1876), 76; from then on he used *Candollea*. (See T.B. Muir, 'An index to the new taxa, new combinations and new names published by Ferdinand J.H. Mueller', *Muelleria*, 4 (1979), 123-68).

86 For limited authority of Congress, see Mueller to J.D. Hooker, undated letter, but probably February 1893, KJH, f.1; for resignation from Le Mans, Mueller to ?, 3 November 1893 [?, dated in another hand; letter is incomplete], KM4, f.53; for justice to author, Mueller to J.D. Hooker, 16 March 1880, KM2, f.260; for justice to person honoured see geographic names manuscript, op.cit. (n.84); for absolute priority, Mueller to J.D. Hooker, 16 March 1880, KM2, f.260.

87 For example, Mueller to W.J. Hooker, 27 May 1854, KAL, f.139.

88 RGS, certificates of nominations, and Mueller to Shaw, 15 May 1858. RGS.

89 Mueller to W.J. Hooker, 12 May 1858, KAL, f.178.

90 LINC.

91 Bufton to Mueller, 22 December 1892, LINC. C.H. French and J.G. Luehmann worked for Mueller and had each been elected FLS, on his nomination, in 1885.

92 Rodway was nominated in 1891, and the quoted annotation to the certificate was made by the Linnean secretary on 28 August 1893. LINC.

93 Mueller to J.D. Hooker, 5 February 1885, KM3, f.139.

94 Mueller to J.D. Hooker, 8 June 1879, KM2, f.234, says the nomination is on Mueller's suggestion alone, but see undated letter McCoy to Mueller, KM2, f.238.

95 Mueller to J.D. Hooker, 8 June 1879, KM2, f.234. For an example of zoological collaboration see Mueller to Owen, 25 August 1864, BMO, Vol.19: ff.370 and 402.

96 J.D. Hooker to Mueller, November 1895, printed in Daley, op.cit. (n.17), p.274.

97 Moore was nominated in 1863, and Maiden 1888.

98 LINC. See also Mueller to Archer, 24 April 1874, University of Melbourne Archives, W.H. Archer papers, 2/137/7.

99 For cash payment see, for example, Mueller to T.B. Moore, 28 September 1889 (MEL); for seeds, see Mueller to Mrs Ryan [of Eucla], 29 August 1895 (copy in MEL).

100 Mueller to J.D. Hooker, 5 February 1885, KM3, f.139.

101 For Burke and Wills, see Alan Moorehead, *Cooper's Creek* (London, 1963), Chap.3; for Ladies Leichhardt Search Expedition, see Mueller to Darling, enclosed in Dispatch of Governor Darling to Cardwell, 25 July 1865, RGS archives, and copy of contract, dispatch of Darling to Cardwell, 21 April 1866, RGS; for death of McIntyre, Mueller, Melbourne *Age*, 23 July 1866; for Elder expedition see Elder to Mueller, 2 July 1890 (South Australian Archives, 1401/197, Elder Scientific Exploration Committee).

102 For example, Jonathan and Stephen Cole, *Social Stratification in Science* (Chicago, 1973).

103 See circular distributed to potential collectors in September 1881 (copy at MEL), and letters to correspondents (e.g. Mueller to Daley, 13 September 1894; (copy of letter, from Royal Historical Society of Victoria, Archives, in MEL).

104 See Muir, op.cit. (n.85), for a list of names.

105 For zoology, see, for example, [Pastor] Hagenauer to Mueller, 4 October 1886, BMO, Vol 14, f.206, apparently enclosed with Mueller to Owen 5 October 1886, BMO, Vol 19, f.391; for Western Australian guides, Mueller to Cpl Jones, 1 January 1880 (Battye Library, Public Library of Western Australia, 2461A/2); for settlers, 'Annie Bellow Macdonald (Mrs Torrens McCann). Brief sketch of her life', *Border Morning Mail* (Albury), 30 September 1927.

106 For examples, see Powell, op.cit. (n.67).

107 I am indebted to the librarians and archivists of the Westminster City Libraries; the Royal Botanic

Gardens, Kew; the Linnean Society of London; The Royal Society; The Royal Geographical Society; the British Museum (Natural History); the Adelaide Botanic Gardens; the University of Melbourne Archives; the La Trobe Library of the Public Library of Victoria; and the National Herbarium of Victoria, especially Mrs D. Sinkora, for the help that they have given me in identifying Mueller material, and for permission to quote from material in their care. Mr Noel Lothian allowed me to see the letters to Otto Tepper, and Mr K.B. Griffiths, Kyogle and Mrs M. Elliston, Latrobe gave permission to use material in their control. I am also indebted to Dr Denis Grundy, of the School of Education, The Flinders University of South Australia, for his interest and advice in the early stage of the study.

Some of the early work on this study was supported by Flinders University research funds, and the later stages were materially assisted by the award of a Visiting Professorship in the Science and Mathematics Education Centre of the Western Australian Institute of Technology.

7

Gorilla warfare in Melbourne:
Halford, Huxley and 'Man's place in nature'

Barry W. Butcher

In July 1863, George Britton Halford, the recently arrived first medical professor at the University of Melbourne, challenged the findings of Thomas Henry Huxley's book, *Evidence as to Man's Place in Nature*. Applying the methods of comparative anatomy Huxley had concluded that 'the differences that separate man from the gorilla and the chimpanzee are not so great as those which separate Man from the lower apes'. Using the same approach, Halford countered that the evidence suggested that there were crucial differences between Man and the apes, which effectively refuted Huxley's conclusions. In the wake of Halford's first lecture on the issue, a public controversy erupted, which embroiled a wide cross-section of the social and intellectual élite in Melbourne.

Historians studying the reception and impact of Darwin's *Origin of Species* in Britain increasingly recognize that the question of 'Man's place in nature' underlay all discussions of evolution. It led to differing interpretations of Man's physiological and genealogical relationship to the 'lower animals', and discussions of these were inextricably bound up in a matrix of debates involving religion, philosophy and political economy. Physiology, anatomy or morphology may have provided the scientific pegs on which various evolutionary arguments were hung, but it was the moral and social implications of those arguments that generated the heat that often accompanied these scientific debates and made them of interest to a wider audience.[1] In this respect, the nature of the debate in Australia was no different. In the case under discussion here, we shall see that the question of whether or not monkeys have feet finds its meaning and significance only because its discussion is embedded in larger cultural values. At the same time, the case of the 'monkey's foot' is illustrative of another aspect of nineteenth-century science — that of scientific colonialism. It represented an opportunity for Halford, an emigré British scientist, to establish his credentials among colonial Melbourne's cultural élites. However, it is suggested here that Halford's success in Melbourne was achieved only at the expense of his international aspirations.

153

Background to the controversy: Huxley and Owen

After publication of Darwin's *Origin*, an evolutionary edge was added to a debate about the correct classification of Man that had surfaced periodically since the seventeenth century. In the 1850s, controversy based on comparative anatomy had flared up in this area between Richard Owen and Thomas Henry Huxley — Owen maintaining that significant anatomical differences could be demonstrated between Man and the 'higher' apes, Huxley denying that there were differences more significant between Man and ape than between ape and monkey. The most significant area of dispute between them lay in Owen's claim that the ape's brain lacked certain structures that were to be found in the brain of Man. By the time his book on Man was published, Huxley had effectively won this particular battle by convincing the majority of his scientific peers that the disputed structures were present in the ape's brain, and he appended a summary of the controversy to the second chapter of his book.[2] There were, however, other areas of dispute between anatomists over the correct classification of Man, and Huxley alluded to one of these, namely the question of whether the hind limbs of the higher apes terminated in a 'true' foot or in a structure better understood as a hand. It was this part of the anatomical debate about Man that Halford carried to Australia, and that provided the grounds of his disagreements with Huxley and his evolutionist supporters in Melbourne.

When *Man's Place in Nature* was first published in Britain its author's thesis did not go unchallenged, and scientific and literary journals such as *The Reader, The Athenaeum* and *The Quarterly Review* devoted considerable space to airing the arguments for and against it. In many ways this dispute 'made' Huxley as a scientific figure of public stature in Britain and was an important factor in his lifetime crusade for acceptance of Darwinism in particular and scientific naturalism more generally. Both issues were seen in some quarters as a threat to the traditional 'common context' of Victorian society.[3]

The cultural matrix of nineteenth-century Australia was British in origin and it is not surprising that Australians should react to perceived threats to it in much the same way as their British contemporaries did. By going to the heart of the evolutionary debate — Man's place in the scheme of things — Halford was initiating for the first time in Australia a full-scale public debate of the wider issues involved. At the same time, like Huxley, he was staking a personal claim to be seen as an authority in science. In the narrower confines of colonial society, that claim was not dependent on the 'truth' (publicly perceived) of his position. Rather it was based on his ability to initiate and sustain debate in a social and intellectual climate that was generally, though not exclusively, favourable to his position. As will be seen, Halford was highly successful in achieving his aims within the colony, but in doing so he undermined his own credibility within the wider circle of British science.

Halford's critique of Huxley

In his discussion of Man's relationship to the ape, Huxley had claimed that in dealing with anatomical structures he would avoid all mention of those that were 'eminently variable'. By this he meant that he would be concentrating on so-called 'primary structures', ignoring those that were sometimes absent and sometimes present within a group. Using this criterion, he rejected the view of some anatomists that the hind limbs of the ape terminated in a hand-like structure rather than a true foot, and claimed that the similarities to a human hand were superficial. He then set down his own reasons for believing them to be true feet — that they shared with the human foot a similar arrangement of the tarsal bones, and had the short flexor and extensor muscles of the digits characteristic of a foot rather than a hand. Finally, and for Huxley conclusively, they possessed the *peroneous longus* muscle, used in balancing the body and flexing the foot.[4]

Halford delivered the first of his four lectures on this subject before an invited audience at the university, which included 'the Chancellor and about fifty other gentlemen'.[5] He began by agreeing that the features designated by Huxley as crucial in determining the status of the human foot were present in the hind limb of the ape. However, according to Halford, one of the chief muscles of the human foot, the *transversalis pedis*, was absent in the hind limb of the 'monkey'. Thus while the 'monkey' possessed some foot-like features, it lacked others; and what is more, it had some features suggestive of hands. For Halford, this was sufficient to prove that Huxley had missed crucial evidence that went against his conclusions and that suggested, when coupled with the fact that the muscles in the monkey's hind limb had greater mobility relative to the same muscle in Man, that the terminal division of the monkey's hind limb was as much hand-like as foot-like. Halford therefore rejected the term 'biped' (two-footed) as applying to the 'monkey' but also the term 'quadrumanous' (four-handed), and suggested instead that these creatures be defined as 'cheiropodous' or 'finger-footed'.[6]

Even at this early stage in the dispute, there were matters about which the two professors would never agree. Huxley had dismissed the greater mobility of the monkey's hind limb muscles as part of the superficial similarity between it and the human hand. As an evolutionist his concern was with primary structural similarities that suggested genealogical relationships, not recent secondary adaptations that could be accounted for, he believed, in terms of the Darwinian principle of descent with modification.[7] For Halford, working within a programme in which he wished to preserve a role for teleology and designing forethought, function was as important as structure and therefore the shape and size of the structures themselves were also of crucial importance. When he concluded his first publication on the subject by drawing attention to the Creator's purposes in designing the limbs of Man and ape to serve different functions, he was highlighting the extent to which he and his adversary were arguing from incommensurable positions.[8]

A second area where problems arose concerned terminology. Halford's direct knowledge of primate anatomy was limited at this stage to dissections carried out on two macaque monkeys and his comments on gorillas were largely extrapolated from these. Critics were quick to point out that the term 'macaque' covered a large genus containing many species, and Halford's failure to identify the species to which his macaques belonged was seen as a

FANCY SKETCH.

Showing the remarkable distinction between a Man and a Monkey.
Respectfully dedicated to MESSRS. HALFORD and OPIFER.

major weakness in his case. More confusing than this, however, was the lack of precision, on both sides of the dispute, when dealing with the great apes, which were, after all, the linch pin of Huxley's argument. 'Monkey', 'monkey-tribe', 'simian', 'simian-tribe', and 'ape' were used interchangeably throughout — on occasion even by Huxley.[9] If nothing else, this supports the view proposed here that the argument at its most meaningful level was not about specific anatomical issues relating to gorillas and Man but about wider concerns with 'Man's place in nature'.

In its own way, this argument between two anatomists over the relative importance of bones and muscles illustrates the complexity of the debate about evolution. On one level, it appears as a dispute over taxonomy and the correct procedures to be followed in zoological classification — what counts as evidence one way or the other — and the roles to be assigned to form and function. But even here it is not possible to separate the arguments from the broader considerations mentioned above; it was always the implications — religious, social and so on — that were paramount. What inferences might be drawn about Man if it were shown that he shared a common ancestry with the apes? And, what would be the result of broadcasting those inferences to the general population? Halford concluded his first lecture by urging his audience to reject any idea that there was any lineal descent from ape to Man. He condemned Huxley's book for 'tendencies' that suggested that it 'might have been written by the devil'.[10]

Consigning the supporters of evolution to the forces of darkness was not unusual in the early responses to Darwinism, but it was rarely spelt out as clearly as this. Although Halford maintained his position throughout, he tempered his rhetoric in later talks, possibly in response to press criticism including some from quarters otherwise sympathetic to his cause.[11]

A month later, during a public lecture in Melbourne, in which he repeated his earlier claims, Halford received support, secular and spiritual, from scientific, religious and social leaders in the city. Frederick McCoy, the anti-Darwinian professor of natural science at the university and director of the National Museum of Victoria, over-riding normal museum practice, had allowed Halford to use stuffed and skeletal material from the museum.[12] During the discussion that followed Halford's talk, the chief spokesman for the Presbyterian Church in Melbourne, Adam Cairns, outlined the weaknesses of the 'development hypothesis' and insisted that it was the 'higher faculties' that separated Man from the apes. This was a view endorsed by most of Halford's supporters throughout the controversy, and represents the closest that the anti-evolutionary forces came to stating their case in an unmediated form, namely their fear of the consequences if Huxley's position were proved to be correct or became accepted orthodoxy.[13] Huxley had, in fact, drawn attention to 'the vastness of the gulf between civilized Man and the brutes' but denied that there was any psychical distinction that could not be accounted for on evolutionary theory.[14]

The Colonial Governor of Victoria, Henry Barkly, presided over Halford's second lecture. Well known for his support for science but antagonism towards evolutionary theory, Barkly closed the meeting with a short speech clearly favouring Halford, thereby ensuring that the new professor had the support and sanction of major figures in scientific, religious and political circles in the colony. Nor was Barkly's the only vice-regal voice raised against the evolutionists. William Denison, a former governor of New South Wales, wrote an anti-Huxley parody from his new diplomatic base in India. Among the general population, however, support was not so complete, and the colony witnessed the rare spectacle of scientific controversy being played out in the local press.[15]

The press reaction

The two Melbourne newspapers, the *Age*, under the control of David Syme, and the *Argus*, with Edward Wilson as its senior proprietor, took opposing sides from the beginning. According to the *Age*, the issues raised by Halford were both complex and important as the subject 'must necessarily have great influence . . . on the minds of the public'. 'Clear and decided opinions' could only be reached when all the evidence was available, but nonetheless the theory propounded by Huxley was 'pernicious' and the logic behind it faulty for there was no reason to suppose that identity of structure implied community of descent, which the *Age* recognized as Huxley's major thesis. This being so the theory could do untold harm if allowed to fall unchallenged into the hands of the credulous, and Halford was to be congratulated for his stand against it.[16] In an article entitled 'The Antiquity of Man', the *Age* professed scepticism as to the likelihood of the transmutation of monkeys into men 'through the operation of laws latent in nature'.[17] Quoting the French naturalist, Armand de Quatrefages, the paper followed Cairns in insisting that the moral and intellectual endowments of Man, which it earlier described as providing 'an impassable chasm' between Man and monkey, should be allowed 'due weight' in any classification. It informed its readers that the palaeontological findings of Boucher de Perthes, which had suggested a long ancestry for Man, had been accepted only after careful scrutiny by geologists 'fully aware of the anti-Biblical uses to which they might be put'.

The *Age* was at pains to point out that it was not suggesting that such enquiries as those of Huxley and de Perthes were illegitimate areas for scientific enquiry; after all, it was 'in speculative researches and in the eagerness to challenge even the mysteries of existence that Mankind infinitely transcends all orders of the lower animals'. According to the *Age*, the human species holds its place at the head of the assemblage of the family of mammalia 'not by instinct but by reason'. In short, for the *Age*, 'it mattered not how near we are to monkeys so long as we are different from monkeys'.[18] Quoting from another Syme publication, the *Leader*, it dismissed Huxley as an unoriginal thinker who had achieved publicity by 'undertaking to carry out the theory of Darwin and of the author of *The Vestiges* as to the development of Man from the inferior animals'.[19]

Clearly the *Age* understood that the dispute encompassed all the issues relevant to the wider context always implicit in discussions about evolution. While supporting Halford's anatomy against Huxley's it was nonetheless concerned to spell out a fall-back position, which effectively ensured that, should Huxley prove to be correct, it would leave the wider question of 'Man's place in nature' unanswered. The moral and intellectual capacities of Man provided the 'impassable chasm' that served as the final arbiter of Man's unique position. The dire consequences that might follow should the 'pernicious' theory fall into the hands of the 'credulous' did not need to be spelt out. In Melbourne, the prevalence of free thought and sectarian disputes were already seen as a potential threat to organized religion.

At the same time, the rapid growth of the colony following the discovery of gold in the previous decade was a cause for concern to those who sought to maintain social and class structures in an effort to ensure social stability. If the religious basis of society were undermined, then the outlook for social stability was bleak. David Syme, the proprietor and driving force behind the *Age*,[20] sustained a lifetime of antagonism to the Darwinian theory of evolution by natural selection, seeing in it an attempted legitimation of the 'law of the jungle' social ethics that he was constantly railing against when opposing the promoters of *laissez-faire* economics and free-trade.[21] In later years Syme proposed his own theory of evolution, which rejected the concept of struggle in favour of internal drives at the cellular level.[22]

Syme's strong anti-Darwinian views allied to his grip on the editorial policy of the *Age* at this time goes some way to explaining why a newspaper with a reputation for promoting liberal if not radical views should be so opposed to a 'radical' new theory ostensibly intended to explain the development of life through the agency of secondary laws.[23]

This antagonism to 'the development hypothesis' did not, however, extend to all sections of the social élite in Melbourne. There were those who saw in evolutionary theory a liberating intellectual idea that posed no threat to true religion or to Man's supremacy in the animal kingdom. Further, the ethical implications of the theory that the *Age* found so disquieting could be interpreted in a manner that supported the opponents of political liberalism and economic protectionism. Even the natural theology that lay at the base of the prevailing social philosophy could be preserved; recent scholarship has shown that the emergence of evolutionary theory did not destroy traditional natural theology but transformed it or, rather, as one commentator has said, 'was a subtle accommodation within natural theology'.[24]

When the *Argus* came out in support of Huxley it provided a powerful vehicle for those taking the more 'optimistic' view in conveying their message to a wider audience. The *Argus* was less explicit than the *Age* in identifying its social philosophy with scientific speculation; initially at least, it was Halford's style and rhetoric that it objected to. The allusion to the Devil it saw as 'vulgar claptrap' and 'imbecile scurrility' designed to imply that Huxley had written with 'a deliberate evil purpose in mind'. If Huxley was guilty of error, then it was scientific error, to be dealt with by recourse to the methods and data of science.[25] Like the *Age*, the *Argus* thought that Huxley's investigations were a legitimate area for scientific enquiry and it condemned Cairns for becoming involved in a dispute that was 'purely scientific . . . its rights and its wrongs must be proved in a rational manner without any reference to anything but the facts'. Reversing the argument spelt out by the *Age*, it warned of the dangers to science if theology were allowed to get involved in scientific investigation. What would have happened if the theologians had had their way with the geologist Charles Lyell thirty years before? The *Argus* had no doubt, reminding its readers of the fate of Bruno and Galileo.[26] While joining the *Age* in believing that it was the 'spiritual and intellectual' abilities of Man that guaranteed his supremacy over the ape, the *Argus* thought that Halford had misunderstood Huxley's position, which was that 'there is less difference between Man and the gorilla than between the gorilla and the lower kinds of monkey and therefore Man has no right to abrogate to himself an exclusive place in the scale of creation'. Halford's continual attempts to show that there was not identity of structure between the foot of Man and ape were therefore pointless, according to the *Argus*.[27]

A number of factors were involved in determining the *Argus's* pro-Huxley position. Its attitude to the medical profession in Melbourne was often hostile, and it appears to have aligned itself to a prominent anti-Halford clique prior to the eruption of this particular dispute. The official journal of the local medical society felt moved on more than one occasion to complain against the paper's attitude.[28] But probably of greater importance was the rivalry between the two newspapers on virtually every issue, especially when related to free trade and protectionism. It seems likely that, once the dispute was up and running, this traditional rivalry came into play.[29]

Edward Wilson, the driving force behind the *Argus*, was an admirer of Darwin and on his return to Britain for health reasons in 1864 became a close neighbour of the Darwin family. In his book *The Expression of the Emotions in Man and Animals* published in 1872, Darwin credited Wilson with using his 'powerful influence' to obtain Australian material used in the book.[30] But, it is unlikely that Wilson played a direct role in the Halford dispute because he no longer had input into the editorial policy of the newspaper — indeed, he appears to have been in conflict with the views of some of his editors on a number of issues.[31] It does seem probable, however, that the general thrust of the *Argus's* policy on social and economic matters would be a strong determining factor in its pro-Huxley stance; it employed the polymath professor William Edward Hearn to write feature articles on law and economics, and Hearn's own social–evolutionary ideas were published in the same year as this dispute began.[32]

Opposition to Halford

In a small community like Melbourne in the 1860s, opposing philosophies could hardly be seen entirely in abstraction. Supporters and opponents of any position, on any subject, were likely to be known by sight and name, a factor that adds a personal edge to this dispute. Halford's opponents in Melbourne provide a good example of this.

The main opposition to Halford came from a group organized around William Thomson, one of the few medical men in Melbourne to achieve an international reputation.[33] Thomson had played some part in the creation of the medical school but soon after Halford's arrival in December 1862, was highly critical of aspects of the proposed medical course. Described by one biographer as 'vain and irascible', he now crossed swords a second time with the new professor, and though the majority of his medical colleagues either

PROFESSOR H———D DRESSING FOR
THE BALL.

MAN'S PLACE IN NATURE.

INTELLIGENT LITTLE BOY.—*Mamma, didn't Mr. OPIFER say the other night that he was descended from a Gorilla?*
MAMMA.—*Yes, dear.*
LITTLE BOY.—*Well, and I think he is right, for there's certainly a family likeness.*

supported Halford or maintained public silence Thomson was not without allies.[34] The most prominent of these was a chemist with the Geological Survey of Victoria, Charles Wood. Wood had studied under Huxley at the School of Mines in London prior to emigrating to New Zealand. There he had worked under James Hector on the Geological Survey before declining health forced him to move to Victoria.[35] Wood was instrumental in eliciting the only public responses from Huxley to Halford's charges, but beyond that remained behind the scenes, apparently for fear of the likely repercussions on his career. Somewhat surprisingly, perhaps, he was on cordial terms with Halford's chief scientific ally, Frederick McCoy. McCoy was deeply interested in the work of the Geological Survey and maintained good relations with both Wood and Alfred Selwyn, the Survey's director, another Huxley supporter. When Charles Aplin, a surveyor with the Survey, spoke up in favour of Huxley at one of Halford's public lectures, he ensured that the colonial geologists were seen as being in the front line of evolutionists. Why they should have been is unclear, but it does accord well with recent studies suggesting that geological surveys were among the first scientific institutions to take up the Darwinian theory.[36]

The antagonism towards Huxley shown by the colonial medical profession remains problematic. Thomson's view was that it stemmed from professional jealousy (presumably aimed towards him) and a fear of falling out of favour with 'the prevailing disposers of patronage' at the university.[37] As his own career prospects at the university disappeared, at least in part through this dispute with Halford, the claim cannot lightly be dismissed.[38] However, there are no convincing reasons for doubting that many Halford supporters were genuine in their statements rejecting the evolutionary point of view. A letter to the *Argus* signed 'Medicus' was highly critical of Thomson, and drew on Richard Owen's writings in support of Halford while at the same time emphasizing that it was the 'moral tendency' of Huxley's book that was the target of Halford's allusion to the Devil, not the nature of the investigation itself. 'Medicus' invoked the same natural theology as Halford when telling Thomson that he should examine all the facts of the case 'before sitting in judgement upon the works of the Creator'. Inadvertently perhaps, 'Medicus' had managed to encapsulate in a few lines the wider context of the debate about evolution, and point up the incommensurability of the positions taken by the opposing sides.[39]

Thomson's defence of Huxley came in the form of a critique of Halford's anatomical dissections. Writing in the *Argus* under the pseudonym 'Opifer', he rejected Halford's claim that the *transversalis pedis* was missing from the hind limb of the gorilla and chimpanzee. Further, he charged Halford with failing to recognize that the so-called *extensor metacarpi* muscle in the ape was in fact the homologue of the *tibialis anticus* in Man.[40] This last point provides a good illustration of the difficulties involved in taxonomic argument and it is difficult now to decide how legitimate the respective positions were. A biographer of Thomson has described his criticisms as 'informed', and they were the basis of later attacks on Halford from sections of the British medical press, most notably *The Lancet* and the *Medical and Surgical Review*. Huxley accepted Thomson's interpretation, and a writer in the British scientific journal, the *Reader*, believed that Thomson had got the better of the argument.[41] On the other hand, one of Halford's successors in the medical school has described *The Lancet's* criticism as 'hostile and none too fair'.[42] Empirical enquiry could not settle the matter then or now, for in essence the argument at the scientific level can be understood only within the context of taxonomies shaped by fundamentally different views of the natural world.

Taxonomic systems are now recognized as being conventional and dependent on data always open to alternative interpretation.[43] Thus when Thomson accused Halford of 'breaching the doctrine accepted by all competent naturalists known as the "unity of type"' in claiming that the *extensor metacarpi* was present in the hind limb of the ape, he was doing so from a standpoint that assumed a particular evolutionary explanation for that doctrine. In short he was assuming, as Darwin had, that closely allied species shared a common ancestry and therefore shared a common anatomical structure.[44]

However, at face value, the doctrine was little more than an observation statement based on the common experience of naturalists, and explanations for it — or, more accurately, interpretations of it — were dependent on a naturalist's theoretical, philosophical or even metaphysical convictions. When Halford concluded his first publication with the claim that 'Surely the intricacies of the monkey's foot was planned, as was the comparative simplicity of man's! They could never run the one into the other, or to use a fashionably scientific term to be "developed" the one from the other', he was in fact proposing an alternative view of 'the unity of type', one based on the time-honoured tenets of natural theology.[45]

Perhaps this aspect of the dispute can be better illustrated by looking at a not unrelated example. In 1873, the Catholic non-Darwinian evolutionist, St George Mivart, published a response to Huxley's book in which, after comprehensively analysing the comparative anatomy of the entire primate order, he claimed that there was no single group of morphological characters that allowed an investigator to place any two (or more) species closer together than they were to any other. By ignoring Huxley's division of the anatomical characters into 'primary' and 'secondary' he was able to offer an alternative explanation, or at least, a refutation of Huxley's.[46] Mivart's attack on Huxley's position has been echoed in this century, most notably perhaps by the neo-Lamarckian Frederic Wood Jones.[47]

Much of the 'scientific' content of the dispute described here provoked a similar multiplicity of possible, usually incommensurable, positions. To account for the apparent 'supererogation' of nature in providing the ape with both a *tibialis anticus* muscle and Halford's *extensor metacarpi* (which was one way in which Halford supporters suggested that his findings could be reconciled with Thomson's critique), a writer in the *Argus* appealed to Owen's law of vegetative or irrelative repetition of parts'.[48] Owen's highly confusing definition of this law proposed that it could account for 'the numerous instances in the animal kingdom, of a principle of structure prevalent throughout the vegetable kingdom exemplified by the multiplication of organs in one animal performing the same function'.[49] Thomson commented acidly that the claim was 'nothing more than evidence that there were still men around who reject the great law of homology'. But, as in the case of the 'unity of type', the law of homology was open to a variety of interpretations, each dependent on a host of prior commitments. Far from the dispute being settled by 'an appeal to facts' as called for by the *Argus*, it had become increasingly bogged down in confusion stemming from incommensurable philosophical positions.[50]

When Halford published his first pamphlet on the dispute, the response was mixed. Entitled *Not like Man, Bimanous and Biped, Nor Yet Quadrumanous But Cheiropodous*, it comprised sixteen pages of text and four plates prepared by two of his students, one of whom, James Edward Neild, was then editor of the *Australian Medical Journal*.[51] Not surprisingly, the *AMJ* reviewed the pamphlet favourably, Neild almost certainly being the reviewer. Reaction to the work in Britain, however, was extremely unfavourable. *The Lancet*, drawing heavily on Thomson's writings, criticized both Halford's findings and the intemperate style of his attack on Huxley. The *Medical Times and Gazette* was rather more restrained in its treatment of Halford, but clearly preferred Huxley's position.[52]

Huxley's response

Through Wood, Huxley was kept informed of events in Melbourne, and in February 1864 the colonial pro-evolutionary group received a boost when Huxley responded with a point-by-point rebuttal of Halford's major criticisms. According to Wood, this was 'a real floorer for Halford', causing a 'sensation' in the medical school where (unnamed) Huxley supporters urged that it be published.[53] In a letter to Alfred Selwyn, Huxley consented to publication, and Wood passed the article to Thomson who promptly sent it to the *Argus* where it appeared on 4 February 1864.[54] There was little new in the piece, Huxley merely reiterating what Thomson had already reported and his reasons for eliminating 'variable structures' from the discussion. In replying, Halford appealed again to those very structures as valid taxonomic markers.[55]

Three weeks later, Halford published his second pamphlet. Entitled *Lines of Demarcation between Man, Gorilla and Macaque*, it was designed to counteract a paper Thomson had delivered to the Medical Society of Victoria. Thomson's paper appeared in the society's *Journal* for February 1864 under the title 'The *Transversalis Pedis* in the Foot of the Gorilla', and in it he drew on the researches of Isidore Geoffroy St Hilaire in asserting that the muscular form and structure of the gorilla's hind limb clearly suggested that it terminated in a true foot, although modified to suit the specific lifestyle of the creature.[56] Halford called on expert witnesses, including Owen, Duvernoy and Geoffrey St Hilaire, when tabulating evidence in his own pamphlet in a manner favourable to his own position.[57]

The *AMJ* was again impressed, but the British journals remained unmoved, *The Lancet* suggesting that Halford's experience of primate anatomy was inadequate to allow him to be seen as an expert on the subject. *The Lancet* did, however, make an important concession when it pointed out that in the final analysis the division of organisms into orders, sub-orders, classes and so on was largely dependent on the weight allotted by particular naturalists to the classificatory value of 'certain characters'. This is precisely the sort of problem evident in the argument between 'Medicus' and Thomson discussed earlier.[58]

The publication of these pamphlets produced different reactions, and had widely different effects on Halford's reputation in Australia and Britain. In Melbourne they increased his public profile, and with the favourable response of the *AMJ* behind them, imbued their author with great prestige. In Britain, however, they further damaged Halford's reputation, which had already been tarnished following the report of a clash between himself and Thomson. After his first lecture and in response to Thomson's initial attack, Halford had placed his macaque dissections on public display at the university. Thomson, together with James Keene, the editor of Melbourne's second medical journal, the pro-Huxley *Medical and Surgical Review*, attended. A heated argument broke out when Halford discovered that Thomson was 'Opifer', which ended with Halford, brandishing a scalpel, hustling Thomson onto the street. Thomson promptly published an account of the incident in the *Argus* and Keene did likewise in the *M and SR*.[59] One of these accounts (probably Thomson's) found its way to Britain where it appeared in a garbled form in the *Medical Times and Gazette*, which severely rebuked Halford for his behaviour.[60] This report could have done little for Halford's aspirations for international recognition later in his career. Before discussing that, however, it is necessary to cover the last act in the saga under discussion here.

Enter the gorillas

The pro-Huxley forces suffered a setback in May 1864 when Wood died of tuberculosis.[61] The dispute disappeared from view for over a year, but re-emerged in June 1865 when McCoy announced in the *Argus* that he had procured a group of stuffed gorillas from the explorer Paul du Chaillu, the first time these creatures had appeared in the colony. McCoy urged his fellow citizens to view this group at the museum to judge for themselves

> how infinitely remote the creature is from humanity, and how monstrously writers have exaggerated the points of resemblance when endeavouring to show that Man is only one phase of the gradual transmutation of animals; which they assume may be brought about by external influences and which they rashly assert is proved by the intermediary character of the gorilla between the other quadrumana and man.[62]

Public interest in the dispute revived: the following week, attendance at the Museum more than doubled.[63] Halford gained access to this new material and encouraged a student, Patrick Moloney, to submit a short paper to the *AMJ* intended to support the 'cheiropodous' nature of the gorilla.[64] The attempt was too enthusiastic, however, and a month later another of Halford's students, William Carey Rees, wrote to the journal pointing out that Moloney's claim that the gorilla was 'four-handed' put him at odds with the professor, who had merely designated it as 'finger-footed'.[65] Halford re-entered the dispute in his own right in July with a lecture to the Royal Society of Victoria in which he once again urged his audience to accept the anatomical evidence in favour of his position. The discussion that followed became animated, with *Charles*

Aplin of the Geological Survey defending Huxley against an audience that was overwhelmingly in favour of Halford. The broad boundaries of the dispute once again surfaced when McCoy claimed that 'the world could now rest easy, free from any belief that the passage between Man and the inferior mammals was to be bridged over by a creature like the gorilla'. Huxley's book was 'a swindle' according to the president of the society, the Reverend John Bleasdale, who now added a Roman Catholic voice to that of the Presbyterianism of Cairns. Bleasdale left no doubt as to whom the 'credulous' readers of the book earlier referred to by the *Age* might be when he claimed that the reasoning of the book was 'only suited to the half-educated intellect fashioned in Mechanics Institutes'. Halford, abandoning his reformed tone, called the book 'disingenuous' and 'calculated to mislead the class of persons to whom it was addressed'.[66] Thus two years of argument had not moved the issue forward, each revival of interest simply emphasizing the extent to which the dispute was bound up with the wider context of the debate about evolution.

The repercussions

It was suggested earlier that Halford's reputation suffered in Britain as a result of his involvement in this dispute. The evidence for this claim emerges from examination of his attempts to gain recognition for his work from his British counterparts. These attempts were made at a time when British science was undergoing major changes, and it was clear that Halford appeared to have been left behind. In 1867, Halford wrote to Richard Owen seeking advice on the procedures to be followed in applying for fellowship of the Royal Society of London. To support his claim he cited two areas of his scientific work, his investigations into the physiology of the heart, carried out prior to his appointment to the Melbourne chair, and his research into a cure for snake-bite through injection of ammonia.[67] Perhaps in recognition of the antagonisms it had aroused in Britain, he avoided mentioning his dispute with Huxley, who, together with John Tyndall and others of the 'scientific naturalism' school, was rapidly becoming a major power-broker in British science. Increasingly, scientific institutions were being moulded in the image of this group, who were taking control of professional appointments and setting the agenda for science policy. As a consequence, scientists identified with the older school of British natural theology were increasingly frozen out from appointments and power. In retrospect, Huxley's reference to Halford in his 1864 letter to Wood as 'your eccentric professor of medicine at the University of Melbourne' has an ominous ring.[68] Halford's desire to become an FRS was never realized, and his career became entirely bound up with the Melbourne medical school.[69]

Some of his supporters fared better, illustrating the distinction between being rewarded for endeavour and being shut out for 'political' misjudgement. Frederick McCoy was elected FRS in 1880, the reward for decades of diligent correspondence with major British scientists such as the geologists Adam Sedgwick and Roderick Murchison. McCoy was generous in exchanging geological and zoological material with overseas museums, and his control of the colonial museum kept him within the imperial network of science.[70] It was another 'network' scientist, the government botanist Ferdinand von Mueller, who canvassed support for McCoy's FRS application. Years of correspondence with William and Joseph Hooker at Kew Gardens and collaboration with George Bentham on the *Flora Australiensis* gave Mueller access to the new scientific power-brokers in Britain. He wrote to both Huxley and John Lubbock on McCoy's behalf, temporarily at least overcoming his distaste for the Darwinian doctrines that both enthusiastically espoused; indeed, when his own position at the Botanic Gardens was under attack he even wrote to Darwin seeking support.[71] While both McCoy and Mueller spoke out publicly against 'the theories of Darwin and *Vestiges*', neither engaged in the type of head-on controversy undertaken by Halford, and both made solid, if unspectacular, contributions to colonial and international science.[72]

Conclusion

This chapter has sought to show how an apparently trivial argument about comparative anatomy can be seen as a major debate when connected with the complex question of 'man's place in nature'. By the unravelling of some of the major components of the debate, it has been shown how social, religious and philosophical factors play a role in 'scientific' controversy. Evolutionary theory was perceived as a threat by some sections of Melbourne's social and intellectual élites while, in contrast, other sections were happy to welcome evolutionary theory in the belief that it could be used to reinterpret and validate ideas in social, religious and political spheres.

The colonial background to this particular debate allows us to witness the fluctuating fortunes of individual careers. It was suggested that in this area Halford was successful in achieving a high degree of prominence within Melbourne public life, and that this success was not dependent on the correctness of his position but rather on his ability to initiate and sustain a debate for over two years. In defending a position that had broad support from social and intellectual élites within the colony, Halford was guaranteeing his place as a prominent spokesman for science in Melbourne. However, the rapidly changing power structures in British science in the years following the publication of the *Origin of Species* was important among the factors that led to his failure to achieve the international scientific honours he sought.

When placed in this wider social and cultural context and seen as part of the debate on 'Man's place in nature', the dispute between Thomas Henry Huxley and George Britton Halford cannot be dismissed, as one recent commentator on the affair has suggested, as 'a trifling argument on comparative anatomy'.[73]

Notes

1 The best discussion of these and other matters relating to the debates about evolution in the nineteenth century is to be found in R.M. Young, *Darwin's Metaphor: Nature's Place in Victorian Culture* (Cambridge, 1985).

2 T.H. Huxley, *Evidence as to Man's Place in Nature* (London, 1863), pp.113-8.

3 Young, op.cit. (n.1), especially Chap 6.

4 Huxley, op.cit (n.2), Chap 2, especially pp.89-91.

5 The *Age*, 17 July 1863; The *Argus*, 17 July 1863.

6 The *Age*, op.cit. (n.5).

7 Huxley's views are spelt out in Chapter 2 of *Man's Place in Nature*. That he deliberately eliminated from his discussion 'all eminently variable structures' is made clear in a letter to C.S. Wood dated 23 October 1863 and published in the *Argus* on 4 February 1864.

8 G.B. Halford, *Not Like Man, Bimanous and Biped Nor Yet Quadrumanous, But Cheiropodous* (Melbourne, 1863), p.16.

9 Criticism of Halford's use of the term 'macaque' in a too general way can be found in *The Lancet*, 12 December 1863, pp.681-3.

10 The *Argus*, op.cit. (n.5).

11 The *Age*, 30 July 1863.

12 McCoy to Halford, 29 July 1863; National Museum Letter Book, 1863, Letter No. 225.

13 The *Argus*, 1 August 1863.

14 Huxley, op.cit. (n.2), p.104.

15 The *Argus*, 1 August 1863. Denison's involvement in the affair is detailed in the *Age*, 4 August 1863.

16 The *Age*, 18 July 1863.

17 The *Age*, 30 July 1863.

18 The *Age*, 4 August 1863.

19 The *Age*, 30 July 1863.

20 Two biographies of Syme are A. Pratt, *David Syme, the Father of Protection* (London, 1908) and C.E. Sayers, *David Syme, a Life* (Melbourne, 1965).

21 Sayers, op.cit. (n.20), p.260.

22 David Syme, *On the Modification of Organisms* (Melbourne, Robertson, 1891).

23 Sayers, op.cit. (n.20), Chapters 7 and 8.

24 R.M. Young, 'Darwinism is Social', in David Kohn (ed), *The Darwinian Heritage* (Princeton, N.J., 1985), p.615.

25 The *Argus*, 20 July 1863.

26 The *Argus*, 3 August 1863.

27 Ibid.

28 *Australian Medical Journal,* Vol. 8 (1863), 135; Vol. 9 (1864), 244-7, 347-9.

29 C.E. Sayers, op.cit. (n.20), pp73, 87. A biographer of Syme's publishing rival, Edward Wilson, quotes Wilson referring to Syme as his 'arch-enemy . . . that wretched beast and imposter'; G. Serle, 'Edward Wilson', *Australian Dictionary of Biography*, Vol. 6, p.415.

30 Charles Darwin, *The Expression of the Emotions in Man and Animals* (London, 1872), p.19.

31 Serle, op.cit. (n.29).

32 W.E. Hearn, *Plutology, or the Theory of the Efforts to Satisfy Human Wants* (Melbourne, 1863). Hearn's career is documented in J.A. La Nauze, *Political Economy in Australia* (Melbourne, 1949).

33 B. Gandevia, 'William Thomson', *Australian Dictionary of Biography*, Vol. 6, p.271.

34 Ibid., p.272.

35 An obituary of Wood can be found in the *Medical and Surgical Review*, 7 May 1864, p.48.

36 J. Secord, 'The Geological Survey of Great Britain as a research school, 1839-1855', *History of Science*, 24 (1986), 223-75.

37 Thomson to Huxley, 26 November 1863; Imperial College of Science and Technology Archives, London.

38 Gandevia, op.cit. (n.33), p.271.

39 The *Argus*, 20 July 1863.

40 The *Argus*, 17 July 1863.

41 Gandevia, op.cit. (n.33), p.271; *The Lancet* (12 December 1863), pp.681-3; *The Medical Times & Gazette* (5 December 1863), p.596. The *Reader* article, dated 28 November 1863, was reproduced in the *Argus* on 19 February 1864.

42 W.A. Osborne, 'George Britton Halford: his life and work', *Medical Journal of Australia* (19 January 1929), p.69.

43 For a discussion of the methodological and philosophical problems associated with taxonomy and classification see J. Dean, 'Controversy over classification: a case study from the history of botany', in B. Barnes and S. Shapin (eds), *Natural Order: Historical Studies of Scientific Culture* (London, 1979), pp.211-30.

44   For Darwin's attitude see *On the Origin of Species by Means of Natural Selection* (London, 1859), pp.53-4.

45   Halford, op.cit. (n.8), p.16.

46   St George Mivart, *Man and Apes: An Exposition of Structural Resemblances and Differences Bearing upon Questions of Affinity and Origin* (London, 1873).

47   F. Wood Jones, *Structure and Function as Seen in the Foot* (London, 1944). Wood Jones does not mention Halford's critique of Huxley.

48   The *Argus*, 20 July 1863.

49   R. Owen, 'Darwin on the origin of species', *Edinburgh Review*, 11 (April 1860), 487-532; reprinted in D. Hull, *Darwin and his Critics* (Chicago, 1973), pp.175-215. The quotation is on p.187 of the latter.

50   The *Argus*, 3 August 1863.

51   Halford, op.cit. (n.8).

52   *Australian Medical Journal* (October 1863), pp.307-12; *The Lancet* (12 December 1863), pp.681-3; *The Medical Times and Gazette* (5 December 1863), p.596.

53   Wood to James Hector, 9 August 1863, Wellington Museum Records, Old Colonial Laboratories Box Files. I am grateful to Miss Joan Radford for copies of Wood's letters to Hector.

54   Huxley's letter to Selwyn has not been sighted; its existence is known only from a comment in Wood's second letter to Huxley, dated 24 February 1864 (Imperial College of Science and Technology Archives, London).

55   The *Argus*, 6 February 1864.

56   Thomson sent a copy of this paper to Huxley, see letter from Thomson to Huxley (note 37).

57   Halford, *Lines of Demarcation between Man, Gorilla and Macaque* (Melbourne, 1864).

58   *The Lancet* (18 June 1864), pp.700-701.

59   The *Argus*, 22 July 1863; *The Medical and Surgical Review* (July 1863). I have only seen the portion of this latter report reprinted in the *Argus*, 22 July 1863.

60   *The Medical Times and Gazette* (3 October 1863), p.362.

61   See n. 35.

62   The *Argus*, 20 June 1865.

63   The *Argus*, 28 June 1865.

64   P. Moloney, 'On some salient points in the anatomical structure of the gorillas lately placed in the National Museum', *Australian Medical Journal* (July 1865), pp.215-17.

65   W.C. Rees, 'The gorillas at the museum', *Australian Medical Journal* (August 1865), pp.271-2.

66   The *Argus*, 25 July 1865. Halford's lecture was published in the *Proceedings of the Royal Society of Victoria*, 8 (1865-1866), 34-49.

67   Halford to Owen, 22 September 1867; Owen Papers, British Museum (Natural History).

68   Huxley to Wood, op.cit. (n.7).

69   K.F. Russell, *The Melbourne Medical School, 1862-1962* (Melbourne, 1977). In 1869 Halford's name was put forward for the fellowship, his application being supported by Richard Owen and James Paget among others. These two had been his original supporters for the Melbourne professorship.

70   For McCoy's career see R.T.M. Pescott, *Collections of a Century, The History of the First Hundred Years of the National Museum of Victoria* (Melbourne, 1954), and A. Moyal, *'A Bright and Savage Land': Scientists in Colonial Australia* (Sydney, 1986).

71   For Mueller's career see M. Willis, *By Their Fruits: A Life of Ferdinand von Mueller, Botanist and Explorer* (Sydney, 1949). The letter to Darwin is dated 10 June 1874 (Darwin Papers, Cambridge University Library).

72   Moyal, op.cit. (n.70).

73   K.F. Russell, 'George Britton Halford', *Australian Dictionary of Biography*, Vol. 4, p.322.

I would like to thank Monica MacCallum, Rod Home, Jan Sapp, Richard Gillespie and Miranda Hughes for their assistance and encouragement in writing this paper. I should also like to thank Miss Joan Radford, Mr Peter Gautrey of the Manuscript Room, Cambridge University Library, Mrs Pingree at the Imperial College of Science and Technology Archives, and the staff of the British Museum Library, for their help in locating or supplying letters; also the director of the National Museum of Victoria for allowing me access to the museum records.

8

'Sweetness and light': Industrial research in the Colonial Sugar Refining Company, 1855–1900

George Bindon and David Philip Miller

The Colonial Sugar Refining Company (CSR), although conceived in London, was born at the farthest reaches of the Empire, in 1855. Australia was a mere seventy years on from the landing of the First Fleet. Yet CSR, almost from its beginning, placed the appropriation and production of scientific and technical knowledge at the very centre of its corporate strategy.

This chapter is a case study in the social history of industrial science in a peripheral economy. Here, are described the growth, and growing influence, of a cadre of chemists within the corporation, which imparted a character to CSR that parallels in both structure and time the development of the 'benchmark' enterprises at the metropolitan centres of the 'second industrial revolution' — especially in Germany and the United States. This is not supposed to happen in a decidedly colonial setting. The role of chemists, and the view that the company developed about the place of scientific and technical expertise in the enterprise are discussed. The degree of creativity and control that CSR's management displayed in competing through the use of 'state of the art' knowledge and technique is also examined. Technical details are discussed only insofar as they are germane to these broader considerations.

The most notable feature of the science systems of peripheral countries is the relative paucity of industry-based or industry-financed research and development.[1] Richer, more developed non-metropolitan communities such as Canada and Australia might succeed in creating a significant scientific enterprise as measured by indicators such as the number of scientists or research papers per capita. But this presence is achieved almost exclusively by the development of government research establishments and publicly financed universities. Such countries continue to be characterized by technological dependence, low industrial productivity and inordinate emphasis on primary resource extraction.

Science indicators that display expenditures on research and development by both sector of performance and sources of funds illustrate the peculiar weakness of the private sector in Australia, and also show that the various performers and supporters of research operate in almost hermetically sealed enclaves. On these criteria Australia occupies a position somewhere between the highly developed economies (i.e. the OECD group of which Australia is a part) and the less developed nations (to which some commentators suggest Australia is moving).[2]

Although there is a growing literature on the history of science and technology in countries other than the major industrial economies, the work has, for the most part, concentrated on high-profile individual scientists and those found in government establishments and universities. Historians of science have naturally tended to approach the problem of these minor scientific communities in terms that value what they observe according to the degree of involvement in, and contribution to, the contemporary scientific discourse at the metropolitan centres.[3] While some recent studies of the imperial dimension of Australian science pursue a 'top down' approach in a rather different way, more attention is now being paid to relations between what Inkster has called the local 'scientific superstructure', the 'cultural-institutional infrastructure' and the 'socio-economic base of support'. And it is here that studies of Australian science and technology begin to converge in a sectorally focused historiography delineating the relations of science, technology and production in late nineteenth- and twentieth-century Australia.[4]

The approach taken in this chapter begins with the question of production and examines the nature of the social agents that contribute the knowledge embodied in production processes. From this viewpoint, the creation of wealth, the distribution of surplus, the conscious intention of the local élite to maximize its dominion, are the standards by which the procurement, production, and deployment of knowledge are valued. While understanding that the participation of the periphery in 'high' scientific discourse is important, confining attention to this part of the problem has meant that a number of major questions have not been seriously addressed by historians of Australian science and technology: why, in the late nineteenth and early twentieth centuries, when an epic alliance was taking place in the metropolis (particularly the USA and Germany) between institutions that produced organized knowledge and those involved in the production of goods and services, did the relationship apparently fail to 'jell' in peripheral societies? Why, despite sometimes quite dramatic success in developing institutions in which scientific knowledge and manpower are created, did the industrial sector not appropriate that competence to create innovative and competitive enterprises? Why did industry, rather, generally continue to draw on the central economies for industrial know-how? It is unlikely that these questions will be answerable without careful examination of the formative period of scientific institutionalization *and* industrial development in the peripheral economies.

CSR was chosen as a case study because this company, since the late nineteenth century, *has* seen the production and appropriation of scientific and technical knowledge as a major pillar of its corporate strategy. By the middle of the twentieth century CSR was one of the world's major raw sugar manufacturing companies and its research operations were the largest private scientific organization in the South Pacific.[5] In the Australian industrial and R&D landscape, CSR stood out as an anomaly.

The story of CSR appears to parallel that of many of the industrial giants of the metropolitan centres, which seized the possibilities presented by the emergence in the nineteenth century of formalized, organized knowledge —

particularly in chemistry and engineering — and contributed to the creation of this knowledge. While 'colonial' enterprises (colonial being defined as a structural relationship to metropolitan centres rather than a jurisdictional relationship) are often characterized by the application of the latest techniques to virgin resources, the initial establishment of these techniques usually takes place in the metropolitan centres. The peripheral companies, at best, concern themselves with minor adaptations to local conditions. The style of management is operational rather than innovative, and generally the end results are industries dominated by foreign interests. By contrast, CSR, which came to dominate the Australian sugar industry, remained almost totally Australian owned, its management was innovative almost from the beginning, and its R&D activities involved more than minor adaptations to imported technological 'packages'. Thus the case of CSR allows investigation of the peculiar conditions and difficulties, if any, of a colonial enterprise developing an aggressive innovative strategy, and the degree to which there were fundamental historical constraints that restricted the industrialization of innovation in a peripheral economy.

This chapter restricts itself to an examination of the early history of innovation in CSR. In Schumpeter's terms, the company evolved from a classical technically entrepreneurial venture to a 'trustified' capitalist corporation in which progress was 'automated' as innovation became institutionally embedded in the corporate structure. Not only was the institution of industrial research established within the company, but it also became an integral part of business strategy.[6] By the charting of this development tentative insights into the ambitious questions posed above may be gained.

Three phases in the company's history between its establishment in 1855 and *circa* 1900 are examined. In the first phase (1855 to the late 1860s) the company was involved in the refining of raw sugar imported into Australia. The difficulties experienced induced a decision to move into milling, and efforts were nade to control Australian-grown cane as its feedstock. During the second phase (late 1860s to *circa* 1880), CSR established the centralized mill system. Management engaged in technical entrepreneurship and supported a rush of experimentation and innovation. In the final phase (1880 to 1900) the company moved towards a 'trustified' corporate structure with innovation embedded in the company. In particular, a cadre of chemists was established, which increasingly commanded the management process through the enthronement of a system of chemical control.

Early history, 1855–1868

The establishment of the Colonial Sugar Refining Company in 1855 was the outcome of protracted efforts to develop sugar refining in Australia on a stable footing.[7] The very earliest ventures were dogged by the slump of the early 1840s, under-capitalization, dissension among the partners and competition from importers. It was against this background that ten Sydney investors, including a number of former directors of the Australasian Sugar Company (founded 1842, dissolved 1854) formed the Colonial Sugar Refining Company.

One of those former directors, Edward Knox, was to take the leading role in CSR's early years.

Edward Knox (1819–1901) was born and educated in Denmark. After a thorough commercial training at Lübeck in Germany, at sixteen he was attached to his uncle's merchant house in London. This commercial training, his European contacts, and his command of English, Danish, French and German were later to prove invaluable in the conduct of CSR's affairs. Knox came to New South Wales in 1840 with the intention of making his fortune as a pastoralist. But he soon turned to the greater attractions of Sydney. There he was appointed manager of the Australasian Sugar Company, a director of the Commercial Banking Company and to other positions in Sydney's commercial life.[8] Knox's personal enterprise in transforming himself into a prosperous and influential member of the capitalist class of the colony while not yet thirty was matched by the recognition he gained as an astute accountant and business manager with a flair for eliminating competition.

CSR was formed as an unlimited liability partnership with a capital of £150 000. As the chairman of the company, Knox carried one-third of the liability but only 40 of the 352 votes. Although apparently established on a firmer base than its predecessors, in its first years CSR still exhibited the somewhat 'hit and run' tactics of enterprise at the periphery. In its first year, despite heavy debts, a dividend of 50 per cent was declared. In 1857 Knox sold his home and £20 000 of his stock and returned to Europe with his fortune. However, an international fall in the volatile sugar market, an Australian depression and an attempt by one of the shareholders to split from CSR to establish a rival concern, brought a badly shaken Knox back to Sydney with a determination to save the business and re-establish his reputation.[9]

Knox realized that he had entered into an unstable enterprise, which not only faced competition from direct importers but was also vulnerable to uncertain raw sugar feedstock markets. The sugar-from-cane business was shaken by the explosive success of the subsidized ('bounty-fed') and innovative European beet-sugar industries. The production process was also unreliable and complex. The emphasis of the new corporate strategy that was forged out of this initial experience was one of unending efforts to reduce uncertainty in every way possible: by eliminating the debt and financing growth internally through retained earnings; by reaching back to control the sources of supply and forward to the uncertain markets; and, most importantly, by maximizing in every way possible the understanding and control of the full spectrum of the manufacturing process through acquisition of scientific and technical competence.

Knox's correspondence with the banker Donald Larnach reveals major features of CSR's problems and progress in the 1860s. The company at this stage was still engaged only in the refining of imported raw sugar. In September 1865, Knox recognized the pressures to improve the quality of the local product as a consequence of foreign competition:

> Improvements in the qualities of sugars imported from Java and Mauritius has called for similar improvement in the qualities we manufacture and the consequence is that our 3rd quality sugar is now far superior to what our 2nd used to be . . . The largest Profit attaches naturally to the highest qualities . . . All I can

say is that I am doing my utmost to produce satisfactory results and setting aside
that it is my duty to do so, my stake is so large as to render any other course out of
the question.[10]

Thus technical improvements in the refining process assumed considerable
importance at this stage and were closely identified with Knox's self-interest.
However, the company's problems were not easily solved and 1865–66 was a
bad year. Larnach wrote to Knox on 20 April 1866 that he could not conceal his
'very great disappointment with the result of your operation for the past year'.
Larnach was prepared to write off his investment (£6000) rather than send
good money after bad. He suggested withholding dividends to finance the
operation back to health.[11] While Knox went along with this policy, he did so in
full awareness of the need for innovation. Edward Knox was above all an
astute financial manager and entrepreneur. He had described himself some
years before as only a 'theorist in sugar boiling'[12] and this he remained. These
early experiences were sobering. Self-financing and tight management with a
view to underwriting investment in innovation became a part of the company's
culture.

As an entrepreneur, the next major step taken by the elder Knox (Knox's
son, E.W. Knox, had also joined the company) was to take CSR into the
milling of sugar cane — a vertical integration of the company's operations
designed to give greater control over the supply of its feedstock of raw sugar.
Together with this strategy, Knox's awareness grew of the importance of
maximizing productivity through systematic consideration of all stages of
milling and refining with the aid of whatever organized knowledge could be
obtained from external sources or the encouragement of experimentation
within the company. Initially this was to mean management becoming more
directly involved in the details of innovation and experimentation. Later it
was to mean that an integral part of the company's operations would include
both the acquisition of specialist expertise, notably chemists, from outside,
and then their training within the company itself.

The move into milling and the apprenticeship of E.W. Knox, 1868–1880

The growing and milling of sugar cane to produce raw sugar was established in
the Australian colonies in various forms prior to CSR's move into milling in
the 1870s and 1880s. Commercial production began in earnest in Queensland
from about 1863, and the next twenty years saw rapid expansion. The
structure of the industry there followed the traditional model of tropical cane
production, with plantations combining the operations of growing and milling
and relying on indentured Melanesian (or Kanaka) labour supplied by 'black-
birders'. 'Black-birding' was the business of 'recruiting' labourers from the
islands of the South Pacific under conditions that ranged from legitimate
contractual arrangements, to kidnapping and out-and-out purchasing of what
could, without too great an exaggeration, be called slaves. The Queensland
industry exported the bulk of its raw sugar to the Melbourne market, the rest
supplying the colony's own needs.

Although CSR was to have a major impact on the Queensland sugar industry when it established mills there in the 1880s, the company's earliest ventures were in the Northern Rivers region of New South Wales. CSR's technical entrepreneurship transformed the sugar industry in that region and gave it a secure commercial basis. The strategy was to establish large central mills with the cane supplied by independent 'yeoman' farmers. Centralized mills were to provide CSR with a competitive advantage through considerable economies of scale that facilitated use of the latest and best in sugar-milling technology.[13]

The initial reasons for CSR's move into milling in New South Wales were defensive. Cane growing had expanded in New South Wales in the 1860s, and by 1869 nine mills erected by growers were in operation. There was a consumer market for the raw sugar produced by these mills so it could be sold without recourse to a refinery, and it also avoided the £5 duty levied on CSR's imports of raw sugar. So cane growing in New South Wales presented a threat to CSR's position. Thus the company's efforts were aimed at minimizing this threat and maintaining stable prices for the basic feedstock of its refineries. It seems that in the late 1860s the company's management did not foresee that milling would prove to be a major source of future profits. By adopting a central mill system, rather than establishing its own plantations, CSR sought to capture the production of independent growers while avoiding the additional problems of actually growing the cane.

In 1868 CSR employed an experienced West Indian planter, Melmoth Hall, to assess the cane-growing areas of the Northern Rivers and the supplies of cane that could be secured from independent growers. On the basis of Hall's reports, it was decided in 1869 to erect mills on the Macleay and Clarence rivers. In 1870 the first cane was crushed at the Darkwater Mill on the Macleay and at the Southgate and Chatsworth Mills on the Clarence. But, there were many problems. The Macleay venture was plagued by an inferior quality of cane and much of the cane supply was captured by independent mills. As a result, the Darkwater Mill was moved to the Clarence in 1873. But on the Clarence, the Southgate and Chatsworth Mills already had cane supply problems due to the competition from small mills. In the 1880s this would mean the consolidation of the older Clarence River mills at Harwood and the erection of mills in other locations — the Condong Mill on the Tweed River (1880) and the Broadwater Mill on the Richmond River (1881). Cane supply at the Condong and Broadwater Mills relied for a time on partial use of plantation modes of control. By that time CSR was not so single-mindedly devoted to the central mill strategy because, to the defensive concerns that had spawned that system there was now added a positive desire to expand milling and hence to develop new sources of supply of cane. As early as the second season of operations in New South Wales, CSR's profits from milling had exceeded those from refining. In the late 1870s and 1880s CSR continued its move into milling by establishing mills in Queensland and in Fiji.[14]

Implementation of a new technology on a large scale stimulated innovation and experimentation in the company. In this respect the early career of Knox's son, Edward William (1847–1933), is of great interest. In 1869, at the age of twenty-two, E.W. Knox was appointed superintendent of the Northern Rivers mills. During the next ten years he mastered most aspects of sugar manufacture and refining technology before assuming the post of general manager in 1880. It was through his activities that innovation and experimentation became a vital and sustained concern of management.

At first E.W. Knox found his task rather daunting. The mill technology, imported mainly from Britain but also from France, represented a massive capital investment; and it was by no means easy to render it operational. There were breakdowns and fatal accidents. But from the start Knox appears to have been convinced that a major task was to increase the efficiency of the recovery of sugar from the cane at all stages of production.[15] This is clear from his regular reports to Joseph Grafton Ross, the general manager of CSR from 1870 to 1880.

Initially Knox had to rely on the knowledge of the company's engineers and mill managers. For example, in 1872 he reported to Ross that he was at the Chatsworth mill and the manager had told him of 'his Experiments with taking back the Syrup and how by this plan he increases the yield of sugar:- I cannot quite make out why there is a heavier yield . . . '[16] At this stage Knox displayed an acute awareness of the possibilities for technical improvements, but a certain vagueness about how to proceed, especially in matters outside the particular competence of the company's engineers and mill managers.

The introduction of the polariscope or polarimeter forced the company to look outside for help of a new kind. This instrument was designed to give a quick and accurate measure of the sugar content of solutions by measuring the degree of rotation of a beam of polarized light passed through the solution. While it first came into use in the sugar industry in Germany and the United States in the 1840s, it was slow in replacing hydrometric determinations of specific gravity.[17] CSR was by no means unusually tardy in acquiring a polariscope in 1873, but they needed outside help to use it. Sir Daniel Cooper, one of CSR's original shareholders, wrote to Edward Knox from London:

> I hope by the aid of Professor Smith that you will work the polariscope, and with the help of the other chemist inform yourself as to the component part of the juice and syrup, so you may know exactly what enemies you have to contend with. If you once know this I feel certain your work will be greatly facilitated and success be sooner attained.[18]

Thus we see the first glimmerings of recognition within the company of the possibilities offered by the competence of chemists.

Another factor came into play as a result of a visit to the West Indies by E.W. Knox in 1876, during which he investigated the local sugar industry. On his return he introduced 'double crushing'. This was based on the realization that a significant amount of usable sugar remained in the cane after one squeeze.[19] The return on investment in the additional plant involved could be increased through accurate measurement of the remaining sugar in the cane, and, based on this information, striving to adopt the most effective means of ensuring that as much of this residue as possible was recovered. Thus the engineering decision to install a second set of rollers led to an enhanced interest in sugar chemistry. As we shall see, physical analysis in turn presented possibilities for increasing intelligent control of the whole production process from selection and planting of the cane through to the marketing of the final products. But the important point at this stage is that by 1876 E.W. Knox was initiating innovation. He wrote to J.G. Ross in 1876:

> I am having some experiments carried out . . . in clarifying to see if more juice can be used by mixing more sulphur . . . As regards the double squeeze . . . my ideas as to what is wanted are now coming into shape. — By the middle of next week I hope to have everything settled . . . [20]

Activity was feverish throughout the decade, with the time between experimentation and application measured in days, if not hours. Often in a single day Knox would receive experimental results from the mills and fire off a number of letters to mill managers specifying further studies.[21]

By 1880 CSR already had well-developed commercial and technically-innovative strategies. Convinced of the profitability of milling, especially when conducted on a large scale and with due attention to continued technical improvements, CSR was already committed to establishing mills in Queensland and had for a number of years been seriously considering establishing operations in Fiji. This expansion was now predicated on profits expected from milling and a defensive strategy of anticipating the competition. While Edward Knox had been merely a 'theorist in sugar boiling', his son had acquired extensive practical knowledge of sugar technology and chemistry, which had been put to good use. And yet both were aware of the wider scene and in particular of the rise of the beet sugar industries of Germany and France. They recognized that the strength of those industries and their increasing superiority in the international sugar market derived not only from the bounty system but also from the organized application of scientific principles and scientific manpower to sugar production.[22] It was to such organized application that the company turned, stumbling a little at first, in the 1880s.

Growth of the chemical staff and some early work

CSR's investment in specialist scientific expertise began in 1879 when Andrew Fairgrieve from Greenock in Scotland was hired. He was described as a chemist with 'large experience in refineries in England and France' — something of a novelty in Australia. The Company did not know how much to pay him, writing to London to ascertain the going rate for such an employee.[23] And there was uncertainty as to his proper function. Most of his work was done in connection with the mills, rather than at the refineries as had initially been intended. Judging by E.W. Knox's postmortem on this first attempt at using a chemist, the experiment was a failure. Speaking in 1890, he recalled:

> About ten years ago, we were led by the great attention then being paid to chemical research in connection with the beet-sugar industry to commence such investigations into our mode of work. We did not know clearly what we wanted, nor did the man we engaged, so the first start was not a success; but it showed us that we were on the right track . . . [24]

But this was said with the benefit of hindsight. At the time there were still sufficient hopes invested in Fairgrieve to justify giving him some more resources. A decision was made to engage a junior. Knox wrote to the company's London agent that 'we find that we have derived so much benefit from the researches of our chemist during the past few months, that we have decided on engaging a junior to assist him in the routine work in Sydney'. The chemist's prospective assistant was envisaged as being about twenty years of age with a good knowledge of the chemistry of sugar, familiarity with French practice and of 'good character'. The man engaged was in fact Thomas Utrick Walton (1850-1917).[25]

Walton, like Fairgrieve, was a product of Greenock, the Scottish 'saccaropolis'. But he was far from being the stuff of which juniors are made. He was educated at Greenock Academy and at the age of sixteen entered Glasgow University. There he studied natural philosophy in the laboratory of Lord Kelvin and engineering with Professor Macquorn Rankine. Graduating as a Bachelor of Science and of Civil Engineering and obtaining the major prizes of his year, Walton next studied technical chemistry at Anderson's University. He was subsequently elected a Fellow of the Chemical Society of London and of the Institute of Chemistry. Evidently intent on a career in industry in his home town, Walton was employed as a refinery chemist, first by Crawhall, Alison & Co. and then by the Scott Co. where he gained considerable experience. Although Walton was undoubtedly an attractive recruit, the mismatch between his qualifications and experience and those initially envisaged by CSR for their 'junior' is indicative of continuing uncertainty within the company about its precise requirements.

Walton's primary expertise was in refinery chemistry, and he oversaw innovation and control in CSR's refinery operations. But he also played an important part in devising and instituting a system of chemical control. Yet another task that Walton undertook during his long tenure of the post of chief chemist was the training of young chemists within the company.

The next man to be engaged was Thomas Steel (1857-1925) who became the first chemist to be specifically stationed at one of the company's mills — Condong on the Tweed River. Meanwhile, Fairgrieve was suggesting further recruitment. Knox replied that he 'should have no objection to getting one or two out' and suggested that a 'first rate German chemist' would be ideal. The reasoning here was that extensive manuring of the cane fields would have to be done at Mackay in Queensland and a German chemist would ensure that this was done 'scientifically'.[26] Two German chemists were engaged in early March 1883. But once again the recruitment process was quite haphazard and ideas about how these gentlemen were to be employed were continually shifting. While the appointees, Dr Gustav Kottmann and Ernst Marquardt, were on the high seas, E.W. Knox and Fairgrieve discussed what was to be done with them. It was not until late April that Knox received the testimonials of the new chemists. He wrote to Fairgrieve with evident surprise:

> The Elder of the two (Kottmann) appears to be a man of very varied experience, and as he could not have held the situation he filled without being a good Chemist he should be of great service to us. The knowledge he possesses would appear to point to his fitness for the post of Sub-Inspector, but of course much must depend upon his possession of other qualities which are necessary for such an officer.[27]

Kottmann (1852-1936) was a man of impressive credentials, with a PhD in chemistry and experience as assistant to the director of the Institut für Zuckerindustrie in Berlin. The latter was a co-operative research institute established by German beet sugar manufacturers in 1867. Given the remarkable innovative powers of the German industry, encouraged by the bounty system and by the close relations between the universities, technical schools and industry in that country, Kottmann had been at the very centre of things. However, Knox's initial attitude towards Kottmann, respectful but cautious, indicates that the German's credentials and experience were not to give him any automatic authority in CSR. On meeting Kottmann for the first time Knox considered him to be 'intelligent' and was impressed to learn from some of Kottmann's fellow passengers from Europe that the chemist had gained his knowledge of English from scratch during the sea voyage. Nevertheless, Knox's closest technical advisers were asked to form a judgement of Kottmann's capabilities. One such judgement, by Fairgrieve, reveals both a hard-headed concern about Kottmann's awareness of practical realities in the company's operations and a rather defensive one-upmanship. Certainly Kottmann had to earn his position within the company. But subsequently, as inspecting chemist from 1885, he supplanted Fairgrieve. The latter was given notice apparently because of his failure to create an ordered programme of work with clear relevance to the operations of the company, and his inability to draft reports in a style that was consistent and accessible to the management.[28]

Kottmann became CSR's main technical 'troubleshooter'. For a number of years he was to live a peripatetic existence, visiting the mills, instituting, co-ordinating and rendering routine a system of chemical control, performing an increasing number of elaborate and protracted experiments on such problems as carbonatation and the diffusion method of sugar extraction, devising the 'P.O.C.S.' formula, solving the technical problems associated with CSR's establishment of its Fiji mills, and performing field experiments on the culture of sugar cane.

The chemists quickly established a major presence as a recognized group within the company. The growth in the total number of chemists and analysts employed by CSR is shown in the graph below. This expansion was a consequence of the recognition of the need for, and successful definition of the role of, chemists and analysts. This was reinforced by a steady growth in the number of mills and refineries operated by the company. The disposition of chemical staff among these operating units reveals an orderly increase in the number of such staff in the mills, refineries, and at the head office laboratory.[29]

As suggested by the title 'chemists and analysts', the chemical staff were not a homogeneous group in terms of their expertise or their functions. There was a definite hierarchy among the chemists. Walton and Kottmann emerged from the small pool of early overseas recruits to become the leading chemists in their positions of chief chemist and inspecting chemist respectively. Their status is indicated by salary level, proximity to 'general management' in the line of command, and by their responsibility for the initiation and oversight of innovation. The next stratum was what might be called the 'better sort' of chemist — men with significant training and/or experience, who could take

Young chemists at a CSR sugar mill in 1899. (Courtesy of CSR Limited)

general charge of the chemical work at a mill or refinery, be responsible for chemical reports from that unit, and be able to conduct experiments either independently or under instructions from the leading chemists. Finally, the largest group of rank-and-file chemists functioned entirely as routine analysts. E.W. Knox, writing in about 1884, described the situation concisely: 'The Chemists are Walton, Kottmann and Marquardt. Steel, Flint & the two Helms' are rather better than analysts. The others have been mainly trained by us.'[30]

Through the 1890s and the early 1900s a small number of chemists were recruited in Britain and Germany. A larger number of Europeans by birth and training already resident in Australia were engaged, as well as a significant body of 'Australians'. It appears that chemists of the 'better sort' came from the first two of these categories. There was considerable stability of the chemical leadership, but when, in 1902, an assistant chief chemist was sought he was recruited from overseas. (That job, incidentally, attracted some 250 well-qualified applicants!)[31] All but the most senior chemical staff were subordinate to mill and refinery managers. The eventual destination for many of the men who succeeded in working their way up from analyst to assistant chemist to head chemist at a mill was general management of a mill or refinery.

In characterizing what the chemists did it is useful to examine first the early activities of the chemist from whom perhaps most was expected — Dr Gustav Kottmann.

It was decided to send Kottmann to Fiji to tackle teething troubles in the new mill there. But before going he visited the Clarence River mills in New South Wales. At the Harwood Mill he began a series of experiments on the carbonatation process. Essentially this was one of a number of stages whereby juice from the mills was clarified. It involved the addition of an excess of lime, which was subsequently precipitated by the action of carbon dioxide gas. The resulting calcium carbonate mechanically carried down many of the impurities in the juice. Described by Deerr as one of 'the most important steps forward in the clarification of sugar juices, both beet and cane', carbonatation had been in use in the beet industry for some time but was introduced in the cane industry only in the 1870s. The expectation was that Kottmann, with his knowledge of the process in the German beet sugar industry, could improve the process in CSR's mills. Writing to the general manager from the Harwood Mill, Kottmann noted that he had had 'very important and surprising results' from his investigations of the nature of the precipitates obtained in the different stages of the process, indicating that there was evidence that some of the precipitates were again decomposed by the carbonic acid gas. Head office had directed Kottmann to proceed to the Fiji operation but he asked for a delay as he had made 'some advance in the knowledge of the carbonatation process' that suggested it would be worthwhile for him to remain. This first potential conflict between Kottmann's 'on-line' duties and experimentation was decided, albeit temporarily, in favour of the latter.[32]

Kottmann next wrote requesting advice about the point at which increases in clarified juice obtained using greater quantities of lime would lead to diminishing financial returns. He noted that in the European beet industry,

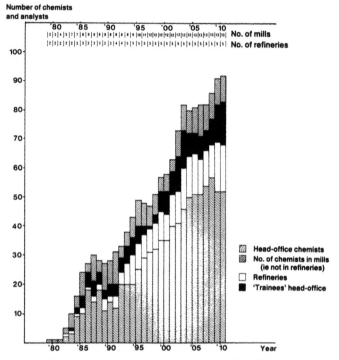

Number and location of 'chemists and analysts' employed by CSR, 1879–1910.

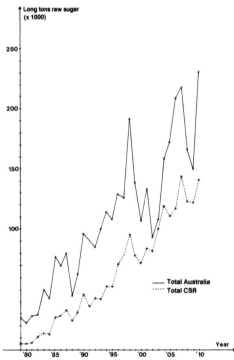

Raw sugar production: total Australia and CSR (Source:A.G. Lowndes (ed.), South Pacific Enterprise: The Colonial Sugar Refining Company Limited *(Sydney, 1956). Appendix II, pp. 435–7)*

2 to 3% of lime of the weight of the beet is used ... but that even 5% are applied'. Unfortunately, as head office pointed out, the economics of the process were very different in the South Pacific due to the high cost of lime. Here was a fine illustration of the type of problem faced when adapting a process employed in the European beet sugar industry to cane sugar in the South Pacific. But Kottmann was not deterred: within a week, using the differences between the processes for producing cane and beet juices to his advantage, he had obtained a good result using only a small amount of lime.

Later in the same year, while in Fiji, Kottmann conducted experiments on the application of another process. This was the diffusion method of juice extraction in which the cane was sliced and pulped and the sugar dissolved from the unruptured cells in closed vessels at 90°C. Diffusion was in extensive use in the beet sugar industries of Europe by the late 1860s. It offered a more elegant, and potentially more efficient process than the brute force of multiple crushing. The cane industries in a number of countries sought to emulate what they took to be the advanced methods of the Germans and the French. Major investment in this technology was made, for example, in Louisiana during the 1870s, particularly at the Belle Alliance plantation. And diffusion was one of the methods that Dr Harvey W. Wiley, appointed head of the Division of Chemistry in the US Department of Agriculture in 1883, looked to in his attempts to place the US sugar industry on firm scientific foundations.[33] The work undertaken on diffusion by Kottmann and CSR was thus concurrent with its attempted application to cane at the metropolitan centres.

In July 1884 Kottmann, then stationed at the Nausori Mill in Fiji, wrote to E.W. Knox outlining a scheme for operating diffusion. Worried at the time about the low recovery rates of sugar from the cane in the Northern River mills, Knox replied that he was interested in anything offering any chance of success, and gave the go-ahead. A model of Kottmann's suggested apparatus was rapidly constructed and tested at the Harwood Mill. And, through consultation between Kottmann, company engineers, and draughtsmen, designs for a large machine were prepared. Knox proceeded with these designs 'because if the results promise well, I should be disposed to fit the apparatus to each of our mills — the reduction by one half of the present loss of sugar in the megass (the fibrous cane residue after crushing) meaning a saving next year of 1500 to 2000 tons of sugar'. Satisfied that they had developed a unique design, plans and specifications were drawn up in order to apply for a patent. Detailed drawings were also sent to England together with an order for a number of machines. E.W. Knox reported on the project in glowing terms to his father, then in England, saying that he thought the thing 'too good to be made public', and that his present idea was that 'we shall keep the thing in our own hands here and let Kottmann have any royalty obtained if the system be adopted in other parts of the world'.[34]

But all did not run smoothly. The following year W.H. Rothe, a company officer acting for E.W. Knox who was then in England, was forced to write persuasively to Kottmann, now struggling to work-up the diffusion process at the Harwood Mill. The crushing season was upon them and Rothe asked whether it would not be advisable to revert to the 'old system of work' if the diffusion apparatus were not soon operational. Kottmann resisted. Noting the chemist's strong objections Rothe assured him that it was not proposed to abandon the method but merely to suspend its trials so that the mill could be worked steadily. Telegrams flew between Harwood and head office for days. Finally, faced with the urgency of getting the cane into production, Rothe laid down the law:

> I must say that though I am as anxious as you are to give the apparatus a thorough trial it seems to me that we cannot go on experimenting with it much longer . . . as the cane in its present damaged state must be taken off as quickly as possible . . .[35]

A year later, diffusion still held out its possibilities and a watching brief was kept on the technology.[36] The method was employed for a while in some of CSR's mills but, as elsewhere in the world, it was gradually abandoned. While technically elegant and capable of offering significant increases in sugar extracted under controlled conditions, diffusion proved uneconomical for the cane industry, in particular because of its high energy costs. However, for CSR and the industry more generally, it was not a total loss. As Deerr argues, the cane sugar industry ultimately benefited because the experience with diffusion helped to draw attention to many shortcomings of milling as then conducted.[37]

Carbonatation and diffusion are just two examples of the way in which the arrival of the chemists transformed the innovative climate within CSR. But their major impact lay not in particular discrete innovations, though these were undoubtedly very important. It lay, rather, in a system of chemical control that affected all stages of production and significantly altered management practices and patterns of authority within the company.

Chemical control, managerial control and the chemists

The essence of chemical control lay in a sophisticated, continuous and comprehensive monitoring process giving important quantitative information about the cane, the megass, the juice and so on at all stages of production and at different parts of each stage. It provided criteria of optimum efficiency given a particular technology, indicated 'black spots' at or within particular stages of production and provided both incentive and information conducive to technical improvements designed to bring overall efficiency closer to the theoretical maximum possible. Apart from the conceptualization of the process of chemical control, its introduction depended on devising monitoring, sampling and analytical techniques at numerous points in the production process. Besides major innovations, these were the kinds of problems that exercised Walton, Kottmann and the 'better sort' of chemists in their early years with CSR.

The increasing attention paid by head office to the rigorous improvement of the production process and the contribution that professionally trained scientists could make also led to important changes in management practices. Such changes were eventually to affect the whole corporate structure. Kottmann, as inspecting chemist, touring the mills and suggesting improvements, precipitated the changes.

While visiting the Broadwater mill on the Clarence River in September 1885, Kottmann had a nasty altercation with the manager, a Mr Wyness. In an urgent telegram to the general manager in Sydney, Kottmann wrote in a style that probably reflected his agitation as much as his imperfect English:

> After several days apparent good agreeing with Mr Wyness when speaking this morning over losses at the second presses Mr Wyness told me that I might take immediate run of the place immediately afterwards he showed me a telegram about his resignation directed to you. He told me that I had to leave the mill as he was the manager and had no intimation that I was sent by the Company . . . I ask you what to do.[38]

This was followed the next day by a letter in which a nervous employee is seen walking a tightrope. What exactly was his status in the company? What should his role be in relation to a mill manager? He made a detailed case that there was significant and unnecessary loss of juice for various technical reasons, and that this was why Mr Wyness 'went off'. Wyness' behaviour was a personal affront and also an insult to head office and the company. He had 'full confidence in the Company which has sent me to Broadwater Mill, that this matter will be settled in a way which will give satisfaction to me'.[39]

The general manager sympathized but suggested that Kottmann tactfully pacify Wyness. The inspecting chemist acknowledged the suggestion but again made clear his view of the problem:

> The difficulty is that suggestions even when expressed in an unpretentious manner, are sometimes looked at much more from a personal than a business point of view . . . When dealing with a totally unbusiness-like ambition it may even happen that the anger of the other party will be the greater, the better the suggestion has been. Another trick played with more or less nonchalance . . . is to make only the plans sent or the instructions given, responsible in case of bad working. This is sometimes next to hum bugging him who — this being his duty — proposes or suggests to work the plant in a better manner. Should I not be allowed in such a case to point out that the plant is not worked properly?[40]

'Unbusiness-like' was, for Kottmann, analogous to unscientific. The dispassionate technical man had a duty to both rational efficiency and the company to call a spade a spade. The traditional mill managers had had their day, but they did not yet know it. Such strains and tensions are familiar to the historian as examples of the conflicts of authority often experienced when science begins to establish its presence in an industrial setting.[41]

Over the next year the assertion of technical control over the factory floor became complete. Kottmann, Walton and the chemists emerged as the keystone of this new dimension of managerial mastery. A battery of communications calling for experiments, reporting of results, together with directives establishing well defined standards of operation, flowed between Kottmann and the various mills. When the mill manager was unsympathetic or unfamiliar with the emerging style of management, as at Broadwater, Kottmann dealt with someone else. During 1885–86 Kottmann devised the P.O.C.S. (pure obtainable cane sugar) formula which was a clever combination of elegant chemical analysis of the sugar content of the juice combined with a practical assessment of the capabilities of the physical plant. The systematic use of the polarimeter to determine the sugar content of the juice through all stages of the manufacture made it possible to institute a system of 'chemical bookkeeping'. 'Scientific' management could be based on an exact measurement of the amount of sugar recovered or lost at each stage. Chemical control became the basis for evaluating innovations in, and adjustments to, the production process, and use of the P.O.C.S. formula indicated the 'unconquered margin' — the difference between the percentage of sugar in the cane and the percentage actually recovered — which the efforts of all company personnel should be directed towards reducing.[42]

As early as 1884 E.W. Knox had announced to the mill managers the intention of instituting a system of chemical bookkeeping. By June 1886 Kottmann was in a position to send to all mills a table that identified thirty-one stages in the production process and was organized on the basis of a six-day week. It indicated the tests to be done at each stage and their frequency. These tables were to be completed and returned to head office each week.[43] Thus, although the system of chemical control would be subsequently revised, it was established in its essentials by 1886. The ramifications were many.

Chemists were now ensconced in all the mills, with laboratory space set aside for their work and their tasks duly specified. The troublesome question of their authority and that of the mill managers was also resolved in an unambiguous and decisive way that marks a watershed in the evolution of the company. In a circular to all managers in May 1886, E.W. Knox noted that some dissatisfaction had been caused because the position and duties of the chemical staff had been ill-defined. Drawing attention to the 'enormous quantity' of sugar lost in manufacture, he stated:

> Now, the experience of the past few seasons in which we have utilized the Services of the Chemists has shown very clearly that a diminution of this serious loss can only be obtained by a close and constant watch on the manufacturing processes, and such a watch can always be kept best by men trained to observe carefully and possessing a knowledge of the chemical changes which the juice etc. may from time to time undergo. I therefore wish that to the chemical staff be entrusted the oversight and, to a certain extent, the control of the manufacture from the rollers to the packing of the sugar. The Managers will, as hitherto, have the regulation of the work entirely in their own hands . . . but they will be expected to give effect to any reasonable requests of the chemists, and be guided by their advice in all questions relating to the treatment of the juice, sugar and syrups.[44]

Although there remained some resistance to the new authority of the chemists — Mr Wyness in particular proved 'still unwilling to avail himself' of their services — Knox quickly put it down. The new system also required adaptation by senior management. Previously E.W. Knox had been master of most of the scientific and technical developments in the company. However, the full impact of chemical control overwhelmed him. He found it impossible to process the chemical reports quickly enough. He also began to find his expertise insufficient: 'even if I could spare the time it is quite likely that I would miss many points to which reference should be made'. In future the entire oversight of the chemical reports and the assessment of the action to be taken in the light of them would lie with the chemists at head office. Knox would merely sanction the chemists' recommendations.[45] CSR had now become chemist-dependent — a kind of corporate 'junkie' that could not do without its technical fix.

The new role and authority given to the chemists and the continuing investment in such specialist personnel were based on some evidence, and considerable confidence, that such measures would pay off. The first consideration of a system of chemical bookkeeping by CSR coincided with the drastic international slump in sugar prices of 1884. This slump was precipitated by the massive production of the bounty-fed and technically innovative European beet sugar industry. In 1886 E.W. Knox visited Germany and on his return reported to the CSR Board of Directors that:

> [t]he extraordinary success achieved in the German beet sugar industry is almost entirely due to the union of science and practical skill which is manifest in every detail ... We cannot do better than follow the example of our German competitors in making chemical research their main standby in all their operations.[46]

This decision lay at the base of a successful competitive strategy.

When we examine CSR's changing competitive position in the Australian colonies and Fiji in the 1880s and 1890s it is clear that they survived the crisis of those years and emerged from it in a stronger competitive position. In the Northern Rivers district numerous mills were forced to close, leaving the field open for CSR and its rationalized, large-scale and highly efficient central mill system. In Queensland, too, the industry experienced a 'shake-down' in which CSR's competitive position improved. But the support given to Queensland planters and millers by the Queensland government, occasional punitive measures against CSR, the encouragement of milling co-operatives, and efforts culminating in 1900 with the establishment of the Queensland Bureau of Sugar Experiment Stations, kept the competitive situation more fluid there. In Fiji, the fall in sugar prices accentuated the entrepreneurial and technical shortcomings of both small, and larger but more diversified, companies. CSR by contrast, with considerable financial resources thanks to its conservative financial management, specialization in sugar, and heavy investment in milling elsewhere 'was more able to finance the required research than proprietors with fewer mills over which to spread the cost. The result was that CSR developed skills that far surpassed those of other millers in the colony'.[47]

CSR's rise as a producer of raw sugar is charted in the graph (p. 182). The clear trend was for the company to grow steadily by contrast with the 'boom and bust' character of the Australian raw sugar economy taken as a whole. And it can be seen that CSR's output quickly came to dominate Australian raw sugar production. A precise determination of the contribution that CSR's investment in scientific and technical expertise and its innovative strategy made to the company's growth and emergence as *the* major power in the manufacture and refining of sugar in the South Pacific would be very complex. The company's early adoption of central milling was crucial. The world-wide pattern of development in the cane sugar industry reveals the move from plantation forms of organization towards central milling as a widespread adaptive response to the massive price falls of the late nineteenth century. That CSR pioneered this form in Australia was a prime ingredient of its success. On the basis of this organizational form, new milling technology was acquired and its introduction financed from within. (Heavy indebtedness due to the expense of new equipment accounted for the demise of many operations in other sugar economies during this period.)[48] However, it seems likely that such rationalization of production and updating of technology would not in themselves have been an adequate response in the context of the massive price falls in the last decades of the century. Continued cost reductions through concerted innovation and control within the company may well have provided the margin that gave CSR its competitive edge. Certainly E.W. Knox considered that the chemists had been the vital ingredient, and that not only chemical control but also the innovations resulting from experimentation during the 'down' season had proven a highly satisfactory investment.[49]

Conclusions

The foregoing account may be considered on several levels. First, there is the question of whether — or the extent to which — CSR became a 'trustified' capitalist enterprise as a consequence of having successfully 'automated' or 'industrialized' the innovative process. Second, did CSR face any particular difficulties in doing this that are attributable to its functioning in a peripheral economy? And, if so, can any broader conclusions be drawn from this?

At first sight the answer to the first question would be unequivocally 'yes'. CSR took concerted and definite steps to acquire and train men of specialist expertise who could engage in sustained experimentation and innovation and whose activities would make a considerable contribution to the company's market position and financial success. Moreover, managerial decisions and policy were increasingly predicated on an innovative strategy. The company succeeded in achieving the scale of operation and dominance in its market necessary to support continued innovation and it came to see competition as taking place at the heart of the enterprise through technical innovation.

But a question that faces this account, and all others dealing with the emergence of industrial research, is: at what point can we say that the era of industrial research has 'arrived'? One response is to date it to the employment of specialist expertise within a company or industry. The problem here relates to what such employees do. They may be engaged in entirely routine, 'cut and try' improvements in product or process. Historians have been understandably reluctant to dignify such work with the title of 'research'. Yet it can be argued that trying to draw distinctions is not simply difficult but futile. Hence, rather than try to evaluate the work performed and locate it on some spectrum of activity, we should return to an institutional mode of analysis. On this view, industrial research can be said to have truly arrived if specialists are employed, if they are granted space (such as laboratories) in the company or industry structure, and, more importantly, if they acquire some degree of autonomy from day-to-day 'on-line' problems of production, and concern themselves with research problems of a more strategic nature.[50] How do the activities of CSR sketched in this chapter relate to such interpretations?

CSR did not have a separate research department in the period under study, although one was subsequently established in the 1920s. It did have a central laboratory and smaller ones in the mills and refineries. However, the plant of the mills and refineries remained the real laboratory — generally the case in process industries of that period. The majority of chemists employed by CSR were engaged almost entirely in routine 'on-line' testing and analysis. But the work of a few of them was of a genuinely innovative character, not least in devising and implementing the routines to be followed by their chemical colleagues. It was not in any sense fundamental research. Nor would one expect this since that kind of research function was not securely established in any industrial concern, even in the United States or Germany, until the early twentieth century, and then only in a handful of corporations. And there, too, by far the majority of scientists in industry were engaged in routine testing and analysis.[51] So, on balance, we can say that CSR did launch a programme of industrial research that was a central and important part of its managerial and competitive strategy in a way that parallels, in both style and timing, the initial emergence of industrial research functions in the metropolitan centres. Given the usual view of the development of science and industry in Australia, and present-day realities, this is surprising to say the least: and it invites questions for further research. The most obvious is whether CSR was unique or whether any other Australian companies followed similar early trajectories.

The second major issue is whether the enterprise launched by CSR in any way bears the marks of its peripheral economic and cultural context. The most obvious difficulty that faced CSR was the need to import scientific expertise. This was done quite successfully, but it was an expensive and risky business. The question of the recruitment of scientifically trained personnel raises the broader issue of the supply of such individuals in the Australian colonies. Those colonies certainly had their universities (Sydney, f. 1850; Melbourne, f. 1853; Adelaide, f. 1874; Tasmania, f. 1890; Queensland, f. 1909; Western Australia, f. 1911) where the sciences were taught and some research conducted. But the relations of the universities with industry remain unclear and virtually unstudied.[52] Nevertheless, it can be seen that, insofar as this case study is concerned, no major symbiotic relationship existed between the teaching of science and research skills in institutions of higher education and the demand for such manpower in CSR. This contrasts with the corporate-university relations that did develop in Germany and the United States in the late nineteenth and early twentieth centuries. After importing scientific manpower directly and employing emigrés already in Australia but trained overseas, CSR turned to training its own men. Walton took the leading role here. Classes were at first informal but from 1904 were placed on a formal basis. A few of the recruits to the company's training scheme had some scientific training at the tertiary level, but by far the majority were acquired from local grammar schools. Indeed, E.W. Knox publicly expressed a preference for such individuals, implying that special training before entering the company's service could be more of a hindrance than a help.[53]

This attitude suggests that although Knox was highly innovative by the standards of the 1880s in his perception of the need to utilize organized science, by the early twentieth century he was more conservative. A further indication of this in a broader context comes from Knox's participation in discussions prior to the establishment of CSIR (the precursor of CSIRO). Knox was disposed to encourage applied science in Australian industry but was evidently unsympathetic to industrial research of a less directed kind. He pursued his *idée fixe* that 'it was chemical control which the industries of Australia required and not research'.[54] It is by no means clear that Knox, in his waning years, still represented the company's view on such matters. Indeed, it seems highly probable that the establishment of CSR's separate research department in the 1920s indicated powerful countervailing attitudes within the company, which were further encouraged by, and exemplified in, CSR's increasing diversification. Nevertheless, it appears that there were obstacles to the whole-hearted adoption of a longer term research strategy, and that these may have owed a great deal to the absence in Australia of that nexus between academic research and corporate strategies that proved so powerful a stimulus to the continuing evolution of industrial research in the metropolitan countries. The extent to which this was a result of inevitable limitations to the 'discursive', or academic, element of the research community as a consequence of its isolation, small size, and focus on the metropolitan centres, or a result of limitations imposed by a colonial political order, or perhaps restricted vision of the local capitalist élite, only further research will tell.

Inkster and Todd suggest that the systematic application of science to production in Australia first occurred in the context of the late nineteenth-century depression and that it typically involved attempts to improve the efficiency of extractive processes.[55] CSR, with its concern to improve yields in sugar production, is perhaps assimilable to this process. And yet there are some important differences between the examples that Inkster and Todd offer (the cyanide process of gold production, and uses of microbiology) and the case of innovation in CSR. In the former, applications of science to production emerged through a combination of individual initiative and government policy with sustenance from the local 'infrastructure' and 'superstructure' of science. In the sugar industry a similar pattern of development might be discerned in the history of the state-supported Queensland industry where co-operative research became the norm. But CSR does not appear to have drawn to the same extent on local support structures in line with a corporate ethos of self-sufficiency, perhaps reinforced by proprietorial concerns. A strategic research site is appearing for those interested in the historical relations between science, innovation and production in Australia. That is the nature of the nexus between an innovative corporate culture and the support structure of Australian science. Could it be that, insofar as the former did emerge, it did so despite the latter?[56]

Notes

1 The term 'peripheral' is used here to denote 'new' countries with colonial backgrounds, but with higher per capita incomes than those countries generally referred to as 'developing', 'less-developed' or 'underdeveloped'. From a structural point of view, peripheral countries can be characterized as being on the mid-point of a spectrum between the metropolitan centres and the less developed countries. The case examined here, Australia, has a few characteristics that differentiate it from most other countries: continuity of colonial relationship and a shared language with a 'Mother Country' and an indigenous (Aboriginal) culture that was 'neutralized' at virtually no cost to the colonizers. The idea of a periphery in science has been discussed in a variety of ways, notably by George Basalla, 'The spread of Western science', *Science*, 145 (1967), 611–22; C.K. Vanderpool, 'Centre and periphery in science: conceptions of a stratification of nations and its consequences', in S.P. Restivo and C.K. Vanderpool, eds, *Comparative Studies in Science and Society* (Columbus, Ohio, 1974), pp.432–44.

2 See P. Stubbs, *Technology and Australia's Future* (A.I.D.A., 1980), pp.81, 104 and J. Nevile, 'The disaster of private sector research and development in Australia', in *Science, Technology and the Economy*, The University of New South Wales, Occasional Papers No.11 (Kensington, 1986). For a statement of the increasingly common view that even what Australia has gained historically in this connection is now being rapidly lost, see R. Johnston, 'Australian science policy: now we can steer, where do we want to go?', *Current Affairs Bulletin*, 59 (1982), 20–30.

3 A. Mozley Moyal (ed.), *Scientists in Nineteenth Century Australia: A Documentary History* (Melbourne, 1976) for all its value is informed by the 'high profile' approach. L. Pyenson, *Cultural Imperialism and Exact Sciences: German Expansion Overseas, 1900–1930* (New York, 1985) is important in showing the role that the implantation of centres of research in Argentina, Oceania and China played in German imperialism, especially the non-economic dimensions of that role.

4 R. Macleod, 'On visiting the "moving metropolis": reflections on the architecture of imperial science', *Historical Records of Australian Science*, 5 (1982), 1–16; I. Inkster, 'Scientific enterprise and the colonial "model": observations on Australian experience in historical context', *Social Studies of Science*, 15 (1985), 677–704. Much valuable work has also been done on thematic lines, for example, G. Blainey, *The Tyranny of Distance: How Distance Shaped Australia's History* (South Melbourne, 1966); K.T.H. Farrer, *A Settlement Amply Supplied: Food Technology in Nineteenth Century Australia* (Melbourne, 1980); C.B. Schedvin, 'Environment, economy and Australian biology', *Historical Studies*, 21 (1984), 11–28; and especially A. Moyal, *Clear Across Australia: A History of Telecommunications* (Melbourne, 1984). A more general work that takes seriously innovation in Australian manufacturing industry is G. Linge, *Industrial Awakening* (Canberra, 1980).

5 A.G. Lowndes (ed.), *South Pacific Enterprise: The Colonial Sugar Refining Company Limited* (Sydney, 1956), p.401. Note also the claim that 'In the unindustrialized period in the South Pacific preceding World War I, C.S.R. must have looked like a mountain in the industrial landscape. Among business organizations only one or two banks could compare in financial strength'. (*Idem*, pp.296–7).

6 Recent accounts of such developments at the centre include Georg Meyer-Thurow, 'The industrialization of invention: a case study from the German chemical industry', *Isis*, 73 (1982), 363–81; J.L. Sturchio, 'Chemists and industry in modern America: studies in the historical application of science indicators' (PhD dissertation, University of Pennsylvania, 1981), Chap. 4 (on E.I. du Pont de Nemours and Company). For the perspective on colonial industrial enterprise outlined here see Raymond Vernon, 'International investment and international trade in the product cycle', *Quarterly Journal of Economics*, 80 (1966), 190–207; J. Schumpeter, *The Theory of Economic Development* (Cambridge, Mass., 1949); A. Heertje, *Economics and Technical Change* (London, 1977). For Schumpeter's characterization of the effect of technical change on the structure of the corporation and economy see J. Schumpeter, 'The instability of capitalism', *Economic Journal*, 38 (1928), 361–86, reprinted in Nathan Rosenberg (ed), *The Economics of Technological Change* (Penguin, 1971), pp.13–42, on pp.39–41. And on the integration of industrial research and business strategy see: A.D. Chandler Jr., *The Visible Hand: The Managerial Revolution in American Business* (Cambridge, Mass., 1977), esp. Chap. 13; L. Galambos, 'The American economy and the reorganisation of the sources of knowledge', in A. Oleson and J. Voss (eds), *The Organization of Knowledge in Modern America 1860–1920* (Baltimore, 1979), pp.269–82; and L.S. Reich, 'Research, patents and the struggle to control radio: big business and the uses of industrial research', *Business History Review*, 51 (1977), 230–48.

7 The following is based on Alan Birch, 'The origins of the Colonial Sugar Refining Company 1841–55', *Business Archives and History*, 5 (1965), 21–31 and the same author's historical introduction to Lowndes, op.cit. (n.5), pp.11–53.

8 On Knox see A.G. Lowndes, 'Sir Edward Knox', *Australian Dictionary of Biography*, 5, pp.38–9; Lowndes, op.cit. (n.5), p.17; B. Dyster, 'Prosperity, prostration, prudence: business and investment in Sydney 1838-1851', in A. Birch and D.S. Macmillan (eds), *Wealth and Progress: Studies in Australian Business History* (Sydney, 1967), pp.65–6.

9   The rival concern was Messrs Robey and Co. established in 1857 by the brothers James and Ralph Meyer Robey. The episode was finally resolved to CSR's advantage when the Robeys were forced to sell out to CSR at a loss in 1859. See S. Edgar, 'Ralph Meyer Robey', *Aust. Dict. Biog.*, 6, pp.47-8, and Lowndes, op.cit. (n.5), pp.18-9.

10 Edward Knox to Donald Larnach, 21 September 1865, Knox Family Papers, Mitchell Library, Sydney, ML.MSS. 98, Vol. 16, ff.443-5. Larnach appears to have been a representative of the Bank of New South Wales in London and also a shareholder in CSR.

11 Larnach to Knox, 20 April 1866, ML.MSS. 98, Vol.16, f.467; Larnach to Knox, 20 September 1866, ML.MSS. 98, Vol.16, ff.487-93.

12 Lowndes, op.cit. (n.5), p.17.

13 R. Shlomowitz, 'The search for institutional equilibrium in Queensland's sugar industry 1884-1913', *Australian Economic History Review*, 19 (1979), 91-122, on pp.93-4, 100-101 and A. Graves, 'Crisis and change in the Queensland sugar industry', in B. Albert and A. Graves (eds), *Crisis and Change in the International Sugar Economy, 1860-1914* (Norwich and Edinburgh, 1984), pp.261-79. Linge, op.cit. (n.4), pp.538-43. Though the basic trend was towards a central mill system, there were various organizational forms in the region. See B.W. Higman, 'Sugar plantations and yeoman farming in New South Wales', *Annals of the Association of American Geographers*, 58 (1968), 697-719.

14 See Higman, ibid., pp.706-8. Lowndes, op.cit. (n.5), pp.22-5. On CSR in Queensland see G.C. Bolton, *A Thousand Miles Away: A History of North Queensland to 1920* (Canberra, 1970), pp.137-8, 236-8, and in Fiji, M. Moynagh, *Brown or White? A History of the Fiji Sugar Industry, 1873-1973* (Canberra, 1981).

15 On E.W. Knox see M. Rutledge, 'Edward W. Knox', *Aust. Dict. Biog.*, 9, pp.626-8; Lowndes, op.cit. (n.5), pp.24-5.

16 E.W. Knox to J.G. Ross, 26 November 1872, C.S.R. Deposit, Archives of Business and Labour, Australian National University, Canberra (hereafter ABL), 142/2324, p.245.

17 N. Deerr, *The History of Sugar*, Vol. II (London, 1949), p.588.

18 Quoted in Lowndes, op.cit. (n.5), p.36. The 'Professor Smith' referred to was almost certainly John Smith (1821-1885) a graduate of Marischal College, Aberdeen, and Professor of Chemistry and Experimental Philosophy at Sydney University from 1852 to 1881. Smith was a prominent member of the Royal Society of New South Wales, of which E. Knox, E.W. Knox and J.G. Ross were also members. On Smith see M. Hoare and J.T. Radford, 'John Smith', *Aust. Dict. Biog.*, 6, pp.148-50. We have been unable to identify the 'other chemist' mentioned by Cooper.

19 Moynagh, op.cit. (n.14), p.42. On double crushing see Deerr, op.cit. (n.17), Vol. II, pp.537-46.

20 E.W. Knox to J.G. Ross, 18 September 1876, ABL 142/2329, p.57.

21 See for example, E.W. Knox to E.W. Hayley, 26 September 1876, ABL 142/2329, p.69; Knox to W.A. Poolman, 26 September 1876, ABL 142/2329, p.71; Knox to Poolman, 30 September 1876, ABL 142/2329, p.74.

22 On the rise of the German and French industries see Deerr, op.cit.(n.17), Vol. II, p.502. On research in the German sugar industry see C. Winner, 'Forschung als Wegbereiter des Rubenbaues: Ein Ruckblick auf die Geschichte des Zuckerrubenforschung in Deutschland', *Zucker*, 25 (1972), 75-80. John Perkins, 'The political economy of sugar beet in imperial Germany', in Albert and Graves op.cit. (n.13), pp.31-45, argues that the great competitiveness of the German industry did not derive directly from export bounties, but from the technical efficiency the bounties stimulated. The prescience of CSR management matched that of officials of the US Department of Agriculture. See W. Lloyd Fox, 'Harvey W. Wiley's search for American sugar self-sufficiency', *Agricultural History*, 54 (1980), 516-26; O.E. Anderson, *The Health of A Nation: Harvey W. Wiley and the Fight for Pure Food* (Chicago, 1958), Chap. 3. Also, on developments in Louisiana, see J.A. Heitmann, 'Scientific and technological change in the Louisiana sugar industry, 1830-1910' (PhD dissertation, Johns Hopkins University, 1983).

23 J.G. Ross to F. Parbury & Co., 2 March 1880, ABL 142/2342, pp.52-3. Parburys were CSR's London agents.

24 E.W. Knox, 'An application of chemical control to a manufacturing business', *Report of the Second Meeting of the Australasian Association for the Advancement of Science*, Melbourne 1890 (Sydney, 1890), pp. 372-9, at p.372. A very useful descriptive source on chemists and chemical control in CSR is E.C. Hood, 'The originators and development of chemical control in C.S.R. Sugar Mills', CSR Ltd, Mill Chemical Department, (1969?). (Typescript held in CSR Library, Sydney.)

25 E.W. Knox to Parbury & Co., 3 August 1880, ABL 142/2342, pp.117-21 on p.121. On Walton see Anon. 'The Scot we know', *The Scottish Australasian* (October 1913), pp.2121-32.

26 E.W. Knox to A. Fairgrieve, 31 August 1882, ABL 142/2855, pp.5-19. On Steel as the first mill chemist see Hood, op.cit. (n.24), p.126. We can glean little information on Steel's origins save that, like Walton, he came from Greenock, he applied for the appointment with CSR and that his previous employer was probably the company of Mirrlees, Tait and Watson. (See E.W. Knox to Parbury & Co., 13 February 1882, ABL 142/2342, pp.380-3 on p.382.)

27 E.W. Knox to A. Fairgrieve, 24 April 1883, ABL 142/2855, pp.246-79 on p.258.

28 A. Fairgrieve to E.W. Knox, 23 May, 1884, ABL 142/1047, pp.109-19. See also E.W. Knox to J.M. Knox, 28 May 1883, ABL, 142/2855, pp.319-26, on p.325. On Kottmann see Lowndes, op.cit. (n.5), pp.36-40; Hood, op.cit. (n.24), pp.124-6. On Kottmann's linguistic prowess, Knox to Fairgrieve, 9 August 1883, ABL, 142/2855, pp.372-84, on p.382. For Fairgrieve's demise and Kottmann's elevation see Hood, op.cit. (n.24), pp.121-2, 150.

29 The graph drawn in the first figure and the following characterization of the chemical staff are based on information gleaned from early staff books held in the Personnel Department of CSR, 1 O'Connell St, Sydney. From 1892 staff books give a systematic annual record of name, position and salary. They are organized by category of employee, one category being 'Chemists and analysts'. Before 1892 the records are less systematic. A volume labelled 'C.S.R. Co. Staff' contains a roughly alphabetical listing of staff with details of positions held, dates and salaries for each entry. The entries go up to 1891. A second volume, less systematically organized, labelled 'Staff revision 1879-188[?]' contains notes (by E.W. Knox?) on salary revisions of staff and also on disposition of chemical staff among the Company's various operational units. Not all entries are dated.

30 Memo in E.W. Knox's hand headed 'Chemists and analysts' pay' in volume labelled 'Staff revision 1879-188[?]', CSR Ltd, Personnel Department (see note 29). By internal evidence this memo can be dated to late 1884.

31 William P. Dixon to General Manager, Sydney, 29 May 1902, ABL 142/2909. The person appointed as assistant chemist was T. Guthrie. References in correspondence give some insight into recruitment. Thus Paul Seeliger was engaged in Berlin in 1885 and T.O. McMillan and G.M. Matheson in Britain in 1890. (See E.W. Knox to Parbury & Co., 14 January 1885, W.H. Rothe to Parbury & Co., 24 October 1889, 23 January 1890, 30 January 1890. ABL 142/2343, pp.20-21, 95-96 and 142/2344, pp.369-71, 422-423, 426-427.) In 1890, E.W. Knox stated that the chemical staff then consisted of 9 Scots, 2 Germans, 3 Danes, 1 Belgian, 1 Swiss, 2 Englishmen and 14 Australians. See E.W. Knox, op.cit. (n.24), p.414. There is a slight discrepancy between Knox's count and our own, as represented in the graph (see n.29).

32 Deerr, op.cit. (n.17), p.580; idem, *Cane Sugar*, 2nd edition (London, 1921), pp.280-8. Kottmann to General Manager, 9 April, 15 April 1884, ABL 14/2938.

33 On quantities of lime for carbonatation and Kottmann's experiments see Kottmann to General Manager, 18 April 1884, and 26 April 1884, ABL 14/2938. For a brief account of the diffusion process see Deerr, op.cit. (n.17), p.547 and idem op.cit. (n.32), pp.251-6. On the American efforts with diffusion see Fox, op.cit. (n.22), pp.517-22 and J. Carlyle Sitterson, *Sugar Country: The Cane Sugar Industry in the South, 1753-1950* (Louisville, 1953), pp.279-81.

34 E.W. Knox to Kottmann, 7 August 1884, ABL 142/2003, pp.28-31; E.W. Knox to Kottmann, 19 November 1884, ABL 142/2003, pp.99-100; E.W. Knox to Edward Knox, 4 August 1884 and 19-21 October 1884, Knox Family Papers, Mitchell Library, Sydney, ML. MSS. 98, Vol. 20, ff.49-84, 129-45.

35 W.H. Rothe to Kottmann, 1 August 1885, ABL 142/2003, p.286 and W.H. Rothe to Kottmann, 28 July 1885, ABL 142/2003, pp.265-6.

36 E.W. Knox to Kottmann, 28 June 1886, ABL 142/2004 pp.79-80.

37 Deerr, op.cit. (n.32), pp.251-2; Deerr, op.cit. (n.17), p.547.

38 Kottmann to General Manager, 17 September 1885, ABL 142/2940 (Kottmann Personal).

39 Kottmann to General Manager, 18 September 1885, ABL 142/2948 (Kottmann Personal).

40 Kottmann to General Manager, 22 September 1885, ABL 142/2940 (Kottmann Personal).

41 See J.J. Beer and W. David Lewis, 'Aspects of the professionalization of science', in K.S. Lynn (ed.), *The Professions in America* (Boston, 1967), pp. 110-30. S.B. Barnes, 'Making out in industrial research', *Science Studies*, 1 (1971), 157-75 offers a solid critique of the clash of value systems theory. Barnes' more flexible and comprehensive approach would seem to apply best to the case in point in which the tensions were between 'new' and 'traditional' lines of authority rather than between some supposed scientific value system and supposed corporate one.

42 On the P.O.C.S. formula see Lowndes, op.cit. (n.5), pp.452-4.

43 E.W. Knox to Managers of Mills, 19 June 1884, ABL 142/3505 (Mill Circulars); Kottmann to All Mills, 8 June 1886, ABL 142/2004, p.56.

44 E.W. Knox to Managers of Mills, 6 May 1886, ABL 142/3505 (Mill Circulars).

45 E.W. Knox to Kottmann, 6 August and 20 August 1886, ABL 142/2004, pp.107-8, 115-9.

46 Quoted in Lowndes, op.cit. (n.5), p.37. Knox's confidence is illustrated by his response to a report of one of CSR's chemists, Paul Seeliger, on the yields obtained at the Brunswick factories in Germany: '. . . the particulars given are enough to make every cane sugar manufacturer shiver with fear were it not that we know that by improving our own processes we can extract more sugar from the cane' (Knox to Seeliger, 19 June 1886, ABL 142/2004, p.77).

47 Linge, op.cit. (n.4), pp.541-2; Moynagh, op.cit. (n.14), p.32. On Queensland government measures in support of the industry see Shlomowitz, op.cit. (n.13), pp.105-6 and Graves, op.cit. (n.13), pp.276-8; Linge, op.cit. (n.4), pp.692-3; Queensland Bureau of Sugar Experiment Stations, *Fifty Years of Scientific Progress* (Brisbane, 1950).

48 B. Albert and A. Graves, 'Introduction', in Albert and Graves, eds, op.cit. (n.13), pp.1-7.

49 E.W. Knox, op.cit. (n.24), pp.376-7.

50 Beer and Lewis, op.cit. (n.41), pp.112-5; Sturchio, op.cit. (n.6), pp.72-3.

51 Enthusiasm for pioneers of industrial research of a more fundamental kind (such as General Electric and AT&T in the United States) sometimes obscures this point. A useful corrective in the American case is provided by Galambos, op.cit. (n.6), p.277.

52 There has been some disagreement on the nature of the scientific work done in Australia's universities revolving around the issue of 'quality'. See D. Fleming, 'Science in Australia, Canada and the United States: some comparative remarks', *Proceedings of the Tenth International Congress of the History of Science* (Ithaca, 1964), Vol. 1, pp.179-96, and compare A.M. Moyal (ed.), *Scientists in Nineteenth Century Australia: A Documentary History* (Melbourne, 1976), pp.221-2, 227-8. J. Radford, *The Chemistry Department of the University of Melbourne: Its Contribution to Australian Science, 1854-1959* (Melbourne, 1978) does indicate considerable interaction between academic science and industry in the Melbourne department. Similar studies are urgently needed.

53 On Knox's policy and attitude see E.W. Knox, op.cit. (n.24), p.379. Our characterization of the nature of the 'intake' is based on casual references in letters regarding apprentice chemists (for example, Kottmann to Marquardt, 12 April 1887, ABL 142/2005, p.43); an analysis of Walton's Class Lists from 1904, which gives some information on educational background (ABL 142/2948-50); and on the age of junior chemists taken into the company revealed by Personnel Records to be typically 17 or 18.

54 Quoted in G. Currie and J. Graham, *The Origins of C.S.I.R.O.. Science and the Commonwealth Government 1901-1926* (Melbourne, 1966), p.45.

55 I. Inkster and J. Todd, Chapter 5 this volume.

56 We thank CSR Ltd for access to, and permission to quote from, the company's archives. Mr Joe Daniels, Manager of Information Services at CSR, and the staff of the Library and Personnel Department there welcomed us and gave us every assistance. We are also grateful to Maureen Purtell and Michael Saclier of the Archives of Business and Labour at the Australian National University for their help in using CSR archives held there. For their helpful criticisms of an earlier version of this paper, our thanks to Ian Inkster and David Oldroyd.

PART III:
Passage to modernity

9

Cancer, physics and society: Interactions between the wars

H. Hamersley

Challenge and opportunity:
cancer, physics and radiotherapy in the 1920s

Within a few months of their discovery by physicists at the end of the nineteenth century, ionizing radiations were known to have effects on living tissues that might be put to therapeutic use. The first experiments in radiotherapy soon followed, including one in Australia in late 1896 that was widely reported locally at the time.[1] By the outbreak of the First World War some success was being achieved in the treatment of skin malignancies, but ionizing radiation had not yet proved to be an effective weapon for attacking cancers deep within the body. Early hopes remained largely unfulfilled until the 1920s. By then new methods of deep radiation therapy were coming into use, which, according to an editorialist in the *Lancet*, had all at once made radiotherapy 'the treatment of first choice in the majority of cases of cancer'.[2] This over-enthusiastic assessment rested on some definite advances made in the previous few years in the methods, science and technology of radiation therapy. These included new methods for using radium (or its emanation, the radioactive gas radon) for intracavitary and interstitial therapy developed in France, the United Kingdom and the United States,[3] and the introduction in 1914 of Coolidge's hot-cathode X-ray tube.[4] This tube gave the X-ray therapist hard, penetrating radiation at previously unattainable dose rates.

But much of the optimism of the early 1920s about the role radiotherapy might play in the treatment of cancer was based on advances gained in Germany during the war by the collaboration of physicists and radiotherapists. They achieved an integration of radiotherapy, radiobiology and the physics of the absorption and scattering of radiation in matter of a kind not seen previously, which according to one Australian commentator in 1922 gave radiotherapy for the first time a 'rationale based on logical and scientific principles'.[5] Although many important refinements of detail remained to be made, the German school had significantly advanced the solution of the central problem of radiotherapy — the delivery of defined, therapeutically effective doses of radiation to sites within the body — and the treatment planning and dosimetric concepts it introduced are still the basis of those used today.

The 'new' radiotherapy arrived in Australia in 1922, when there were 'enthusiastic discussions' of the subject at a number of meetings of the British Medical Association branches in Sydney and Melbourne.[6] At this time Australia was still one of the most affluent Western societies.[7] By this time, also, it had the kind of population structure that ensured growing public concern with the 'cancer problem' and a demand for modern radiotherapy services. But providing these in Australia at this time posed some problems, which were not solved by the importation of a few sets of modern deep-therapy equipment for use in private medical practice. The new methods of radiotherapy depended critically on accurate measurement of radiation dose. This required specialist physicists to make equipment calibrations and dosimetric calculations in radiotherapy clinics, and, in national standards laboratories or elsewhere, to establish and maintain units of radiation measurement and methods of calibrating clinical instruments in terms of them.

Radiotherapy was on its way to becoming a medical high technology, and its efficient practice required an infrastructure of supporting technical and scientific services. But while this infrastructure could be taken for granted in countries with established traditions of support for research in physics and biomedicine, in Australia in the early 'twenties, most of it was still lacking. At times, research of high quality had been done by physicists and biomedical scientists working in Australian universities. But, by and large, the record was one of discontinuous and episodic achievement in an unsympathetic and unsupportive environment.[8] Most Australians did not yet realize that if they were to continue to share in advances in the quality of life being won elsewhere through science and technology, a much more supportive environment for local research and development in fields such as physics, which apparently had only marginal contributions to make to the nation's economic life, would eventually have to be created.

The 1920s brought some of the first opportunities to change ideas here, and one of the vehicles of opportunity was the need to have in Australia the scientific infrastructure required for the practice of the new radiotherapy. Partly on the promise of the new therapy, extraordinary success was achieved in the 1920s in mobilizing public support for the cause of cancer research and treatment in Australia. The initial impetus came from Sydney, where the university's Cancer Research Committee brought its 'Cancer Campaign' to raise funds for cancer research and treatment to a very successful conclusion in 1927. Similar appeals were then launched in all mainland states except Victoria. The Commonwealth government also entered the field, and from the late 1920s came to assume leadership of the nation-wide effort.

For physicists this national ferment created some unique opportunities and challenges. Their participation in the national cancer effort encouraged the perception — new for Australians — that physics, traditionally one of the abstract sciences most remote from them, had something to contribute in matters of central human and societal concern. A limited number of new career openings for physicists also appeared, as a nation-wide scheme for providing physical services for radiotherapy, based on the physics departments of the state universities, gradually evolved. Finally, for physicists in the major centres of Sydney and Melbourne, there was the prospect of obtaining from cancer funds support for basic and applied research in radiation physics and radiobiology; if this support eventuated, it could bring significantly nearer realization emerging ambitions to establish physics as a viable research discipline in Australia.

Such ambitions were particularly in evidence at the University of Melbourne, where T.H Laby held the chair of physics (or 'natural philosophy', as it was called there at the time). No physicist in Australia was better placed to see and, it might be thought, to seize, the opportunities of the situation. He was a seasoned investigator of ionizing radiation with useful contacts, from his time at the Cavendish Laboratory as a research student, with some of the leading figures in Britain at the research frontiers of the field. First in New Zealand, and from 1915 in Melbourne, he assiduously tried to build up the departments he led as centres for significant physical research.[9] During the war he was the leader in his department in Melbourne of work on problems in diagnostic radiology of importance to the war effort. This work convincingly showed that physicists had useful contributions to make to medical radiology.[10]

By the early 1920s Laby had an authority and prestige with medical users of ionizing radiation unique in Australian physics. His department was also becoming the recognized centre in Australia for research on X-rays and their applications, which was a tribute both to the quality of the work its staff were doing and to Laby's talents as an entrepreneur of research who could attract outside sources of funding to supplement modest university subventions for research. However, Laby's talents as a leader of scientific research and an entrepreneur failed to win him support from his university's Cancer Research Committee, established in 1915. Nor did anything come of attempts made by Laby and medical allies in the 1920s to establish a radiotherapy or cancer institute in Melbourne[11] from which support might have come for research in his department on medical radiation physics. As it turned out, Laby's acknowledged interest and authority in this field, and his department's growing research reputation in radiation physics, only became valuable assets after the Commonwealth government decided in 1927 to make a major purchase of radium to be used for cancer treatment in Australia. This led to Laby's appointment as the Commonwealth's consultant radium physicist, and to the location, in Laby's department, of the Commonwealth's radium laboratory. We shall later see how, through this Commonwealth association, Laby was to play an important part in the 1930s in the events that finally put in place in Australia the scientific infrastructure needed for advanced radiotherapy.

In Sydney there was no-one comparable to Laby as an investigator of ionizing radiation, or, perhaps, as committed as he was to the values of professionalism in scientific research and the vision of creating a sustainable tradition of Australian physical research. But it was here, in the 1920s, that some of the opportunities Laby saw were first seized. The University of Sydney's Cancer Research Committee was not only a spectacularly successful fund-raising body; as we shall shortly see, it also went on to support ambitious research on the physics and radiobiology of radiotherapy. As we shall also see, however, the story of this research illustrates not only some of the novel opportunities that were opening to Australian scientists in the 1920s, but also some of the novel challenges and pitfalls faced by those who, themselves trained in a tradition where research had been a desultory pursuit on the margins of academic science, suddenly had the responsibility for managing a major research enterprise thrust upon them.

Opportunity seized: the formation of the University of Sydney Cancer Research Committee

The spectacular career of the Sydney Cancer Research Committee began in 1919, when the University of Sydney received a bequest of £7000 from the estate of J.F. Archibald of the *Bulletin*, to be added 'to any funds possessed by the University for the prosecution of research into the problem of cancer'.[12] The university had no such funds, but promptly created some by a small appropriation from other sources to satisfy the conditions of the bequest. Eventually, in November 1921, a senate committee was established to consider how the funds should be used, and it was this committee that recommended the scheme proposed by F.P. Sandes, the professor of surgery. The Cancer Research Committee was formally established in late April 1922 to carry through the scheme.

The Sandes scheme envisaged modestly funded academic research, using existing university and teaching hospital laboratories and staffs. The cancer funds were, in effect, to be used to nurture in a small way the idea that research should have a part in the training and careers of scientists. The funds produced an income of only about £400 a year. The senate committee thought £1000 to £1500 a year would be needed to place the scheme 'on a proper working basis',[13] and urged action to raise funds to bridge the gap. At this time, however, no larger ambitions were evident. The only remarkable feature of the Sandes scheme, which continued to influence funding and research priorities even when the scale of operations had greatly expanded, was that it proposed to apply the 'cancer funds' to a range of disciplines, which included the physical sciences.

Instead of proposing exclusive support for projects in the biomedical fields, from where advances in the understanding of cancer had come in the past, Sandes argued in his memorandum that the time had come when 'workers in different sciences, such as mathematics, physics, chemistry, botany, zoology, histology, physiology, pathology and even medicine and surgery may all be able to contribute something'. Further, he took the view that 'team work is essential in any scheme of cancer research'. To direct research until a permanent director of research was appointed he proposed a collegiate committee of direction the membership of which should be determined on the assumption that 'the biophysical and biochemical part of the foundation . . . should be the first to be laid'.[14] As members he proposed J.A. Pollock (professor of physics), J. Read (organic chemistry) and H.G. Chapman (physiology). Sandes himself was to act as secretary.

It was an unusual committee to nominate to direct research on cancer at that time, but it was duly established in May 1922, constituted as Sandes had proposed except for the addition of the pathologist, D.A. Welsh. Two meetings in May 1922 produced little except a research project on the juice of *Euphorbia pehlus*, a popular cancer cure.[15] This was an unimaginative choice given some of the ambitious proposals for possible lines of research set out in Sandes's memorandum. But many of these were probably unworkable at the time, showing more evidence of Sandes's wide reading in the recent speculative scientific literature on cancer than knowledge of the current research frontiers of the subject. The committee did not meet again until September 1923. By then Read had left for a chair in Scotland and Pollock had died, depriving the committee of its only members with claims to more than local scientific reputation or significant achievement to that time in scientific research. O.U. Vonwiller, long Pollock's assistant and trained under him, eventually succeeded to the physics chair, and replaced him on the committee. It was this committee that, over the next few years, would have to manage the transition from small-scale (and desultory) academic research — the only kind its members knew — to something much more ambitious and professional, funded at levels unprecedented in Australia.

For in September 1923 the Cancer Research Committee made the abrupt change of course that was to make it the major entrepreneur of cancer research and treatment in Australia in the 1920s. Plans now rapidly evolved to approach various prominent Sydney businessmen and other notables to serve on a special finance committee of the Cancer Research Committee and 'draw up a scheme for raising £100 000' to establish a 'Cancer Research Institute'.[16] A major effort was promised 'to make provision for radiotherapeutic research', and a special department of the proposed research institute was projected, which would have 'facilities . . . for the treatment of cancer by special radiation methods, and for the estimation of the value of such treatment'.[17] The interest now shown in 'radiotherapeutic research' was undoubtedly a tribute to the recent arrival in Australia of the new radiotherapy.

While this event had no effect on the Melbourne Cancer Research Committee, local circumstances and personalities made it decisive in the career of the Sydney one. Some credit for this should go to H.M. Moran, the Sydney surgeon who introduced to Australia radium 'needles' for interstitial therapy, and subsequently became one of the severest critics of the Cancer Research Committee. According to Moran, it was only after he and the X-ray therapist E.H. Molesworth had convened a meeting between 'several professors from the . . . University' with a businessman 'active in philanthropic work and in hospital affairs' that the Cancer Research Committee 'widened its activities so as to take in . . . the treatment side'.[18] Moran's account leaves unclear whether this meeting took place in 1923 or later. Whether or not it was Moran's businessman, by late 1923 there was someone (his identity is not known) outside university circles who had plans of his own for a 'cancer research institute' in Sydney that would undertake 'X-ray therapy, pathological research, general physical, biochemical, biological, physiological and anatomical research in cancer'.[19]

In early 1924 an approach was made to the university, proposing a joint venture. But the promoter had already advanced arguments to potential donors why 'an institute should be founded free of University or public hospital control . . . governed by a body of scientific men who would devote their whole time and energies to its advancement'.[20] He was also critical of both university plans to have 'students working in the laboratories' and of the tendency 'in University organization . . . to select amongst those workers only men already attached to the University'.[21] Not surprisingly, his terms for collaboration were unacceptable to the university. After late April 1924 no more was heard of the scheme.[22] But it seems clear that it significantly influenced the final form and scope of the university's own plans, although in the end nothing came of the proposal in 1924 that the university would move to establish its own 'cancer institute'. Instead, under the Cancer Research Committee a 'cancer organization' evolved that had its own staff and major items of equipment paid for from the cancer funds. But they were located in university departments or the teaching hospitals, and their control remained with the Cancer Research Committee and its scientific executive, the Committee of Direction.

After some elaborate preparations to secure wide community support, the university's public appeal was launched in April 1925.[23] By the end of June 1926, total receipts, including grants from the Commonwealth and New South Wales governments, amounted to £23 000.[24] This was enough to evoke the interest of T.H. Laby, who wrote from Melbourne for 'information in relation to the constitution of the Cancer Research Committee and . . . as to its future proposals in regard to research'.[25] But much more spectacular results were achieved when the decision was taken to re-launch the appeal in July 1926, emphasizing that the Cancer Research Committee intended to support cancer treatment as well as research. By December 1926 'the required amount was reached'.[26] When the appeal finally closed in July 1927 the total 'cancer funds' — the Archibald bequest, the government grants, and the donations from the public — exceeded £130 000, equivalent perhaps to $3 000 000 today.[27]

Cancer and radiobiological research at Sydney, 1924–1930

Although some account had to be taken of the view, always potentially influential in the Cancer Research Committee from the time it set out on the path of expansion, that priority should be given to the development of radiotherapy services,[28] in fact the Committee used the larger portion of the cancer funds for research. In October 1926, when the success of the big appeal was already clear, it boldly resolved that research would be 'concentrated and intensive', and that the fund 'would be expended, if necessary, within the next ten years'.[29] This decision may have been the easier to take because the committee by this time already had in hand a major research project from which great results were expected.

This project was concerned with a novel (and controversial) radiobiological hypothesis, first proposed in 1924 by Warnford Moppett, who had just graduated in medicine at Sydney but wished to pursue a career in research. He was, in many ways, the kind of young graduate whose career in research Sandes's scheme partly aimed to foster. The hypothesis was that 'homogeneous' (nearly monochromatic) X-rays of certain frequencies were specially efficient in producing changes in living cells, some of a kind that led to cell death, others of a kind that promoted cell proliferation. When rays of both types were present (as they would be in the usual X-ray beams used for therapy) their biological effects were mutually inhibited, a process Moppett termed 'antagonism'.[30] The novelty of this hypothesis was that it postulated an abrupt and discontinuous 'selective action' at specific places in the X-ray spectrum, in contrast to earlier concepts of selective action which had referred either to supposed differences in biological action of whole regions of the X-ray or gamma ray spectrum, or to a general 'quality effect' which became apparent as harder radiation was used.

Moppett came to Sandes's notice and was given the opportunity to put his ideas to the test in 1924, using laboratory space and equipment in the physics department.[31] To make a systematic search of the X-ray spectrum to identify the specially potent frequencies, Moppett used a Bragg spectrometer as a wave selector'. Because this unavoidably gave him homogeneous X-rays of very low intensity, he used the sensitive allantoic membrane of the embryo chicken as his experimental tissue.[32] The membrane was exposed to the rays through a window cut in the shell, although this exposed it to the air for up to two hours during the experimental treatment. This hazardous procedure was used because preliminary experiments showed there was no reaction under intact shells. This, Moppett thought, was owing to 'the relatively high absorption and scattering properties of the shell'.[33]

Although use of this risky window technique apparently could not be avoided, preliminary experiments in 1924 showed, according to Moppett, that 'opening or closing of the shell . . ., if done with sufficient care, gives rise to no visible reaction which might be confused with the X-ray effect'.[34] By early 1925 Moppett had apparently strong experimental confirmation of his ideas. There was, he reported in the account of the work published later in 1925, 'complete atrophy of the membrane . . . at a wavelength of 0.11, 0.53, and 0.79 Ångström unit . . . The effect is sharply restricted to those wavelengths and the greatest is at 0.11 and the least at 0.79'.[35]

In March 1925 Moppett, his ideas, and his experimental results were brought to the Committee of Direction, meeting for the first time after nearly 18 months. They arrived at a critical time for the committee, just when it had in prospect large funds to support research, some ambitious ideas deriving from Sandes's memorandum about the kind of research it would like to support, a broad commitment from 1924 to support 'radiotherapeutic' research, but on its record to this time no very good ideas of its own for research that reconciled these various ambitions and commitments. In Moppett's work the committee had, ready-made, a research project that did reconcile them. His hypothesis had revolutionary implications for radiotherapy, since it implied that useful therapeutic effects might be obtained even at very low doses if homogeneous rays were used. It also had some startling physical and biophysical implications.

On the model most physicists accepted by 1925, the biological effects of radiation were ultimately the result of ionization along the tracks of the beta particles emitted when photons were photoelectrically absorbed or Compton-scattered by atoms of the biological material.[36] But, as Moppett noted in his published account of his first experiments, highly frequency-dependent effects of the kind he expected (and found) were difficult to reconcile with this model.[37] 'Selective action' was thus a problematic idea. But, just possibly, it was also a discovery that would entail a major revision of accepted models. The committee was not well equipped to make a rigorous technical assessment of Moppett's experimental work, and at this point, perhaps, the views of experienced workers in radiobiology elsewhere should have been canvassed. But its members lacked the kinds of contact with the international research community that would have made this exercise easy and possible for someone like T.H. Laby. Instead, the committee decided there should be replications of Moppett's experiments, and investigations of the physical factors (beam quality and spectral distribution) of the X-ray equipment he had used. But, less cautiously, it also proposed 'investigations upon the effect of homogeneous X rays upon the nucleus in contradistinction to the electrons of the Nitrogen atom to ascertain whether Dr Moppett's suggestion is correct that the Moppett effect is produced by disintegration of the Nitrogen atom in living tissue',[38] but showed no interest in Moppett's equally radical suggestion that X-ray absorption edges corresponding to lines in C.G. Barkla's phantom 'J series' might be implicated.

After a period in London at the Lister Institute, Moppett returned to Sydney in April 1926 as cancer biologist in the research team the Committee of Direction was assembling to explore the Moppett effect. His time in London was mainly spent in learning tissue culture techniques. He either did not have, or did not take, the opportunity to discuss the ideas, methods or findings of his Sydney work with any of the experienced investigators of radiobiological problems working there at this time. This may indicate either remarkable confidence or remarkable diffidence, but in either case the intellectual isolation in which the Sydney work was proceeding remained intact. In Sydney, Moppett worked on the replication and extension of his radiobiological experiments.

Meanwhile, parallel investigations aimed at a physical explanation of his radiobiological data had already begun in 1925. These included microchemical investigations into 'cell architecture', to identify atomic receptors in cells that, excited in a highly frequency-sensitive way by radiation, might bring about the 'selective' action Moppett thought he had demonstrated.[39] The microchemical work led to a staining technique for locating potassium in tissues and then, in February 1927, to the discovery of the heavy elements lead and uranium in trace amounts in egg tissue. The significance of this discovery was that these elements, unlike light elements such as potassium or nitrogen, which it was plausible to think on biochemical grounds should be implicated, have critical X-ray absorption edges near positions in the X-ray spectrum where Moppett detected maximum biological action. So it seemed that the discontinuities in the X-ray biological action spectrum that were the Moppett effect could be explained in terms of the standard physics of photon absorption, and the speculations about J-series and nuclear disintegrations were discarded by everyone except Moppett himself. Lead and uranium became the preferred candidates for receptor 'key atoms' in the cell architecture, although it was obscure what their biochemical role might be.[40]

Between 1925 and 1927, the Moppett effect came to dominate the Sydney research programme. Although the Committee of Direction co-opted some able new members, including V.A. Bailey, who had joined the Sydney physics department from Oxford in 1925, the commitment increased in the following years. Yet there were some early indications that Moppett's radiobiological work might be fundamentally flawed, notably in late 1926 when there were meetings of the committee that revealed that some of the Sydney physicists had reservations about the validity of Moppett's explanation of why there were no radiation effects on egg membranes under intact shells. Experiments then showed that Moppett's explanation (attenuation of the X-rays by the shell) was untenable. But in April 1927 Moppett came up with an alternative explanation invoking his concept of 'antagonism'. This was deemed satisfactory and the issue was not pursued further.[41]

The system for the direction of research by a part-time committee had, of course, been designed for very different circumstances from those prevailing after the ambitious plans of 1924 had been realized. When, as a result of those plans, the Committee of Direction acquired additional responsibilities for the organization of radiotherapy services (which frequently embroiled it in heated technical and medico-ethical controversies),[42] the collegiate system became unworkable. In October 1926 moves to appoint a full-time Director of Research began. The sub-committee that drew up the terms of appointment stipulated that the Director should already be a 'proper researcher'.[43] A strong overseas committee, which included the Secretary of the Royal Society and C.J. Martin of the Lister Institute, was appointed to short-list canidates.[44] Later, the physicist W.H. Bragg was included.[45] But sound though the decision to seek a director of this calibre might have been in the changed circumstances, there were some problems inherent in it. The successful candidate would probably owe his appointment to training and achievement in one of the conventional disciplines of cancer research. But how such an appointee would accommodate to the existing research programme, stressing linked radiobiological and chemico-physical investigations, was problematic. So also was the question of the scope he would have to shift resources to other lines of investigation, because it was intended that the Committee of Direction would continue to have a significant role in determining research policies and priorities.

Only in late 1927 were some of these questions addressed, when Sandes 'reported that criticism had been levelled at the plan of research followed by the Sydney cancer organization and that the team system had been decried'. It was decided that 'pending the appointment of a permanent director, the present plan of research should be discussed and the opinion of some authoritative body obtained'.[46] But these plans for a review were overtaken by events. In November the appointment as director of G.M. Findlay, an experimental pathologist working at the Imperial Cancer Research Laboratories in London, was announced.[47] He then promptly declined the position when conditions he wanted, impinging on the unresolved questions of the future role of the Committee of Direction, were rejected.[48] Soon, there was pressure to conclude the whole matter by appointment of a local candidate.[49] The choice fell on H.G. Chapman. Although he had won a Syme Prize in 1910, by 1928 Chapman was not the 'proper researcher' on whom sights had first been set. He had, however, been a successful chairman of the New South Wales Technical Commission on Miners' Diseases in 1921,[50] and had been a member of the Committee of Direction from its creation in 1922. He also had a reputation for brilliance, acknowledged later by both the very hostile H.M. Moran[51] and Vonwiller.[52] But according to Vonwiller, 'relatively new members of the University generally . . .' were against his appointment.

Under Chapman and the Scientific Advisory Committee, which replaced the Committee of Direction, research continued to be dominated by work on the Moppett effect. Interest continued in the possible role played by heavy elements in producing the effect, and some success was reported in demonstrating their presence, not just in bulk egg tissue but in the allantoic membrane, by both X-ray spectroscopy and microchemical analysis.[53] As a working hypothesis it was assumed that they functioned as catalysts in cell chemistry, and that their catalytic action was somehow disrupted when radiation was absorbed. Accordingly, an investigation was approved in July 1929 of how 'radiation might change the catalytic action of lead', and a chemist was recruited to carry it forward.[54] By the end of 1929, Chapman was directing a minor research empire, which included two physicists, four chemists, a cytologist, an experimental physiologist, an anatomist, a pathologist and a 'cancer biologist'. While some members of this team were working on departmental projects supported by the Cancer Research Committee, the investigations ramifying from the Moppett effect still received the major share of resources.

The Commonwealth radium purchase and first steps towards a national scheme of physical services for radiotherapy

Possibly as a result of its tendency to internal controversy, the Sydney Cancer Research Committee contributed less to solving the technical and organizational problems of radiotheraphy in Australia in the 1930s than the charter it had given itself and the resources it controlled would have justified. However, an initiative it took early in its career indirectly did make an important contribution to the eventual solution of some of these problems. In collaboration with the New South Wales government, in July 1926 the committee commissioned a report on modern methods of cancer treatment from a British radium expert, Arthur Burrows, who was making a private visit to Australia at the time.[55] Burrows, who had been director of the Manchester Radium Institute since 1915, was a substantial figure in British radiotherapy. Armed with his report, from early 1927 the committee began lobbying the Commonwealth government to establish a national 'Radium Bank'.[56]

In late February 1927 J.H.L. Cumpston, the permanent head of the Commonwealth Health Department, accepted co-option to the committee,[57] thus putting himself in the extraordinary position of joining the council of a body lobbying for a policy on which he would be his government's principal bureaucratic adviser. The explanation is that the Commonwealth was already favourably inclined to purchasing some radium. Its interest had begun in October 1926, when D.C. Cameron, the Federal member for Brisbane, had written to the Federal Treasurer, Earle Page, urging 'the establishment of a

Radium Institute in Australia and the despatching of one or two Australian graduates abroad to secure information.[58] At the time Cameron wrote, a 'cancer campaign' on the lines of the visibly successful Sydney one, was being launched in his own state. Page was receptive, and the proposal was referred to the Commonwealth's recently formed Cancer Research Advisory Council and then to the Federal Health Council. From these bodies emerged the proposal that the Commonwealth government should establish a radium bank that would lend radium to hospitals or state cancer treatment organizations.[59]

It could be argued that the Burrows report and the submissions based on it from Sydney only gave more definite form to Commonwealth intentions that were already there. But this was still an important contribution, as their ultimate effect was to persuade the Commonwealth to take a more active role than as the passive owner and lender of radium first ideas suggested it might be. After the decision to purchase radium was finally made by Federal Cabinet in July 1927,[60] an ad hoc advisory committee (consisting of Cumpston, Laby, Sandes and Bailey) was formed to advise the Commonwealth on 'the conditions under which radium should be purchased'. The committee took a broad view of this brief and proposed various steps the Commonwealth should take to ensure the use of its radium under the most efficient conditions.[61] One outcome was that in August 1927 Burrows himself was placed under contract to act as radium treatment adviser to the Commonwealth for three years.

Burrows, however, was strongly committed to the use of radon as a versatile alternative to the use of radium 'salt' itself. It was used at the London Radium Institute where Burrows was trained, and at his own institute in Manchester and other British radiotherapy centres. On his advice, it soon became settled Commonwealth policy that radon for therapy would be produced at the treatment centres in Sydney and Melbourne which Commonwealth plans called for.[62] In Sydney the Cancer Research Committee agreed, after the inevitable internal controversy, to set up a radon plant and to appoint a physicist to operate it and dispense the radon in the forms required for therapy.[63] But in Melbourne, where it also maintained the central reserve of radium and a 'central research department' to investigate improved methods of mounting and screening radium and radon, the Commonwealth had to maintain its own physicist, who, under an agreement reached with Melbourne University in May 1928, worked in space provided by Laby's department, under Laby's general supervision.[64] The physicist, A.H. Turner, was recruited from Laby's department, where he was one of the research group developing an X-ray spectroscopy method of chemical analysis. Laby himself was appointed consultant radium physicist to the Commonwealth. The arrangement worked smoothly until 1936, when some serious frictions developed between Laby and Turner's successor C.E. Eddy.[65] It set a pattern for co-operative arrangements with universities in other states, under which their physics departments produced and dipensed radon from Commonwealth radium for local hospitals and medical practitioners.[66]

The Commonwealth's concern with radiological physics initially extended only to custody and management of its radium. However, both technical developments in radiotherapy and the Commonwealth's policy (as shaped by Cumpston) of developing a co-operative 'national cancer research and treatment scheme' — of which the system of local radon services was one expression — conspired to widen the scope of its concern. To further the policy, in 1930 a new consultative body to replace the Commonwealth's Cancer Advisory Committee was formed. It had become obvious, Cumpston said, 'that a larger conference offered greater possibilities of successful discussion'.[67] The new body, the Australian Cancer Conference, met every year from 1930 to 1939. The brief of the conferences was to 'discuss three principal topics . . . what has been done, what is being done, and what can be done to improve national activities against cancer'.[68] In numbers varying from year to year physicists took part in all the conferences, and Laby, in particular, was able to use them to show how new Commonwealth initiatives to support radiological physics could improve 'national efforts against cancer'.

As it was the focus of Commonwealth interest at the time, the physics of radium and radon therapy was the main concern of physicists at the early conferences. But as early as 1931 a new field for technical contributions from physicists emerged: protection measures for radiation workers. Those most at risk were physicists and others who prepared radon and radium sources, and 'protective measures in the laboratory' was one of the topics covered in the review of work at the Commonwealth Radium Laboratory at the second conference.[69] It turned out there was more cause for concern about the hospital radium departments where the sources were used. A study of their 'arrangements for protection of personnel' revealed that 'the precautions taken are in many instances ineffective'.[70] This provoked the formation at the second conference of a protection committee, chaired by Laby, to report back to the conference on what should be done. Among other things, Laby's committee recommended 'that ionization or photographic measurements be made of the intensity of radiation in all rooms in which radium or X rays are used, in order to discover whether the radiation exceeds 10^{-5} roentgen units per second'.[71] This value of 'tolerance dose' in the new roentgen units adopted provisionally by the committee was the one on which several estimates, variously derived in the recent literature, converged.

The precise number accepted for 'tolerance dose' by Laby's committee was less important than the fact that a number could now be quoted. 'Methods for making uniform determinations of the exposure of workers in order to ascertain whether at any centre the protection measures taken are insufficient in protecting the personnel' were now feasible.[72] By the time of the 1932 conference, 'film badge' dosimeters were already in routine use at the Commonwealth Radium Laboratory in Melbourne, and experimental issues of them were being made to the treatment centres.[73] Subsequently, this became a routine service for radiation users. The film badge dosimeter had first been used in the late 1920s to monitor X-ray exposures, but its use in Australia from 1932 to monitor gamma radiation exposures as well appears to have been in advance of practice elsewhere. The protection problem in the treatment centres aside, the explanation is that Laby's department, where the Commonwealth laboratory was located, was expert in photometric methods of X-ray measurement, which it continued to use when experimental physicists elsewhere had turned to ionometric methods. Then in 1930 J.S. Rogers of Laby's department extended the photometric method to gamma radiation, showing in a pioneering investigation that for gamma rays, as for X rays, the blackening of photographic film depends only on the exposure (i.e. the 'quantity' of the radiation) and is independent of the intensity.[74] Some uncertainties about the definition and realization of the roentgen unit for gamma radiation aside (of minor importance for radiation protection measurements where high precision was not required), the technical basis for the photometric measurement of gamma radiation exposure had been laid.[75]

Perhaps exceeding its brief, at the time it reported in 1931 Laby's protection committee declared that 'the Commonwealth Department of Health, now that it has overcome the initial difficulties of providing cancer treatment by radium, could with great advantage assist in improving the facilities for the treatment of cancer by X rays'.[76] The most significant improvement within technical reach by 1931 was in the specification and measurement of X-ray dose in terms of its 'quantity' and 'quality'. Unless both these physical factors of treatment were well defined, the dose obtained at depth by X-ray therapists was uncertain. From 1928, when the second International Congress of Radiology agreed on a conceptual definition for an ionization unit of dose, the roentgen, there was, in theory, a reproducible unit for 'quantity' of radiation, in terms of which doses delivered at the skin (one of the factors determining dose at depth) could be measured. Subsequently this unit had been realized with large ionization chambers at National Standards laboratories in Germany, Britain and the United States, and dosimeters for clinical use calibrated in terms of the unit had become commercially available.[77]

But in 1931 the Sydney radiotherapists A.T. Nisbet and Leila Keatinge found that, in a comparison of dosage practices, only five X-ray therapy installations of the ten examined had ionization dosimeters in use. The instruments were of at least four types, variously calibrated in one or another of the transitional ionization units in use before the international roentgen was defined and realized.[78] Further, as one colleague commented to Nesbit and Keatinge, 'each instrument we try out gives different results at different times and from each other'. He therefore preferred 'relying on constant factors giving constant results, ... constant factors which I know will give an E.D. [erythema dose]'.[79] That is, he preferred to rely on combinations of tube milliamperage and kilovoltage known by trial and error to give erythema doses. But this method, which radiologists could use after the advent of the Coolidge tube for lack of a satisfactory 'direct' method of measuring dose, ignored the real possibility of changes in machine adjustment with time that could affect both the quantity and quality of the X-rays for given combinations. It also made it difficult to compare methods and results on different installations or at different clinics.

Before the situation revealed by Nesbit and Keatinge could be remedied, some means was required in Australia for keeping clinical dosimeters in calibration. This point was taken by the 1931 cancer conference, which urged that 'steps be taken in each state ... to secure the calibration of instruments'.[80] As a result of this resolution, Vonwiller was asked in June 1931 by the Cancer Research Committee in Sydney 'to consider some means of calibrating dosimeters'.[81] But the local calibration of dosimeters implied the establishment in Australia of a standard for the international roentgen, and as

Turner pointed out at the 1932 cancer conference, although 'the definition of the international "r", in terms of which most dosage meters record, is well known, . . . to realize the unit in practice is a matter of some experimental difficulty'. There was at this time no counterpart in Australia of the national standards laboratories that had taken on this work overseas. But, Turner noted, the essential requirement was 'a laboratory possessing a staff experienced in experimental X-ray technique'.[82] Turner did not spell out the point, but such a laboratory existed in Melbourne, in the physics department where he had been trained and where his Commonwealth laboratory was located. He offered no view on who should fund the work, but the 1932 cancer conference did. It recommended that 'an Australian standard and state sub-standards be prepared for calibrating recording instruments . . . and that the Commonwealth Department of Health be asked to prepare these standards'.[83]

By the time of the next cancer conference in March 1933 the Commonwealth already had in place equipment 'at the Physics Department, Melbourne University . . . for the calibration of measuring instruments used in X-ray therapy',[84] and Turner had an assistant physicist, T.H. Oddie, to take over the radium work. The instrument establishing the roentgen in Australia, however, was not one of the large standard chambers that had been in view in 1932, but a portable 'transfer' standard of the type designed by Taylor and Singer for the US National Bureau of Standards. This design of instrument was chosen as the Australian primary standard because of its 'simplicity of construction and demonstrated accuracy'.[85] By 1933, therefore, the technical solution to the problem of measuring clinical X-ray doses in terms of the international ionization unit existed in Australia, although institutional arrangements for applying it in the clinics and hospitals had yet to be devised.

There still remained the problem of specifying and determining the quality (hardness or penetrating power) of the X-ray beam, the property critical in determining the dose at depth relative to the dose at the skin. Overseas work related to this problem was regularly reviewed at the cancer conferences from the early 1930s by Laby, Turner, and C.E. Eddy. Eddy, a research physicist in Laby's department until 1935 when he succeeded Turner at the Commonwealth Radium Laboratory, became known for his 'lucid and logical' lectures at the conferences, expounding difficult points in the physics of clinical radiation measurements.[86] Work was started in both Sydney and Melbourne on methods for measuring the spectral energy distribution of therapy machines.[87] The import of the physicists' contributions to the conferences was that beam quality, and therefore depth dose, were related in a complex way to the voltage applied to the X-ray tube, the wave form of the voltage, and the filtration used on the beam, and thus optimization of equipment for deep X-ray therapy would regularly require physical measurements of some complexity. Accordingly, 'in these days of ultra-specialization, the services of physicists . . . should be available for the standardization of . . . radiations . . . used in therapy.[88]

How the services of specialist physicists could be provided, and who should fund their work, became issues from 1932, with Laby taking the leading part in promoting them. In 1932 he convened with W.G. Cuscaden a 'Hospitals Conference' in Melbourne in which one of the topics discussed was an inspection service for X-ray plants used in therapy or diagnosis, to embrace not only electrical and radiation safety but also 'the performance of equipment from a physical point of view'.[89] Laby suggested that this 'physical research service' could be provided by his department, given some hospital funding. Two Melbourne hospitals showed some interest. By 1933 Laby or members of his department had made 'preliminary inspections of the equipment at the Melbourne Hospital and the Women's Hospital' and 'spectroscopic observations' had been taken.[90]

But funding from the hospitals did not follow, and it was soon in doubt whether Laby could proceed further, as 'the research staff . . . is at present not large enough to take up this service and the appointment of an additional physicist is urgently desirable'.[91] To fund the physical research service Laby turned to the University of Melbourne's Cancer Research Committee. But it controlled only modest resources, many of which were already committed. When it refused to fund work in his department on 'applied X-ray research in relation to X-ray therapy and X-ray equipment', Laby initiated a public controversy, in which he attacked the support the committee was giving to Thomas Cherry's investigations of his controversial theory that cancer was related to 'immunity to tubercle'. In the course of the controversy Laby weakened his case by candidly revealing that one concern was the impact of threatened cuts in 'the University's grant for research from the State government' on his department's 'fundamental research . . . in X-ray physics'. He thus brought on his head the charge that he wanted to divert 'funds intended for cancer research to the promotion of research in physics'.[92] In Melbourne the only immediate result Laby achieved was the formation of a university committee under his chairmanship to report to the University Council 'as to whether research into the physics of X-rays was needed and, if it were, how it should be financed and organized'.[93] This was not the end of his campaign, but its successful conclusion was only reached in 1935.

Even less was achieved in Sydney, although given its resources and the role it had assumed in developing local radiotherapy services, the Sydney Cancer Research Committee might well have achieved rather more. But no senior figure associated with the Committee had the combination of entrepreneurial drive with the scientific authority in radiation physics to give the scientific leadership required. Even so, the Committee might have achieved more but for some fortuitous circumstances. From 1928 the committee had its own radium physicist, and in time his work might have diversified into the physics of X-ray therapy. This possibility was removed, however, by his long illness and eventual resignation in 1933.[94] No successor was appointed since the demand in Sydney for radon for clinical use was small, and the work was taken over by members of the physics department.[95] Again, during the critical years when a lead from Sydney might have been decisive, W.H. Love, the Cancer Research Committee's research physicist, was in Cambridge. After his return to Sydney in 1932, he was able to carry out some experimental work on photometric methods for measuring the spectral distribution of X-ray therapy beams,[96] and by 1934 had an ionization chamber set up in Sydney for the calibration of clinical dosimeters.[97] This work prepared the technical ground for a routine physical measurement service for the hospitals, but the committee made no move to establish one.

By this time the Cancer Research Committee had other, more urgent, matters to occupy it, namely the suicide of its director of research, H.G. Chapman, in May 1934 while being 'pressed for information about the state of the funds of the Royal Society and the National Research Council',[98] and the collapse of the research programme built round the Moppett effect.

Shattered dreams: the end of a radiobiological hypothesis

In late 1929 Moppett published a key paper in Section B of the *Proceedings of the Royal Society of London* reporting his 1924 experiments on the effects of homogeneous radiation on the allantoic membrane and replications and extensions performed in 1926 and 1927-28.[99] This was possibly in response to pressure from V.A. Bailey in 1928 for early and full publication of the work.[100] A warning of troubles ahead was the comment of one of the Royal Society referees who reported on a draft of the paper, that it was known to him that 'extensive effects on the embryo may result from apparently small causes'.[101]

In the next three years Moppett published five more papers in the *Proceedings*.[102] These attempted to confirm the findings reported in the key paper in other types of tissue, but no results were reported that provided independent evidence of selective action. Such evidence was elusive, as W.H. Love found in Cambridge in 1930 when, on leave from Sydney on a Rockefeller Fellowship, he began work on 'the effects of homogeneous X-radiation on the division of cells in tissue culture in vitro'.[103] Love hoped to obtain quantitative data demonstrating selective action. But after some reports to Sydney of encouraging preliminary results, no more was heard about this investigation, and he turned to another topic for his Cambridge research dissertation.

The first published work that brought into serious question the validity of Moppett's techniques and findings was done in London by Daphne Goulston who, ironically, was one of the Sydney team. On leave from Sydney, during 1931 she worked under J.C. Mottram of the Radium Institute and Mount Vernon Hospital. Here she proposed to continue work begun in Sydney on the effect of radium beta radiations on the allantoic membrane exposed, following Moppett's technique, through a window cut in the shell. But in London, under Mottram's supervision, she first investigated the changes induced in the membrane by this technique. Goulston and Mottram reported two disturbing findings in early 1932: 'first, that changes in the allantoic membrane, exactly similar to those described by Moppett following irradiation, occur with eggs dealt with by the usual technique, but not exposed to radiation'. Secondly, 'all the effects appear to be due to injury', in which the presence of fungi or spore-forming bacteria might be implicated.[104]

Further questions about the reality of the Moppett effect were posed by experiments performed by C.M. Scott in Edinburgh, reported early in 1933,[105] and by E.S. Duthie in Dublin, reported early in 1934.[106] Duthie's experiments, unlike Scott's,[107] reproduced all the essential features of Moppett's experiments exactly. But Duthie found no evidence for selective action, while 'the eggs used as controls . . . showed an extremely high percentage of damage', which further experiments indicated was 'in all probability the result of cooling the tissue'.

By early 1934, therefore, there was strong pressure on Moppett and the Sydney team either to refute the doubts that had been raised or retract. Re-appraisal in Sydney of the evidence for selective action was delayed, however, by Chapman's suicide. No successor to him as director of research was appointed. Instead, under arrangements confirmed in August 1934, Vonwiller and D.A. Welsh became joint supervisors of research for the Cancer Research Committee. The new regime resolved 'to have Dr Moppett's work and his conclusions submitted to rigid tests'.[108]

By the time of the 1935 cancer conference, 'experiments designed with a view to testing the correctness of the fundamental fact raised by Dr Moppett' had not reached a 'definite result'. However, by then Moppett, in collaboration with N.E. Goldsworthy, had begun work on 'infective reactions in the allantoic membrane of the chick embryo' to ascertain whether 'the effects observed in the radiation experiments could be obtained by infection'. The discouraging finding was that they could, as also by 'the effect of impact of air at various temperatures and of varying degrees of humidity'.[109] After so much had been invested in it, the Moppett effect was now on its way to the limbo reserved for phantom effects and irreproducible results.

Consolidation and partnership: the national scheme of physical services for radiotherapy from 1935

The 1935 cancer conference saw, however, a successful conclusion to T.H. Laby's campaign for 'physical research services' After his rebuff in Melbourne, his proposals were elaborated in a paper for the 1934 cancer conference.[110] In the course of 1934 the proposals Laby had made 'became the subject of careful consideration by the Commonwealth Department of Health', which soon began preliminary negotiations with the University of Melbourne on the question of jointly providing facilities for research on the problems Laby had defined as being within the scope of a 'physical research service'.[111] Negotiations were then suspended so that the whole question of 'the essential services necessary, physical and otherwise, for the scientifically accurate application of X-rays and radon in the treatment of disease' could be referred to a special 'preliminary conference' to be held the day before the 1935 cancer conference.[112] The meeting included delegates from the radiotherapy departments of the major hospitals and 'the Departments of Physics at the several Universities'.[113] It recommended, among other things, that 'there should be provision for the testing of X-ray equipment in all its characteristics essential for its effective operation', and that 'facilities should exist in several centres' for 'the testing, examination, and calibration of scientific apparatus and instruments used in radiotherapy, and for carrying out of scientific investigations connected with (a) the purely physical aspects of radiotherapy (b) the biophysical aspects (c) the performance of x-ray equipment'.[114] The main conference accepted these recommendations.

The recommendations determined the final arrangements negotiated between the Commonwealth and the University of Melbourne in June 1935.[115] Under these arrangements the enlarged Commonwealth laboratory was intended to be the national keystone of a system of local physical services to be established in each state. As C.E. Eddy saw it, 'the Commonwealth laboratory will maintain standards of dosage, develop methods of carrying out the physical measurements necessary for the accurate application of X-rays and radon, and conduct research into physical problems of treatment', while the local services would carry out physical measurements in the hospitals. Their work would therefore include

> Standardisation of clinical dosimeters against the local sub-standard; determination of . . . quality of X-ray beams used clinically; determination of the doses given at different depths under various conditions of treatment; investigation of the most efficient types of filter; measurement of peak kilovoltage; [and] investigation of dosage given by different distribution of radium needles.[116]

As far as possible, Eddy thought, 'these local services should be based upon the Physics Department of the local university, thus providing a self contained system within each state'. As many of the physics departments were already operating radon services, the new system of local physical services was thus to build on the existing pattern of collaboration between the Commonwealth and the academic physics community.

Sandes, on the eve of retiring as director of cancer treatment for the Sydney Cancer Research Committee, expressed some scepticism about how long the new system would work. 'The system of relying on University departments', he observed, 'will soon break of its own weight.' Instead, going on his recent observations overseas, he saw a future in which 'each hospital in possession of X-ray machinery and radium . . . will have its own physicist and regard him as an essential member of its staff '. The universities, he thought, 'should confine their attention to the higher branches of research, and to the instruction of hospital physicists in the use of apparatus for testing and standardizing purposes'.[117]

By 1937 the system was established and working in most states. In his review of activities at the Commonwealth laboratory at the 1937 cancer conference, C.E. Eddy noted that 'there will shortly be at least eight physicists throughout Australia wholly engaged in radiological physics'.[118] Ten years before there had been none. These physicists included A.F.A. Harper, appointed hospital physicist to the Cancer Research Committee in Sydney in December 1935,[119] and H.C. Webster, who in 1937 moved from the Radio Research Board to a lectureship in biophysics at the University of Queensland, with responsibility for operating the local radon and physical services.[120] As with Harper and Webster, the physicists were usually associated with university departments, though only exceptionally as regular members of academic staff. Their work was funded by 'State anti-cancer funds', either direct state government or hospital board subventions, or funds controlled by local 'anti-cancer organizations', many of them formed on the model of the Sydney Cancer Research Committee's 'cancer campaign' organization. None of these anti-cancer organizations attempted, however, as Sydney's Cancer

Research Committee had, to mount, control and direct its own ambitious programme of research.

As Sandes predicted, the 'system of relying on university departments' turned out to be a transitional one, although elements of it survived into the era when the radium and radon, through which academic physicists had been drawn into the national effort against cancer in the 1920s, were passing into obsolescence as therapeutic agents. On the whole the system was a successful consummation of Commonwealth ambitions from the late 1920s 'to focus the interest in cancer research and treatment which had been steadily growing in Australia for some time upon certain definite tangible aspects of the work',[121] and an illustration of the thesis that, even in matters in which its constitutional powers to shape events are limited 'the Commonwealth can take the initiative if it tries hard and persistently, if it has something substantial to offer, and if the wind blows in the right direction'.[122]

The Sydney experiment: some concluding comments

In contrast, the Sydney Cancer Research Committee, which in the 1920s had much to do with turning the wind in the right direction for the Commonwealth, saw many of the high hopes and ambitions on which its career had been launched come to very little by the end of the 1930s. In the aftermath of the collapse of the research programme built round the Moppett effect, and faced with the consequences of its policy of spending the capital as well as the income of the cancer funds it controlled, the committee recommended to the University Senate in late 1937 that its research organization should close down in April 1938.[123] The appointments of research staff were terminated. Warnford Moppett's research career ended, and he entered private medical practice in Sydney. The physical service for Sydney hospitals became a charge on university appropriations, pending consideration of the use of the residual cancer funds for 'the development of a general scheme of anti-cancer work in which adequate provision is made for the maintenance and extension of physical services'.[124] The committee had already divested itself of responsibility for the X-ray therapy machines in the hospitals in 1933,[125] leaving only vestigial responsibilities for cancer treatment in Sandes's hands until he retired from the position of director of cancer treatment in 1935. With his departure, only Vonwiller and D.A. Welsh remained of the men who had guided the early career of the enterprise, and it was Vonwiller who wrote the Cancer Research Committee's apologia that appeared in the last issue of its journal.[126]

By the time Vonwiller wrote this, much had occurred that, in many ways, vindicated those who had urged in 1924 'an institute ... free of university and Public Hospital control ... governed by a body of scientific men who would devote their whole time and energies to its advancement'.[127] According to such critics as H.M. Moran, the public hospital affiliation was fatal to the committee's attempts to build an efficient radiotherapy treatment organization.[128] There was, however, probably no realistic alternative open to the Cancer Research Committee except to work with the hospitals and their honorary appointment system, for better or, as Moran thought, for worse. But there can be less argument that some of the misgivings expressed in 1924 about the way the university proposed to prosecute its ambitious plans for cancer research were justified. The experiment in the collegiate direction of research that prevailed at Sydney in the formative years of the Cancer Research Committee's research programme failed to provide scientific leadership of the calibre required, as also did H.G. Chapman's appointment as director of research in 1928.

Much of the blame for the eventual debacle was placed on Chapman by Vonwiller in his apologia. But Vonwiller's assessment must be viewed sceptically. Chapman inherited, and did not himself create, the flawed research programme that collapsed in 1934. Vonwiller himself, with other members of the Committee of Direction, played as important a part as anyone in the events and decisions that led to major commitments of resources being made to investigations of the Moppett effect. If there was folly, it was collective folly, in part explained by the difficulties of managing the transition from a tradition in which research had been on the margins of academic science to a new one, for Australia, that emphasized research in the training of scientists.

Notes

1 F.J. Martell, *Scientific Australian*, December 1896, 22.
2 *Lancet* (2 July 1921), 5.
3 J.T. Case, 'Radium therapy and the American Radium Society', *Am. J. Roentgenol.*, 82 (1959), 578–80.
4 W.D. Coolidge, 'A powerful Roentgen ray tube with pure electron discharge', *Phys. Rev.*, 2 (1913), 430.
5 L.J. Clendinnen, 'The possibilities of irradiation in malignant disease', *Med. J. Aust* (29 April 1922), 457. For an account of the achievements of the German school, see D.P. Serwer, *The Rise of Radiation Protection: Science, Medicine and Technology in Society, 1896–1935* (Springfield, Va., 1977), Chap.5; also E. Quimby, 'The history of dosimetry in Roentgen therapy', *Am. J. Roentgenol.*, 54, (1945), 696. For a classic text from the German school, see B. Kroenig and W. Friedrich (trans. H. Schmitz), *The Principles of Physics and Biology of Radiation Therapy* (London, 1922).
6 *Med. J. Aust.*, 27 January 1923, 104–5.
7 *The Australian Economy: a View from the North* (Sydney, 1984), 4–5.
8 For some examples of this phenomenon in biomedicine, see F.C. Courtice, Chap.12 this volume; in physics, R.W. Home, 'Origins of the Australian physics community', *Historical Studies*, 20 (1983), 383.
9 Home, ibid., 387–8.
10 T.H. Laby to G.W.C. Kaye, 1 April 1919 (draft), 1–3 (Laby Papers, University of Melbourne Archives); T.H. Laby, *Rep. 5th Aust. Cancer Conf.* (1934), 37; Anon., 'Radiography in medicine', *Med. J. Aust.*, 2 (1917), 121–2.
11 Wettenhall Papers, AMA Victorian Branch Library, W 1548 and W 1538/3; T.H. Laby, 'Report to Victorian Minister of Health with reference to proposal to establish an Institute of Malignant Diseases', copy in Australian Archives CRS A1928, 125/36.
12 University of Sydney. Senate. Minutes, 22/12/19.
13 University of Sydney. Senate. Minutes, 8/5/22.
14 F.P. Sandes, 'Memorandum for the Committee of the Senate', attached to minutes of meeting of Committee of Direction, 5/5/22 (Committee of Direction 1922-1928, University of Sydney Archives, G 3/15).
15 Committee of Direction minutes, 5/5/22 and 8/5/22 (n.14).

16 Cancer Research Committee minutes, 23/11/23 (Cancer Research Committee 1923-1961, University of Sydney Archives G 3/14).

17 Paragraph 9 'Preliminary scheme ... for presentation to the Minister', attached Cancer Research Committee Minutes, 15/8/24 (n.16).

18 H.M. Moran, *Viewless Winds* (London, 1939), p.263.

19 L. Utz to Acting Warden, University of Sydney, attached to Cancer Research Committee minutes, 29/4/24 (n.16).

20 Ibid.

21 Ibid.

22 Cancer Research Committee minutes, 29/4/24 (n. 16).

23 University of Sydney. 1926 Calendar, 834.

24 Cancer Research Committee minutes, 8/7/26, Treasurer's report.

25 Ibid., Correspondence item (a).

26 University of Sydney, 1927 Calendar, 838.

27 Sandes, *Trans. Australas. Med. Congr.* (1929), 14.

28 This view was strongly held by the businessmen who had organized the public appeal for the Cancer Research Committee. In February 1927 they demanded representation on the Committee of Direction 'as one means of keeping faith with the subscribing public'. See Cancer Research Committee minutes 4/2/27. Later, this group also pushed hard for the Cancer Research Committee to make its own radium purchase (Cancer Research Committee minutes 19/5/27 (n.16)).

29 Cancer Research Committee minutes 10/9/26 (Notice of motion by D.A. Welsh), and Cancer Research Committee minutes, 7/10/26 (n.16).

30 For a concise account of Moppett's theories, see W. Moppett, 'Recent work in cancer research', *Med. J. Aust.* (24 July 1926), 103; for an extended account, W. Moppett, *Homogeneous X-radiation and Living Tissues* (Sydney, 1932).

31 O.U. Vonwiller, 'Cancer research in the University of Sydney', *J. Cancer Res. Comm. Univ. Sydney* (1 October 1938), 62; F.P. Sandes, 'The Cancer Campaign', *J. Cancer Res. Comm. Univ. Sydney*, (1 May 1929), 19.

32 W. Moppett, 'An investigation of the reactions of animal tissues to monochromatic X-rays of different wavelengths', *Med. J. Aust.* (11 April 1925), 366.

33 Ibid.

34 Ibid.

35 Ibid.

36 See, for example, A. Bachem, *Principles of X-ray and Radium Dosage* (Chicago, 1923), pp.174-5. Bachem was one of a number of German radiological physicists who found career opportunities in the United States in the 1920s.

37 W. Moppett, *Med. J. Aust.* (30 May 1925), 564.

38 Committee of Direction minutes, 5/3/25; ibid 26/3/25 (n.14) Cancer Research Committee minutes, 3/4/25 (n.16).

39 Committee of Direction minutes 22/4/26 (n.14) Memorandum, 'Expansion of Cancer Research', and attachment, J.M. Petrie, 'Report to the Committee of Direction'.

40 'Report of the meeting of research workers' 13 June 1927, in Committee of Direction minutes, 16/6/27; ibid. 28/7/27 (n.14).

41 Questions about the validity of the shell-window technique were first raised in the Committee of Direction at its meeting on 22 September 1926, possibly by V.A. Bailey attending his first meeting for the physics department. Vonwiller was not present on this occasion. There was further reference to the shell-window problem at a meeting between the committee and the research workers (reported in Committee of Direction minutes, 7/3/27 (n.14)) at which it was resolved that Moppett and Love should carry out 'further egg control experiments', and that 'diffusion experiments on egg shells be conducted'. An account of these experiments, and an interpretation of the results, is given in Moppett, op.cit. (n.30), pp.58-9.

42 For the physicists, the most important of these controversies was the long-running technical dispute with members of the Radiotherapy and Hospitals Sub-Committee about the choice of radiotherapy equipment of the Royal Prince Alfred Hospital. For the beginning of the controversy, see Cancer Research Committee minutes, 23/9/27; for its conclusion, ibid, 4/10/28 and 28/10/28 (n.16).

43 Ibid., 29/10/26 (n.16), 1st report of the Sub-Committee on Research Directorship.

44 Ibid., 21/12/26 (n.16).

45 Bragg was added to the advisory committee on the motion of Sandes, Cancer Research Committee minutes, 4/2/27 (n.16).

46 Committee of Direction minutes, 10/11/27 (n.14).

47 Findlay was elected to the directorship by the University Senate on 14/11/27. The election, with information about Findlay, was reported in the *Sydney Morning Herald*, 17 November 1927.

48 An indication of the reasons that led Findlay to refuse the appointment is given in Committee of Direction minutes, 17/11/27 (n.14), 'Conditions of Dr Findlay's appointment'. The Cancer Research

Committee was formally advised on 15/12/27 that Findlay had finally declined the directorship.

49 Cancer Research Committee minutes, 16/2/28 (n.16), Motion of D. Benjamin that the question of the early appointment of a research director be placed on the agenda for the next regular meeting.

50 *Sydney Morning Herald*, 26/5/34 (Obituary of H.G. Chapman); B. Kennedy, *Silver, Sin and Sixpenny Ale* (Melbourne, 1978), pp.166–74.

51 Moran, op.cit. (n.18), pp.267–8.

52 Vonwiller, op.cit. (n.31), p.55.

53 W.H. Love, The X-ray absorption spectrum of the chorio-allantoic membrane of the chick embryo and the emission spectrum of the ash of the egg of the domestic fowl', *Med. J. Aust.* (21 September 1929), 396.

54 Cancer Research Committee minutes, 18/7/29 (n.16), Report of Director of Cancer Research.

55 Ibid, 28/4/26; Committee of Direction minutes, 8/7/26 (n.14), Correspondence item (k).

56 Committee of Direction minutes, 20/1/27 (n.14), draft memorandum for Commonwealth government and Acting Director's report.

57 Cancer Research Committee minutes, 17/2/27 (n.16), under heading 'Radium Bank'.

58 As reported in J.H. Cumpston to Minister of Health, 18/8/27 (Australian Archives CRS A1928, 126/2).

59 Cumpston, ibid., and attached 'Proposals for consideration'.

60 The Commonwealth government's intention to purchase radium was reported in the Melbourne press on 19 July 1927. Provision for the purchase in the estimates was probably finally made at the Cabinet meeting on the same day. The press reports brought importunate inquiries from the Australian Radium Corporation for the contract to supply the radium, beginning with a letter to Cumpston from its Manager on 23 July 1927 (Australian Archives CRS A1928, 126/4). But Cumpston and his principal adviser T.H. Laby were justifiably sceptical whether the Corporation could supply the quantity required (10 grams)in the time limits they had in mind from its Mt Painter mine (J.H. Cumpston, 'Notes of an interview with Prof. Laby', 9/11/27, Australian Archives A1928, 126/4). The contract went to Union Minière of Belgium.

61 Laby, op.cit. (n.11), section headed 'Action by Commonwealth government' (Australian Archives CRS A1928, 125/36).

62 Ryrie to Prime Minister, 12/9/27 (Australian Archives CRS A1928, 126/5).

63 Committee of Direction minutes, 14/2/28 (n.14), motion by Chapman that 'steps be taken to obtain the services of a physicist to take charge of the proposed emanation station'. For the subsequent controversy, ibid., 17/5/28; ibid., 31/5/28.

64 J.F. Richardson, *The Australian Radiation Laboratory: a Concise History* (Canberra, 1981), Appendix 1.

65 Ibid., p.18.

66 The first radon service outside Melbourne or Sydney was established in Adelaide in 1931, the second in Perth in 1934, and the third in Brisbane in 1937.

67 J.H. Cumpston, *Rep. 1st Aust. Cancer Conf.* (1930), 4.

68 Ibid.

69 M.J. Holmes, *Rep. 2nd Aust. Cancer Conf.* (1931), 16.

70 Ibid. 14.

71 'Report of the Committee on "Protection" ', Appendix F, *Rep. 2nd Aust. Cancer Conf.* (1931), 43.

72 *Rep. 2nd Aust. Cancer Conf.* (1931), 15.

73 *Rep. 4th Aust. Cancer Conf.* (1933), 25.

74 J.S. Rogers, 'The photographic effects of gamma rays', *Proc. Physical Soc.*, 43 (1931), 59–67.

75 O.U. Vonwiller, 'The conference of physicists, mathematicians and astronomers, Sydney, August 1931', *J. Cancer Res. Comm. Univ. Sydney*, 3 (1931-2), 158.

76 'Report of the Committee on "Protection" ', Appendix F, *Rep. 2nd Aust. Cancer Conf.* (1931), 43.

77 T.H. Laby, 'Measurement of X-ray dosages', Appendix E, *Rep. 2nd Aust. Cancer Conf.* (1931), 41–2.

78 A.T. Nisbet and L. Keatinge: 'Some observations on the problem of the "erythema dose" ', *J. Cancer Res. Comm. Univ. Sydney*, 3 (1931), 17.

79 Ibid., 16.

80 *Rep. 2nd Aust. Cancer Conf.* (1931), 7, Resolution 8.

81 Cancer Research Committee minutes, 18/6/31 (n.16).

82 A.H. Turner, 'The standardization of dosage meters and kilo-volt meters', *Rep. 3rd Aust. Cancer Conf.* (1932), 39.

83 *Rep. 3rd Aust. Cancer Conf.* (1932), 9, Resolution 4(b).

84 Appendix B, Rep. *4th Aust. Cancer Conf.* (1932), 16; Appendix C, ibid., 28.

85 A.H. Turner, 'Standardization of instruments used in X-ray therapy', *Rep. 4th Aust. Cancer Conf.* (1933), 57.

86 *The Bulletin*, 22/7/36, 14.

87 W.H. Love, 'The photometric determination of X-ray quality', *J. Cancer Res. Comm. Univ. Sydney*, 5(1934), 11–15; *Rep. 4th Aust. Cancer Conf.* (1933), 16; Ibid., 26.

88 C.E. Eddy, 'Recent developments in the production and measurement of X-rays', *Rep. 5th Aust. Cancer Conf.* (1934), 44.

89 Appendix C, *Rep. 3rd Aust. Cancer Conf.* (1932), 31.

90 Appendix C, *Rep. 4th Aust. Cancer Conf.* (1933), 26.

91 Ibid., 28.

92 T.H. Laby (ed.), *Cancer Research in the University of Melbourne* (n.p., 1934), pp.7, 11.

93 T.H. Laby (Chairman): 'Report on radiation therapy to the Council of the University of Melbourne' (Melbourne, 1934).

94 Cancer Research Committee minutes, 16/11/34 (n.16).

95 Appendix A, *Rep. 6th Aust. Cancer Conf.* (1935), 23.

96 W.H. Love, op.cit. (n.87).

97 Love, ibid. p.211; Appendix B, *Rep. 5th Aust. Cancer Conf.* (1934), 23.

98 *Sydney Morning Herald*, 20/6/34, 14; ibid., 2/6/34, 15; ibid., 6/6/34, 14.

99 W. Moppett, 'Differential action of X-rays on tissue growth and Vitality', *Proc. Roy. Soc.*, B, 105 (1929), 407.

100 Committee of Direction minutes, 26/4/28 (n. 14).

101 W.J.E. Barnard, Referee's report, Royal Society of London, received 20/4/29.

102 *Proc. Roy. Soc.*, B, 107 (1930), 293-302; ibid., 302-7; ibid., 308-12; ibid., 108 (1931), 503-10; ibid., 110 (1932), 172-9.

103 Cancer Research Committee minutes, 18/3/30 (n.16), Report of Director of Research (18/3/30).

104 D. Goulston and J.C. Mottram, 'On the technique of exposing the chorio-allantoic membrane ...', *Br. J. Exp. Pathol.*, 13 (1932), 181.

105 C.M. Scott, 'The biological action of homogeneous and heterogeneous X-rays', *Proc. Roy. Soc.*, B, 112 (1933), 374.

106 E.S. Duthie, 'Experiments on the chorio-allantoic membrane', *Br. J. Radiol.*, 7 (1934), 238.

107 W. Moppett, 'X-radiation on the allantoic membrane', *Nature* (23 November 1933), 483.

108 Cancer Research Committee minutes, 30/8/34 (n.16), 'Future Organization', para. 6(b).

109 'Review on research carried out by the Cancer Research Committee of the University of Sydney', *Rep. 6th Aust. Cancer Conf.* (1935), 33.

110 T.H. Laby, 'The need for a physical research service in relation to X-ray and radium therapy', *Rep. 5th Aust. Cancer Conf.* (1934), 37-9.

111 Appendix B, *Rep. 6th Aust. Cancer Conf* (1935), 32.

112 Ibid.

113 *Rep. 6th Aust. Cancer Conf.* (1935), 5.

114 Ibid.

115 J.F. Richardson, op.cit. (n.64), Appendix 2.

116 C.E. Eddy, 'Review of the activities of the laboratory during the year 1935', *Rep. 7th Aust. Cancer Conf.* (1936), 40.

117 F.P. Sandes, 'Report to the Federal Government of Australia', Appendix D, *Rep. 6th Aust. Cancer Conf.* (1935), 36.

118 Eddy, op.cit. (n.116) p.41.

119 Harper's appointment was recommended by a Special Meeting of the CRC, 31 October 1935. For an account of his work, see his 'Physical services to hospitals', *J. Cancer Res. Comm. Univ. Sydney*, 8 (1938), 108-14.

120 *Rep. 8th Aust. Cancer Conf.* (1937), 21; H.C. Webster, *A History of the Physics Department of the University of Queensland* (n.p., 1977), pp.4-5.

121 A. Burrows, *Trans. Australas. Med. Congr.* (1929), 21.

122 M. Roe, 'The establishment of the Australian Department of Health: its background and significance', *Historical Studies*, 17 (1976), 153.

123 Appendix A, *9th Aust. Cancer Conf.* (1938), 14.

124 Rep. 10th Aust. N Z Cancer Conf (1939), Appendix B, 6.

125 Cancer Research Committee minutes, 16/11/33 (n. 16).

126 O.U. Vonwiller, 'Cancer research in the University of Sydney', *J. Cancer Res. Comm. Univ. Sydney* (1 October 1938), 52-73.

127 Utz, op.cit. (n.19).

128 Moran, op.cit. (n.18), pp.270-2.

10

Science on service, 1939-1945

R. W. Home

The unprecedented impact of science on military affairs during the Second World War has long been recognized. The sudden ending of hostilities with the capitulation of the Japanese following the dropping of atomic bombs on Hiroshima and Nagasaki was but the most dramatic (and chilling) illustration of this. The development of radar, of jet engines and V2 rockets, of plastics and synthetic rubber, of penicillin, DDT and vastly improved long-distance radio and telephone communications, all testify to the same point. Less visible manifestations of war-orientated scientific developments — for example operational research, codebreaking, and blood transfusion services for the wounded — have also been described at length in journals, books and television series such as 'The Secret War'.

Australian scientists played a role in several of these developments. Their principal achievement, however, was cumulative rather than specific, namely the part they played in the remarkable transformation of Australia's technical and industrial capacity during the war years. Here, their contribution has been admirably documented by D.P. Mellor in his impressive 738-page volume, *The Role of Science and Industry*, in the Australian War Memorial's official history of Australia in the Second World War.[1] Mellor describes, among other things, the work of Australian chemists in developing production facilities for a range of commercially and strategically important substances previously available only through imports; of engineering scientists in building up a local aircraft industry; of biochemists in developing an Australian capacity for manufacturing vitamins and drugs; and of physicists in creating an optical industry out of nothing, a working system of national standards of measurement (essential for the successful manufacture of munitions but also for large-scale secondary industry of every other kind), and a radio industry capable of designing and manufacturing the most advanced electronic systems of the day such as successful new radar equipment.

Such striking achievements inevitably redounded on the work of the scientists themselves, placing new demands on them but also creating exciting new opportunities. This chapter will chart the way in which they responded. Its concern will, however, be different from Mellor's. Rather than further documenting the contribution of Australia's scientists to the war effort, it will focus on the inverse question of the effect of the war on the Australian scientific community.

That effect was substantial and of long-term significance because it was felt differently in different scientific fields. While the physical sciences expanded dramatically during the war, the agricultural sciences, which had in the pre-war period dominated the Australian scientific scene, were held in check. At war's end, matters did not simply revert to the *status quo ante*. On the contrary, the changes that the war had brought about in the Australian economy rendered permanent the shift that had taken place in the balance of Australian scientific power.

This shift is most clearly revealed in the structure of the Federal government's principal engine of scientific research, the Council for Scientific and Industrial Research (CSIR). Prior to 1939, CSIR was, in effect, an organization devoted to research on behalf of the nation's primary industries. None of its six full-scale divisions was concerned either with the physical sciences or with problems relating to secondary industry. CSIR's only engagement in this area was through its responsibility for the Radio Research Board, but even here much of the funding came from the Postmaster General's Department rather than CSIR's budget.[2] By 1947, CSIR had acquired a further six divisions, all of them concerned with the physical sciences; and the two largest of these, Industrial Chemistry and Radiophysics, had grown to be by far the largest divisions in the entire organization.

Australian science was affected in other ways as well. Significant numbers of established scientists joined the armed services where, in contrast to what had often happened during the First World War, the value of their technical skills was generally recognized and put to use. The entire meteorological service, considerably expanded to meet military requirements in the South-West Pacific region, was in 1940 transferred to Air Force control for the duration of hostilities. Many raw young graduates (or even, later, undergraduates) in physics and engineering were recruited to operate the new radar equipment as it came into widespread use. Many other new graduates were pressed immediately into war-related industrial or laboratory work instead of first acquiring the research experience that in more peaceful times would have been their lot. Especially in the national emergency following Japan's entry into the war, an acute shortage of trained scientific manpower outside the existing research establishments, a reflection of Australia's previously low state of industrial development, meant that, perforce, the research laboratories themselves frequently had to take on production as well as research and product-development responsibilities. Inevitably, their capacity to maintain a programme of more advanced scientific work was disrupted in the process. Even if the production lines were kept out of the laboratories, the scientific work that was required to be done was often fairly low-grade problem-solving. The skills of scientists were for the most part pressed into meeting immediate or short-term needs. Projects without obvious direct application or that could be expected to yield practical fruit only in the longer term were either abandoned or drastically curtailed.

The new skills and interests that Australian scientists developed during the war were not lost during the peace that followed. To be sure, the pace of advance thereafter was not so furious or so obvious, as scientific skills that had been sharply focused during the war on a relatively small number of problems of military significance were dispersed into a multitude of peace-time occupations. The net result, however, was that in all sorts of fields, the level of scientific and industrial sophistication in post-war Australia was dramatically above its 1939 level. A new base level of scientific expertise was established in the nation. The beginnings of a modern industrial society had been laid.

The Julius Committee

The establishment by the Australian government in July 1936 of a committee of inquiry known as the Secondary Industries Testing and Research Committee, was a major turning point in Australian scientific life.[3] In fact, this was the beginning of the transformation of CSIR. The committee, chaired by CSIR's Chairman, Sir George Julius, drew up a blueprint for the implementation, at long last, of the power the Council had always had under its statute to undertake research of benefit to secondary as well as primary industry. Previously, funds had not been made available for such work. Now, with the worst of the Depression over, the government had recognized that an expanding population would not be able to find employment in the increasingly mechanized rural industries but could only be absorbed by broadening the nation's economic base. A further consideration in the minds at least of some was the looming prospect of renewed military conflict. Should war come, Australia would need to be much more self-sufficient than it had been in 1914. A consortium of Australian industrialists was pressing for the establishment of home-grown aircraft and automobile industries. Yet even in the 1930s aeroplanes and engines were very complicated pieces of precision engineering, the successful mass production of which was beyond the capacity or experience of existing Australian firms. The government had therefore agreed to help.

Julius' committee went to work with a will and by February 1937 was able to submit its report. The central feature of this was a recommendation that a National Standards Laboratory be established as soon as possible. Substantial costs were envisaged — a capital investment of £80 000 spread over three years and annual expenditure to maintain the laboratory rising to £10 000 per annum by the end of that time — but without such a facility, the committee argued, 'which can only be given by those in authority', modern mass production techniques simply could not be deployed by Australian industry:

> The outstanding feature of modern industry is the development of mass production, in which interchangeability of component parts is essential . . . In all mass production of components, individual fitting together is a thing of the past. Any one of tens of thousands of pistons must be a true and satisfactory working fit with any one of the similar number of cylinders for the same make of engine. This is only possible if manufacture is controlled rigidly within very fine limits of accuracy . . . Such accuracy is much beyond that of the ordinary measuring scale, and can only be achieved by the use of very carefully designed and accurately made 'limit gauges' . . . Without the facility for securing guaranteed gauges in the first place and for maintenance of accuracy by periodical calibration . . . interchangeability is quite impracticable.[4]

Similar considerations also applied, the committee pointed out, with regard to electrical, optical and thermometric measurements. Here, too, there was a need for basic reference standards. The proposed laboratory and associated legislation to establish national standards of measurement of physical quantities were thus 'fundamental to the effective development of Australian secondary industry'. So, too, was the establishment of a National Testing Authority to organize an authoritative system of testing and certification. In confirmation of its views, the committee stressed the 'unanimity and emphasis' with which industrialists had endorsed these proposals.

The committee also recommended that a research service be provided to assist secondary industry. Following a wide-ranging survey, it pointed to a number of areas, chiefly chemical and metallurgical, in which such a service would be of value. Linked with this proposal was a further recommendation that a properly staffed scientific information service be established in order to make 'the results of investigation, both abroad and locally', available to Australian industrialists. A recurring theme was Australia's dependence, total in many cases, on imports of manufactured goods, and her consequent vulnerability should supplies for any reason be cut off.

The industrialists whose lobbying had led to the formation of the committee would have been well pleased with what it had to say about the 'particular urgency' for an aircraft and engine testing and research laboratory. In view of 'the national and commercial importance of the projected industries', the committee reported, and of 'the specialized conditions to be observed', it was imperative that such a laboratory be established at the earliest possible date. An appendix to the report[5] discussed in greater detail the requirements of such a laboratory and the kinds of work it might be expected to undertake, using Britain's Royal Aircraft Establishment at Farnborough as a model.

In other respects, however, the report was much less definite as to how a research service for secondary industry might be organized. The range of problems that might be addressed was vast. So, too, was the scope (and hence the cost) of the investigations that might be called for. Moreover, the relationship between the proposed research service and the manufacturers who were to benefit from it remained unclear. The committee recommended a cautious and exploratory approach to begin with:

> It is essential . . . that attention be concentrated in the first instance on work that is likely to be productive of early results and that will give experience on which a sound and more extensive organization may gradually be developed . . .
> The most effective procedure for the initiation of a research service would therefore be the immediate appointment of senior research officers in those branches of science most directly associated with the problems of secondary industry, the examination of the position in consultation with them, and the preparation of a programme of initial activities.

In short, in the committee's view, 'a beginning should be made only on a small scale and . . . the work should be developed as experience is gained'. At the same time, there was no question as to the 'urgent importance' of such a beginning being made, with the actual work being done initially in existing institutions.

Finally, the committee pointed to Australia's 'dearth of men adequately trained in certain branches of science' essential to the work of the proposed new laboratories. Industry, too, would as time went on need to hire greater numbers of scientifically trained staff. Yet the facilities available in the Australian universities for post-graduate research training were generally inadequate. The committee therefore recommended that until these facilities could be expanded, CSIR's capacity under the Science and Industry Endowment Act to send promising young investigators abroad for advanced training be extended to cover the demand for research workers in these additional fields. Indeed, the committee stated, all the proposed initiatives should become the responsibility of CSIR, though not at the expense of its existing activities.

As has been noted, strategic considerations, and in particular the urgent need to render Australia less dependent in times of emergency on overseas manufacturers, were a recurring theme of the report of Julius' committee. No similar sense of urgency is evident in the government's response. The report was submitted, as we have seen, in February 1937. It included recommendations for an initial allocation of funds in the 1937/38 fiscal year and much larger allocations, especially to build and equip the proposed standards laboratory, in the following two years. (The special status of the recommendations concerning the aircraft and engine testing laboratory is revealed by the committee's not feeling it necessary to provide cost estimates in relation to them.) In fact, matters proceeded at a much more leisurely pace than the committee had envisaged, and not until more than twelve months later, on 28 April 1938, was the report formally tabled in Parliament and ordered to be printed.

Of the more substantial proposals, that concerning aircraft and engine research moved fastest, but even here it was almost two years after the lodging of the report before the first research workers were in place. In July 1937, H.E. Wimperis, recently retired as Director of Scientific Research at Farnborough, arrived in Australia to prepare further detailed advice on the setting up of the envisaged aeronautical research facilities. He submitted his report in December of the same year, it was adopted by the government in April 1938, and the positions of Officer-in-Charge of the proposed laboratory and Chief Assistant were advertised soon afterwards by CSIR. Wimperis was retained following his return to England as a consultant to recommend on the purchase of equipment, which was to include such expensive items as a wind tunnel and an engine test house. He also interviewed British applicants for the positions that had been advertised. Late in 1938 the appointment was announced of a research officer from Britain's Air Ministry, L.P. Coombes, as Officer-in-Charge. Coombes arrived in Australia early in 1939 to set up the new laboratory on land that had been reserved for it at Fishermen's Bend, Melbourne.

As far as the other recommendations of the committee were concerned, matters remained largely at a standstill throughout 1937. As late as December, Julius could still do no more than report to the Council for Scientific and Industrial Research that

> there was not very much to report in this connection as all action has been held up
> for months past owing to the fact that the Estimates had not been passed, and
> thus no funds had been available. It looked as if the Government would agree to
> the proposals . . . regarding a fundamental standards laboratory; there was also

general recognition that an Information Service would be of considerable value ... The cost of such a service would not be large, and in the first instance all that would be required would be the appointment of two or three senior men. As regards other possible activities, it was difficult to say in what direction developments would take place.[6]

The Information Service was, in fact, the first of the new activities proposed by Julius' committee actually to be set in place, albeit on a very small scale. CSIR's Secretary, Gerald Lightfoot, was sent overseas during the latter part of 1937 in order to investigate various models for such a service, and early in 1938, soon after his return, a unit was established under the leadership of the long-time CSIR functionary, G.A. Cook, and including, as well, two chemists.

The standards laboratory was a much more far-reaching and expensive proposition, but the importance of the functions it would perform was clearly understood. A CSIR committee under the chairmanship of Sydney University's professor of electrical engineering, J.P.V. Madsen, ten years earlier had looked into what would be involved, and had obtained an extensive report from the National Physical Laboratory in England on the equipment

CSIR Executive Committee, date uncertain (late 1930s?); left to right: *G.A. Julius, A.C.D. Rivett, A.E.V. Richardson (*Courtesy of the Basser Library, Australian Academy of Science*)*

that would be needed. Planning began again in earnest early in 1938. Negotiations were begun with the University of Sydney for a site for the laboratory within the university grounds, and a steering committee, chaired once again by Madsen, was appointed. By April, it had been decided not to bring out a consultant from the National Physical Laboratory, as had at first been proposed, but rather to advertise for three section heads, to take charge respectively of metrological, electrical and other physical standards; and then to send these officers, once appointed, to England for training. A suggestion that they should be accompanied when they returned to Australia by a senior officer from NPL was quietly shelved soon afterwards. Not until July, however, was ministerial approval obtained to place the advertisements.[7] In December 1938, three studentships were authorized for training in the work of the respective sections. The appointments of the three section heads, N.A. Esserman, D.M. Myers and G.H. Briggs had been approved shortly before, and all three left for England as soon as they could, where they were joined later in 1939 by their three junior officers. Construction of the laboratory itself began just before the outbreak of war on 3 September 1939, but the building was not ready to be occupied until September 1940. By that time, the first group of staff had returned from England and another group of trainees had taken their places there.

At least the nature and function of a standards laboratory were clear. Such was far from being the case with respect to the proposed research service for secondary industry and, as a result, this in due course became the subject of lengthy discussions at Executive Committee and Council meetings of CSIR. In particular, though Julius was anxious to proceed, the Council's Chief Executive Officer, A.C.D. Rivett, long remained worried by the political cost to CSIR of raising hopes within the manufacturing sector for assistance that the Council could not possibly provide 'unless very huge sums were available for the establishment and staffing of laboratories'.[8]

Still more basic questions remained to be answered, too, as became clear when Julius introduced the subject for discussion at a meeting of the CSIR Executive Committee on 25 March 1938, once the long delays in Parliament over the passing of the Estimates had at last been resolved:

> Sir George Julius said that what is necessary is to crystallise the issue as to how a beginning can best be made, what problems should first be tackled, where the work should be done, to what extent, if any, existing facilities can be utilised, and what will be necessary in the way of the erection and equipment by C.S.I.R. of laboratories of its own.[9]

By July, after further discussions at several CSIR Executive Committee meetings, thinking had crystallized sufficiently to raise the matter with the Minister, R.G. Casey.[10] The creation of separate physical and chemical laboratories was proposed. Casey agreed that 'it would be necessary to proceed with both chemical and physical research if we are going to make any more than a pretence of secondary industry research' and that 'the first step would be to get men to take charge of chemical and physical research work and to explore the whole position'. Unfortunately, before this decision could be translated into action the Minister issued a countermanding instruction to the effect that, 'owing to heavy commitments in respect of Defence and National Insurance, combined with a certain shrinkage of revenue', CSIR should proceed neither to appoint the proposed senior officers nor to embark on any substantial capital expenditure in relation to secondary industries research. However, while the Executive Committee was still exploring other ways of moving ahead, the government changed tack again and Casey in February 1939 authorized an advertisement being placed (as usual, in England and Australia) for 'an officer to take charge of investigations into chemical problems', and also an additional £20 000 being spent on 'accommodation and equipment at the Standards Laboratory for research on physical problems'.[11] By May, a short-list of three applicants for the chemistry post had been arrived at, but not until August was it decided to appoint I.W. Wark to the new position. Even then, no provision had been made for laboratory facilities, beyond ascertaining that temporary accommodation could be found in the newly completed chemical laboratory at the University of Melbourne.

Wark began duty in October 1939 and quickly formulated ambitious plans for the new section that envisaged the appointment of a further five senior chemists, 25 assistants and juniors, and sundry laboratory assistants and clerical staff.[12] His first real task, however, was to convince the Minister that there was a need for a laboratory at all, and he accordingly carried out an extensive survey of the kinds of research industrialists wished to see undertaken. With war-time problems already beginning to make themselves felt, Wark and his assistant, the engineer E.J. Drake, were able to draw up an impressive list. In response, the government in March 1940 committed large sums of money to erect and equip a substantial laboratory for a new Division of Industrial Chemistry, alongside the Aeronautical Laboratory at Fishermen's Bend.[13]

The onset of war

There is a striking contrast between the leisurely approach adopted in government circles when it came to building up the facilities discussed so far — even where, as in the case of aeronautical research, the project had government backing from the start — and the speed of the response to two further initiatives that contributed to the war-induced reorientation of CSIR. The dates alone provide an explanation here. By 1939, when the new proposals emerged, the international situation had deteriorated drastically. War had come to be seen by many as inevitable. And both proposals had obvious and important implications for Australia's defence.

The dramatic story of the founding of the secret Radiophysics Laboratory has been told many times.[14] In February 1939, S.M. Bruce, the Australian High Commissioner in London, under conditions of great confidentiality, cabled an invitation from the British Air Ministry to the Australian government to send a top-ranking physicist to Britain to be informed about recent developments in relation to a new weapon that came to be known as radar, that located enemy targets by means of reflected high-frequency radio signals. The British had already begun installing the radar network along the English coast that, eighteen months later, was to play a crucial role in the Battle of Britain. An officer of CSIR's Radio Research Board, D.F. Martyn, was selected and despatched at once to England. He returned to Australia in August of the same year, less than a month before the outbreak of war, laden with reports, blueprints and specimen electronic valves. In the light of his report, the government decided to establish a laboratory under the aegis of CSIR to undertake radar research in concert with what was being done in Britain and, perhaps more importantly, to develop a cadre of Australian workers familiar with the new techniques. Martyn was appointed officer-in-charge. In addition, the Radiophysics Advisory Board was established to develop policy under the chairmanship of J.P.V. Madsen and including Rivett, the Director-General of Posts and Telegraphs, Sir Harry Brown, and the chiefs of staff of the three armed services. By September, the government had committed the large sum of £80 000 to the project, advertisements had been placed for the first scientific staff, and arrangements had been completed with the University of Sydney for those appointed to find temporary accommodation in the university's Department of Electrical Engineering. As to a more permanent home, plans were quickly developed for an extension to the proposed National Standards Laboratory about to be constructed on the Sydney University campus. Erection of this additional section of the thus enlarged building was pressed ahead with the utmost speed, with the result that it was ready for occupation long before the part being built for the NSL. By March 1940, staff began to move in.

The other new development was less spectacular but came to pass even more rapidly. It emerged as a result of a visit to Australia during the northern summer of 1939 by an expatriate Australian physicist, F.P. Bowden, who had become a lecturer at the University of Cambridge and a leading member of a research group working for a number of years on the physics of surfaces. As the war clouds darkened over Europe, Bowden in August 1939 offered his services to the Australian government as an expert on lubricants and the wear of metal surfaces. To be sure, there was already a strong commitment on the part of the government to fostering aircraft and automobile industries in Australia, industries to which Bowden's skills and knowledge were highly relevant but otherwise in exceedingly short supply. Nevertheless, it is astonishing how rapidly Bowden's offer was accepted and laboratory space found at the University of Melbourne for the new Lubricants and Bearings Section formed under the CSIR umbrella with him as Officer-in-Charge. By mid-October, that is, only a few weeks after the proposal had first been mooted, the basic arrangements were settled and the process of building up the section had begun in earnest.

During the First World War, on account of the low technical state of the nation's secondary industries Australia had been unable to make systematic use of its scientific resources, sparse though these were. In the years that followed, little thought was given to how best to deploy such resources in a future emergency. As international tensions built up in the late 1930s, some areas of Australian industry had become more technically sophisticated and the first, albeit desultory, signs appeared that the possible strategic significance of scientific work was beginning to be recognized. The formation of the Julius committee on research for secondary industry was one early straw in the wind. Another was a memorandum from the Prime Minister's Department to CSIR in November 1938 seeking 'advice as to how the services of C.S.I.R. officers would best be utilized in time of emergency'. In response, 'schedules of 156 officers of C.S.I.R., with detailed statements of their training, qualifications and experience, etc.' were sent to the Minister in March 1939.[15]

By then, several of the nation's professional scientific organizations including the Australian Chemical Institute, the Institution of Engineers Australia and the Australian Veterinary Association were also drawing up registers of their members, which could be laid before the government when the need arose. The creation in August 1939 of a formally constituted Australian Branch of the (British) Institute of Physics meant that there was at last a professional body in the country with sufficient status to undertake a similar survey for physics. Preparing a register of Australia's physicists was one of the earliest tasks taken up by the first Secretary of the Branch, Professor A.D. Ross of the University of Western Australia. In most other sciences, however, it was not until long after the war was over that analogous national professional organizations emerged. Hence, at the outbreak of the war, no comprehensive data were available on either the size or the training and competence of much of the Australian scientific community. Such figures as are available for the chemists and physicists indicate an alarming shortage of scientific manpower. There were, for example, only fifty-six Fellows of the Australian Chemical Institute in the whole country, and only fifty-five 'corporate' members (i.e. Fellows or Associates) of the Institute of Physics.[16]

Not until mid-1939 did systematic thought begin to be given, even within the government's own scientific organization, to the question of how best to make use of the scientific manpower available. In June of that year, less than three months before war actually broke out, the CSIR Executive Committee was confronted by a request from one of its officers that it 'define the policy of C.S.I.R. in regard to service by research and administrative officers in combatant rather than technical units'. It decided that leave could and should be granted to administrative staff who wished to join combatant units, but that leave could not be guaranteed to scientific and technical staff 'whose services would obviously be needed in other than combatant units'.[17] Inevitably, problems of interpretation arose almost at once; an engineer in the Division of Forest Products who also held a commission in the Royal Australian Engineers asked whether, in the light of the Executive Committee's decision, he should resign his commission. Since the only directive that had been received from the government's manpower planning authorities had been to the effect that 'in cases in which officers of CSIR wish to enrol in the militia, all practicable facilities are to be granted to them to enable them to do so'[18] — as Mellor has noted, 'the need to mobilise science was more apparent to scientists than it was to administrators'[19] — the Executive Committee adopted as liberal a stance as it could. Yet its members were by this time acutely conscious of the urgency and importance of the tasks likely to be imposed on the Council in the event of war, and no longer felt able to abide by the letter of the instructions they had been given. The Executive Committee had no objection, they resolved, to the officer in question retaining his commission in the R.A.E., 'particularly as the latter is a technical unit'. As to the more general issue, however,

> it was felt that where a Research Officer already held a commission in a Defence unit, the Executive Committee would not suggest his resignation, leaving to the individual to decide where and how he could be most usefully employed in time of emergency, having regard to all the circumstances, and particularly the function and nature of the Defence unit in which he is interested. However, ... the Executive Committee feels it is sound in principle to assume that, in time of emergency, the great majority of Research Officers can be of more value in the special work for which their training and experience fit them than in the front line with combatant units.[20]

When war was declared, individual Australian scientists were anxious to place their expertise at the service of the nation. For those already within the government service, whether at the Defence Department's Munitions Supply Laboratories at Maribyrnong, Victoria, or in CSIR, this would involve, at most, a redirection of effort within the existing structure. As far as CSIR was concerned, for example, on 4 September 1939, following Prime Minister Menzies' dramatic radio broadcast the previous night announcing that Germany had invaded Poland and that, with Britain, Australia was therefore at war, the Council's Secretary, Gerald Lightfoot, wrote to the chief of each division and the officer-in-charge of each section, 'asking that consideration be given to the question of re-orientation of their work with a view to accelerating investigations which are likely to be of value in connection with problems arising out of the war'.[21]

For scientists employed elsewhere, however, most notably in the universities, it was much more difficult to find useful war-related work, even though the promulgation, at last, of a list of reserved industries and occupations (in which restrictions were placed on who could be accepted for military service) provided reassuring evidence that the value of scientific skills was at last beginning to be recognized.[22] The newly elected president of the Australian Branch of the Institute of Physics, T.H. Laby from the University of Melbourne, had written to the Prime Minister a week before war broke out, offering the services of members of the Branch in an emergency, but when Menzies replied seeking specific suggestions of ways in which physicists might be able to help, Laby was unable to formulate a satisfactory response.[23] The Australian and New Zealand Association for the Advancement of Science (ANZAAS), which, as the senior national scientific organization, might have been expected to play a role in marshalling the country's scientific talent, was in fact powerless to do so because it was so constituted that it virtually ceased to exist between biennial congresses. The Australian National Research Council, comprising about 200 of Australia's leading scientists and dominated by university people, recommended the formation of a scientific advisory committee after an existing British model. To be sure, this would have allowed university scientists a larger public voice than they enjoyed on the Council of CSIR; but in other respects the proposed committee would largely have duplicated the Council's functions and the government thus rejected the idea. However, it did grant the ANRC a lesser, consultancy role in some matters.[24] Despite this Rivett, for one, long remained sceptical about using university scientists in war work, as may be seen from a letter he drafted for CSIR's new Minister, Harold Holt, early in 1940:

> I know that complaints are to be heard on occasion that the Government is not using Australian scientific men as fully as they desire. Such complaints at times come from persons who have not shown great competence in scientific work in the past; but frequently from capable men who are already in full time employment and are able only to give spare time to Government work. While in many cases we can use, and are using, such men (particularly in Universities) one has to recognize that war work nowadays is usually of such a nature that it requires the whole time of the scientific staff engaged upon it. Further, a great deal of it is too confidential to be handled in any but carefully controlled Government laboratories.[25]

A number of university scientists sought to bring their talents to bear on war-related work through the Australian Association of Scientific Workers, which had been formed at the ANZAAS Congress in Canberra in January 1939 with the aim of promoting a greater social role for science.[26] With the onset of

COMMONWEALTH OF AUSTRALIA

SCIENCE ON SERVICE

A DIRECTORY OF SCIENTIFIC RESOURCES IN AUSTRALIA

Prepared by
THE SCIENTIFIC LIAISON BUREAU
under Direction from the Hon. J. J. Dedman

Melbourne	- M 2639*	Sydney - -	B 6224
Adelaide	Central 3355	Brisbane -	B 2501
Hobart	- 6038	Perth - -	B 3439

Newcastle - - Hamilton 203

* To change in April, 1944, to MU 7639.

SECOND EDITION

SYDNEY
AUSTRALASIAN MEDICAL PUBLISHING COMPANY LIMITED
1943

Title page, Science on Service *directory (1943).*

war, the association's working committees turned their attention to such questions as the provision of essential drugs and industrial chemicals, and potential shortages of manpower for scientific work. Initially, their activities met with official indifference but in due course they managed to play an important role in these matters. At a later stage in the war, in 1942, the association agitated successfully for the creation, despite opposition from CSIR, of a scientific liaison bureau to act as a bridge 'to ensure that scientific problems which arise in the Services, government departments and war industries are promptly brought to the notice of suitable scientific men'.[27]

Many physicists outside the government laboratories eventually found a rôle for themselves in the work of the Optical Munitions Panel (later the Scientific Instruments and Optical Panel), established in July 1940 under the chairmanship of Professor T.H. Laby of the University of Melbourne. Some chemists, most notably Professor E.J. Hartung of the University of Melbourne, also made a crucial contribution to this work by initiating the local manufacture of optical glass. Later, when the war spread to tropical regions and the growth of moulds on optical equipment became a problem, a number of botanists and biochemists likewise became involved.[28]

A few specialist groups found their skills in demand from the outset. Having access to a reliable meteorological service, for example, was of paramount importance to the armed services. The Commonwealth Bureau of Meteorology, the focus of all meteorological work in the country, had already been considerably expanded in the late 1930s to meet the growing demands of civil aviation. A conference in July 1939 established a programme of further expansion to meet the needs of the military, and laid plans, as mentioned above, for the transfer of the entire meteorological service to the Air Force should war break out. This new growth led to a severe shortage of trained personnel — the first department of meteorology in an Australian university, at the University of Melbourne, had been created only that year — and caused the bureau to expand its in-house training scheme and to recruit numbers of science graduates, chiefly school teachers, into the service in the first months of the war. Not until 1940, however, did the proposed transfer of the bureau to the Air Force take place.[29]

Radio physicists and engineers were perhaps in greater demand than any other group. This was due partly to the recruiting of staff for the Radiophysics Laboratory, but, in addition, radio communication services in general were of major strategic significance. Fortunately, Australia was relatively well supplied with radio scientists, chiefly as a result of the activities of CSIR's Radio Research Board. The board, established in 1927 under the chairmanship of J.P.V. Madsen, had quickly established active research groups under Madsen at the University of Sydney and Laby at the University of Melbourne. These were built around small cadres of permanent Radio Research Board staff, but quickly developed the practice of engaging research students to work on related topics for their thesis projects. The work done was generally of a very high standard, and D.F. Martyn in particular developed a considerable international reputation for his studies of the propagation of radio signals in the ionosphere.

By 1939 a considerable number of young Australian physicists had by this means gained significant research experience in radio science. Some had then gone on to work in Britain under leading figures in this field such as E.V. Appleton and J.A. Ratcliffe. One of this group, J.L. Pawsey, subsequently joined the research laboratory of the British electronics group, EMI Ltd, where he was an important early contributor to research on television systems. Another, J.H. Piddington, while in England did research related to radar, and following his return to Australia, CSIR in November 1938 sought to continue this, requesting additional funds for 'certain special and confidential work' in which the Radio Research Board was interested, to do with 'long-range detection of aeroplanes'. However, *in a manner long characteristic of Australian attitudes on sensitive matters, the Minister, Casey, felt that before* authorizing this, he should gain clearance from London for what was proposed. This, of course, was not forthcoming. Madsen, however, undaunted, found money elsewhere and Piddington was shortly experimenting with a large and powerful radar set located in the basement of Madsen's department.[30] A year later, after Britain had decided to let the Dominions into the secret, Pawsey and Piddington were among the first recruits into the Radiophysics Laboratory. They were joined by most of the staff of the Radio Research Board.

Radio had been through a boom period in Australia, as elsewhere, during the 1920s and 1930s, and a number of those who worked for a time under the auspices of the Radio Research Board had subsequently found positions in the radio industry. By far the largest private employer of radio scientists and engineers — it is almost impossible to draw a line between engineers and physicists at this time in the field of radio research — was Amalgamated Wireless (Australasia) Ltd which, under the dynamic leadership of its managing director, E.T. Fisk, dominated the Australian market and was to play a key role in the nation's war effort. The Australian government owned 50 per cent-plus-one of the company's shares and had granted it exclusive rights to operate both the coastal radio service for merchant shipping in Australian waters and Australia's international radio links. The company also owned several commercial broadcasting stations and manufactured both broadcasting equipment and household wireless receivers. It was a major supplier of telephone equipment to the Postmaster General's Department. With the formation of a subsidiary, the Amalgamated Wireless Valve Co., AWA in 1933 had begun manufacturing, under licence from RCA in America, many of the electronic valves required by its other operations.

In 1931, AWA had established research laboratories at its new works at Ashfield, in Sydney. Part of the function of the laboratories was to maintain a variety of physical sub-standards and associated measuring instruments for quality control purposes in the valve-manufacturing plant. They were also intended to keep the company abreast of the latest developments in applied electronics research and to undertake research of their own with a view to strengthening the company's position with respect to patent rights in this rapidly growing field. The laboratories expanded steadily throughout the 1930s. As we have noted, they recruited a number of Radio Research Board staff; others transferred to the laboratories from elsewhere in the company's operations. By 1939 they were one of the largest employers of physicists in the country.[31] There were other electronics manufacturers in Australia by 1939, most notably local subsidiaries of the European conglomerates Standard Telephones & Cables (STC) and Philips. They, too, were to make an important contribution to the war effort.[32] Much more than AWA, however, they had tended to rely on assembling imported components. None had laboratories or research staffs to compare with those of AWA.

Like the Radio Research Board, the AWA laboratories proved a fertile recruiting ground for the Radiophysics Laboratory, once this began hiring staff. However, a careful balance had to be kept in this case, for the radar project was as dependent on AWA's manufacturing capacity (and that of the company's commercial rivals) as it was on any research that might be done at the Radiophysics Laboratory. Matters came to a head only weeks after the outbreak of war, when O.O. Pulley, a senior AWA engineer and former Radio Research Board research worker, applied for and was appointed to one of the recently advertised senior positions in Radiophysics. AWA's managing director, Fisk, protested vigorously, first to CSIR and then to the Minister, stating that losing Pulley would seriously disrupt his company's production work. Madsen responded on behalf of Radiophysics that it was essential that the laboratory secure 'a first-class engineering man with a sound training in physics' such as Pulley.[33] Given the absolute priority being given to building up the Radiophysics Laboratory at this time, Madsen's argument prevailed, though in the event, Pulley's stay at Radiophysics proved an unhappy one. Despite this, the incident well illustrates how, even at this early stage of proceedings, Australia was being forced to give more careful thought to the husbanding and deployment of its all too scarce scientific resources. The new attitude was confirmed a month later when CSIR took prompt action to obtain exemptions for several of its scientific officers who had been called up for service in the militia.[34]

'Phoney' war

The above developments fell far short, however, of the full mobilization of Australian science that was to come later. After the initial alarms associated with the declaration of war and the setting up of the Radiophysics Laboratory, there ensued a period in which little of consequence occurred on the scientific front. Though several new laboratories were soon under construction for CSIR — the Radiophysics–National Standards Laboratory complex in Sydney, the Aeronautics and (by early 1940) Industrial Chemistry laboratories in Melbourne — the council as a whole had been given 'strict instructions that the utmost economy should be exercised' in its day-to-day operations.[35] Only in the Division of Forest Products, where two shifts were being worked per day on plywood and structural timbers for aircraft, was there any real intensification of activity. Moreover, as we have seen, it was some time before the nation's university scientists found ways of contributing usefully outside their classrooms. Some, such as T.H. Laby, in their frustrated enthusiasm to become more directly involved, developed a considerable animus towards CSIR. This led in Laby's case to a remarkable anti-CSIR outburst, which Rivett attributed to pique at not being involved in the

Radiophysics developments, when a deputation of physicists waited on a group of Ministers in Canberra in June 1940.[36] Even in aeronautical research, previous patterns of work changed very little. The attitude of L.P. Coombes, the newly appointed Chief of CSIR's Division of Aeronautics, seems to have reflected the situation at this time in Australian science more generally. According to his report to CSIR's Council meeting in April 1940,

> on the outbreak of war, his Laboratory had been asked how best it could help in the national effort. In war, however, emphasis is given to increase of production rather than to improvement of type, and actually there is not much more aeronautical research in war time than in peace. The Aeronautical Laboratory would, therefore, just carry on, although it would accelerate its progress as much as possible.[37]

Rivett in a letter to an English friend evinced a similarly relaxed (and resigned) attitude at about the same time, though even in advance of Laby's outburst in Canberra, he was clearly finding the rumblings from the university scientists rather tiresome:

> As to the national effort here, I mean the national scientific effort, I daresay that if one put all the large and small odds and ends together, the sum might be bigger than one's non-quantitative impression of it all. There has been the inevitably large amount of drawing up lists of so-called scientific workers and laboratories and all the rest of it. It all seems to end in polite thanks and the pigeon-hole. Apart from the growing numbers of full-timers with scientific training in munitions establishments, etc., a certain amount is being done in Universities. I suspect, however, that the time has come when the University teacher, in his odd hours of freedom, counts for relatively little in the research world — at any rate in these under-staffed Dominion Universities. Many of the older men grumble because, perforce, they are not handed out nice little problems in their own particular lines, and prophesy that the Government's neglect of science will mean the destruction of the Empire. Poor Government! Poor Empire![38]

There had been considerable talk early in the war, especially in connection with radar, of co-ordinating the research being done in Britain and Australia so as to avoid needless duplication of effort; and to that end the Australian authorities established a scientific liaison office in London in early 1940 with a senior Radiophysics scientist, G.H. Munro, in charge. In the event, however, Munro's work and that of his successors amounted to keeping themselves informed about British developments and channelling the latest information and equipment to Australia. There was little traffic in the opposite direction. No doubt, this was partly due to the much smaller scale of the Australian operation. Partly, too, however, it doubtless reflected the same traditional British tendency to underestimate colonial capabilities that the expatriate Australian physicist H.S.W. Massey reported in connection with some work he had been doing for the British Navy:

> I was somewhat surprised to find . . . that the authorities here thought that it would be necessary to send scientists and technicians from here to take over the work concerned in Australia and the other Dominions, an opinion of which I quickly disabused them.[39]

Given the physical separation of the two countries, however, and the resulting delays in communication between them, it would have been inevitable anyway that as the war situation became more critical as far as Britain was concerned, all notions of an orderly parcelling out of research problems should have been forgotten. Matters were far too pressing for that to be feasible any longer, even if it might once have been possible.

Throughout 1940 and 1941, there was a slow but steady build-up of Australia's scientific resources, the main thrust of the work being to render the country less dependent on imports of essential specialized materials and to support the expanding munitions industry. The staff of the Munitions Supply Laboratories grew considerably, the new CSIR laboratories opened one by one, and university-centred groups, such as the Optical Munitions Panel and AASW's Drugs Sub-Committee, became established features on the scientific landscape.

The fall of France in May 1940 gave rise to a greater sense of urgency, but the war was still a long way away and the concern that was felt was for Britain's safety, not Australia's. Rivett captured the mood perfectly in a letter to another British colleague:

> What the future holds it is impossible even to guess at this distance: but I can assure you that we are all of us thinking very much of you people and of the black days that are ahead and only wish that we were able to send our good wishes in a [f]orm that could be applied with effect against the enemy.
> Ordinary affairs seem hardly to count at such times; here as with you there has been such a rapid awakening to the general peril that the whole direction of National effort seems quickly to be changing.[40]

Even so, in many areas Australia's scientists still had a long way to go in making the nation's industrialists aware of what they could contribute. On the very same day that Rivett wrote this letter, CSIR's Executive Committee discussed the need to draw the attention of manufacturers to the existence of its information section, now more than two years old! A press notice had recently been issued highlighting the facilities that the section offered. Now, the Executive Committee approved the sending of a follow-up letter to 'selected manufacturing establishments', 'drawing attention to the establishment of the Information Section and to the opportunity afforded of obtaining technical information'. In the course of the discussion, Rivett indicated that he had recently also felt it necessary to write to Essington Lewis, Director-General of Munitions, 'summarising facilities which C.S.I.R. might provide, e.g., with respect to the Standards Laboratory, Aeronautical Laboratory, Forest Products Division, Information Section etc'.[41]

There was still some feeling, too, that the best way for Australian scientists to contribute to the war effort might be, as in the First World War (when a contingent of Australian chemists had volunteered to work in British munitions factories), to work in Britain rather than in Australia. At the height of the Battle of Britain, CSIR's Head Office, prompted by a discussion at the Radiophysics Advisory Board, sent a cable to London inquiring 'whether a team of Australian chemists, biologists and physicists would be useful in Britain'. The British authorities, however, on this occasion poured cold water on the notion: 'As a result of the advice given it had become clear that Britain did not need any considerable personnel from Australia for scientific work.'[42]

Meanwhile the establishment of a national system of standards had been proceeding very slowly. Gerald Lightfoot on behalf of CSIR had written to the Minister in June 1939, shortly before war was declared, about the legislative action that would be needed, but nothing had eventuated. In November, CSIR's Executive Committee took up the matter again but decided that 'the present time was inopportune to press the matter'. In April 1940, it was decided to commission J.M. Baldwin, Victorian Government Astronomer, to survey the existing state Weights and Measures Acts (under which the states exercised their responsibility for standards of measurement used in trade) and advise on both the legislative and administrative aspects of the Federal government's accepting responsibility for the maintenance of basic metrological standards. In the event, it was Lightfoot who carried out the survey but, early in 1941, the matter was quietly shelved, not to be taken up again until the end of the war was in sight.[43] Negotiations likewise stalled over the question of co-ordinating testing services for manufacturing industry throughout the Commonwealth, the state governments in New South Wales and Victoria, where the need for sub-standards and testing services was greatest because of their greater degree of industrialization, proving extremely reluctant to commit any funds to the scheme. This problem, too, was soon set aside as the now fast-growing munitions industries began to press both the Munitions Supply Laboratories and the newly opened National Standards Laboratory for assistance of a more urgent and short-term nature.[44]

As early as December 1939, Madsen in his capacity as chairman of the National Standards Laboratory advisory committee had with his usual foresight drawn the CSIR Executive Committee's attention to 'the desirability of someone taking immediate action to provide a service for the checking of gauges' needed for manufacturing components with precise tolerances, especially in building the aircraft engines for which the government had just placed an initial order. Since the states were doing next to nothing, he suggested that the Commonwealth be advised that it should place orders immediately, on its own behalf, for equipment that he estimated would cost £15 000. A month later — perhaps in response to Madsen's initiative — the Commonwealth Department of Supply and Development convened a conference to consider the same question at which it was agreed that the calibration of the gauges to be used by the sub-contractors would be handled by the Munitions Supply Laboratories.[45]

The problem continued to escalate, however, and by May 1940 CSIR faced a request to take a share of the load by calibrating gauges for Sydney firms, by advising firms 'in the selection and purchase of equipment, in principles of metrology and the technique of measurement', and by testing and if necessary calibrating measuring instruments in factories. Julius, the engineer, at first demurred, saying that 'it was not necessary to utilise the services of trained metrologists to calibrate testing machines'. Within a few more months, however, the pattern of work at the National Standards Laboratory was to shift much further away from the normal perception of what 'trained metrologists' were supposed to do. Now, with Julius' concurrence, CSIR accepted requests from the Department of Munitions actually to manufacture twenty-five sets of Johannsen gauge blocks and 20 screw thread profile apparatus. In addition, albeit reluctantly since 'difficulties will probably arise owing to lack of equipment and personnel', NSL agreed that workshop gauges should be sent directly to them for periodical checking, rather than to one of the network of NSL-licensed testing stations it had been intended to establish but which had not yet come into existence. By early 1941 an agreement had been forged between CSIR and the Munitions Supply Laboratories at Maribyrnong, which had been expanding their gauge-testing staff for some time, that the latter would take responsibility for mechanical testing in Victoria, South Australia and Western Australia, leaving New South Wales and Queensland to NSL. (Apparently no demand for such services was anticipated from Tasmania!) NSL had already had to recruit a number of new workshop staff from the University of Sydney in connection with the Laboratory's newly acquired manufacturing responsibilities.[46] Now it was also found necessary to take on a considerable number of non-scientifically trained 'observers', many of them females, to handle a growing volume of routine testing of instruments and gauges.

The Munitions Supply Laboratories had always been intended to provide, as part of their function, advice and routine standards control services for the munitions industry.[47] Yet to be so directly involved in the production process was a new experience for most civilian scientists in Australia. It was to become almost the norm in the dark days following the entry of Japan into the war in December 1941.

State of emergency

War tension had already been building up within the scientific community, as elsewhere within Australian society, prior to the Japanese attacks on Pearl Harbour and Malaya. In November 1941, a letter went to the chiefs of all CSIR divisions calling them to a meeting early the following month — fortuitously, on the very day Australia awoke to find itself at war with a new enemy. The chiefs were asked to consider the implications for their divisions of a Ministerial instruction to CSIR 'to give careful consideration to its programme of work in order to ascertain what investigations could be held in abeyance during wartime with a view to concentrating energies on problems connected with the war effort'.[48] On 12 November, the Minister in charge of CSIR in the new Labor government, J.J. Dedman, addressed a meeting of the full council:

> Mr. Dedman said that, under existing conditions, it was not easy to carry on all activities associated with peace time ... He realised that certain peace-time activities had already been set on one side by CSIR and this must continue progressively for some time. Some investigations might have to be discarded for the time being. It was essential to devote all activities to winning the war. He was more than sympathetic with peace-time projects, but sacrifices must be made.[49]

At the meeting of chiefs these strictures led to a further significant reallocation of CSIR's resources away from most of the old-established agriculturally orientated divisions — Animal Health and Nutrition, Plant Industry, Economic Entomology, Soils, Fisheries — to 'urgent war requirements' in the

Divisions of Food Preservation and Transport, Forest Products and Aeronautics, in the National Standards Laboratory, and in the Lubricants and Bearings Section.[50] By early the following year, Rivett felt able to claim that most of these units, as well as Industrial Chemistry and the separately funded Radiophysics Division, were 'practically full-time on direct war work'. 'I sometimes wonder', he mused, 'how on earth we are going to face the situation that will arise when peace comes; but apparently there is no great need to worry about that for some little time yet'.[51]

Both within CSIR's laboratories and elsewhere, 'direct war work' came more and more to follow the pattern of direct involvement with routine testing and actual production that had been set at the National Standards Laboratory. In October 1941, for example, Philip Bowden reported that his Lubricants and Bearings Section had been approached by the Aircraft Production Commission to work out methods of manufacturing certain bearings and to train workers in the relevant procedures. The commission proposed then to set up its own 'annexe', or factory, using Bowden's group as consultants. However:

> The immediate production of a number of bearings was ... essential, and it would take some time to construct the annexe and obtain the necessary equipment. The equipment available for casting bearings at the University of Melbourne would be suitable as a small production unit. He had therefore offered to begin production on a small scale.[52]

At the Division of Aeronautics, there had been long delays in procuring necessary equipment and staff. In May 1941, when a report was prepared for a CSIR council meeting, two of the planned four sections of the laboratory were still inoperative for one or other of these reasons; indeed, not until early 1942 did the division reach what was regarded as 'working strength'. In the meantime, and contrary to the intentions Coombes had expressed a year earlier, it had come to be chiefly engaged, as it would be for the remainder of the war, on 'many problems of aircraft production' submitted to it by the manufacturing authorities. At this early stage, these included the design of a local gun turret for the Wirraway machine and studies of the forces operating on the bomb doors of a Beaufort bomber, the use of improved wood for aircraft, corrections to sighting of guns, and gas welding of alloy steel tubes. In addition, the division became responsible for giving 'an independent opinion as to the airworthiness of different types of aircraft'. Much work was done in this connection on rationalizing specifications and, in conjunction with a specially formed Air Force flight, on flight testing Australian-produced aircraft.[53]

Similar close involvement with actual production was quickly the norm at Wark's Division of Industrial Chemistry. Here, most effort went into developing processes using Australian ingredients to pilot-plant stage to produce essential chemicals that were in short supply. By mid-1942, pilot plants were either operating or under construction to produce tri-cresyl phosphate, required as a plasticiser in synthetic resin nose caps for bullets and shells, from phosphorus oxy-chloride; phosphorus oxy-chloride from rock phosphate; ethylene from alcohol; and ethylene oxide, ethylene chlorhydrin and ethylene dichloride from ethylene. In the case of the ethylene chlorhydrin, 'required for the synthesis of the essential drug, novocaine', the Division's plant was capable of producing the whole of Australia's requirements. Similarly, the mineral flotation process on which Wark and his colleague K.L. Sutherland were world authorities was successfully adapted to separate the source of another essential drug, ergot, from rye, in sufficient quantities to meet the whole of the nation's needs. In almost none of the work undertaken was fundamental new research required. Instead, the problem was usually either to adapt existing processes to Australian ingredients and operating conditions — for example, the key to the problem of producing ethylene from alcohol was finding 'a suitable clay catalyst of Australian origin' — or to identify and 'scale up' the best means of obtaining substances that were in demand from materials available at the time. In other cases, the Division advised government departments or industrial concerns on chemical and metallurgical problems, without itself becoming responsible for actual production. A wide range of examples is listed in the secret 'War Reports' prepared by CSIR in 1942, 1943 and 1944.[54]

As has been noted, many university scientists became involved in the work of the Optical Munitions Panel. Technically, the panel's role was to advise the Ordnance Production Directorate of the Ministry of Munitions on the production in Australia of the large range of optical equipment needed by the armed forces. As the panel's secretary/historian, J.S. Rogers, records, however, as time went by it was also asked 'to keep an eye on production'.[55] Yet even in its advisory role, the panel was closely connected with the problems of production. Its Glass Committee, for example, was responsible in the early stages of the work for deciding how much optical glass, and of which types, should be ordered from overseas, and later, once optical glass was being made in Australia, for recommending how many melts were needed of each type of glass. The task of the panel's physicists was for the most part to analyse the existing British-made instruments on which most of the armed forces' orders were based, and then to redesign their optics for Australian glass to meet the same performance specifications. Later, the panel also took on the task of developing prototypes of new instruments for which there were no British counterparts to serve as models.

The grinding of precision lenses and prisms was a specialized skill possessed by practically no-one in Australia at the outbreak of war. One of the panel's principal achievements was the training of substantial numbers of operatives for the contracting firms. Both university and Munitions Supply Laboratories physicists were active in helping contractors to overcome problems as they arose.[56] Others became much more directly involved in the production process, as both the Physics Department at the University of Tasmania and the Mt Stromlo Observatory were, in effect, converted into factories for producing large numbers of high-quality optical components. Many of these components were subsequently individually tested in the Natural Philosophy Department at the University of Melbourne.

Difficulties frequently arose in attaining the high degree of mechanical precision specified for many of the made-up instruments for the production of which the panel was responsible. For some time there was an acute shortage of the necessary machine tools, of skilled technicians to operate these, and of instrument makers able to assemble the various parts into satisfactory working instruments. Prototypes were routinely constructed in the co-operating university laboratories and, in some cases, full production lines had to be set up.

Once Japan entered the war and the focus of Australia's military effort shifted to the islands to her near north, a new problem arose, namely the rapid growth of fungi under tropic conditions on the glass surfaces of the instruments. 'Tropic proofing' thus became an urgent task. Once again, the burden fell chiefly on university laboratories. It took some time, however, to find a solution. Not until late 1943, following tests on a number of substances in Melbourne University's departments of botany and biochemistry, was the decision reached to recommend the use of lacquer impregnated with the fungicide 'Merthiosal' (sodium-ethyl-mercuri-salicylate), which could be painted on the inner metallic surfaces of the instruments. The minutes of the meeting at which this decision was made starkly reveal the pressure on Australian scientists during these crucial years of the war, even within university laboratories, to produce immediate solutions to practical problems of production at the expense of thoroughly worked out research. The use of Merthiosal was proposed by V.M. Trikojus, professor of biochemistry at Melbourne, it at least having been demonstrated by then that this substance prevented the growth of fungi. Trikojus stressed, however, that it was still not known how it acted, that is, whether it killed the spores or merely prevented them from germinating. The committee noted, too, that other possible fungicides remained to be investigated. Its deliberations were cut short, however, by the Army's representative:

> [I]t was necessary at the present time to take some action immediately with regard to instruments on hand. The Army was not particularly interested at the moment whether the germicide would remain affective (*sic*) for weeks or for months so long as it caused an improvement on present conditions ... Mr. Sims ... point[ed] out that whether finality was reached or not regarding what was the best fungicide, some interim recommendation would have to be made at once so that stores could be treated as rapidly as possible and sent to replace instruments in New Guinea affected by fungus.[57]

Binoculars were the main problem. The committee therefore agreed that 'a fairly large number of binoculars be sent at once to the Botany School [at the University of Melbourne] with instructions that they are to be treated at once'. When staff there had gained enough experience to be able to recommend 'a good practical procedure', they were to forward instructions to the other OMP co-operating laboratories.

In the meantime, facilities at other universities were used to recondition thousands of binoculars impressed from the civilian population for military use. The physics departments at the universities in Sydney, Adelaide and Perth and the botany department at the University of Melbourne were all active in this work, which for the most part involved university scientists training and supervising large numbers of women operatives hired specially for the task.[58]

There was a similar shift towards a direct involvement with production at

AWA physicist-engineers J.G. Downes and D. Dane discussing a new transmitter-receiver for civil aircraft (Reprinted from AWA and the War *[Sydney, 1946(?)].* Courtesy AWA Ltd)

the AWA research laboratories. In the early stages of the war, the overwhelming emphasis in the work of the laboratories continued to be on questions of developmental design. By early 1942, however, the Communication Receiver Section was delivering fully made-up units — telephone links, UHF equipment, etc. — to the company's communication services, not just designing them. Likewise, the Test Instruments Section, staffed by two engineers and a few technicians in December 1940, by mid-1942 had become 'a complete unit for the design and manufacture of test instruments' with its own machine shop, assembly section, progress store, planning section, test rooms and self-contained development and design section, employing forty-six people and occupying 2500 square feet of floor space. Even more spectacular was what happened to the Crystals Section, a small specialist group, which in December 1940 produced twenty-four carefully cut quartz crystals for radio receiving sets, and held outstanding orders at the end of that month for a further sixty-eight. During 1941 the section tooled up to mass-produce crystals. An old grinding machine was converted into a precision cutting engine — the munitions factories had first call on all new machines of this kind — a range of physical sub-standards was acquired for quality control purposes and a team of young women trained in the cutting, grinding and polishing techniques involved. By June 1942, the section was able to produce in a single month no fewer than 351 precision crystals and another 847 mass-produced crystals made to somewhat less exacting but still rigorous standards under contract for the Air Force. In September 1942, to meet an emergency order from the company's Commercial Engineering Department, 200 crystals were produced in five days, at which stage the supply of quartz temporarily ran out. (In fact, the company spent a considerable sum on mining operations at Glen Innes, NSW, and elsewhere, in a largely unsuccessful attempt to secure a reliable local supply of quartz.) Under pressures such as these, production problems for a time swamped all other activities in the laboratories.[59]

As far as the radar project was concerned, the initial arrangements envisaged the basic research and design work being done in the Radiophysics Laboratory, with subsequent development and production in the hands of the Postmaster General's Department and its research laboratories. Relations between the two organizations soon deteriorated. In addition, there were disagreements between Madsen and D.F. Martyn over the management at Radiophysics. In 1941, Martyn was replaced as chief by the New Zealander, F.W.G. White, and some months later left the laboratory to develop an Army operational research group focusing particularly on problems associated with the use of radar.[60] Developmental research necessarily continued to be a central feature of the work at Radiophysics, but even here, as Rivett remarked in a letter written in early 1942, 'it is production that counts now with us in the work with which [the Laboratory] is concerned. Research in the ordinary sense, while still highly desirable, begins to wear a too distant appearance.'[61]

Rivett doubtless had in mind here the first Australian-made air-warning radar sets that were sent to Darwin, Port Moresby and Port Kembla in early February 1942, which had been partly constructed in the laboratory. A more typical arrangement had before this seen a much larger number of 'shore defence' sets, built in the PMG's Department to an Australian design, installed at strategic points around the Australian coast. In this as in most other cases, only experimental sets and the first prototype models were constructed in the Laboratory, while the responsibility for carrying matters forward from prototype stage to actual production was left with the PMG's Department or, from early 1942, a commercial manufacturer.

Magnetron manufactured at the School of Natural Philosophy, University of Melbourne, for the Australian radar project.

Enemy in retreat

By 1943, there was a growing sense that the tide of war had turned and that Australia and its allies would eventually emerge victorious. As early as March of that year, the question of post-war reconstruction reappeared on the agenda of a CSIR Executive Committee meeting,[62] and it recurred with increasing frequency from then on. (An elaborate system of government committees with CSIR representation had, rather optimistically, been established much earlier, in February 1941, to lay plans for the reconstruction that would be needed at the end of the war. However, all such planning had ceased with the sudden Japanese onslaught in December 1941.)

Though much of Australia's scientific effort continued to go into problems directly associated with war production, there were other signs, too, that the crisis had passed. The dramatic expansion of physical science laboratories that had accompanied the growth of war-related industries, slowed. Now, it was rather a matter of maintaining and, in due course, gradually redirecting the enhanced level of activity achieved in the early years of the war. There is, for example, a striking contrast in the tone of CSIR Executive Committee minutes dealing with the National Standards Laboratory between 1940 and 1941, when the major build-up was taking place, and from 1943 onwards. Before, it had been a matter of major decisions concerning the laboratory — its building, its staffing, its equipment, its mode of responding to the war-driven demands being placed on it — needing to be taken at almost every meeting. Now, NSL was up and running on its war-related tasks, and such decisions as were required of the Executive Committee were of a much more routine, administrative character. The same applied, more or less, to the Division of Aeronautics, though here some problems remained concerning relations between the division and the manufacturers. Radiophysics had already assembled a first-rate scientific staff. Under its new management and with both a clearer vision of its own role and better defined relations with the Ministry of Munitions and the manufacturers, the laboratory settled down to the task of developing new and improved radar systems suited to Australia's strategic situation and the nature of the war being fought in the Pacific. The Optical Munitions Panel's production laboratories and the AWA research establishment maintained large through-puts of work, as did various other scientific laboratories throughout the country, where production or near-production activities had likewise become the norm.

The earliest concrete signs of a redirection of scientific work towards post-war tasks came later in 1943. There was an extensive preliminary discussion of 'C.S.I.R. and Post-war Reconstruction' at a full council meeting in May of that year. Then in August and September, the first practical problems were addressed by the Executive Committee. In response to a request for assistance from the Rural Reconstruction Commission, the committee considered the role of the Division of Soils in relation to plans that were beginning to be developed for closer agricultural settlement in post-war Australia; it also discussed in a preliminary way a Department of Post-War Reconstruction proposal to establish an experimental building station with CSIR involvement — a scheme that was eventually to give rise to the whole new CSIR Division of Building Research.[63]

During 1944, the pace of change gradually picked up as the sense of planning for a peaceful future strengthened. In January it was agreed that the scientific liaison offices in London and Washington should continue after the war as part of a larger British scheme of 'maintaining permanent scientific and technical representation in London and in other capital cities of the English speaking world'. In March, it was decided to seek from the government 'a definite policy for the development of the Council's work' in connection with post-war reconstruction. In May, there was a request from the Director-General of Munitions that 'the National Standards Laboratory should retain a nucleus staff after the war for the manufacture of slip gauges'. This was agreed to because the expected demand was small and staff who could do the job would have to be retained for other purposes anyway. It was formally noted in this connection, however, that 'C.S.I.R did not wish to enter the manufacturing field'. The question of patenting some of the products of Radiophysics research came up at the same meeting. In July there was discussion of a major development in wool research, which was to become a feature of CSIR work after the war, and, in response to a request from the Department of Post-War Reconstruction, of the Council's likely post-war needs for new buildings.[64] Towards the end of the year, redeployment of staff began as university scientists who had been seconded to CSIR divisions (especially Radiophysics) during the emergency started returning to their departments and the first of the young science graduates who had been pressed into war-related work immediately on graduation were released to pursue more normal patterns of post-graduate training.

Preparations for peace

By early 1945, the stream of matters coming before the Executive Committee concerning the return of CSIR to peace-time working conditions had turned into a flood. More general issues were aired at two meetings of chiefs and section heads, each extending over several days, that were held in September 1944 and March 1945, and at subsequent meetings of the full CSIR Council. Several major threads can be discerned in the discussions that took place and in the decisions made.

As the production of munitions became of less pressing concern and increasing food production received greater emphasis, the work of the old agricultural divisions gradually began to pick up. In general terms, at least, their rôle was clear. This was far from the case, however, with the new physical-science laboratories. All of these had been established under the immediate threat of war and, as we have seen, much of their war-time work had been of a fairly routine problem-solving kind, which had from time to time shaded over into actual production. They had never previously had the opportunity to build up orthodox scientific research programmes. In some cases, most notably Radiophysics, they had still to find themselves a peace-time rôle of any kind. Everywhere there was a strong feeling that, having been immersed for so long in immediate, practical problems, it was now time to get back to (or to take up) more fundamental questions. The issues to be resolved were complex and to handle them F.W.G. White was transferred from

Radiophysics to CSIR headquarters, to a newly created position as Assistant Executive Officer with special responsibility for the physical science divisions. He proved a veritable powerhouse in this role, and under his leadership substantial planning documents were produced by the divisions concerned.[65]

The relationship between CSIR's new activities and the nation's secondary industries was of particular concern. Even before the war, it had been generally recognized that the way in which secondary industry was structured, with different firms competing against each other in the local market, created difficulties that did not arise in the case of research conducted on behalf of the export-orientated primary industries.[66] Nearly half a century later, these difficulties have still not been overcome. In the later years of the war, they were discussed repeatedly within CSIR.[67] There was general agreement that the emphasis within CSIR ought to be, as its name implied, overwhelmingly on research; that it ought not to become bogged down in providing routine testing services for industry; and that an extension service was required that would provide a means both for disseminating scientific and technical information to the secondary industries and for bringing problems that industry could not solve for itself to the attention of appropriate scientific research workers. Opinions differed, however, as to whether providing such a service should be part of CSIR's responsibility, or whether CSIR should be left responsible for carrying out the research only and an independent extension service created to provide the links with industry.

The role of the National Standards Laboratory was another matter. Here, there was a general agreement as to what was required. Now, the question of Commonwealth legislation for a national system of standards was revived, and also that of establishing certified testing authorities to serve the day-to-day needs of industry. The laboratory itself, its three sections of metrology, physics and electrotechnology now classified as separate CSIR divisions, was able at last to take up the more fundamental work originally planned for it.

What was clear to everybody was that, while the war had seen an extraordinary increase in Australia's scientific workforce, it had also created severe distortions in the distribution and training of individual scientists, which would have to be redressed with the coming of peace. The AWA laboratories, for example, had recruited a number of promising young men during the war years despite the competing attractions of the Radiophysics Laboratory and the Air Force. Under the pressures of war, however, new graduates had been rushed straight into jobs without any opportunity to undertake higher degree studies. By 1945, a number of these people began to seek their release in order to return to university. At the same time, the production side of the laboratories' work on which many of them had been employed began to be scaled back. The remaining scientific staff were reorganized in September 1945 so as to enable the laboratories to handle the by then rapidly increasing proportion of non-defence research projects — many of them, however, utilizing new skills developed in response to war-time needs — being directed to them from within the company.[68]

The same thing had happened on a much larger scale within CSIR and consequently the problems of adjustment were much more severe. Easiest to deal with were university scientists who had been seconded to CSIR for war work. Most of these returned to their universities, where their services were becoming desperately needed on account of rapidly increasing enrolments, in the first months of 1945. The real problem lay with younger staff who, as at AWA, had often been recruited immediately on graduation. Rivett put the matter in a nutshell at the March 1945 meeting of Chiefs:

> He did not feel happy about some of the recent appointments to the C.S.I.R. staff; the Council had been forced to take the best offering rather than to select the type of man it knew it required. Before the war it was usual to appoint only those people who had had at least one year's training after graduation. Now C.S.I.R. had been forced to take people who had not even completed their graduate courses. He realised that it would be impossible to eliminate all these people, nor did he feel it desirable, but something would have to be done to train them.[69]

Though Rivett did not say so, the problem had been compounded by the nature of the war-time work on which many of these raw young graduates had been engaged, because this had often not been of a kind suitable for training them in the ways of research. The same applied to those who had been recruited into the Munitions Supply Laboratories or into the armed forces (for example as Air Force radar operators). It even applied to those who had remained in the universities since there, too, as we have seen, most ordinary research activities had been supplanted by war-related tasks much more closely connected with problems of production. Members of this group were, however, better placed than others to revert to the status of research student once the demands of war production abated somewhat.

Before the war, except for an isolated department here and there, the Australian universities had not been active in research. Chronically starved for funds by parsimonious and short-sighted state governments, and with their small staffs weighed down by heavy undergraduate teaching loads, they had traditionally sent their best students overseas, to Britain, for their advanced training, rather than building up graduate schools of their own. No Australian university offered the PhD degree. Now, as the end of the war approached, the universities faced the prospect of being flooded with large numbers of new government-assisted undergraduates and demobilized servicemen, and there was every likelihood that their capacity to undertake research would be reduced even further as a result. So, too, would be their already small capacity to train research workers.

Inevitably, therefore, if the young scientists who had been recruited for war work were to undertake advanced training, many of them would have to go overseas. As in the past, most eyes turned to England. The brightest talents had already attracted university 'travelling scholarships' on graduation, which they were now free to take up. Others found their names on a list of over fifty young CSIR staff in line for grants from the Science and Industry Endowment Fund to send them abroad in the way that much smaller numbers had been sent before the war.[70] Others again scrimped and saved and obtained support from one source or another, and also went away. The pre-war trickle turned into a flood of young Australian scientific talent heading for Great Britain. Many of those involved eventually returned to Australia, but many others did not. Within Australia, the conviction spread that the nation could no longer afford to be so totally dependent on others for the training of its future scientific leaders. While CSIR experimented with a scheme to appoint 'research associates' who would undertake up to three years of training in one of the Council's laboratories,[71] a move to create a national research university in Canberra gained increasing support. At the same time, the existing universities were at last empowered, one by one, to award the PhD degree.

It was not just the younger scientists, however, who found their careers interrupted by the war. At the height of the emergency, virtually every laboratory and every scientist in the country was diverted to war work for a period of several years. For some of those involved, this constituted no more than a hiatus — for example, many scientists from the state departments of agriculture were diverted from research to administrative duties associated with wartime controls[72] — for others it was an occasion to apply their particular skills for the common cause; for others again it was an opportunity not to be missed to build (or reshape) a career.

As we have seen, the Radiophysics Laboratory was in a somewhat special category, able to maintain a strong programme of developmental research throughout the war years while others were being swamped by problems associated with production. Those working there developed great expertise with high-frequency transmitting and receiving systems and remarkable insight into the behaviour of radio waves. Many stayed at the laboratory after the war, one group quickly carrying Australia to the forefront of the emerging science of radio astronomy while others tackled the equally promising but ultimately less successful field of cloud physics and more immediately practical problems such as developing radar systems for civil aviation.[73] They were joined by others who, during the war, had honed their skills elsewhere, for example W.N. Christiansen, who had spent the war years at Rockbank, Victoria, in charge of AWA's programme of systematically up-grading its beam radio links with Britain and America and in the process developing the extraordinary mastery of antenna systems that subsequently distinguished his career as a radio astronomer.

The more production-orientated laboratories could also offer opportunities for those prepared to grasp them. At the University of Tasmania, A.L. McAulay and F.D. Cruickshank, as part of their commitment to the production of optical munitions, developed a powerful new technique for analysing optical systems, which led to a notable series of publications and an on-going tradition of optical research in their department after the war.[74] W.H. Steel, a young mathematics graduate from the University of Melbourne, was hired by AWA in 1943 in an effort to salvage an abortive attempt by the company to break into the optical manufacturing industry. Though his labours did not save the programme, they launched him into a career that later, at NSL, saw him become one of Australia's leading optical physicists.[75] In the Optical Munitions Panel's factory at Mt Stromlo Observatory, a number of young scientists learned skills at manipulating optical components that contributed significantly to the remarkable efflorescence of Australian optical astronomy in the post-war period.[76] Several young physicists recruited into the Commonwealth government's war-time Minerals Survey thereby learned the techniques of geophysical prospecting and after the war became the nucleus of a large Geophysics Division within the newly formed Commonwealth Bureau of Mineral Resources.

Such examples could be multiplied many times over. By the end of the war, and because of the pressure it had created, the Australian scientific community was not only much larger than it had been before the war, it also boasted expertise in a much wider range of scientific skills. These skills were not lost at war's end. The war brought about a remarkable growth and increase in the technological sophistication of Australian industry; and even though so much of the war-time scientific work that contributed to this was of a fairly routine nature, a new base level of scientific expertise was established. The dramatic post-war expansion of the universities ensured that this was maintained in the peace that followed.

To be sure, the beginnings of this transformation were evident prior to the war, for example in the rise of an Australian electronics industry and in the decision to implement the recommendations of Julius' Secondary Industries Testing and Research Committee. However, the war undoubtedly hastened the process and at the same time greatly increased the demand for skilled scientific workers. In addition, as we have seen, it created new opportunities, that many scientists were quick to grasp, for enhancing their longer-term research capabilities. Hence it was, indeed, a turning point in Australia's scientific history.[77]

Notes

1 D.P. Mellor, *The Role of Science and Industry* (Canberra, 1958) (Australia in the War of 1939-1945: Series 4 [Civil], Vol. 5).

2 W.F. Evans, *History of the Radio Research Board, 1926-1945* (Melbourne, 1973).

3 G. Currie and J. Graham, 'G.A. Julius and research for secondary industry', *Records of the Australian Academy of Science*, 2(1) (1971), 10-28; Mellor, op. cit. (n.1), pp. 21-2.

4 'Secondary industries testing and research ... Report of Committee appointed to report thereon', *Australia. Parliamentary Papers*, Session 1937-40, iv, 941-85.

5 Ibid., Appendix 4, pp. 984-5.

6 CSIRO Archives, Council minutes, 8-10 December 1937, p. 15.

7 CSIRO Archives, Executive Committee minutes, 20 April, 12 July, 27 July 1938.

8 CSIRO Archives, Council minutes, 5-7 April 1937, p. 7.

9 Ibid., Executive Committee minutes, 25 March 1938.

10 Ibid., 27 July 1938.

11 Ibid., 16 September and 14 October 1938; 24 January, 17 February, 17 May and 16 August 1939.

12 Ibid., 13 December 1939.

13 CSIRO Archives, Council minutes, 30 April - 2 May 1941. The circumstances surrounding the creation of the Division of Industrial Chemistry have been described by Wark, 'The CSIRO Division of Industrial Chemistry, 1940-1952', *Records of the Australian Academy of Science*, 4(2) (1979), 7-41. See also A.L.G. Rees, 'Ian William Wark, 1899-1985', *Historical Records of Australian Science*, 6(4) (1987), 533-48.

14 Mellor, op.cit. (n.1), Chap. 19; W.F. Evans, *A History of the Radiophysics Advisory Board, 1939-1945* (Melbourne, 1970).

15 CSIRO Archives, Executive Committee minutes, 21 November 1938 and 20 March 1939.

16 Registrar, RACI, to D.P. Mellor, 22 March 1957 (Australian War Memorial, AWM 74, Box 4); R.W. Home, 'Origins of the Australian physics community', *Historical Studies*, 20 (1982-83), 383-400, p. 398. That matters were not quite as bleak as these figures suggest, at least as far as chemistry was concerned, is perhaps indicated by the fact that besides the 56 fellows, there were 882 associates of the ACI at this time. On the other hand, the huge difference between the numbers of fellows and associates

suggests that significant numbers of the latter group were in fact under-trained men who could never aspire to fellowship, who had been admitted to the Institute in its early years on the ground that they were already practising 'chemists' at the time the Institute was formed, and not because they satisfied the qualifications requirements subsequently established for professional certification. This in turn highlights the more general question of whom one should count as a qualified scientist in compiling historical manpower statistics. Mellor (op. cit. (n.1) p. 57), after recording a contemporary estimate that there were between 3000 and 4000 scientists in Australia at the beginning of the war, notes that this 'probably refers to science graduates' and adds: 'Not everyone would agree to call a graduate in science a scientist. Some writers prefer to restrict this term to those who are or have been engaged in scientific research.' However, since the figure he cites probably includes the many diploma holders from the technical colleges and other non-graduate associates of bodies such as the ACI, even the number of bachelor-level science graduates in the country (apart from school teachers, who were not encouraged to join such organizations and so tended not to be counted) was very likely considerably smaller than this. The number who would have had experience of research would have been very much smaller again.

17   CSIRO Archives, Executive Committee minutes, 16 June 1939.
18   Ibid., 14 December 1938.
19   Mellor, op. cit. (n.1), p. 57.
20   CSIRO Archives, Executive Committee minutes, 18 August 1939.
21   Ibid., 19 September 1939.
22   The list of reserved occupations included the following:
     Laboratory Staffs, to include: Chemists, Engineers, Physicists and other Scientific Workers, qualified by Degree, or Diploma, or Students in 3rd year at the University or Technical College.
     Students (University): Medical and Dental. All students in 2nd and later years of Physical Sciences, Chemistry, Physics, Metallurgy and Engineering. All students in 2nd and later years of Biological Sciences, Bacteriology, Bio-Chemistry, Zoology, Botany and Agriculture. Last Year Students in Economics, Commerce, Mathematics. 3rd Year Honours Students in Languages. Post Graduate and Research Students. University Teaching Staffs.
     Scientific and Research Workers: Physicists. Scientific Research Workers (full time) in a University or Technical School.
     (Australian Institute of Physics files, Basser Library, Australian Academy of Science, box 86/14).
23   Ibid.
24   Mellor, op. cit. (n.1), p. 60.
25   CSIRO Archives, Series 9, M14/40/2.
26   J. Moran, 'Scientists in the political and public arena: A social-intellectual history of the Australian Association of Scientific Workers, 1939–49', M. Phil. thesis, Griffith University, 1983.
27   Mellor, op.cit. (n.1), p. 63; also *Science on Service: A Directory of Scientific Resources in Australia* (Sydney, 1943).
28   J.S. Rogers, 'The history of the Scientific Instruments and Optical Panel (initially Optical Munitions Panel), July 1940–December 1946' (Australian Archives MP 730/11, Boxes 3 and 4); also Mellor, op. cit. (n.1), Chap. 12.
29   Mellor, op. cit. (n.1), Chap. 22; also L.J. Dwyer to Mellor, 4 May 1956 (Australian War Memorial, AWM 74).
30   CSIRO Archives, Executive Committee minutes, 21 November 1938; J.H. Piddington, personal communication.
31   Details about AWA and its research laboratories have been drawn from the company's records, now in the Mitchell Library, State Library of New South Wales, Sydney.
32   Mellor, op. cit. (n.1), Chap. 21.
33   CSIRO Archives, Executive Committee minutes, 18 October 1939.
34   CSIRO Archives, Executive Committee minutes, 16 November 1939.
35   Ibid., Council minutes, 22 April 1940.
36   Ibid., Series 9, M14/40/6 (Rivett papers).
37   Ibid., Council minutes, 23 April 1940.
38   Rivett to B.H. Wilsdon, 30 April 1940; CSIRO Archives, Series 9, M14/40/4.
39   Massey to Rivett, 28 August 1940; CSIRO Archives, Series 9, M14/40/8.
40   Rivett to J.E. Nicholls, 19 June 1940; CSIRO Archives, Series 9, M14/40/6.
41   CSIRO Archives, Executive Committee minutes, 19 June 1940.
42   Ibid., 5 September 1940.
43   Ibid., 15 November 1939, 14 February, 19 April and 14 May 1940, and 11 February 1941. The subject reappears in the minutes in April 1944, when it was again resolved to ask Baldwin to look into the matter.
44   Ibid., 18 October and 17 November 1939; 14 February, 6 March, and 4 September 1940; 11 March, 15 April, 30 April and 15 May 1941. Soon after the end of the war, in November 1945, CSIR convened a representative conference as a means of re-opening the whole question.

45 Ibid., 13 December 1939 and 14 February 1940.

46 Ibid., 14 May and 4 September 1940; 11 February, 15 April and 15 May 1941.

47 Mellor, op. cit. (n.1), pp. 9–13.

48 CSIRO Archives, Executive Committee minutes, 11 November 1941.

49 Ibid., Council minutes, 12 November 1941.

50 Ibid., Executive Committee minutes, 9 December 1941.

51 Rivett to B.H. Wilsdon, 29 April 1942; CSIRO Archives, Series 9, M14.

52 CSIRO Archives, Executive Committee minutes, 7 October 1941.

53 Ibid., Council minutes, 1 May 1941; also CSIR's 'War Report to June 30th, 1942', pp. 32–5.

54 CSIRO Archives, Series 50. See also Wark, op. cit. (n.13), pp 14–19.

55 Rogers, op. cit. (n.29), p. 5.

56 H.C. Bolton, 'J.J. McNeill and the development of optical research in Australia', *Historical Records of Australian Science*, 5(4) (1983), 55–70.

57 Scientific Instruments and Optical Panel, Specifications Committee, minutes of meeting, 29 December 1943; University of Melbourne Archives, Trikojus papers, 2/83.

58 Rogers, op. cit. (n.29), pp. 70–2; also C.A. Ramm, personal communication.

59 These details have been drawn from the monthly reports from this period prepared by A.L. Green, director of the AWA laboratories (AWA papers, Mitchell Library, Sydney).

60 Mellor, op. cit. (n.1), pp. 433, 656–60; also A. Moyal, *Clear across Australia: A History of Telecommunications* (Melbourne, 1984), p. 167.

61 Rivett to H.E. Wimperis, 7 March 1942, CSIRO Archives, Series 9, M14.

62 CSIRO Archives, Executive Committee minutes, 9 March 1943.

63 Ibid., Council minutes, 5 May 1943, Executive Committee minutes, 10 August, 22 September and 23 September 1943.

64 Ibid., 12 January, 14 March, 23 May, 24 May, 12 July and 13 July 1944.

65 See this volume, p. 309 for a discussion of the plans developed by Radiophysics.

66 CSIRO Archives, Council minutes, 19 April 1939.

67 Ibid., 5 May 1943 and 31 October 1944; Executive Committee minutes, 9 January, 16 March, 18 April, 11 July and 23 October 1945.

68 A.L. Green, monthly report, September 1945; AWA papers, Mitchell Library, Sydney.

69 CSIRO Archives, Series 9, M13/20 (16), p. 10.

70 Ibid., Executive Committee minutes, 12 September 1945.

71 Ibid., Series 9, M13/20 (16), and Executive Committee minutes, 12 September 1945.

72 A.E.V. Richardson to J.G. Crawford, Department of Post-war Reconstruction, 26 April 1945; ibid., M13/20/27/2.

73 See this volume, Chap. 14.

74 In 1945 alone, at the start of their publication programme, seven papers (five by Cruickshank, one by McAulay and one by Cruickshank and McAulay jointly) appeared in the *Proceedings of the Physical Society of London* (vol. 57).

75 A.L. Green's monthly reports, AWA papers, Mitchell Library, Sydney; also W.H. Steel, 'Optics in Australia', *Australian Physicist*, 24 (1987), 242–5.

76 See this volume, p. 346.

77 Research for this chapter was supported by grants from the Australian Research Grants Scheme, the Australian War Memorial, Canberra, and the University of Melbourne. Much friendly assistance was provided by the CSIRO Archivist, Mr Colin Smith, and his staff, and by the Honorary Archivist of the CSIRO Division of Radiophysics, Miss Sally Atkinson. Mr J.A.L. Hooke, Chairman of AWA Ltd, kindly granted me access to the company's records.

11

Plant introduction in Australia

R.L. Burt and W.T. Williams

The beginning: 1788–1830

New plants for old

Australia is probably the only country in the world in which, with a single exception (*Macadamia*, the Queensland nut), every edible cultivar and every crop plant is, or has been derived from, an introduction, as indeed is true of most of the ornamentals, lawn grasses and sown pasture species. The First Fleet and its colonist successors were to provide Australia not only with a new human population, but also with an almost completely new economic flora. Pike has summarized the reports of the early explorers in these words: 'No local plant could be coaxed to make food; no indigenous animal gave milk for human use; no native tree yielded edible fruit.'[1] Nor was the countryside inviting. Sir Joseph Banks, during his voyage of 1768-69 in the *Endeavour*, wrote that the country 'resembled . . . the back of a lean cow . . . where her scraggy hip-bones have stuck out . . . accidental rubs and knocks have entirely bared them'.[2] Perry, paraphrasing Lieutenant-Captain Tench's statements of 1793, has noted: 'The grasses grew in clumps and tufts, not in a sward, and the soil was either a dry sand or a heavy clay'.[3] He also quotes an anonymous author as writing in 1790: 'The country . . . is past all dispute a wretched one'.[4] Britain had been faced with a similar problem when she settled Barbados in 1627. In that case, however, there were the friendly Arawak Indians, who provided, *inter alia*, cassava, maize, yams, bananas and pineapples. Unfortunately, the early Australian explorers lacked the communicatory skills that would have enabled them to discover how the Aboriginal population had maintained itself for thousands of years on a vegetation that appeared to the newcomers to be completely inedible. They were convinced that, if they wanted plants to eat, plants — in some form — must be taken with them.

The First Fleet, under the command of Captain Arthur Phillip, RN, sailed from Portsmouth on 13 May 1787. Details of its voyage have been provided by Collins.[5] Its eleven ships carried almost 1500 human beings (though surprisingly no botanist or even gardener) and over 400 assorted animals. It has not proved possible to find any record of the plants taken on board; such a record may never have existed, since the arrangements had been placed by the Admiralty in the hands of junior officers, who pointedly ignored the advice of Banks. However, Wilson notes that the First Fleet 'brought seed from

England'[6] and Wrigley and Rathjen believe that this included English wheats.[7] It must also have carried hay and stock-feed for the animals, which may well have been the source of some of the so-called 'English grasses' that were soon to become naturalized; certainly 'corn and hay for the stock' were purchased at the Cape of Good Hope.[8]

There appears to be no record that plant material was taken aboard at Tenerife, the first stop; collection began in earnest on 4 August 1787, when they reached Rio de Janeiro. Here they procured coffee, cocoa (in the nut), cotton (as seed), indigo, bananas (as living plants), oranges of various kinds (both as seed and plants), lemons (seed and plants), guava (seed), tamarind, 'prickly pear-plant with the cochineal on it', *Eugenia* (probably *E. jambos*), a jalap (probably a Mexican species of *Ipomoea*), and three sorts of ipecacuanha[9] — probably *Cephaelis (Uragoga) ipecacuanha*, *Asclepias curassavica* (later to become a troublesome weed) and *Ionidium ipecacuanha*.

The fleet reached the Cape of Good Hope on 13 October 1787. Here they had the assistance of Francis Masson, the King's botanist, the first of the collectors sent out by Kew. The plants taken aboard included a fig-tree, bamboos, sugar-cane, various vines, quince, apple, pear, strawberry, myrtle and 'all sorts of grain'.[10] The fleet left the Cape on 12 November, and landed in Botany Bay on 26 January 1788. Deliberate plant introduction into Australia had begun.

The early years

European-style agriculture began immediately. Governor Phillip, whose plaintive request to Lord Sydney for a botanist was not to be fulfilled until the arrival of Caley in 1800, established a government farm in Sydney, which by July possessed '9 acres in corn' (wheat). The soil proved infertile, and in 1789 the farm was moved to Parramatta under the charge of Henry Dodd. The governor, who had earlier expressed misgivings as to whether the Australian environment could support European-style living, then decided to use James Ruse as an experimental 'first settler'. Ruse, a convict whose sentence had now expired, had formerly been a farmer, and Phillip gave him 2 acres to cultivate on his own. Fifteen months later, Ruse 'having got in his first crop of corn . . . said that he was able to support himself by the produce of his farm' and 'declared himself desirous of relinquishing his claim to any provisions from the [government] store'.[11] The experiment was evidently successful; in fact by December 1791, a total of 920 acres of land were under cultivation — Collins enumerates maize, wheat, barley, oats, potatoes, vines and turnips.[12] The colony was nevertheless becoming conscious of deficiencies; in 1798 P.G. King, himself to become governor in 1800, wrote to Banks asking for medicinal plants such as fennel, marshmallow and thistles, English grasses, and broom seed 'as substitute for Hops'.[13]

In 1801 the *Investigator* arrived from England under the command of Captain Matthew Flinders, with the distinguished Scottish botanist Robert Brown and an excellent gardener, Peter Good. Flinders suggested that seed of European vegetables should be sown at successive anchorages for the use of future expeditions. This task was allotted to Good, and it is known that he produced a 'List of Seed Sown'.[14] As governor, King continued to request plants. In a letter to Lord Hobart he enclosed, for the attention of Banks a list of Fruit Trees, Plants, Herbs and Forest Trees, which would be very useful to the inhabitants hereafter, and beg leave to suggest, that if a Plant Cabin was to be fitted up on board one of His Majesty's ships, coming to this Colony, and a person sent with the director of it, many useful plants might be brought . . . [15]

Meanwhile, some at least of the introductions were making themselves at home in their new country, for in 1804 Robert Brown noted twenty-nine non-crop alien species growing wild in the Sydney area.[16] In 1806, Bligh — who had been a notable collector of breadfruit material — arrived, and promptly resumed the land of the old government farm. By 1810 this area was in use as a nursery with two functions: to prepare native plants for their journey to England, and to cultivate introduced plants, in the hope of establishing them in the new climate and soil. In fact, it was the first of many 'acclimatization gardens' to be established in Australia. Not that all went well. The difficulty of growing crops in infertile soil without adequate manure; the onset of various unspecified forms of 'blight'; deleterious changes in the native vegetation brought about by the trampling of the hard-footed European cattle; these and other less botanical vicissitudes have been amply documented in standard histories. It is necessary only to note that by about 1820 the colony appeared to promise a stable future, based on a substantially European agriculture. Moreover, during all this period ships were arriving, bearing cattle, sheep, sometimes convicts, and an assortment of stores for a colony that was all too often short of food. It is possible that many also brought with them seeds of their favourite plants, but no records of this have been found.

The undisciplined years: 1830–1930

The problem of documentation

The nineteenth century is a 'nightmare' period for anyone interested in the history of plant introduction, for documentation — where it exists at all — is both meagre and scattered. In the early years of the century Australia was not a single national entity but a mosaic of largely isolated and autonomous colonies, each of which looked after its own affairs, including its plant introduction. Much material was undoubtedly brought in by individual settlers, but nurserymen were also very active. This is made clear by, for example, the early catalogues of W. MacArthur of Camden,[17] of the Sydney nurseryman T.W. Shepherd,[18] and of the Government Botanic Garden, Sydney.[19] This last alone lists over 3000 species and varieties obtained from many parts of the world, though only rarely is it known who brought them in and when. Perusal of these lists suggests that there were two separate driving forces behind the constant stream of introductions. The first was, of course, simply a need for food and medicines; the second seems to have been sheer nostalgia. The settlers, faced as they were by an alien vegetation under Australia's 'pitiless blue sky',[20] longed for the sights and sounds of home; for roses, daffodils and, regrettably, the London sparrow.

New settlers were not the only source of plant material. There was also correspondence, of which occasional tantalizing glimpses can be seen. For example, the *Sydney Morning Herald* (then simply the *Sydney Herald*) in its first year records the arrival of ships carrying 'four boxes of plants for Mr

Jones' and '1 cask of hop plants, 2 boxes of seeds' for Archdeacon Broughton.[21] There were yet other sources; during the 'blackbirding' period, kanakas brought in to work in the Queensland cane-fields brought with them a variety of tropically-adapted vegetables, including the soon to be familiar sweet potato, taro and snake beans.[22]

There was, too, increasing dissemination of plant material within the country. The Brisbane Botanic Garden (created in 1855 under the directorship of Walter Hill) in 1872 'supplied 50 000 cuttings of sugar-cane of some 14 varieties, 5000 of white mulberry, 5056 coffee plants, 1020 tea plants, 1060 ginger roots, 300 papers of . . . tobaccos'; but in the same year 'the demand for Cinchona, cloves, nutmeg, allspice, cinnamon, cocoa, mangoes and oil palm exceeded the garden's supply'.[23] It did not take long to discover that not all tropical plants can be successfully cultivated in Queensland. Early failures were the durian and the mangosteen; and Dr Dunmore Lang claimed that rice would not germinate in Queensland.[24] Since he used polished rice obtained from a grocer, this is hardly surprising; but he blamed the climate.

The lack of documentation was not due simply to a lack of botanists;[25] the problem from our point of view is that most of the documentation is in the wrong direction. Almost without exception the then botanists were not particularly interested in plants they already knew well; they were fascinated by the native flora, which they assiduously collected, pressed, annotated, and sent overseas, usually to Kew. The magnitude of this operation can be demonstrated by the fact that in England George Bentham (admittedly with considerable help from Baron von Mueller, the director of the Melbourne Botanic Gardens) was able, during the years 1863 to 1878, to write his monumental *Flora Australiensis* without ever feeling the need to visit the country.

The botanists of the early days were all European-trained, and knew a European weed or cultivar when they saw one. If on their travels they encountered such a plant apparently established among the native flora, they recorded it as a naturalized alien. As a result, although for many introductions it is not known when or by whom they were first introduced, we do at least know when and where they were first recorded as established. Surprisingly, the plants of which we have most knowledge are those that, it was discovered all too late, should never have been introduced at all.

The blackberry (*Rubus fruticosus* agg.) was undoubtedly introduced for its edible fruit, though as usual it is not known precisely when. It was first recorded wild in Victoria in 1843, and was spread as a result of the misguided enthusiasm of von Mueller. During the millennia of isolation, Australia had never evolved an invasive shrub; as a result, the blackberry found an ecological niche that was not being filled, and promptly proceeded to fill it. An exactly comparable situation arose in the north with *Lantana camara*. This was first introduced — possibly even by Governor Phillip — as an ornamental hedge-plant. It was first recorded wild in Queensland in 1869 and has been a nuisance ever since. *Echium plantagineum* ('Paterson's curse' or 'salvation Jane') is first recorded as an ornamental in the Camden gardens of John MacArthur in 1843; it began to spread explosively in the Albury region in 1896. The prickly pear, *Opuntia* sp., was probably brought in by a Dr Carlisle in 1839 as a hedging plant. Captain Phillip has sometimes been blamed, but the *Opuntia* brought in by the First Fleet was not the species that later became a pest.

Some introductions were accidental. *Xanthium spinosum*, the Bathurst burr, first appeared in 1840, and was believed to have been carried in the tails of horses imported from Chile. *X. pungens*, the Noogoora burr, first recorded around 1860, is suspected of having arrived as a contaminant of cotton seed.

Some limited information is available concerning the origin of some of the more northerly pasture grasses. *Paspalum dilatatum* was probably introduced by von Mueller about 1870; *Brachiaria mutica*, 'para grass', by Dr J. Bancroft at about the same time; *Axonopus affinis* by a Mr Shipway in 1891; and seeds of *Cenchrus setigerus* were sent in 1920 by Field-Marshal Lord Birdwood to his son-in-law in Western Australia. *Axonopus compressus* was first recorded in Cairns in 1887; the now ubiquitous *Panicum maximum*, 'guinea grass', was first recorded in Queensland in 1867, though it had almost certainly been in the country for years. One other species of *Cenchrus*, *C. pennisetiformis* or 'Cloncurry buffel', appears to have come from rotting saddles of Afghan camels. The most interesting legume record is of the plant now known as *Stylosanthes humilis*, 'Townsville stylo', first recorded in the neighbourhood of Townsville in 1904.

The better known temperate pasture plants generally present greater difficulties. Grass genera such as *Festuca*, *Lolium* and *Poa*, and the many species of legumes such as *Medicago* and *Trifolium*, have been in Australia for a very long time. There seems little doubt that many of them were introduced accidentally, possibly with feed for horses and cattle; but a comprehensive study of these plants would be a major research project, which writers in northern latitudes cannot hope to undertake. The studies on *Lolium*,[26] *Medicago* and *Trifolium*,[27] and *Trifolium subterranean* would provide an admirable starting-point.[28] A more recently developed species, *Phalaris tuberosa*, was probably first introduced into Australia at the Toowoomba Botanic Gardens in 1884; it arrived in a packet of seeds from the American Department of Agriculture but probably originated in Italy.

The beginning of organization

Amid the morass of unorganized and idiosyncratic introduction characteristic of this period, the beginnings of planned introduction are discernible. Four aspects call for brief consideration. These are a search for improved varieties of crop-plants; the rise of the 'acclimatization societies'; a growing concern over the dangers of uncontrolled introduction; and a growing realization of the need for some form of inter-state integration.

The search for improved varieties of crop plant included such examples as wheat,[29] sugar-cane and vines. The first wheats were probably mixtures of seed obtained from England, and (*en route*) from Capetown; most were unsuited to Australian conditions, and there were many failures. The Agricultural Society of New South Wales was founded in 1822, and immediately embarked on a scheme of introduction and testing selection and breeding of wheat varieties. Outstanding successes were the introduction of White Essex in 1850 and the selection of Purple Straw in 1860.

Sugar-cane had also been brought out by the First Fleet. In 1820 an itinerant missionary and sugar-cane planter, John Gyles, visited the Port Macquarie area with Oxley and suggested that this would be an appropriate area for the development of a mill and a commercial cropping area. Some 50 tons of sugar were produced in 1823, but frosts were a problem and sugar-cane cultivation moved north. The first successful cultivation was the 50 acres grown by Louis Hope at Ormiston. His success caused what Herbert has called 'a band of public spirited colonists' to join forces and introduce (in 1863) a large collection of cane varieties.[30] The Queensland Acclimatization Society grew out of this effort, and promptly embarked on a vigorous programme of exchange with other parts of the world.

Vines, too, came with the First Fleet but in 1831 James Busby visited Europe specifically to bring in new varieties. He obtained 433 from the Botanic Gardens of Montpellier, 110 from Luxembourg, 74 collected personally in France, 17 collected personally in Spain, and 44 from Sion House near Kew Gardens — the latter 'through the instrumentality of Mr R. Cunningham of the Kew Botanic Gardens, who undertook to pack the whole collection for despatch to Sydney'.[31]

History often has its lighter side. In 1876, the then Governor of Queensland demanded that the acclimatization society pay more attention to the provision of street shade trees, which 'would have to be protected from noxious animals such as larrikins and goats'.[32] But the oddest introduction was surely that of the irrepressible Dr Dunmore Lang, who imported '600 virtuous Presbyterians' in three boat-loads for cultivation of cotton.[33] The project, like his earlier attempt to germinate rice, was a failure; the government of the day refused to provide the land he wanted, and the virtuous Presbyterians had to seek other employment.

It must be remembered that in the early years of the nineteenth century, botanists — including von Mueller — believed implicitly in 'special creation' and the fixity of species. Nevertheless von Mueller evidently believed that species had been created with some degree of innate variability, since in his capacity as director of the Melbourne Botanic Gardens he avers that 'an endless number of plants of the whole temperate zone and ... many from the warmest parts of the globe can be acclimatized in our latitudes'.[34] His detailed suggestions were documented in *Select Extra-tropical Plants*, the Australian content of which drew on his earlier work, notably the volumes of *Fragmenta Phytographiae Australiae*.

However, it soon became clear that adaptability varied greatly between species. Some acclimatized readily, others never survived the transfer to the new conditions. Some form of clearing-house was needed for the exchange of information concerning already-known successes and failures of acclimatization, in order to avoid the disheartening experience of repeated failures; and this was the primary purpose of the acclimatization societies. These were not an Australian invention, for they already existed in Europe. The Société Impérial d'Acclimatation was set up in Paris in 1854; the date of formation of the Acclimatization Society of Great Britain is uncertain, but it was definitely in existence before 1860. The first such Australian society was the Acclimatization Society of Victoria, set up in Melbourne in 1861. Its published 'Rules and Objects' are scientifically irreproachable even by today's standards; they include the statement that it was set up 'for the purpose of spreading knowledge of acclimatization, and inquiry into the causes of success or failure'.[35] The idea quickly spread; New Zealand — where such societies still exist — set up its first in 1864. It is difficult to establish the date of formation of such bodies, since they tended to arise gradually out of local groups of

SELECT PLANTS

READILY ELIGIBLE FOR

Industrial Culture or Naturalisation

IN VICTORIA,

WITH INDICATIONS OF THEIR NATIVE COUNTRIES
AND SOME OF THEIR USES,

BY

BARON FERD. VON MUELLER,

C.M.G., M. & PH.D., F.R.S.

"Omnia enim in usus suos creata sunt."—Syrach xxxix, 21, 26.

Printed for the Government of Victoria by
M'CARRON, BIRD & CO., FLINDERS LANE WEST, MELBOURNE.
1876.

Ferdinand von Mueller played a major role in the distribution of useful plants. This book was especially important. It went through several editions and was translated into other languages.

enthusiasts, and might not immediately dignify themselves with the appropriate title. As an example, three authorities differ as to the date of formation of the Acclimatization Society of Queensland, giving the date as 1862, 1865 and 1868.[36]

A contemporary botanist may be forgiven for feeling that the concepts underlying the acclimatization societies were unduly Lamarckian. But, surely, this does not excuse the extraordinary attack on the Victorian society made by Kynaston.[37] He writes of the 'almost Goon-show-like fantastic comedy of the acclimatization of the species' and of 'this amiable-seeming cloud-cuckoo-land that was to do so much damage'. Such societies undoubtedly made some serious mistakes, but they represented a brave and commendable attempt to bring some degree of order into a chaotic situation. They have now disappeared from Australia, not because they were useless, but for two quite different reasons. First, their work has been taken over by CSIRO and the state departments of agriculture and/or primary industry; secondly, with modern facilities for long-distance travel, it is now more expeditious to seek plants likely to be already substantially adapted to the climate for which they are needed.

By the middle of the nineteenth century the inadvisability of uncontrolled introduction was becoming apparent; it was clearly unwise to allow anybody to bring in anything and grow it in his garden. As a result, 'noxious weeds' legislation began. The first general legislation of this type seems to have been that of South Australia in 1851; the first specific case appears to be the 'thistle bill' which was introduced in Victoria in 1856, 'An Act to make provision for the eradication of certain thistle plants and the Bathurst Burr'.[38] Still in Victoria, first the blackberry (1874) and then St John's wort (*Hypericum perforatum*) in 1892 were duly declared to be noxious.

Nevertheless the practice died hard. As late as 1912, Turner, in an article in the *Sydney Morning Herald*, was criticizing the amateur introduction of plants without sufficient thought to their possible ill-effects.[39] The case in point was *Emex australis*, introduced as a garden vegetable, which has indeed become a serious weed throughout much of Australia. Ornamentals such as *Eichhornia* and the rubber-vine have become pests; the little hybrid water-fern *Salvinia molesta*, first recorded wild in Brisbane in 1953, is now all too familiar to those who know Lake Moondarra near Mount Isa, or the Ross River in Townsville. Even in 1985 George Diatloff, Acting Director of the Biological Division of the Queensland Department of Lands, found it necessary to repeat Turner's warning when opening the Tropical Weed Research Centre in Charters Towers.

In this period it was also becoming increasingly clear that introduced plants were apt to bring their diseases with them. Plant pathology came later;[40] although pathologists were working in Victoria in 1890, the first professional course was given by Waterhouse, in Sydney, in 1922.

The need to investigate problems of plant disease was at least partly responsible for the setting up in 1927 of what was then the Division of Economic Botany in CSIR, later to become the Division of Plant Industry in CSIRO. The first Chief of the Division, Dr B.T. Dickson, was himself a plant pathologist, and rapidly built up an appropriate research team.

Dickson was to do more. During the whole of this period, introduction was still the responsibility of the individual states. There was no federal instrumentality capable of addressing the problem, for although the Advisory Council of Science and Industry was set up in 1916, CSIR did not come into existence as such until 1926. It is impossible, in limited space, even to summarize the remarkable progress made by the state departments. Attention should be drawn, however, to the work of N.A.R. Pollock and J.L. Schofield in Queensland; to the devoted work by E. Breakwell in New South Wales; to the influence of von Mueller in Victoria; to the farmer Amos Howard in South Australia, the first to realize the potential value of subterranean clover, a plant that later was to become the special interest of Western Australia. All were working to some extent in isolation, a significant factor even today.

Dickson foresaw the advantages that could accrue from the establishment of some form of national body to co-ordinate the activities of the individual states and act as a botanical clearing-house in much the same way as Kew had done for the Empire. Thus, as one of the original units of the Division of Plant Industry, he set up a Plant Introduction Section, designed not to supplant, but to supplement, the activities of the state departments of agriculture. Moreover, he realized that, years earlier, the United States had faced a similar problem, with many more states and a climatic range that, while not as extreme as that of Australia, was considerable. He therefore sent A. McTaggart to America to establish appropriate contacts, then brought him back to head the newly formed section. The stage was now set for the development of a plant introduction network that, if not federally controlled, was at least federally integrated.

Order from chaos: 1930–1980

As a result of the sheer size of Australia, and because the country is divided into a series of separate states, each with a considerable measure of autonomy, a nation-wide procedure cannot be created by a simple stroke of a pen. Some degree of interstate rivalry is inevitable. In 1940, the Hon. Frank Bulcock, the Queensland Minister of Agriculture and Stock, remarked sadly that

> the co-ordination of agricultural experiment and research programmes was by no means easy of accomplishment, for it was found that personal factors and State considerations were frequently a bar to the pooling of the sum total of knowledge available . . . As a result, each State tended to make its work a partial secret. This led to duplication of effort . . . [41]

Nevertheless, basically there was goodwill, and a widespread realization that the largely chaotic and undisciplined introduction procedures of the previous century must somehow be brought into an integrated system. Historically, this is an extremely complex period, with progress proceeding independently on a number of different lines at different speeds. Four main trends can be distinguished: an improvement in documentation; the gradual replacement of individual and idiosyncratic introduction by scientifically planned expeditions engaged in seeking new plant material for specific purposes; the development of improved facilities for the testing and assessment of introduced material; and some (but perhaps not quite enough) rationalization of quarantine procedures. Superimposed on these basic trends were two contingent events. The first of these was a growing interest in the pastoral development of the tropical north, an interest that influenced expedition priorities; the other was the Second World War.

Documentation

The earliest lists of plant introduction were probably those of the Government Botanic Gardens in Sydney and a number of contemporary nurserymen. As the years went by, lists were independently produced by state departments of agriculture and by such organizations as the Waite Institute in Adelaide, the Soil Conservation Service of New South Wales and the Parks and Gardens Section of the Department of the Interior. By 1930 the records of introduced plants were enshrined in many independently numbered and overlapping lists. This undoubtedly resulted in much duplication of effort; anybody wishing to discover whether a specific plant had already been introduced, and if so by whom, was faced by what was virtually a major piece of bibliographic research. A national register was needed, but there is no evidence that any of the existing organizations was anxious to undertake the task of preparing it. Some form of Commonwealth intervention seemed to be the only hope.

We have seen that in 1929 Dickson, the Chief of the CSIR Division of Economic Botany, had set up a Plant Introduction Section. Its Officer-in-Charge was McTaggart, and unquestionably its most illustrious founder member was William Hartley. At a meeting of the Commonwealth Council's Standing Committee on Agriculture in November 1929, Dickson suggested that an inventory be made of all introductions to date, 'with brief notes regarding source, date introduced and tested, success or failure of the introductions ... and any other pertinent information'.[42] The Council approved the proposal, promised that state departmental officers would supply the necessary information, and suggested that the records go back twenty-five years. The dream of a national register had at last attained formal recognition; but almost half a century was to pass before the dream was to become a reality.

Nevertheless, Dickson was determined to ensure that his own house was in order. By 1930, the Division of Economic Botany had acquired 270 cereal varieties, 85 varieties of grasses, over 50 leguminous forage plants, and a miscellaneous collection of crop plants. In December 1930 McTaggart and Hartley published the now historic *Plant Inventory No. 1*, in which all introductions made by the division were listed and given Commonwealth Plant Introduction (CPI) numbers; the *Inventory* was issued 'to all Departments and Institutions concerned'.[43] In the ensuing years, so many scientists took part, directly or indirectly, in the activities of the Plant Introduction Section that it is impossible to list them all, but mention must be made of the outstanding contributions of C. Barnard (who later became officer-in-charge), E.T. Bailey (who joined in 1934) and C.A. Neal-Smith (from 1946). To these should certainly be added the work of J.F. Miles of CSIRO in Queensland (between 1936 and 1958) and that of state counterparts such as Pollock, Schofield and Dorothy Davidson, also in Queensland.

By 1963 the *Plant Inventory* had become a quarterly publication and the CPI numbers had passed 30 000; but there was still no national register. To take only two examples: the State Department of Primary Industries in Queensland had an independent list of over 6000 'Q' numbers, and New South Wales a list of over 5000 'P' numbers. It is true some members of these and other lists also appeared in the Commonwealth lists with CPI numbers, but this was not yet general policy. However, in 1963 *Plant Inventory* fused with *Plant Introduction Notes* (the first number had been published in 1949) to become the *Plant Introduction Review*; and its first number carried, for the first time, a request for other organizations to provide lists of accessions for inclusion in the publication. This time the request did not go unheeded. In 1965 Barnard, who had been appointed officer-in-charge of the Plant Introduction Section in 1962, was able to publish a 'Supplementary Accession List' with the comment: 'We are glad to record the co-operation of other plant introducing agencies, in particular those from Australian State organizations, in making available their lists of introductions for inclusion.'[44] Integration was not yet fully achieved, since readers were advised that 'Enquiries concerning these lists should be directed to the appropriate State Departments or other introducing agencies'.[45]

In March 1974 the Plant Production Committee of the Standing Committee on Agriculture recommended that the CSIRO Division of Plant Industry look into the possibility of preparing a nation-wide list of plant introductions. The Plant Introduction Section (now re-named the Genetic Resources Section and headed by Dr D.R. Marshall) stated that this would be feasible 'provided the relevant information could be provided in computerized form'.[46] But it was decided for the present to exclude horticultural species, wheat, fungal and bacterial cultures, and 'lines continuously imported for resale as cultivars'.[47] An issue of the *Plant Introduction Review* for 1974 carried the statement that 'with the cooperation of all relevant bodies the goal of a national list has now been realised'.[48] Finally, in 1975, the same publication carried the statement that 'commencing at list No. 98, the scope of the CPI list has been extended to include all agricultural plants brought into Australia by the major seed importing agencies'. The dream of 1930 of an all-embracing national list had for most practical purposes at last been realized, some forty-five years later; appropriately the *Plant Introduction Review* became the *Australian Plant Introduction Review* at that time.

The tropical north

The colony had begun in Sydney, where the summer warmth and winter rainfall were reminiscent of the Mediterranean, with which some at least of the colonists were familiar. Moreover, it would support familiar crops such as wheat and the common vegetables, and familiar forage plants such as the grasses *Festuca* and *Lolium* and the legumes *Trifolium* and *Medicago*. The earliest large-scale agriculture and horticulture were therefore essentially temperate; but from the earliest days the eyes of the colonists turned wistfully towards the tropics, as a possible venue for more exotic crops and fruits.

At this point the history of Australian plant introduction falls into two sharply contrasted aspects. From early days the south had possessed crops, pasture grasses and legumes that were at least adequate. There was always the hope that species would be found capable of giving higher yields, or of flourishing in more extreme ecological situations. Over the years there was a steady stream of temperate introductions, the single most important source being the justly celebrated 'reconnaissance mission' of C.M. Donald and J.F. Miles in 1951; this mission visited a remarkable number of Mediterranean countries in search of plants, many of which had not yet been domesticated. In contrast, the problem in the tropics was not that of finding plants that would effect modest improvements; it was, all too often, that of finding anything that would grow at all.

This chapter concentrates on the tropical problem. The temperate story is already very well documented,[49] whereas documentation for the north, where it exists at all, is scattered and often difficult of access.

Many tropical introductions were brought in by the First Fleet, and a list of the tropical fruits introduced during the first half-century contains many surprises. It includes a number that only today are being virtually rediscovered as potential crops by organizations such as the Rare Fruits Council and the CSIRO Division of Horticultural Research. Moreover, England expected her colonies to be self-supporting, and it was at least conceivable that tropical crops might form the basis for an export trade.

But export had its own problems. It might have been true that Britannia ruled the waves, but at that time, for all practical purposes, the British East India Company ruled the seas, and was opposed to any newly founded colony's engaging in activities that might interfere with its own trade — particularly if the colony might be able to undercut its prices by the use of cheap convict labour. In the orders given to Governor Phillip he was told that 'every source of intercourse between the intended settlement at Botany Bay . . . and the settlements of our East India Tea Company, as well as the coast of China, and the islands in that part of the world . . . should be prevented by every possible means'.[50] Specific prohibitions on the export of seal oil and skins, and of sandalwood, were to follow.

However, such prohibitions did not prevent the colonists from opening up the tropics for their own use. Queensland was particularly attractive; the cooler, drier parts might support wheat, the warmer wetter parts might support sugar-cane, and it might be possible to use the subcoastal areas as pastures. Precise documentation is elusive, but it is clear that by the 1850s oats were already widely grown and by about 1870 cultivation of both wheat and cane had been firmly established in Queensland. Sheep had been tried but, except in western districts, were not a success; burning pastures to encourage growth resulted in the invasion by spear-grass (*Heteropogon contortus*), the spirally-twisted fruiting inflorescences of which had disastrous effects on both the wool and the sheep themselves. So, if pastures were to be utilized, it would have to be for cattle; and by 1920 there was a retinue of stock and brand inspectors throughout much of the state.

However, the cattle were to present plant introduction with its most formidable challenge to date. Beef cattle would survive on the northern pastures, but to fatten them for market, or to replace them with high-yielding dairy cattle, appeared to be impossible. The notoriously low phosphorus status of almost all Australian soils could be redressed by the addition of superphosphate; trace-element deficiencies were common, but this was a field of study in which Australian plant physiologists were to become noteworthy pioneers; but the apparently irremediable shortage was of protein nitrogen. The first agriculturalist to place this in a practical context seems to have been N.A.R. Pollock of the Queensland Department of Agriculture and Stock, who wrote in 1925: 'A defect in northern pasturage, both native and introduced, is the absence of a legume to take the place of the clover and trefoils of the more temperate Southern parts'.[51] He also noted that stocking pressure on the native pastures was resulting in the loss of native legumes such as *Rhyncosia*,

Glycine and *Atylosia*. A survey was initiated in 1927 by the Waite Institute, under the aegis of the Empire Marketing Board in collaboration with CSIR and the University of Adelaide; it showed conclusively that the protein content of Australian native grasses would not sustain cattle on a year-round basis. A legume was needed for Queensland, but none was to hand. The CSIR Annual Report for 1934/35 carries the statement:

> One of the outstanding problems in connexion with plant introduction work is the discovery and acclimatization of a legume which will thrive in the summer rainfall regions of Australia, similar for example to subterranean clover which has been introduced so successfully into the winter rainfall areas of Southern Australia.[52]

The first published constructive suggestion probably came in 1933 from William Davies, in Australia on a visit from the English Grassland Research Station. He outlined three alternatives: the breeding or selection of indigenous plants, the possible extension of the use of temperate species of known value into the tropics, and the use of imported plants from tropical and subtropical areas. Pollock, however, had reservations. Also in 1933 he wrote: 'Attempts to introduce grasses and legumes peculiar to cold or temperate climates into the tropics are obviously doomed to failure and the soil temperatures and general climatic conditions are very far from favourable'.[53] From this time onwards the literature, which until then had been strangely lacking in discussion of the problem, was dominated by the quest for new legumes. J.F. Miles, for example, wrote in the CSIR *Journal* for 1939 of 'The Need for a Legume in Northern Queensland'.[54] F.W. Bulcock, the then Queensland Minister for Agriculture and Stock, was writing in 1938 to the effect that some introduced grasses were doing extremely well, but 'the absence of a legume in the pasture constitutes a source of worry'.[55]

Meanwhile, enthusiasm for exploiting the tropical north might easily have been dampened by a series of articles entitled 'The Tropics and Man', published by Dr D.H.K. Lee, professor of physiology in the University of Queensland.[56] He warned his readers of the dangers that beset the white man in the tropics, listing among others boredom, lack of concentration, circulatory failure, the 'mañana' attitude, irritability and excessive drinking. Sir John Russell, out from Rothamsted in 1939, would have none of this. He is quoted as saying, of North Queensland, that 'the children looked very well . . . It was clearly possible for men of British and North European stock to run the country.'[57]

There was published in 1941 what was to prove a vitally important paper, 'Introduced legumes in North Queensland' by J.L. Schofield, the director of the Queensland Bureau of Tropical Agriculture. It was a test of the second and third suggestions of William Davies, a comparison under tropical conditions of a suite of introduced temperate legumes with such few tropical legumes as were already available. The results were unequivocal: 'Results indicate that certain tropical legumes are satisfactory under coastal conditions in North Queensland, but temperate legumes are markedly unsuccessful.'[58] This was the first formal intimation that legumes for the tropics must be sought in the tropics.

A few introduced tropical legumes were already in Australia. Townsville

The native Australian grassland of the dry tropics . . .
. . . may deteriorate rapidly when exposed to hard-footed animals, and the animals in turn suffer.
(Courtesy of R. Rebgetz, P. Gillard and the CSIRO Division of Tropical Crops and Pastures)

stylo (*Stylosanthes humilis*) was first recorded on Townsville Common in 1904; it flourished under semi-arid conditions, and was already in use as a legume in northern pastures. From 1919 onwards its use had been enthusiastically promoted by the indefatigable Pollock. So convinced was he of its all-round efficacy that it was his habit to spread the seed from train windows during his journeys round north Queensland. This must surely represent what has been called 'one of the largest and least conventional species adaptation studies ever conducted'.[59]

However, Townsville stylo had its limitations. It would not withstand the frosts that in inland Australia extend well up into the tropics and, as an annual, it competed poorly in many situations. Two others, *Centrosema pubescens* and *Stylosanthes guianensis*, were both introduced about 1930; but both are essentially plants of the wet tropics, and quite unsuitable for arid or even semi-arid areas. Two other well-known species were already established around Darwin as early as the 1860s — the plants now known as *Macroptilium atropurpureum* and *Leucaena leucocephala*. However, *Macroptilium* may fail under heavy grazing, and *Leucaena* is hardly a pasture plant — if left to itself it grows into a small tree. There were still no legumes suitable for the hotter and drier areas, yet these areas were now due for exploitation. In 1947 we are told: 'The North Australia Development Committee ... has the responsibility of determining the best usage of great areas in the Northern Territory, and, to this end, plans first to obtain facts from scientific surveys.'[60] The first such survey set out in May 1946, under the leadership of C.S. Christian of the CSIR Division of Plant Industry; it was to report on the ecology, vegetation and soils of the Katherine-Darwin-Coast area.

Although it was not realized at the time, plant introduction had entered a completely new phase. Previously, requirements had not been specified in advance. Introductions were largely by correspondence, and Australia happily accepted almost anything thought to be interesting, then investigated its possible utility. Now, for the first time, Australia could define a specific requirement for a particular type of plant for a particular purpose; a type of plant, moreover, that was unlikely to arrive as a result of casual correspondence. It was time for Australia to go out into the world and seek the plants that were so badly needed.

Expeditions

The need for a northern legume was not the only reason why Australian thoughts began to turn in the direction of expeditions. From personal communications it is clear that in the late 1930s some botanists were becoming increasingly concerned about loss of genetic variability, the phenomenon now known as 'genetic erosion'. Hartley, who was undeniably one of the foremost pioneers of the concept of 'genetic resources', urged the need for expeditions into the wild in search of new germplasm material. He received little tangible encouragement, and in 1938 resigned from CSIR and went to work at Kew. However, he rejoined CSIR in 1940; in 1944 information was received that the Division of Exploration and Introduction of the United States Department of Agriculture was planning an expedition to subtropical South America, primarily to seek material of *Arachis* to improve their peanut strains, and of the forage grasses *Paspalum dilatatum* and *P. notatum*. Australian official opinion was becoming increasingly sympathetic to the concept of expeditions. South America had already provided Australia with the only two species of *Stylosanthes* she possessed, both very valuable, and no doubt there could be others. Moreover, such material might be of use to other countries of the British Commonwealth. Therefore, representations were made to the Executive Council of the Commonwealth Agricultural Bureau, suggesting that it would be advantageous if the American expedition could be organized as a joint venture by the addition of Australian personnel. Permission was sought and granted, and two plant collectors were included: J.L. Stephens of Georgia, the officer-in-charge, and William Hartley of Australia.

The expedition left Miami on 12 November 1947; it visited Argentina, Uruguay, Paraguay and Brazil, returning home in April 1948. Hartley, whose detailed report on the expedition was published as *CSIR Division of Plant Industry Divisional Report No. 7*, then returned to Australia, followed shortly after by his 236 accessions.[61] These included particularly the legumes *Stylosanthes*, *Arachis*, *Desmodium* and *Adesmia*, and the grass *Paspalum*. There are three points of interest concerning this expedition that deserve brief consideration. First, Hartley, like his noted predecessor Banks, regarded plant collection as a rigorous scientific discipline, and the expedition was planned with meticulous care to ensure the greatest possible chance of success. Floras of the areas to be visited were studied in advance, and two independent systems of climatic classification were used to select climates homologous with those for which the new material was required. Secondly, the later history of the plants illustrates the time-scale involved in commercial release. The first accession from the expedition to be released commercially was a grass, *Paspalum plicatulum* cv. Hartley, which was released in 1963, fifteen years after its collection. Lastly, we note the opportunist nature of the venture. It was not an expedition planned and financed from Australian sources; it was an expedition planned and organized by another country, to which Australia attached herself.

Australia now embraced the concept of exploration with ever-increasing enthusiasm and this phase has been admirably documented.[62] During the years 1947 to 1960 there were, in addition to Hartley's, five expeditions; during 1960 to 1967 there were twenty; during 1967 to 1971 there were thirty-four; and between 1971 and 1976 there were a further thirty-six. The sponsors were varied, as were the destinations. Of the ninety-six recorded expeditions, thirty-two were primarily to Europe, twenty-seven to Central and South America, twelve to Africa, twenty to India and South-east Asia, and five involved visits to North America, Russia and/or Papua New Guinea. For the period for which we have data, the numbers of tropical and temperate accessions collected for trial as pasture plants have been about equal (47 and 53 per cent respectively).

In some of the earlier expeditions a substantial proportion of the material collected was obtained from existing overseas research establishments; Africa should certainly be mentioned, with its excellent collections, lovingly assembled by the earlier 'colonial botanists', that have provided a number of valuable Australian grass cultivars. However, the practice has its drawbacks, since such material may already have begun the process of genetic erosion; and later expeditions seem to have concentrated to a much greater extent on material obtained directly from the wild. Moreover, it is in wild material that the possibility of serendipity, the happy accident, reaches its highest level. A single example will suffice. In 1965 W. Atkinson was collecting, primarily, in high-altitude areas in the South American tropics, hoping to find plants suitable for New South Wales. When leaving for home he embarked at Maracaibo, in the semi-arid lowland tropics of Venezuela, and noticed a small legume, growing near the airport, that looked interesting; he added it to his collection. It proved to be a tetraploid, *Stylosanthes hamata*, and became the celebrated accession 'Caribbean stylo', cv. Verano, perhaps the most accommodating legume yet naturalized in the northern, hotter and more arid, parts of Australia.

The earlier 'introduction by correspondence' approach had one outstanding weakness. Seeds of a new species might be brought into the country, but nothing whatever might be known about the climate or soils of which it was native. Even in the earliest days Banks stressed the advisability of collecting as much information as possible concerning the soils and climate of origin — what is now colloquially known as 'passport information'. The problem entered a new phase in 1966. By that year CSIRO had installed its second-generation computer, the Control Data 3600, and although this was not yet provided with drums or visual displays, it was well provided with magnetic-tape stations. In that year, L. Albrecht pointed out the enormous advantages of storing passport information on magnetic tape instead of on laboriously hand-written cards.[63] However, two years later, L.F. Myers (then in charge of the Plant Introduction Section) also pointed out that to be successful such a system required that descriptions of the site of collection should be compatible for all collectors. As he put it: 'In the past, some collection sites would be described by geology, others by land-use and altitude, but no information was available for all sites for even one common attribute, say, altitude.'[64]

Myers therefore introduced, and circulated widely, a 'collectors' card', which required the collector to code the area under the eight headings of parent rock, aspect, agriculture, slope, texture of surface soil, depth of surface soil to clay, chemical reaction of surface soil, and the water-status of the area. Later experience showed that the card needed some modification; in particular, botanists with only limited training in geology or soil science could hardly be expected to identify the parent rock, and their assessments of soils were not always reliable. In fact, the expedition of R.L. Burt and R.F. Isbell to Central and South America in 1971 seems to have been the first in which a senior soil scientist (Isbell) was professionally involved. But such minor shortcomings, inevitable in a pioneer situation, do not detract from the importance of Myers' contribution.

Meanwhile an old problem was re-asserting itself. Legume seeds do not carry with them the *Rhizobium* bacteria they require to form effective nitrogen-fixing nodules; they acquire these from the soils in which they grow. Moreover, some legumes are 'promiscuous', and will form effective nodules with a wide range of *Rhizobium* strains; others are 'élitist', and have very precise *Rhizobium* requirements. The *Rhizobium* strains themselves similarly exhibit a wide range of co-operative ability. As a result, many introduced legumes failed to nodulate effectively, and simply died. In any case, Australian soils are not rich in *Rhizobium* strains, so it is necessary to provide them artificially. As early as 1896 agriculturalists were adding crude *Rhizobium* cultures to soil in the hope of improving nodulation.[65] *Rhizobium* can be cultured *in vitro*; D.O. Norris, who joined the Cunningham Laboratory in Brisbane in 1953, realized the potential advantages of a reference collection of *Rhizobium* strains. He initiated the collection in 1956; its members were given 'C.B.' (CSIRO Brisbane) numbers by which such members, now numbering more than 3000, are known. The first strains to be included were from indigenous and already-naturalized legumes, but Norris urged collectors to bring back not only seeds or fruits, but also nodules. The first overseas rhizobial accession was from a *Stylosanthes* that had been introduced into New Guinea — the most famous is CB 756, collected by Mr Norman Shaw in Zimbabwe from a herbaceous legume *Macrotyloma africanum*, which has proved to be the most promiscuous of all known *Rhizobium* strains.

The first cultivar of Stylothanthes hamata, *Verano, is now widely used throughout the world and has spread throughout north Queensland; here it is in association with Guinea Grass (**Panicum maximum*), an African species. (*Courtesy of P.C. Pengelly and the CSIRO Division of Tropical Crops and Pastures)

But there was another, and even older, problem. Fruits and seeds may carry with them unwanted passengers in the form of bacteria or fungal spores on their surfaces. Moreover, although seed-borne viruses are uncommon, they exist. It has therefore always been necessary to ensure that the introduction of a desirable plant does not also involve the introduction of an undesirable plant disease. The problem had of course been for long all too familiar to those who were concerned with the introduction of new animals such as sheep and cattle, and the first state quarantine laws for animals were passed in 1871. The botanists were to follow only four years later.

Quarantine

Quarantine must always be a compromise, since too rigorous a system might end by denying agricultural scientists the germplasm material they need for their experiments. Moreover, the complete policing of every conceivable point of entry — including the seeds innocently sent in correspondence by well-meaning relatives overseas — would be logistically impracticable. Nevertheless, it had become clear towards the end of the nineteenth century that something had to be done.

As so often in Australian history, the first legislation was enacted by the states. The first Plant Quarantine Act was passed in Victoria in 1875, to be followed by New South Wales and Queensland in 1887. An earlier embargo on a specific plant was the attempt by the Queensland Plant Acclimatization Society to prevent, in 1876, the importation by Count Franceschi of grape-vines from the *Phylloxera*-ridden region of Tuscany; subsequently, in 1877, an act was passed regulating the import of any vines. After federation, the first federal Plant Quarantine Act was passed in 1908. Unfortunately, nobody seems to have taken much notice: White goes so far as to aver that it was not until the appointment of Dr T.H. Harrison as director of plant quarantine in 1946 that 'the quarantine regulations began to be vigorously implemented and upheld'.[66] It was, however, immediately clear that a single federal department could not police every port of entry; and so, although the Commonwealth Department of Health and its director of plant quarantine remained responsible for new regulations, bulletins and leaflets as necessary, the day-by-day work of inspection was delegated to the states.

Quarantine in Australia now operates on three levels. At the top is a list of plants, import of whose seeds is completely prohibited. Inevitably, this list is a summary of wisdom after the event; it is a list of plants already known to be troublesome. Unfortunately, nobody in the early days could have predicted that such charming and apparently harmless ornamentals as *Eichhornia*, *Salvinia*, *Lantana* or the rubber-vine would become pests, but all of them did. Nor in recent years was it predicted that *Rhus*, introduced into New South Wales as an ornamental shrub, would cause serious allergies in Sydney.

The second level is represented by a list of plants, import of whose seeds is restricted. These are mainly plants, taxonomically related to important existing crops or forage plants, that might carry diseases that could be transmitted to existing cultivars, or that might indulge in unwanted hybridization. Also restricted are those cultivars, mainly of fruit or ornamentals, that can only be propagated vegetatively, by corms, bulbs, tubers or cuttings. Permission is needed to import any of the restricted items, which is normally granted only to universities, government agencies, or established nurserymen.

Finally, seeds of any plants not on the prohibited or restricted lists, may be imported but such seeds must be inspected (and if necessary fumigated) on arrival. Any plant on the 'restricted' list must spend a complete growing season in quarantine, either in an insect-proof 'quarantine glasshouse' or, if that is impracticable, in a vermin-proof fenced enclosure. In both cases the plants are required to be inspected at intervals by a plant pathologist. This itself causes difficulties: the distances in, for example, Queensland are such that visits from the state pathologists may be relatively infrequent.

Nevertheless, cumbersome though the system may appear, it works. Moreover, Australia is not alone in such matters; P.H. Gregory, writing of the problem in cocoa, sums up the present situation neatly:

> current opinion indicates that existing procedures do not unreasonably restrict the movement of genetic material. Furthermore, experience shows that insect pests and the main fungal diseases are satisfactorily excluded... More research is needed to improve detection of viruses, which experience shows can still slip through quarantine.[67]

Still, the time spent in a quarantine glasshouse is only the first phase of an introduced plant's career. The extent to which it might conceivably be of value in Australian agriculture is still completely unknown. How is this to be established?

Testing and evaluation

These terms have unhappily caused much discussion, controversy and even animosity. Yet the basic problem is simple: a new introduction, or group of introductions, is to be grown under conditions of soil and climate similar to those for which a new plant is required, and performance recorded. The conditions are important; as Hartley has pointed out: 'Many potentially valuable introductions fail to gain recognition simply because they are tested in the wrong environment'.[68] Alternatively, if it is desired to establish the full potential of a new genetic resource, then it must be tested over a range of environments. In a country such as Australia, with a wide variety of soils and an even wider variety of climates, this implies an extensive series of testing stations. The need seems first to have been generally recognized about 1930, since the early 1930s saw an almost explosive dissemination of both state and CSIR testing stations, widely scattered over most of the agricultural and horticultural regions of Australia, many of them with resident plant-introduction staff. If no suitable station were available, a small area of land might be leased from a co-operative grazier for use in experiments, a practice still common today.

Given the plants and a suitable site, three questions arise. How is the experiment to be laid out? What data are to be recorded, and, how, if at all, are these data to be analysed? All three questions, and particularly the last, have proved controversial. Concerning the first, there has often been an over-academic attitude based on the early work of R.A. Fisher, so that what could be simple experiments are laid out in elaborately replicated randomized blocks.

Fortunately, the statisticians themselves are becoming increasingly impatient of unnecessarily complex designs. S.C. Pearce, whose statistical qualifications and experience cannot be questioned, has recently written:

> Supervisors of research and editors insist on randomized blocks and it is nearly impossible to publish a non-significant difference ... The consequence is that research institutes, which once protested that the Fisherian experiment was impracticable, are now so geared to it that they can conduct nothing except these grand localized 'set-pieces'.[69]

Moreover, this over-statistical approach tends to limit the information recorded; some published trials have used only an estimate of yield, since this is statistically tractable.

However, plants usually possess other attributes that may be of agricultural or horticultural interest, and so these too must be recorded. In these days of rapid exchange of information, it is once again necessary to ensure that different workers record the same attributes. So history is repeating itself. The 1967 'collectors' cards' of L.F. Myers, which had themselves been anticipated by instructions given to Banksian collectors, represented an attempt to rationalize the collection of simple environmental information; but it is now necessary also to collect appropriate morphological and agronomic information and to note the occurrence of 'new' variation. So, on an international scale, the International Board for Plant Genetic Resources (IBPGR) publishes, for every major crop, a list of 'descriptors', which workers in introduction are urged to record. Such descriptions may cover the whole range of attribute types — nominal, ordinal or numeric: mixed-data sets of this type are not amenable to conventional statistics, but modern computer programs do permit their use for numerical methods of classification.

Foremost among the advocates of classificatory techniques have been the wheat-breeders. Krull and Borlaug write: 'The present available world collections of the major crop plants contain a tremendous range of variability. The major hurdle to unlocking their secrets and utilizing the valuable characters has been our inability to satisfactorily classify this variability.'[70] Marshall and Brown, specifically writing of wheat, maintain that, if a data-set is extensive, evaluation is impracticable unless it is preceded by a classification.[71] The purpose of such a procedure is to reduce a very large number of items to a small number of substantially homogeneous groups, each of which can be treated almost as if it were an individual accession. What is possibly the first application of this approach to a forage plant is the work of L.A. Edye on *Glycine* (now *Neonotonia*) *wightii*, which was rapidly followed by a series of papers by R.L. Burt and his colleagues on *Stylosanthes*. These workers, using all the morphological and agronomic descriptors available to them, used the pattern-seeking numerical techniques of classification, ordination and graph-theory to elucidate as precisely as possible the configuration of a given set of accessions. They imposed on this configuration any additional available information — climatic, edaphic, biochemical, cytological and/or agronomic — and contended that they were thereby enabled to predict, more precisely than had previously been possible, the situations in which a given accession might be expected to succeed.[72] They also successfully revived the 'agrostological index' of Hartley, which used the native grasses of the area of collection as climatic and edaphic indicators.[73] The claims of these workers are not universally accepted; some agricultural scientists dispute the validity and utility of the methods.

The international problem

Plant introduction is now only part of the world-wide problem area of conservation of genetic resources. Most familiar crops are extremely inbred and contain little unused variation. If new genes — for example for disease resistance — are needed, they must be sought among the wild relatives, or the 'landraces' of primitive cultivation, all of which are rapidly disappearing. As a result, gene banks for all the major, and many of the minor, crops are being established around the world. However, collection is increasingly difficult; most of Vavilov's 'centres of diversity' are situated within the developing nations, whose increasingly complex political boundaries make them less accessible than was formerly the case. Added to this, the maintenance of a collection under optimal storage conditions is extremely expensive.

A gene bank is of little use without documentation, which is why the IBPGR publishes its lists of recommended 'descriptors'. Unfortunately, the accurate recording of passport information and morphological descriptors requires expertise that is not everywhere available; and evaluation in particular is a time-consuming process requiring much experience. As a result, the documentation of almost all gene banks is desperately inadequate, a sad situation admirably summarized by Peeters and J.T. Williams.[74] Genetic material is normally available to breeders anywhere; examples of suppliers include the International Rice Research Institute (for rice, especially in Asia) and the International Institute of Tropical Agriculture (for a variety of crops in Africa). Yet few breeders make extensive use of such material, in part at least because of the inadequate documentation and analysis.

A problem is also arising concerning the 'ownership' of genetic resources, which may restrict distribution for essentially unscientific reasons. Sun, for example, quotes reports that Ethiopia will not export germplasm of coffee, India of black pepper or turmeric, Taiwan of sugar-cane.[75] Underlying this appears to be a feeling that the developing countries have been in some sense exploited. Sun has elsewhere stated: 'Developing countries ... charge that developed countries take their germplasm, improve it, and then sell back the seeds at an unfair profit.'[76] Mackenzie similarly reports that 'the Third World is growing angry that its plant genes are taken freely to the industrial nations — often to be sold back later as highly bred strains'.[77] The situation is exacerbated by a growing interest in plant breeders' rights. In fact, such strains have usually been bred to serve a particular purpose in a particular place, and may well be of little use in their country of origin; many require high inputs of, for example, fertilizer, which are not everywhere available. Australia's record is good. After the Second World War she made germplasm from her collections — particularly of wheat — freely available to those countries whose gene banks had been dissipated or destroyed by the ravages of war; and the fact that species of *Eucalyptus* are now grown overseas over a far wider area than in their native habitats demonstrates that such plant material has been freely and widely distributed. More recently, 'core collections' of *Stylosanthes hamata* and *Desmanthus* spp. from South and Central America and the Caribbean have been sent to the Forage Legume Agronomy Group at the International Livestock Centre of Africa; other collections, including African grasses, have been despatched to Belize and Antigua (in Central America and the Caribbean) for use in projects sponsored by the International Development Resource Centre of Canada. In such matters, Australia's conscience is clear.[78]

Notes

1 D. Pike, *Australia* (Cambridge, 1962), p.28.
2 C. Lyte, *Sir Joseph Banks* (Newton Abbot and North Pomfret, 1980), p.116.
3 T.M. Perry, *Australia's First Frontier* (Melbourne, 1963), p.6.
4 Ibid.
5 D. Collins, *An Account of the English Colony in New South Wales*, Vol. 1 (London, 1798). Subsequently edited by B.H. Fletcher and republished Sydney, Wellington and London, 1975.
6 E. Wilson, 'The Royal Botanic Gardens, Sydney', *Australian Horticulture*, 80(2) (1982), 24–35.
7 L.W. Wrigley and A. Rathjen, 'Wheat breeding in Australia', in D.J. and S.G.M. Carr (eds), *Plants and Man in Australia* (Sydney, 1981), pp.96–135.
8 D Collins, op. cit. (n.5), p.lxxx.
9 Ibid.
10 Ibid.
11 D. Collins, op. cit. (n.5), p.130.
12 Ibid, p.157.
13 Cited by W.J. Parsons in 'The history of introduced weeds', op.cit. (n.7), pp.179-93.
14 P. Edwards, 'Botany of the Flinders voyage', in D.J. and S.G.M. Carr (eds), *People and Plants in Australia* (Sydney, 1981), pp.139-66.
15 P.G. King to Lord Hobart, 9 May 1803 in *Historical Records of Australia*. Series 1. Governors dispatches to and from England, Vol. 4, ed. J.S. Watson, 1803.
16 J.T. Swarbrick, 'Weeds of Sydney Town, 1802-4', *Australian Weeds*, 3(1) (1984), 42.
17 Anon., *Catalogue of Plants Cultivated at Camden, New South Wales* (n.p., 1843): Anon., *Catalogue of Plants Cultivated at Camden, New South Wales* (n.p., 1845).
18 T.W. Shepherd, *Catalogue of Plants Cultivated at the Darling Nursery, Sydney, New South Wales* (Sydney, 1851).
19 Anon., *Catalogue of Plants Cultivated in the Government Botanic Garden, Sydney, New South Wales* (Sydney, 1857).
20 D. Mackellar, 'My country', in *Poetry in Australia*, Vol. 1 (Sydney, 1964).
21 Anon., *The Sydney Herald*, 3 October 1831 (Vol. 1(25)), Commercial News (Facsimile edition, April 18, 1831 to Jan. 2 1832, Lidcombe, NSW, 1976).
22 D.A. Herbert, 'A story of Queensland's scientific achievement', *Proceedings of the Royal Society of Queensland*, 71 (1959) 1-15.
23 H.W. Caulfield, 'Brisbane's two botanic gardens', *Australian Horticulture*, 81(3) (1983), 91-2.
24 D.A. Herbert, op. cit. (n.22).
25 D.J. and S.G.M. Carr (eds.), *People and Plants in Australia* (Sydney, 1981); D.J. and S.G.M. Carr (eds.), *Plants and Man in Australia* (Sydney, 1981).
26 P.M. Kloot, 'The genus *Lolium* in Australia', *Australian Journal of Botany*, 31(4) (1983), 421-35.
27 P.S. Cocks, M.J. Mathison and E.J. Crawford 'From wild plants to pasture cultivars: annual medics and subterranean clover in southern Australia', in R.J. Summerfield and A.H. Bunting (eds.), *Advances in Legume Science* (Kew, 1980), pp. 569-96.
28 J.S. Gladstones and W.J. Collins, 'Subterranean clover as a naturalized plant in Australia', *Journal of the Australian Institute of Agricultural Science*, 49 (1983), 191-202.
29 C.W. Wrigley and A. Rathjen, op. cit. (n.7).
30 D.A. Herbert, op. cit. (n.22).
31 H.E. Laffer, *The Wine Industry of Australia* (Adelaide, 1949), p. 35.
32 D.A. Herbert, op. cit. (n.22).
33 Ibid.
34 F. Mueller, *Report on the Botanic Garden, 1856-7* (Melbourne, Victoria), p. 8.

35 Anon., *The Rules and Objects of the Acclimatization Society of Victoria* (Published with the report adopted at the first general meeting of the members) (Melbourne, 1861), p. 1.

36 H.W. Caulfield, op. cit. (n.23); D.A. Herbert, op. cit. (n.22); I.M. Mackerras and E.N. Marks, 'The Bancrofts: a century of scientific endeavour', *Proceedings of the Royal Society of Queensland*, 84 (1973), 1-34.

37 E. Kynaston, *A Man on Edge: a Life of Baron Sir Ferdinand von Mueller* (Ringwood, 1981), p. 144.

38 Cited from W.T. Parsons, *Noxious Weeds of Victoria* (Melbourne, 1973), in correspondence from R.H. Groves, 12 July 1985.

39 F. Turner, 'Cape Spinach: An obnoxious plant', *Sydney Morning Herald*, 7 December 1912, p. 27.

40 N.H. White, 'A history of plant pathology in Australia', in D.J. and S.G.M. Carr (eds.), *Plants and Man in Australia* (Sydney, 1981), pp.42-95.

41 F.W. Bulcock, 'Agricultural Coordination', *Queensland Agricultural Journal*, 53 (1940), 264-5.

42 Anon., *Fourth Annual Report of the Council for Scientific and Industrial Research for the Year ended 30th June 1930* (Canberra, 1930), p. 16.

43 Anon., *Fifth Annual Report of the Council for Scientific and Industrial Research for the Year ended 30th June 1931* (Canberra, 1933), p. 17.

44 C. Barnard, 'Australian states supplementary accession list', *Plant Introduction Review*, 2(2) (1965), ii.

45 Ibid.

46 Anon., 'National list of plant introductions. Progress report', *Plant Introduction Review*, 10(1) (1974), 4-5.

47 Ibid.

48 Ibid.

49 F.H.W. Morley, 'Subterranean clover', *Advances in Agronomy*, 13 (1961), 57-123; C.M. Donald, 'Temperate pasture species', in *Australian Grasslands*, R. Milton Moore (ed.) (Canberra, 1970), pp.303-20; P.S. Cocks, M.J. Mathison and E.J. Crawford, op. cit. (n.27); M.J. Mathison, 'Mediterranean and temperate forage legumes' in *Genetic Resources of Forage Plants*, J.G. McIvor and R.A. Bray (eds.) (Melbourne, 1983), pp.63-81.

50 Cited by A.G.L. Shaw in '1788-1810', in F.R. Crowley (ed.), *A New History of Australia* (Melbourne, 1974), pp.1-74.

51 N.A.R. Pollock, 'Northern Division — *Stylosanthes*', in the *Annual Report of the Queensland Department of Agriculture and Stock for 1924-25*, p. 9.

52 Anon., *Ninth Annual Report of the Council for Scientific and Industrial Research for the Year ended 30th June 1935* (Canberra, 1935), p. 13.

53 N.A. Pollock, 'Northern Division — pasture grasses and legumes', in the *Annual Report of the Queensland Department of Agriculture and Stock for 1932-33*, p. 10.

54 J.F. Miles, 'The need for a legume in northern Queensland', *Journal of the Council for Scientific and Industrial Research*, 12 (1939), 289-93.

55 F.W. Bulcock, 'Cattle fattening on coastal pastures', *Queensland Agricultural Journal*, 49 (1938), 460-8.

56 See, for example, D.H.K. Lee, 'The tropics and man. Mental capacity', *Queensland Agricultural Journal*, 48 (1937), 736-8.

57 Anon., 'Event and comment: The future of the far north', *Queensland Agricultural Journal*, 51 (1939), 133-5.

58 J.L. Schofield, 'Introduced legumes in north Queensland', *Queensland Agricultural Journal*, 56 (1941), 378-88.

59 A.G. Eyles and D.G. Cameron, 'Pasture research in Northern Australia — its history, achievements and future emphasis', CSIRO Division of Tropical Crops and Pastures, Research Report No. 4 (1986).

60 Anon., *Twentieth Annual Report of the Council for Scientific and Industrial Research for the Year ended 30th June 1946*, p. 12.

61 W. Hartley, 'Plant collecting expedition to sub-tropical South America, 1947-48', CSIRO Division of Plant Industry, Divisional Report No. 7 (Melbourne, 1949).

62 C.A. Neal-Smith and D.E. Johns, 'Australian plant exploration, 1947-67', *Plant Introduction Review*, 4(1) (1967), 1-6.

63 L. Albrecht, 'Computer use for plant introduction data processing', *Plant Introduction Review*, 3(2) (1966), 27-30.

64 L.F. Myers, 'New procedures for plant introduction through CSIRO, Canberra', *Plant Introduction Review*, 5(2) (1968), 26-7.

65 R.A. Date, 'Rhizobium as a factor in soil fertility', in *Soils, an Australian Viewpoint*, Division of Soils, CSIRO (Melbourne, 1983), p. 711.

66 N.H. White, op. cit. (n.40), p. 49.

67 P.H. Gregory 'Cocoa (*Theobroma cacao* L.)', in W.B. Hewitt and L. Chiaruppa (eds.), *Plant Health and Quarantine in International Transfer of Genetic Resources* (Cleveland, 1977), pp. 119-24.

68   W. Hartley, 'Plant introduction services: an appraisal', in O.H. Frankel and E. Bennett (eds.), *Genetic Resources in Plants* (London, 1970), p. 424.

69   S.C. Pearce, 'Experimental design: R.A. Fisher and some modern rivals', *The Statistician*, 28 (1979), 153-61.

70   C.F. Krull and N.E. Borlaug, 'The utilization of collections in plant breeding', op. cit. (n.68), pp.427-40.

71   D.R. Marshall and A.H.D. Brown, 'Wheat genetic resources', in L.T. Evans and W.J. Peacock (eds.), *Wheat Science — Today and Tomorrow* (Sydney, 1975), pp.21-40.

72   R.L. Burt, W.T. Williams and D.J. Abel, 'A new graph-theoretic technique for the analysis of genetic resources data', *Agro-Ecosystems*, 8 (1983), 231-45.

73   W.T. Williams and R.L. Burt, 'A re-appraisal of Hartley's agrostological index', *Journal of Applied Ecology*, 19 (1982), 159-66.

74   J.P. Peeters and J.T. Williams, 'Towards better use of gene banks with special reference to information', *Plant Genetic Resources — Newsletter*, 60 (1984), 22-32.

75   M. Sun, 'The global flight over plant genes', *Science*, 231 (1986), 445-7.

76   M. Sun, 'Fiscal neglect breeds problems for seed banks', Ibid., 329-30.

77   D. MacKenzie, 'UN takes control of world's food genes', *New Scientist*, 100 (1983), 558.

78   Several members of the CSIRO Plant Introduction Section, have provided both encouragement and information; the contributions of Mr W. Hartley, Mr L.F. Myers, Mr C.A. Neal-Smith and Mr R. Pullen are gratefully acknowledged as are those of Dr D.G. Cameron, of the Queensland Department of Primary Industries, and Dr R.H. Groves, CSIRO Division of Plant Industry. Mrs D. Sinkora and Dr D.M. Churchill, of the Royal Botanic Gardens and National Herbarium, Melbourne, Drs B.T. Roach (of CSIR), J.V. Possingham (CSIRO Division of Horticultural Research) and J.R. Syme (of the Queensland Wheat Research Institute) all furnished information on various topics discussed here.

     I would also thank Drs D. Abel and R. Coventry of the CSIRO Davies Laboratory for many helpful suggestions and Ms Penny and Ms Waite for help with library work and Mrs W. Strauch for providing various typed versions of the manuscript.

12

Research in the medical sciences:
The road to national independence

F.C. Courtice

One of the outstanding achievements in the history of science in Australia has been the success of the biomedical scientists since the Second World War. The fact that a country, geographically so isolated from the hub of the scientific world, produced three Nobel Laureates for Physiology or Medicine in such a short space of time, from 1945 to 1963, has evoked in all Australians a feeling of national pride. In the eyes of the world these awards, which are given 'to those who, during the preceding year, have conferred the greatest benefit on mankind', point to Australia as an environment favourable to original research in the medical sciences. Such an environment was not created overnight; it gradually evolved over the course of almost a century as a result of the efforts of many men and women whose ultimate goal was the establishment of our national independence in this field of human endeavour. It is fitting, in the bicentenary of British settlement in Australia, that we should reflect on the part played by, at least, a few of these dedicated individuals.

The beginnings

When the men of the First Fleet built a hospital on the western shore of Sydney Cove 200 years ago, the treatment of the sick was not based on a scientific understanding of the diseases they suffered.[1] At this time systematic research in the medical sciences was unknown, but Europe was soon to see it emerge. By the time the first medical school was established in the Australian colonies, in Melbourne in 1862, radical changes were being made in medical education in Britain, changes that were stimulated by the growth of the new scientific movement in France and Germany.[2] In these countries clinical medicine in the early nineteenth century was deriving great benefit from scientific discoveries, with the result that medical science was rapidly replacing medical empiricism. Britain responded by introducing the laboratory-based sciences of anatomy, physiology and pathology into medical education, necessitating not only the appointment of professors to teach these disciplines but also the provision of laboratories in which the professors could do their work. These changes inevitably led to a more organized approach to research in the medical sciences and to what has been called 'the university spirit'. This implied that university teachers should not only pass on existing knowledge but also create new knowledge by engaging in original research.

It was in the light of these changes in Britain that the first chair — of anatomy, physiology and pathology — was established in Melbourne's new medical school.[3] The first incumbent of this chair was George Britton Halford who is a central figure in another chapter in this book.[4] The second medical school in the Australian colonies was established in the University of Sydney in 1883 when an Edinburgh graduate, Dr Anderson Stuart, was appointed to the chair of anatomy and physiology.[5] It was in this school that, in the 1890s, a group of enthusiastic young men, fired not only with the university spirit but also with a spirit of adventure, were brought together. In 1891 Anderson Stuart appointed Charles James Martin, then aged 25 and a London graduate in medicine, to succeed Dr Almroth Wright as demonstrator in physiology. At about the same time Dr J.T. Wilson, who had come to Sydney from Edinburgh in 1887 to be a demonstrator in anatomy, was appointed to the newly created chair of anatomy. In 1892 a young zoologist not yet graduated, J.P. Hill, arrived in Sydney from Britain as demonstrator in the Department of Zoology and Comparative Anatomy. This group was joined in 1894 by one of Wilson's most brilliant students, the Australian-born Grafton Elliot Smith.

C.J. Martin, J.P. Hill and J.T. Wilson in their laboratory in the Sydney Medical School in the early 1890s. (Dept of Medical Illustration, Univ. of NSW. Courtesy of Ann Macintosh)

Of these four young men, probably the most outstanding was Martin. He was one of those scientists with rare force of mind who triumph over all difficulties — his early years in London testify to this. Born in Hackney in 1866, the youngest of a large family, at the age of fifteen he became a junior clerk in an insurance company where his father was an actuary. His leanings towards science and his desire to become a doctor drove him, against the wishes of his family, to study in the evenings for his matriculation at London University. Later, as a medical student at St Thomas's Hospital, he had to take some of his pre-clinical studies at Guy's Hospital where he became friendly with a fellow student, Ernest Starling, and a young analytical chemist working in the forensic laboratory, Gowland Hopkins. Martin urged Hopkins to study medicine and devote himself to research in physiological chemistry. There began a life-long friendship of three of Britain's most distinguished medical scientists, who were destined to play leading roles in setting the pattern for research in the medical sciences in the twentieth century. After graduating in science (BSc) with honours in physiology and the university gold medal, Martin went to Leipzig to study with the renowned German physiologist, Carl Ludwig, who was then in his seventieth year. Ludwig soon became Martin's physiological hero. Returning to London, Martin became a demonstrator in biology and physiology at King's College, for he had to earn his living, while at the same time continuing his medical education at St Thomas's Hospital where he gained his bachelor of medicine in 1890.[6] Martin was promoted to senior demonstrator early in 1891 and could look forward to a career in physiology in England. It was not unusual at this time, however, for able young men to accept positions in British universities overseas where the salaries offered were considerably higher than in the universities at home. Martin took advantage of this when he accepted an invitation to replace Almroth Wright as demonstrator in physiology in Sydney; the higher salary enabled him to marry before he departed from London.

Martin's research on snake venoms

With this background it could be expected that the Australian environment in the 1890s would excite Martin's zeal to pursue experimental physiology and that no shortage of funds or materials would daunt him. Scientific apparatus at that time was simple, and medical scientists depended mainly on improvisation, of which Martin was a master. He was quick to take advantage of the somewhat unique Australian environment, showing much wisdom in his decision to study the venoms of two of Australia's most poisonous snakes, the black snake and the tiger snake; snake-bite was prevalent in Australia at the time. Even somewhat later, in the early part of the present century, I well remember that as a small boy the property on which I lived in Queensland was infested with snakes; while attending a small bush school, I took with me each day a snake-bite kit consisting of a ligature and a small metal tube with a lance at one end and a quantity of Condy's crystal at the other. Our teacher drilled us from time to time in the use of the kit; in the event of snake-bite, the ligature was to be applied immediately to stop the blood supply to the affected area, the bite incised with the lance and the Condy's crystal rubbed in to neutralize the poison. Such were the fears of the early settlers, especially those living in country areas. In later years I often wondered how I would have coped had I been bitten walking home from school on a lonely road along which only one or two sulkies would pass each week.

In the 1890s very little was known about the chemistry of the venoms of the Australian snakes, although investigations had pointed to the protein nature of the venoms of the rattle-snake in America and of the cobra in India. Martin embarked on a complete investigation of the venom of the Australian black snake by posing three questions: what is the poison? what is its exact physiological action? and how can this action be best prevented or counteracted? He thought that the production of an antivenene for the treatment of snake-bite should be based on a scientific understanding of the poisonous constituent or constituents in the venom.

The first problem that confronted Martin in this investigation, however, was the collection of the venom. For this he devised an ingenious method by which he was able to obtain pure venom direct from the fang and free from other mouth secretions that are invariably produced when a snake bites. Even though he was able to collect pure venom, he found that in different samples the amount of solids varied enormously, from 12 to 67 per cent. This meant that, in order to compare the toxicity of different venoms, it was necessary first to dry the venom and then reconstitute the powder in known strengths in water or salt solution. Martin's early interest in chemistry, when Hopkins encouraged him to 'potter' in his forensic chemistry laboratory in London, proved of considerable value when he attempted to separate the various constituents of the venom. By using the methods then available for separating proteins in solution — heat, alcohol and various neutral salts — Martin and Smith[7] were able to show that black snake venom contained at least four such substances — small quantities of albumin and of deuteroalbumose and large quantities of heteroalbumose and protoalbumose. Only the heteroalbumose and the protoalbumose were toxic when tested for local or general effects. They postulated that the cells of the venom gland of the snake have a hydrating effect on the albumin supplied to them by the blood, whereby toxic albumoses are formed.

Martin had presented this extensive paper to the Royal Society of New South Wales on 3 August 1892, only a little over a year after his arrival in Australia. In further experiments on the toxicity of the venom he found that, when introduced directly into the circulation by intravenous injection, death was due to clotting of the blood or thrombosis. On the other hand, when the venom was administered beneath the layers of the skin by subcutaneous injection, death was caused by respiratory paralysis. This suggested that the venom contained at least two toxins. The toxin that produced death by thrombosis, the heteroalbumose fraction, could be coagulated by heating whereas the neurotoxin, the protoalbumose fraction, could not.

In order to study these two toxins more thoroughly Martin sought a method whereby he could quickly and simply separate them from the venom. He devised an apparatus in which the venom was subjected to filtration at high pressures of 40 to 50 atmospheres, through a gelatin filter, which he made by filling the pores of a Pasteur–Chamberland filter with gelatin or gelatinous silica.[8] He found that the protein that was precipitated by heat and caused death by thrombosis did not pass through the filter, whereas the neurotoxin component that caused death by paralysis was diffusible.[9] This explained why death was caused by paralysis when venom was administered subcutaneously, the neurotoxin passing more rapidly into the blood stream than the non-diffusible component causing thrombosis.

Having answered the first two questions he originally posed, Martin set his mind to answering the third. Here he came into conflict with Dr Albert Calmette of the Pasteur Institute in Paris. Calmette (now more widely known for his development of a strain of the tubercle bacillus, bacille Calmette Guérin, which he used as a vaccine, BCG against tuberculosis) worked in the Pasteur Institute in Saigon in 1891-93 and had produced an antitoxic serum to cobra venom. He maintained that the various constituents in the venoms of different snakes that produced similar effects were identical; he further maintained that his serum was equally valuable against the venom of all snakes. Martin contested these views.[10] Whereas cobra venom contained toxins similar in their effects to those of Australian snakes, the proportions of each toxin in the venoms varied widely. He showed that Calmette's antiserum, while being potent against cobra venom, was only feebly potent against tiger-snake and black-snake venom and then only when given intravenously either before or immediately after the administration of the venom.

At this time little was known of the chemical nature of antitoxins. Two views prevailed concerning the mechanism whereby these substances antagonized toxins. Calmette adhered to the view that the toxin–antitoxin reaction was an indirect one, the antitoxin operating in some unknown way through the medium of the cells of the organism.[11] Martin and Thomas Cherry, on the other hand, supported the view that the antagonism was of a chemical nature, much as an alkali neutralizes an acid.[12] Using antitoxins to diphtheria toxin and tiger-snake venom they showed that the toxin–antitoxin reaction behaved as a chemical one. This was confirmed by showing that an antiserum to tiger-snake venom, a potent antidote to that venom, was not effective as an antidote to the venoms of other Australian snakes. Martin's researches on snake venoms were of considerable practical significance in showing that an antivenene should be administered intravenously as soon as possible after the bite, and that a single antivenene would not be effective against the bites of all of Australia's venomous snakes.

Monotremes and marsupials:
research on Australia's unique indigenous fauna

Martin's other major research interest concerned the body's heat-regulating mechanisms, an interest he maintained throughout his career. This again was, and still is, an important problem in a country with such great differences in climate. He decided once more to take advantage of the Australian environment by studying the mechanisms used by a unique group of animals indigenous to Australia, the monotremes, *Echidna* or spiny ant-eater and *Ornithorhynchus* or duck-billed platypus. These animals are warm-blooded, lay eggs and nourish their young by milk secreted by special glands situated within a temporary pouch. They had been the subject of much discussion by zoologists concerning their position in the animal kingdom, but Martin was interested to find out whether their reaction to a change in environmental temperature would give any clues to the mechanism of homeothermism (maintenance of a constant body temperature) in higher mammals, and in particular in man. Here was a wonderful opportunity to study heat regulation in animals that in some ways resembled the cold-blooded reptiles and in others the warm-blooded mammals. In his experiments Martin again demonstrated his skill in managing wild animals, his mastery in improvising apparatus, and the orderliness of his mind in tackling problems and presenting the results. He found that in its heat regulation *Echidna*, which possesses no sweat glands and hibernates in winter, is the lowest in the scale of warm-blooded animals. Its reaction to a rise in environmental temperature is a lowering of heat production, but at temperatures above 35–37°C the limit of this compensating mechanism is reached. When its own body temperature, normally about 28°C, has reached 38°C death ensues. *Ornithorhynchus*, on the other hand, possesses abundant sweat glands and makes a better attempt at homeothermism by being able to modify to some extent heat loss as well as heat production. In their heat regulation, Martin found that the monotremes were half-way between the cold-blooded reptiles such as the lizard and the warm-blooded mammals such as the rabbit and cat.[13] He was able to conclude from his experiments that on moving up the scale from the cold-blooded reptiles homeothermic adjustment was achieved by a capacity to regulate heat production, but that in higher mammals control depended to an increasing extent on heat loss from the skin, especially by evaporation of sweat.[14] These basic concepts are of considerable importance in understanding man's adjustment to changes from temperate to tropical climates.

Although a member of Anderson Stuart's department of physiology, Martin preferred to work largely in Wilson's department of anatomy, which he found more congenial. Wilson was born in Scotland where his father was a school teacher. As a medical student at Edinburgh University he was a classmate and close friend of J.S. Haldane, who was to become one of Britain's most distinguished physiologists of the early twentieth century.[15] On his arrival in Sydney the young zoologist, J.P. Hill, was also attracted to Wilson. He, too, came from Scotland where his father was a breeder of prize cattle in the highlands. He studied science at Edinburgh University, where he was most influenced by his lecturer in comparative embryology and vertebrate morphology. Before he finished his course, however, he was recommended by his professor to fill the position of demonstrator in zoology in Professor Haswell's department in the University of Sydney.[16] Arriving in Sydney in 1892 as a young, unmarried man of 19, Hill lived for several years in the household of Professor Wilson's parents where Wilson, himself, also lived with his baby daughter after his wife's death in childbirth.

Being interested in embryology, Hill was naturally excited by the opportunity to study the development of Australia's somewhat unique monotremes and marsupials. Martin sought his help in describing a platypus embryo taken from an intra-uterine egg.[17] Martin also collaborated with Wilson on several projects on the comparative anatomy of the platypus, including the anatomy of the snout and the histology of the peculiar rod-like tactile organs in the integument and mucous membrane of the snout.[18] Wilson

Hunting for platypus along a river in the Blue Mountains, New South Wales, in September 1895; left to right: *J.T. Wilson, J.P. Hill, G. Elliot Smith.* (Courtesy of Dept of Medical Illustration, Univ. of NSW)

and Hill began their classical investigations of the development of a marsupial, the bandicoot, *Perameles*, with a study of the teeth.[19] In 1894 Grafton Elliot Smith joined the anatomy department as a demonstrator after graduating in medicine in 1892. He had been born in Grafton in northern New South Wales where his father was headmaster of the local high school. A brilliant student, Elliot Smith soon attracted the attention of Wilson. He planned to be a surgeon, but he also had a craving for original research in neurology. Martin urged him to work on the anatomy of the brains of Australia's monotremes and marsupials, and Wilson soon put him to work dissecting and sectioning the brains of these animals.[20]

This team of scientists, consisting of a physiologist, an anatomist, an embryologist and a neuroanatomist found that Australia was a favourable environment for research in the biomedical sciences. The classes of students were small, which gave them more time for research. Moreover, they were all young; in 1894 when they were all working together Wilson was thirty-three, Martin twenty-eight, Elliot Smith twenty-three and Hill twenty-one. Full of enthusiasm, they were an ideal team to study development, structure and function in some of Australia's fascinating native animals. Those years in the 1890s were an exciting adventure into the secrets of nature for all of them, and one of the most interesting periods in the history of medical research in the Sydney medical school. In writing a biographical memoir of Wilson in 1941,[21] Hill reflected on the conditions under which he worked in Sydney. 'The salary, though not large, was princely compared with the pittance the Edinburgh Demonstrators received in those days.' He went on to praise the medical school building and the facilities for teaching and research 'which were far in advance of any comparable department in the United Kingdom at the time. These were days of intense activity, often prolonged far into the night. In vacations we went on camping expeditions in search of monotremes and marsupials.'

In 1896 the group began to scatter, but all remained close friends throughout their lives. Elliot Smith was the first to leave. He decided to pursue a career in neuroanatomy, so Wilson thought it best for him to go to England for further study. With a Sydney University travelling scholarship he eventually arrived in Cambridge where he was awarded a research studentship and later a Fellowship of St John's College. From Cambridge he went to the chair of anatomy in Cairo, where he developed a deep interest in physical anthropology; from Cairo he returned to England to the chair of anatomy in Manchester and, finally, in 1919, to the chair of anatomy in University College, London. Elliot Smith, elected to the Fellowship of the Royal Society (of London) in 1907, was the first Australian-born to pursue such a distinguished career in the medical sciences, albeit his work after leaving Sydney in 1896 was undertaken entirely overseas.[22]

Martin moved to Melbourne in 1897 as lecturer in physiology; here he continued his research, especially on thermal exchange in the monotremes and on antitoxins to snake venoms. By this time Halford was seventy-three, but he had been appointed to the chair of physiology for life. In 1901, the year in which Martin was elected a Fellow of the Royal Society for the research he had done in Australia, the university appointed him acting professor, implying that he would be appointed to the chair on the death of Halford. Martin had now been in Australia for ten years; naturally he began to feel the effects of his isolation from scientists in England and Europe. He asked for the opportunity to visit Europe during the long vacation every second year to keep in touch with the scientific work of his colleagues overseas. While the council of the university were sympathetic to his request, and would do almost anything to retain him, the turn of the century was far too early for the concept of paid overseas study leave to be acted on.[23] Eventually, Martin returned to England in 1903, to become the foundation director of the Lister Institute in London, a position he held until his retirement in 1930. Here he retained his tremendous interest in Australia. Several Australians went to the Lister Institute to work under his direction and throughout his life he acted on electoral committees for appointments in Australia. So he continued to have a considerable influence on the development of medical science in Australia.

After he retired in December 1930, Martin accepted an invitation from the Council of Scientific and Industrial Research to become director of the Division of Animal Nutrition in Adelaide, following the death of its first director, Professor T. Brailsford Robertson. He immediately entered into the spirit of the laboratory, carrying out investigations into the dietary requirements and nutrition of sheep as well as into the possible use of the virus *Myxomatosis cuniculi* in the control of rabbit plagues.[24] He again returned to England in 1933, to retire in Cambridge. Martin always showed great fondness of Australia and affection for Australians. Of the students he taught in Sydney and Melbourne he said: 'They all meant business and the experimental method of approach came naturally to them.' To honour Martin the National Health and Medical Research Council established in 1951 the C.J. Martin Fellowships to enable young Australians to undertake medical research overseas.

Martin was undoubtedly an exceptional person. The qualities with which he was endowed were suited for pioneer work in the medical sciences in a young country. He believed very passionately that an understanding of the scientific method was essential in the basic education of medical students, for he was fully aware of the impact that science was making on the prevention and treatment of disease. He was determined to promote a system of medical education in which the basic medical sciences were a prominent feature. In this he succeeded, setting an example by his own scientific work and by infecting others with his enthusiasm. When he left Sydney, Wilson and Hill continued their work on the anatomy and embryology of the marsupials and monotremes.[25] In 1906, however, Hill, who had been in Australia for fourteen years, returned to London to fill the Jodrell Chair of Zoology at University College. He was elected to the Fellowship of the Royal Society in 1913 and in 1921 Elliot Smith who was then professor of anatomy at University College invited him to a personal chair of embryology and histology within his department. These two men who were approximately the same age and who had shared their youthful enthusiasm for medical science in Sydney were again working side by side in London; such are the bonds of friendship in the field of experimental science. Wilson, who was elected to the Fellowship of the Royal Society in 1909, remained in Sydney until 1920, when he returned to Britain as professor of anatomy in Cambridge. When Martin retired to live in Cambridge he and Wilson had many occasions to relive their earlier days in Sydney.

The brain drain to Britain

The exciting events of those early years in the Sydney medical school were followed by a long period of relative inactivity as far as research was concerned. This poses several questions. Why, for example, when Martin left Sydney in 1897 did physiological research in the medical school come to a virtual standstill for over half a century? Anderson Stuart as professor of physiology had never given leadership in research and after his retirement in 1920 subsequent professors of physiology also failed in that respect for various reasons,[26] although some members of the staff from time to time engaged in research; but they were not of the calibre of Martin. Another interesting question posed by the events related here is 'Why did all four of these medical scientists, all Fellows of the Royal Society or destined to become so, and all leaders in their particular fields of research, leave Australia to continue their research activities in England?' There is no doubt that, although scientists may for a while enjoy the isolation that the Australian environment provides, ultimately the pull to the centre of the action is very great. Even at the beginning of the century, medical science had become an international pursuit, and leading scientists found that they needed frequent contact with their fellow researchers to maintain their effort. Martin had felt this need at a time when the pool of medical scientists in Australia was very small and the distance from London and other research centres very great. Moreover, Australia was part of the British Empire and, as we shall see, it was made relatively easy for able young scientists to gravitate from the periphery to the centre. When this move proved rewarding, most of the scientists remained at the centre, since the move against this so-called gravitational pull seemed to become increasingly more difficult as the twentieth century advanced.

At the beginning of this century the basic medical sciences in the universities in Britain were growing in strength, but with little or no overall organization of medical research. There was even a 'curtain of aloofness between the interests of those who were working in different major sections of the scientific field'.[27] But this was soon to change. Arising from the National Insurance Act of 1911, the Medical Research Committee was established in 1913 to promote the development of medical and biological research with money provided largely by an annual parliamentary grant-in-aid. The committee decided that a national institute for medical research should be established. In addition, research in various fields would be encouraged by the maintenance of staff, either as individuals or grouped in units, in other institutions such as universities or teaching hospitals, and by the provision of grants to help research workers not in the committee's employ. This policy had been formulated in 1914, but it was not until after the First World War that it could be fully implemented by the Medical Research Council, which, in 1920, replaced the Medical Research Committee.[28] In Britain, therefore, the concept that the promotion of medical research was a responsibility of the state was recognized in the early part of the present century. It had been found necessary to create an organization outside the universities, in which scientists with a special aptitude for research would be supported and encouraged in areas that might lead to an improvement in the health of the community.

Nevertheless, steps were also taken to ensure that research in universities should not be inhibited but should form the basis of the overall organization of medical research since it is in the universities that potential recruits for research are trained. In 1919 the University Grants Committee was established, primarily to distribute government funds to universities. It was recognized that budgets and staffs of universities should be adequate to support research in realization of the concept that higher teaching and research are naturally supportive of each other. Furthermore, it was thought that research should not be an incidental activity, which staff members might indulge in during their free time, but an essential function of all universities, necessitating the provision of public money for these purposes. By having adequate funds at their disposal, universities could maintain that academic freedom that underlies the development of ideas, the motive force in the progress of all knowledge.[29]

There was created in Britain, therefore, an environment in which research in the medical sciences could flourish, an environment that produced in the interval between the two world wars no fewer than six Nobel Laureates for Physiology and Medicine. It was to this environment that many Australians, interested in pursuing research in the medical sciences, gravitated when they saw little opportunity or encouragement for research in the medical sciences in their own country. As British citizens they were free to move to whatever environment provided the best conditions for carrying out their research. They could apply for and accept positions in Britain; in addition, certain British scholarships and fellowships attracted Australian students. Of these, two in particular have been important as far as Australian medical science is concerned. In 1903, under the will of Cecil Rhodes, scholarships were established to enable students from British Commonwealth countries (excluding Britain itself) and from the United States to spend three years studying at Oxford University. Up to the outbreak of the Second World War thirty-eight of the Australian Rhodes Scholars were interested in medicine; of these, eight decided to pursue a career in the medical sciences. All of these eight remained in Britain after the expiry of their scholarships, among them two of our Nobel Laureates, H.W. Florey and J.C. Eccles. Only three eventually returned to Australia. In 1909 the Beit Memorial Fellowships for Medical Research were founded and endowed by Sir Otto Beit as a memorial to his brother Alfred Beit. These fellowships were designed to promote by research the advancement of medicine and of the allied sciences in their relation to medicine. They were open to graduates of any university of the British Commonwealth (including the universities of Britain itself) to work in recognized places of research in Britain. These very competitive awards have since enabled a series of young men and women to devote a few years wholly to research work and by so doing to equip themselves for careers in which research would at least play an important part. Up to the outbreak of the Second World War, 209 fellowships had been awarded. These included awards to twelve Australians, of whom only half eventually returned to Australia, one being our other Nobel Laureate, F.M. Burnet.

The events of the first part of the twentieth century clearly demonstrated Australia's colonial dependence on Britain in the field of research in the medical sciences. But changes were emerging; the most significant being the establishment by private benefaction of institutes in which research in the medical sciences could be pursued as a career.

Establishment of medical research institutes by private benefaction in Australia

The concept that higher education is most effective where research of high quality is being undertaken became widespread in advanced countries in the twentieth century. In his Linacre lecture delivered in Cambridge in 1958, Sir Charles Harington, at the time director of the National Institute for Medical Research in London, aptly described the role that universities had played:

> Medical research has its roots in the universities. It is quite vital to the intellectual life of any country that the universities should retain the place that

they have hitherto occupied in the vanguard of thought, and in science this means not only inspiration in teaching but also activity in investigation.[30]

Nevertheless, Britain had found that it was necessary, while maintaining and encouraging the research activity of the universities, to create a government agency with an overall responsibility for the adequacy of the national effort in medical research, which was so important in maintaining the health of the community.

In Australia, the three medical schools — Melbourne, Sydney and Adelaide (established in 1885)[31] — were all staffed basically for teaching. Although Martin and his colleagues in Sydney had shown what could be done in these circumstances, such gifted men were rare. Those, such as Elliot Smith, who had Martin's ability and enthusiasm for research preferred to expend their creative energies in Britain at the centre of activity. Up to the outbreak of the Second World War the medical schools — a fourth was added in 1936, in the University of Queensland — developed a reputation for their teaching; research was undertaken in somewhat difficult conditions by those individuals inspired to devote their free time to this endeavour. Notable among these was Frederic Wood Jones who came from London to the chair of anatomy in Adelaide in 1919. In 1927 he went to Hawaii but returned to Australia in 1930 as professor of anatomy in the University of Melbourne where he remained for seven years before returning to England to occupy the chair of anatomy in Manchester. A distinguished anatomist, anthropologist and naturalist, Wood Jones was elected to the Fellowship of the Royal Society in 1925. During his fifteen years in Australia, he carried on the early Sydney tradition of research with classical anatomical studies of the marsupials and other Australian mammals; in addition he engaged in and encouraged the scientific study of the Australian Aborigines.[32]

As in Britain, however, the need arose for conditions that would provide opportunities for medical scientists to devote themselves full-time to research. Men of vision who stressed the importance of research in the maintenance of a high standard of health in the community looked to private benefactors for the establishment of institutes of medical research similar to those that had been so successfully developed on the continent of Europe in the nineteenth century. For expediency these institutes began as extensions of the pathology departments of the teaching hospitals. As Charles Martin will always be remembered for his pioneering research in the medical schools of the Universities of Sydney and Melbourne, so Charles Kellaway will be remembered for introducing the tradition of full-time research in the medical sciences in Australia. This he did as director of the Walter and Eliza Hall Institute, which was the first of the institutes founded by private benefaction.

The Walter and Eliza Hall Institute for Medical Research
The Walter and Eliza Hall Trust was established in 1914 at a time when Dr Harry Allen was professor of pathology and Dean of the Faculty of Medicine in the University of Melbourne. One of the trustees felt that the trust should support cancer research, but Allen, wielding considerable power in the medical community in Melbourne, was in a position to guide the trust in the best way their moneys could be used. The Melbourne Hospital was planning a new

Sir John Eccles

Dr C.H. Kellaway

Sir Charles Martin

Lord Florey

Sir Macfarlane Burnet

*Five noted Australian biomedical scientists (*Courtesy of Dept of Medical Illustration, Univ. of NSW; Director, Walter and Eliza Hall Inst; Director, John Curtin School of Med. Res.*)*

pathology department and Allen saw this as a suitable place for research laboratories. In 1915 the Walter and Eliza Hall Institute for Research in Pathology and Medicine was established in the Melbourne Hospital with a governing body that included the vice-chancellor and the professor of pathology of the University of Melbourne. Owing to the war the first director of the Institute, Dr S.W. Patterson, did not begin duties until 1919. It was, however, the second director, Dr C.H. Kellaway, who was responsible for establishing research in the institute.[33]

Kellaway, like Martin, was endowed with exceptional qualities. Born in Melbourne in 1889, he graduated in medicine in the University of Melbourne in 1911. In 1915, as a captain in the Australian Army Medical Corps, he went overseas in the Australian Expeditionary Force, first to Gallipoli and then to Cairo. It was here in the Middle East during the First World War that Kellaway first met and came under the stimulating influence of C.J. Martin, who was serving with the rank of Lieut.-Colonel in the Australian Army Medical Corps as pathological adviser to the Australian Forces in the Mediterranean Area. At that time Martin was forty-nine and an experienced medical scientist, whereas Kellaway was a young medical graduate of twenty-six. Kellaway received from Martin his first training in research methods in the field of pathology and bacteriology appropriate to military conditions of the war in the Middle East, where one of the main concerns was the prevention and treatment of dysentery.[34] While later serving in France, in 1917, Kellaway was severely gassed with phosgene and was declared unfit for further active service. Accordingly, early in 1918 he went to London to act in an administrative position for the Australian Flying Corps. It was here that he met Dr H.H. (later Sir Henry) Dale when the newly established National Institute for Medical Research was housed temporarily in the Lister Institute. There began a life-long friendship, which was to have a considerable influence on the development of medical research in Australia.

Kellaway returned to Australia in 1919, but in 1920 he again went to England, sponsored by Martin and Dale as a Royal Society Foulerton Research Student, to work in the National Institute's new laboratories in Hampstead where Dale was head of the department of biochemistry and pharmacology. A few years before the outbreak of the war Dale had found that the amine, β-iminazolylethylamine (later known as histamine) produced a condition of peripheral circulatory failure.[35] This shock-like state closely resembled that observed in severely wounded soldiers during the war. In 1919, Dale and his colleagues were reinvestigating the possibility that histamine, which is released in injured tissues, might be the cause of wound shock. It was natural, therefore, for Kellaway to work in this field and he applied himself to the possible role of histamine in the production of anaphylactic shock. Kellaway later worked with Professor T.R. Elliott, head of the medical unit in University College Hospital, on adrenal antagonism to the shock-like effects of histamine. The organization of medical research in London — in a private institute (Lister Institute), in the National Institute for Medical Research and in a medical unit in a teaching hospital — was, therefore, very familiar to Kellaway when he once more returned to Australia in 1923, at the age of thirty-four, as the second director of the Walter and Eliza Hall Institute. In his new venture, he had the support and blessing of one of the most eminent, and one who was to become the most influential, medical scientist in Britain during the period between the two world wars, Sir Henry Dale. At this time the Walter and Eliza Hall Institute was primarily responsible for the routine pathological services of the Melbourne Hospital; Kellaway set about the task of adding research as one of its main functions. He decided to build up a structure similar to that in the National Institute in London, with departments of physiology, biochemistry and bacteriology, each staffed with scientists whose primary responsibility would be basic research.

For the first five years the going was not easy. Without any Commonwealth government funds for research Kellaway had to rely on private donations to finance any development he wished to make. Moreover, he had not developed any important field of research of his own. Then in 1928 his fortunes changed. It was in this year that one of his earlier colleagues, Dr N.H. Fairley, returned to the institute after working in India, where he had gained experience of the value of specific polyvalent antivenene in the treatment of cobra and Russell viper's bites. In Australia, records showed that deaths from snake bites averaged about fourteen per year from 1910 to 1926 and were not decreasing. The number of individuals bitten each year was, however, considerably greater than this since only a few of the seventy or so species of snake classified as venomous can cause death in man. These are the death adder, the tiger snake, the copper-head, the brown snake and the black snake. Moreover, the mortality rate in persons bitten by these snakes varied greatly from 50 per cent for the death adder, 40 per cent for the tiger snake, 9 per cent for the brown snake and only 1 per cent for the black snake. Kellaway and Fairley decided to undertake a wide investigation of the venoms of these snakes, Kellaway to investigate the pharmacological actions of the venoms and Fairley to organize their collection and study the venom yields, the toxicity of the venoms, the mechanism of the bite and the efficacy of forms of local treatment. These investigations would continue where Martin and others had earlier left off.

Fairley experimented on sheep, showing that, compared with cobra venom, tiger-snake venom was twenty-five times, death-adder venom ten times, copper-head venom two-and-a-half times and black-snake venom one-third as toxic. The amount of venom injected when a snake bites was also shown to vary widely in different species, depending not only on the amount of venom in the venom gland but more especially on the effectiveness of the bite. From the standpoint of inoculation and biting efficiency Fairley found that the death adder easily ranks first, followed by the tiger snake, black snake, copper-head with the brown snake last. Although tiger-snake venom was two-and-a-half times more lethal for sheep than death-adder venom, and more potent that the venom of any known snake, the tiger snake injected much less venom than the death adder. When, therefore, both the toxicity of the venom and the amount injected on biting were taken into consideration, Fairley concluded that the death adder ranked as one of the most deadly of all known terrestrial snakes. His findings also indicated that if snakes were capable of inoculating the whole contents of their venom glands into their victims, recovery would be rare. The clinical problem of snake bite must, therefore, be related to the quantity of venom injected into the tissues. Since this is always unknown in cases of snake-bite in man, the efficacy of any form of therapy is difficult to assess.

Fairley again turned to experiments in sheep to study the effectiveness of local treatment. He found that if the snake injects several lethal doses into the subcutaneous tissues, as is frequently the case with death-adder and tiger-snake bites, a ligature has to be applied to stop completely the blood flow in less than two minutes for it to have any beneficial effect. Even so, it was found that the most that could be expected from ligature alone was a prolongation of the death time. When incision or scarification of the bite was combined with ligature, the standard treatment advocated at the time, the results were no better; nor did the packing of the incision with Condy's crystal improve the result. The most effective local treatment was found to be immediate ligature followed as soon as possible by complete excision of the affected area. In the case of finger or toe bites excision could be effectively accomplished by amputation. The importance of ligature, which delays the rate of absorption of the venom, is that it allows time for the effective treatment by excision or by the intravenous injection of an antivenene.[36]

In a long series of experiments Kellaway and his several colleagues were able to recognize in the various venoms several toxic components. In summary, these were: a neurotoxin leading to asphyxia by paralysis of the respiratory muscles: a component causing circulatory failure: a thrombin causing coagulation of the blood, and an anticoagulant, which inhibited coagulation: a haemolysin, which destroyed red blood cells: a haemorrhagin, which by its cytolytic action (dissolution of cells) on the endothelial lining of small blood vessels caused haemorrhagic lesions in many tissues: and a proteolytic substance. The distribution of these different components varied greatly in the several venoms, which made it difficult for the Commonwealth Serum Laboratories, with whom Kellaway collaborated, to make a single antivenene to protect against snake-bite. Two other aspects of snake venom interested Kellaway. The first was the possible use in human therapy of some of the very pharmacologically active constituents. The coagulation component was used successfully to stop bleeding, for example after teeth extraction in haemophiliacs, but in more recent times this has been superseded by a factor derived from blood serum proteins. The second interesting aspect of the venoms was that snakes often bite themselves or other not closely related snakes with little or no ill-effect. The possible mechanisms of protection intrigued Kellaway. He summarized his work on snake venoms in the Dohme memorial lectures which he delivered at Johns Hopkins Hospital in 1936.[37]

The close similarity of some of the effects of venom to those of acute anaphylaxis, which Kellaway had earlier studied with Dale, led to an investigation of the possibility that this might be due to the liberation of histamine by the cytolytic action of the venom. Dr W. Feldberg, an expert in the latest techniques of histamine detection, who was working in Dale's laboratory in London, came to Melbourne in 1937 to study this problem. He was able to show that the venoms could cause the liberation of sufficient histamine to explain some of their actions.[38] Later, Kellaway undertook a more general study of the action of histamine and of other natural cell constituents in response to various physical and chemical injuries.[39]

The other event in 1928 that was to affect medical research in the Institute, and eventually in Australia, was a medical tragedy that occurred incidentally in the early stages of a campaign of active immunization against diphtheria in the town of Bundaberg in Queensland.[40] Using a toxin–antitoxin mixture that did not contain an antiseptic, provided by the Commonwealth Serum Laboratories in a 10 millilitre rubber-stoppered bottle for multiple inoculations, one of the local medical practitioners, who was also the medical officer for health in Bundaberg, inoculated groups of children on 17, 20, 21 and 24 January with material from this bottle without any untoward effects. On 27 January he inoculated with material from the same bottle, a further twenty-one children. Of these, eleven died on 28 and one on 29 January while six others suffered severe symptoms from which they ultimately recovered. Kellaway was appointed chairman, with Drs Peter MacCallum and A.H. Tebbutt as members, of a royal commission of inquiry to investigate the cause of the tragedy. Their report to the Governor-General was presented on 11 June 1928 after the conclusion of very extensive scientific investigations. The findings showed that the toxin–antitoxin mixture had been contaminated, probably on 24 January, by a staphylococcus that had multiplied in the next three days, during which time the bottle was stored in a cupboard at the prevailing sub-tropical summer temperatures.

As a high school student in my home town of Bundaberg, I knew most of the children who died and so could understand the extent of the human tragedy that had so suddenly occurred, while not understanding its scientific basis. The loss of twelve young children, in one instance an entire family of three, was mainly for the local community to bear, but the medical implications were to extend not only throughout the Commonwealth of Australia but to the entire outside world. Campaigns for immunization against diphtheria were about to be launched in many countries, which were naturally interested in the findings and recommendations of the royal commission. One of these recommendations was: 'That the Commonwealth Department of Health should make full and careful enquiry as to whether it is advisable to substitute an atoxin or some similarly modified immunizing agent for toxin–antitoxin.' The newly established (in 1926) Commonwealth Department of Health as well as the community at large were deeply impressed with the efficient and painstaking way that the commission, under Kellaway's chairmanship, had carried out the inquiry. The status and the national importance of medical research and of the Walter and Eliza Hall Institute were raised in the eyes of the public and of the Commonwealth government. A very practical effect of the commission was that Kellaway was given a grant by the Commonwealth Department of Health without which he would not have been able to pursue his investigations on snake venoms. This grant was the fore-runner of a more permanent and more far-reaching scheme for the provision of public funds for medical research, a scheme that was established less than a decade later.

Kellaway's qualities of leadership enabled him to gather a group of young scientists, all born in Australia, who were destined to pursue distinguished careers in the medical sciences. The first of these was Neil Fairley,[41] who worked with Kellaway on hydatid disease and on snake venoms. Fairley was only two years younger than Kellaway and after graduating from the Melbourne medical school in 1915, went to the Middle East in 1916 where he served as pathologist and later as senior physician to the 11th Australian General Hospital. It was here that he, too, met and worked with C.J. Martin; after the war, in 1919, he went from Cairo to London where he worked for some months in the Lister Institute. In Egypt Fairley had used a complement fixation test for the diagnosis of the parasitic bilharzial infection and when he returned to Australia in 1920 as first assistant to the director (Dr Patterson) of the Walter and Eliza Hall Institute he conceived the possibility of using a similar test for the diagnosis of hydatid disease, a common disorder in Australia. Desirous of pursuing a career in tropical diseases, however, he went to Bombay in 1922, where he contracted sprue, a tropical disease he was studying. Returning to Australia he rejoined the Walter and Eliza Hall Institute in 1927, where he worked with Kellaway before going to London in 1929 to work in the London School of Hygiene and Tropical Medicine. Here he remained except during the Second World War when he served with the Australian Army Medical Services in the Middle East and in the Pacific theatre of the war, where he established and directed the Malaria Research Unit to carry out classic experiments on the pathogenesis and chemotherapy of malaria.

The other two members of Kellaway's group at the Hall Institute in the 1920s were G.R. Cameron and F.M. Burnet, both of whom graduated in the University of Melbourne in 1922. Cameron's forebears came from the highlands of Scotland, but he was born in the Victorian country town of Echuca.[42] Like Kellaway, who had a great influence on his career, he was a son of a minister of the church. Soon after graduating in medicine, Cameron decided to become a pathologist for he was a devotee of the professor of pathology, Sir Harry Allen in whose department he was appointed Stewart Lecturer in Pathology. In 1925 Kellaway asked him to be his first assistant and deputy director of the Walter and Eliza Hall Institute. To further his studies in pathology, however, Cameron left Australia in 1927 to work with Aschoff in Freiburg after which, in 1928, he went to University College Hospital Medical School in London to work with Professor A.E. Boycot. As a Graham Scholar in pathology and later a Beit Fellow for medical research, his main interest was in diseases of the liver. He remained at UCH Medical School, succeeding Boycot in the chair of morbid anatomy in 1937. It was mainly through Cameron that I first met Kellaway. Cameron and I were working in the Chemical Defence Experimental Station at Porton on Salisbury Plain during the Second World War when Kellaway, in his capacity of Chairman of the Chemical Defence Board in Australia, visited the station. We both learned a good deal about the progress of medical research in our native land which we had left many years earlier. Kellaway impressed me with his friendly, easy, out-going manner. After the war when he was working in London he was most helpful before I returned to Australia in 1948. In London he naturally remained interested in medical research in Australia; from time to time I wrote to him of my experiences at the Kanematsu Memorial Institute and he invariably replied with valuable and encouraging words of wisdom.

The fourth of the group, F.M. Burnet, was destined to become the most distinguished scientist of them all. His forebears also came from Scotland, but he was born in the small Victorian country town of Traralgon where his father was manager of the local branch of the Colonial Bank. After graduating in medicine he spent three years as an intern at the Melbourne Hospital. Much of this time was spent as a pathologist in the Walter and Eliza Hall Institute where he began a classical series of studies on bacteriophages. In 1925–27, supported by a Beit Memorial Fellowship, he continued these studies in the Lister Institute in London, where C.J. Martin was director. On his return to the Hall Institute early in 1928, his immediate research was influenced by the outcome of the Bundaberg tragedy. For the next few years he worked on staphylococcal toxin. In 1931, however, in the depth of the Depression when the finances of the institute were severely strained, Kellaway arranged with Dale for Burnet to work on animal viruses for two years in Dr P.P. Laidlaw's department at the National Institute for Medical Research. At that time relatively little was known about the somewhat mysterious ultramicroscopic infective agents called viruses. The National Institute had recently isolated the virus of canine distemper and in 1931, with the arrival of an influenza epidemic, was embarking on a programme of isolating and describing the influenza virus.[43]

When he joined the group at the National Institute in London in 1932, Burnet participated in what he later called 'the first whole-hearted attempt to define the nature of viruses and to develop means for their study and control'.[44] It was while studying a virus that affected canaries that Burnet decided to grow it on the membrane of the chick embryo, a technique that had recently been introduced into viral culture. The experiment was successful and he went on to spend two exciting and productive years in London before returning to the Walter and Eliza Hall Institute, where he was able to use the chick-embryo technique for growing the influenza virus. This was the beginning of his distinguished career in the investigation of infectious diseases, especially those caused by viruses, a field that had much public appeal. Techniques to isolate and culture a virus were essential, but Burnet's main interest was in the natural history of a virus and of the disease it caused. He was able to show that influenza viruses from the throat of a patient could be isolated by inoculating the amniotic cavity of the chick embryo, a technique that became the standard method for primary isolation of the influenza virus. The aim of isolating and growing influenza virus was to produce a vaccine that would protect against influenza. At the beginning of the Second World War, Burnet's concern was that another pandemic would occur similar to that of 1918–1919 in Europe, which caused more deaths than the fighting of the 1914–1918 war: 'I aimed to produce a live virus vaccine which would produce a harmless infection of the respiratory tract, without symptoms but producing enough antibody to protect against real flu of the same immunological type'. His vaccine, of human influenza virus strains attenuated by repeated passages in chick embryos, was tested on student and army volunteers with some, but not outstanding, success. During this phase of his career Burnet also made considerable contributions to studies of such disorders as psittacosis, Q-fever, poliomyelitis and simple herpes.[45]

All four of this group of medical scientists were elected to the Fellowship of the Royal Society during the 1940s, Kellaway in 1940, Fairley and Burnet in 1942 and Cameron in 1946. Part of Kellaway's success at the Hall Institute was his leadership of men, which was undoubtedly enhanced by his friendship with two of Britain's most distinguished and influential medical scientists, Martin and Dale. This link between London and the Walter and Eliza Hall Institute was a very close one and of the greatest importance to Kellaway, especially in those early stages when he was trying to establish research in the institute. Dale later described Kellaway's directorship of the Walter and Eliza Hall Institute from 1923 to 1944: 'During that period he had the satisfaction of seeing the Institute, under his stimulating and generous leadership, become the leading Australian centre, and one of high rank among those in all the world, for researches in Experimental Pathology and in the related field of medical science'.[46]

Whereas Dale was able to develop the National Institute in London with the security of government funds, Kellaway was at a considerable disadvantage when he had to depend on private benefactions. The Australian government was slow to recognize the importance of medical research in this country's rise to nationhood, so there was little organization of medical research at a national level. Kellaway welcomed the beginnings of such an organization when in 1936 the Federal Health Council, which had met annually since 1926, was reconstituted as the National Health and Medical Research Council.[47] One of the functions of the new council was to advise the Commonwealth government on the expenditure of money on medical research and on the projects of medical research generally. At its first meeting in 1937 the council stressed the importance of initiating a national research scheme that would encourage young medical graduates to take up medical research as a career and one that would ensure continuity and permanence. From the public moneys at its disposal, the council was able to provide grants to individual research workers in universities and institutes. This was the first step toward the recognition at a national level that it was necessary to provide public money to aid medical research, since private donations were often irregular and insufficient. Kellaway reflected at this time in his Stawell Oration:

> In a young country there is little opportunity for the development of scientific research. The pioneers are fully occupied in the struggle for existence, and many years must elapse before the conditions of life afford to suitable persons the necessary leisure and detachment to set about the business of adding to the world's store of knowledge. A few gifted individuals may in their leisure hours make fresh discoveries, but until the conditions are created for setting aside persons, with special gifts, for whole-time research advance must be slow and difficult.[48]

He went on to say that the establishment of the National Health and Medical Research Council would probably prove to be the most important single step forward in the history of medical research in Australia.

Other medical research institutes

While the Walter and Eliza Hall Institute was becoming firmly established in research, other institutes were also emerging. The Alfred Hospital, the second teaching hospital of the Melbourne medical school, established a clinical biochemistry department in 1923 and in 1924 Dr A.B. Corkill was appointed biochemist. In 1926, however, Thomas Baker endowed the Baker Institute into which the biochemistry department was absorbed. Dr W.J. Penfold became the first director in 1927.[49] Much of the work of the institute up to the outbreak of the Second World War related to the provision of clinical pathological services for the hospital. Research was mainly in biochemistry centering on carbohydrate metabolism, which was at that time a very active field since insulin had only recently been introduced for the treatment of diabetes. Corkill went to England in 1929 and again in the early thirties to work at the National Institute for Medical Research on carbohydrate metabolism in muscle and liver. By the time he became director in 1938 he had gathered a group of researchers to work on the effect of various extracts of the anterior pituitary gland, especially a diabetogenic substance, on carbohydrate metabolism.

In Sydney in the 1930s two institutes were established in which research in the medical sciences was pursued. Like the two institutes in Melbourne, their primary function was to maintain the pathological services of the hospitals to which they were attached. One of these institutes was attached to Sydney Hospital, Australia's oldest hospital. Ever since he became pathologist to Sydney Hospital in 1924, Dr W.K. Inglis had aimed to establish such an institute. With a benefaction from the Japanese firm, Kanematsu (Australia) Ltd, Inglis became the chief architect for the establishment of the Kanematsu Memorial Institute of Pathology in 1933.[50] In his determination to establish research in the medical sciences as one of the main functions of the Kanematsu Institute, he called on Kellaway's experience. Inglis's perseverance together with Kellaway's wise counsel resulted in the appointment in 1937 of Dr J.C. Eccles as director. Within a few months of arriving in Sydney Eccles had established laboratories for research in neurophysiology.

Eccles was a graduate of the Melbourne medical school. As Rhodes Scholar for Victoria he went to Oxford in 1925 to work with Sir Charles Sherrington, one of Britain's leading neurophysiologists. He remained in Sherrington's laboratory until his return to Australia in 1937. Many years later he wrote:

> England in my time there (1925–1937) was a delightful and stimulating place for a young academic, although by present standards the laboratory facilities were primitive. . . . we physiologists of Great Britain were united by our membership in the Physiological Society, which, I think, was then in one of its great periods.[51]

Eccles was to find Sydney a very different place: 'a lovely place to live, but the academic isolation was severe. The Sydney University Medical School was a very dim place, being little more than a teaching institution.'

Eccles brought to Sydney his twelve years experience in neurophysiology together with his apparatus and his technician to work on the mechanism of neuromuscular transmission, a very active but controversial field. Two hypotheses had been put forward to account for the passage of an impulse from a nerve to a muscle at the neuromuscular junction, the chemical and the electrical. The chemical hypothesis postulated that a chemical substance, liberated at the nerve ending was responsible for transmission of the nerve impulse to the muscle fibre whereas the electrical hypothesis assumed that the chemicophysical disturbance constituting the nerve impulse passes on to the muscle. Eccles leaned to the electrical hypothesis and set out to test it. He was joined by Dr S.W. Kuffler, a graduate of Vienna with no previous research experience and Dr Bernard Katz, a graduate of Leipzig who had worked in London before coming to Sydney — 'the wilderness soon blossomed'. In due course their experiments led them to support the chemical hypothesis.[52] Such a basic problem in neurophysiology did not relate specifically to the Australian environment and so did not enjoy the same public understanding as the problems of snake venoms and of infectious diseases that were being studied at the Hall Institute. As Eccles wrote later: 'There was much criticism of our activities from some clinicians because the research on curarized muscles and anticholinesterases seemed so remote from clinical usefulness.' Yet time has proven that the fruits of this research have been of great value in modern therapy, such as the use of muscle relaxants in anaesthesia and of hypotensive drugs for the treatment of hypertension. It was while working at the Kanematsu Institute, in 1941, that Eccles was elected to the Fellowship of the Royal Society.

While Eccles was investigating neuromuscular transmission in a somewhat isolated environment on the top floor of the Kanematsu Institute, Dr M.R. Lemberg, who like Katz was a refugee from Hitler's Germany, was establishing a small group of biochemists in the biochemistry department of Royal North Shore Hospital's Kolling Institute, not far away. This institute had been established in 1923, primarily for the pathological services of the hospital, and took its name from its first benefactor, Mrs Eva Kolling.[53] Lemberg had left Heidelberg in 1933 to work in Cambridge on the role of cytochromes in the complex pathways of biological oxidations. He would have preferred to remain in Cambridge, but some of the many refugees from Germany had to be distributed to other countries. In 1935 Dr W.W. Ingram, who had been mainly responsible for the establishment of the institute, arranged for Lemberg to accept the position of chief biochemist, his main duty being to engage in research. Lemberg reflected 30 years later:

> So I went into the wilderness, for I did not expect inspiration from my Australian colleagues at that time. For many years I worked indeed in an almost complete vacuum with little response . . . Fortunately I had learned to stand on my own feet and to take inspiration from the literature. This is not, of course, ideal, but Australian science is now (1965) on a vastly different level from what it was 30 years ago.[54]

Lemberg remained at the Kolling Institute until his retirement in 1972, developing an active group working on the biochemistry of the haematins and bile pigments.[55] He was elected to the Fellowship of the Royal Society in 1951 for his work on tetrapyrole chemistry and biochemistry.

Although the four institutes in Melbourne and Sydney were primarily responsible for the routine pathological services of the teaching hospitals to which they were attached, all became active in research in widely differing disciplines of the medical sciences. In Adelaide the Institute of Medical and Veterinary Science was created by Act of the South Australian parliament in 1937, not only to provide diagnostic facilities in clinical pathology but also to act as a laboratory basis for research in the medical sciences. Up to the Second World War, it is evident that research in the medical sciences in Australia was struggling to survive, although fresh hope was derived from the establishment of the National Health and Medical Research Council. The effects of this were just beginning to be felt when war erupted on 1 September 1939.

A national institute for medical research in Australia

Throughout history, wars have heralded radical changes in society. After the six years of the Second World War, from 1939 to 1945, the people, not only in Australia but world-wide, were prepared for changes in their attitude to medical research, indeed, to research in all of the natural sciences. One of the greatest benefits to mankind, the development of penicillin to combat bacterial infections, emerged during the Second World War and it was an Australian, H.W. (later Lord) Florey, who was mainly responsible. Towards the end of the war, in 1945, Florey (jointly with E.B. Chain and Alexander Fleming) was awarded the Nobel Prize for his work on penicillin, the first Australian-born to win the Nobel Prize for Physiology and Medicine. These circumstances helped to create in Australia an environment in which the society of the day through its government was prepared to support medical research by the establishment of a national institute.

Born in South Australia, Howard Florey graduated in medicine in the University of Adelaide in 1921. In the same year he won a Rhodes Scholarship and proceeded to Oxford early in 1922 to embark on a career of medical research. He spent the rest of his life working in laboratories in England. After leaving Oxford he went to Cambridge, London and Sheffield before returning to Oxford in 1935 as professor of pathology. Florey tackled a wide range of problems in experimental physiology and pathology, but his main interest was the body's defence against invading organisms. In the late 1930s, well before the war began, he drew up his plans to work on antimicrobial substances. Penicillin was, therefore, not the outcome of a project embarked on because of the war, but the advent of war in 1939 no doubt had an influence on its exciting course, which is told by Dr E.P. Abraham in his biographical memoir of Florey.[56]

In 1940 Florey and his colleagues published in *The Lancet* their first paper on penicillin as a therapeutic agent.[57] It tells a simple but graphic story of how mice infected with a lethal dose of *Streptococcus pyogenes* were restored to normal health if treated with a penicillin preparation. Abraham describes what must have been one of the most exciting week-ends in the history of medical science, a week-end just one month before the fall of France. He writes:

> Florey began a protection experiment on Saturday 15 May 1940, with eight mice infected intraperitoneally with *Strep. pyogenes*. One hour after injection of the streptococci two mice were given 10 mg of a preparation of penicillin subcutaneously and two were given 5 mg, and those doses were followed by three similar doses at intervals of about two hours. These four mice remained well, while four untreated controls died after about fifteen hours.

This dramatic experiment ushered in the antibiotic era in which treatment of bacterial infections and the pattern of medical practice were to undergo radical changes.

Towards the end of the war, in 1944, Florey was invited to visit Australia to advise the government on the production of penicillin. It was at this time that he expressed an interest in medical research in his native country, encouraged by R.D. Wright, professor of physiology in the University of Melbourne who had worked with him in Oxford in 1936–1938. Florey's concept was that of an organization for medical research similar to that which had been so successful in Britain, namely an Australian Medical Research Council with a central National Institute for Medical Research. In 1945, however, the idea of a national institute was linked to that of a national university in Canberra and there emerged the John Curtin School of Medical Research as one of four research institutes in a post-graduate national university, established by Act of Parliament in 1946.

Florey was an outstanding laboratory scientist who demanded from others high standards of excellence; he did not suffer fools gladly. The permanent laboratories of the John Curtin School of Medical Research were completed in 1957 and at the opening of the school on 27 March, 1958, he stressed the importance of the experimental method in the advancement of medical research in Australia.[58] He said:

> Australia produces a high proportion of people of outstanding intellectual attainments, yet in comparison with the more advanced countries in the world it is not considered pre-eminent, as it is, for example pre-eminent in sporting activities. In short Australia has not, so far, been looked on as a land of opportunity for the intellectual. But things are changing, and today we are celebrating an event that will surely go down as a landmark in the endeavours of Australia to foster experimental science in a world that is at last beginning to realize the potency of the experimental approach to the study of natural phenomena . . . In these days we are witnessing the invasion of all branches of medicine by the experimental outlook . . . The opening of this building is a demonstration of the intentions of Australia to take its share in the winning of new medical knowledge by experimental means.

Florey realized that to achieve this, medical scientists in Australia had to compete in a very competitive international sphere. He also realized that Australian universities and institutes had to compete for scientists, themselves, in this same international sphere:

> This School will be judged entirely by the research work that it carries out. Such judgement is not made only locally, it is international and it is not indulgent. The School will have to contend for a long time to come with the pull on young men to go for post-graduate training to the great centres which exist and are rapidly multiplying in the United States, and which exist in Europe. If it is to attract and hold the best people, it will need not only to be good but it must be superlatively good . . . But how worthwhile it will all be if a growing and vigorous Australia can attract the best Australian scientists to work here, and if a constant traffic of men between different parts of the Commonwealth and foreign countries can be established.

The story of the beginnings of the John Curtin School of Medical Research has been told more fully by Professor Frank Fenner in his Victor Coppleson lecture in 1971 and more recently by Professor Robert Porter.[59] Other events towards the end of and immediately after the war contributed to a change in the medical research scene in Australia. In 1944 Kellaway resigned as director of the Walter and Eliza Hall Institute to take up the position of director-in-chief of the Wellcome Research Laboratories in London. He was succeeded by Dr F.M. (later Sir Macfarlane) Burnet. At about the same time Eccles resigned as director of the Kanematsu Memorial Institute to take the position of professor of physiology in the University of Otago, New Zealand. He was succeeded in 1948 by Dr F.C. Courtice who had been working in England since 1933. In Brisbane the Queensland Institute of Medical Research was established by an Act of the State Parliament in 1945 as the result of a proposal by Dr E.H. Derrick for a laboratory devoted to full-time research into problems of special importance to northern or tropical Australia.[60] Dr Ian Mackerras became its first director in 1947. At the Baker Institute Dr Corkill retired as director in 1949 and was succeeded by Dr T.E. Lowe.

In the first decade after the war these institutes were engaged in research in many different medical fields. The Walter and Eliza Hall Institute at first concentrated its work on viruses, with special emphasis on the influenza virus.[61] Later more emphasis was placed on obtaining a better understanding of immunological processes when Burnet put forward his clonal selection theory, which postulated that the antibody pattern was produced by genetic processes in the relevant cells during embryonic life.[62] A clinical research unit was also added to the institute, in which Dr Ian Wood studied diseases of the stomach, liver and the immune system. In the Baker Institute emphasis was placed on cardiovascular research, especially on the physiology of the heart and the mechanisms of blood coagulation; research in diabetes and in metabolic disorders was also pursued in the clinical research unit. In the Kanematsu Memorial Institute research concerned the role of the lymphatic system in the maintenance of fluid balance in the body especially after injury, of lipoprotein transport in relation to the aetiology of atherosclerosis and in various aspects of the microcirculation. A clinical research unit in which coronary heart disease and renal disorders were studied was also established. In the Kolling Institute investigations into the biochemistry of the haematins continued. The Queensland Institute of Medical Research investigated the zoonoses (scrub typhus, tick typhus, Q-fever and leptospirosis), parasitology (hookworm), entomology (mosquitoes and their role as vectors of disease and the epidemiology of Murray Valley encephalitis). Meanwhile, the John Curtin School of Medical Research was getting under way. The first chair was filled in 1948 with the appointment of Dr A.H. Ennor as professor of biochemistry. Three other professorial appointments soon followed, in microbiology (Dr F.F. Fenner), medical chemistry (Dr A. Albert) and physiology (Professor J.C. Eccles). Research in these fields was pursued at first in laboratories scattered widely from Otago to London, then in 1952 in temporary laboratories in Canberra until the permanent building was completed in 1957, after which further disciplines of experimental pathology (Dr F.C. Courtice), physical biochemistry (Dr A.G. Ogsten) and genetics (Dr D.G. Catcheside) were added.

With the establishment of the Australian Academy of Science in 1954, of the University Grants Committee and a new medical school in Western Australia in 1957, and of several specialist national societies, the 1950s were exciting years for the medical scientists. The effects of the National Health and Medical Research Council and of the more recently founded John Curtin School of Medical Research were beginning to be felt. Young graduates saw that there were now facilities in Australia for their training in research. More importantly, they could see that opportunities for a career in research or in teaching and research were increasing, a vastly different picture from that which existed in the 1930s before the war. The 1960s proved to be even more exciting. Three more medical schools were established — in the University of New South Wales, in Monash and in the University of Tasmania — providing more opportunities for research and teaching. Moreover, the medical schools had become fired with the so-called 'university spirit' or concept that research is essential for the highest standards of teaching. The climax of the post-war effort, however, came with the award of the Nobel Prizes for Physiology and Medicine in 1960 and 1963.

Nobel prizes for Physiology or Medicine, 1960 and 1963

Under the will of Alfred Nobel, a Swedish chemist and engineer who died in 1896, Nobel Prizes have been awarded annually since 1901 and have become recognized as the highest awards for human achievement in the specified fields. In 1960 Sir Macfarlane Burnet, Director of the Walter and Eliza Hall Institute was awarded the Nobel Prize for Physiology or Medicine, jointly with Dr P.E. Medawar of Britain, for the discovery of acquired immunological tolerance. They had shown that under certain conditions the body can be induced to tolerate the transplantation of foreign tissue, which normally it would reject.

Burnet was the first Australian to win this award for work that had been carried out solely in the Australian environment. He had resisted many offers to establish himself overseas. Of a somewhat shy and reserved nature, he preferred to bring up his family in Australia and to work on the periphery rather than in the centre of the world of science. As he has often said:

> I have worked in England and visited America frequently and extensively but I have never been wholly absorbed into either scientific universe. I have, as it were, always retreated to the smaller country where I could be less subject to the two pressures of competition and conformity, and from which I could look at what was happening with less sense of direct involvement.

However, whereas Martin at the turn of the century was really isolated from the main scientific centres, Burnet as a full-time research worker with no teaching duties, was able to visit these centres at intervals, visits that were very frequent after the war when international air travel became established.

Burnet built for himself a very distinguished international reputation as a virologist. His output of important scientific papers in this field was prodigious. To anyone so interested in infectious diseases, the body's defences against these micro-organisms naturally became an important subject for thought and study. Burnet realized that the essential basis of this defence is the ability of the body to recognize the micro-organisms as foreign or 'non-self' and so form antibodies for their destruction. Such antibodies, which were shown to be large protein molecules or gamma-globulins, were not formed to the body's own cells, which were recognized as 'self'. In the 1930s when he was working with influenza virus, he noted that when the virus was introduced into the chick embryo it would multiply but no antibodies would be produced. The embryo would hatch and grow into a chick without any antibody to the influenza virus. Burnet was able to formulate an immunity theory whereby the body develops during its embryonic stage the ability to recognize foreign antigens as 'self'. In 1949 he suggested that it should be possible to produce artificial tolerance to an antigen by inoculating the animal with it during the prenatal or embryonic stage.[63]

Medawar and his colleagues, who had been working in Britain since the beginning of the war on skin grafting, were able to show that embryos of mice of a certain pure strain, given cells of mice of a second pure strain, would after birth accept skin grafts from this second strain and retain them indefinitely without rejection. The mice of the first strain had become immunologically tolerant to tissues from mice of the second strain, as Burnet had predicted. This finding stimulated work on ways of producing tolerance in human beings by suppression of the immune response with suitable drugs, thus opening the way for successful kidney transplants.

Three years later, in 1963, Sir John Eccles was awarded the Nobel Prize for Physiology and Medicine, jointly with A.F. Huxley and A.L. Hodgkin of Britain, for his work on the transmission of nerve impulses along a nerve fibre. Eccles had gone from the Kanematsu Institute in Sydney to the University of Otago in Dunedin in 1944, but he returned to Australia in 1952 as professor of physiology in the John Curtin School of Medical Research. The mechanism whereby a nerve impulse is transmitted from a nerve ending to a muscle across the neuromuscular junction or from a nerve terminal to another neurone across a synapse in the central nervous system was the subject of considerable controversy during the first half of the present century. Of two theories, the work of Sir Henry Dale and his colleagues in London supported the hypothesis that a chemical substance was responsible for the passage of the impulse from nerve ending to effector cell. The second or electrical hypothesis, however, had to be tested; Eccles set out early in his career, while still at Oxford, to do this. His arguments and discussions frequently brought him into friendly, but sometimes heated, conflict with Dale and his supporters.

We have seen that while at the Kanematsu Institute Eccles and his colleagues, Katz and Kuffler, tested the electrical hypothesis concerning transmission across the neuromuscular junction. Eventually, they reached the conclusion that the impulse was transmitted by the chemical transmitter, acetyl choline. Eccles wrote later: 'By 1949 the great weight of the evidence had caused me to repudiate electrical transmission at all the peripheral synapses, but with synapses in the central nervous system the problem was still open. I continued to espouse the electrical theory both for excitation and inhibition.'[64] While in Dunedin, Eccles decided to test the electrical hypothesis regarding transmission across synapses in the central nervous system. To do this he introduced in 1951 a new technique of intracellular recording from motoneurones in the spinal cord.[65] This involved the insertion of a glass micro-electrode, 0.5 to 1 micrometers(μm) external diameter, into a motoneurone of the spinal cord, making it possible to record action potentials of a single motor nerve cell.

Eccles tested the electrical hypothesis and tells of a dramatic experiment that took place in Dunedin in August 1951. In the closing address of the Sir Henry Dale Centennial Symposium in 1975 he said:

> Thus it was a clear test. If the quadriceps volley caused the trace to go up it was electrical, if down it was chemical. It went down ... We were momentarily stunned, as well we might be after a long day's experiment, enhanced by some extraordinary obstetrical complications. The wife of one of my two associates (Jack Coombs) was delivered of a baby girl by the other (Lawrence Brock), I meanwhile tending the experiment. It was by then in the early hours of the morning. But on recovering from the shock (the physiological, not the obstetrical) the decision was made. Inhibitory synaptic action was chemically mediated and it was evident that the mirror image response, excitatory synaptic action, was also chemical.

He concluded his address with a tribute to Sir Henry Dale:

> It was a great privilege to have been so closely associated with him (Dale) in those great creative years, first as a sparring partner in opposition and then as a convert. Such great men are infinitely precious in our lives and in our culture. This centennial occasion gives the opportunity for those who knew Sir Henry Dale personally to hand on our memories to those who have come later.

Unlike Martin, Dale never visited Australia; yet, as has been indicated throughout this chapter, his influence in stimulating the development of the medical sciences in this country has been immense.

When Eccles moved to the newly established John Curtin School of Medical Research in Canberra, he took with him 'four magnificent electrical stimulating and recording units designed by Jack Coombs and built in New Zealand. At that time and for many years to come — in fact until the transistor era — they were the best general research instruments for electrophysiology in the world.'[66] He also took Jack Coombs with him, for there were fresh fields to conquer. As master of the technique of recording from individual cells of the spinal cord, Eccles and his many colleagues in Canberra went on to investigate the patterns of organization that exist in the spinal cord, which he described in the Herter lectures at Johns Hopkins Medical School in 1955,[67] and in his book, *The Physiology of Synapses*,[68] which was published on the eve of the Nobel festivities in Stockholm on 10 December 1963, the day he received his Nobel award.

The environment for research in the medical sciences in Australia in the 1960s had changed dramatically from that which existed before the Second World War, a just reward for those who had struggled for so long to reach their goal of national independence. Australian scientists responded when the stage was set for them to work in their own country under conditions that enabled them to compete in the international sphere of medical research. The first light at the end of the tunnel was seen in the late 1930s when the Commonwealth government established the National Health and Medical Research Council, recognizing the importance of medical research and the necessity of providing public funds for its support. This early ray of hope for the medical scientists was further enhanced with the establishment of the

John Curtin School of Medical Research, a national institute that acted as a catalyst for research in the medical sciences. Despite Florey's concern about attracting young scientists to Australia, this school became an international research centre. As Eccles wrote many years later:

> A remarkable feature of the Australian National University was its international orientation. No preference was made to Australians, and there were unrivalled facilities for overseas scholars . . . Australia gained enormously in two respects. First there was a transformation from the academic isolation of the prewar years that I had experienced on arrival in Sydney in 1937, and many Australian scholars returned to Australia, often after long sojourns overseas. Second, the new generation of young Australians had the great advantage of association in Australia with scholars from overseas. The academic renaissance spread through the whole University structure of Australia.[69]

Australia's medical scientists had reached the end of the long road to national independence.

Notes

1   J.F. Watson, *The History of the Sydney Hospital from 1811 to 1911* (Sydney, 1911).
2   C. Newman, *Evolution of Medical Education in the Nineteenth Century* (London, 1957); F.N.L. Poynter, *Medicine and Science in the 1860s*. Proceedings of the 6th congress on the History of Science, University of Sussex, 6-9 September 1967, Wellcome Institute of the History of Medicine (London, 1967).
3   K.F. Russell, *The Melbourne Medical School 1862–1962* (Melbourne, 1977).
4   B. Butcher, Chapter 7, this volume.
5   J.A. Young, A.J. Sefton and N. Webb, *Centenary Book of the University of Sydney Faculty of Medicine* (Sydney, 1984).
6   H. Chick, 'Charles James Martin', *Biogr. Mem. Fellows Roy. Soc.*, 2 (1956), 173-208.
7   C.J. Martin and J.M. Smith, 'The venom of the Australian black snake (*Pseudechis porphyriacus*)', *Proc. Roy. Soc. NSW*, 26 (1892), 240-64.
8   C.J. Martin, Note on a method of separating colloids from crystalloids by filtration, *J. Roy. Soc. NSW*, 30 (1896), 147-9; idem, 'A rapid method of separating colloids from crystalloids in solution containing both', *J. Physiol. London*, 20 (1896), 364-71.
9   C.J. Martin, 'An explanation of the marked differences in the effects produced by subcutaneous and intravenous injection of the venom of Australian snakes', *J. Roy. Soc. NSW*, 30 (1896), 150-7.
10   C.J. Martin, 'The curative value of Calmette's anti-venomous serum in the treatment of inoculations with the poisons of Australian snakes', *Intercolon. Med. J. Australasia*, 2 (1897), 527-36; idem, 'On the advisability of administering curative serum by intravenous injection', ibid., 537-8.
11   A. Calmette, 'A lecture on the treatment of animals poisoned with Snake venoms by the injection of antivenomous serum', *Lancet*, 2 (1896), 449-50; C.J. Martin, 'The curative value of Calmette's antivenomous serum in the treatment of inoculations with the poisons of Australian snakes: A rejoinder to M. le Dr Calmette', *Brit. med. J.*, 2 (1898), 1805-7.
12   C.J. Martin and T. Cherry, 'The nature of the antagonism between toxins and antitoxins', *Proc. Roy. Soc. London*, 63 (1898), 420-32; C.J. Martin. 'Further observations concerning the relation of the toxin and antitoxin of snake venoms', *Proc. Roy. Soc. London*, 64 (1898-9), 88-94.
13   C.J. Martin, 'Thermal adjustment and respiratory exchange in monotremes and marsupials. A study in the development of homeothermism', *Phil. Trans. Roy. Soc. Series B*, 195 (1902), 1-37.
14   C.J. Martin, 'Thermal adjustment of man and animals to external conditions. Croonian Lectures delivered before the Royal College of Physicians on June 12, 17 and 19, 1930', *Lancet*, 2 (1930), 561-7, 617-20, 673-8.
15   J.P. Hill, 'James Thomas Wilson', *Obit. not. Fellows Roy. Soc.*, 6 (1948-9), 643-60.
16   D.M.S. Watson, 'James Peter Hill', *Biogr. Mem. Fellows Roy. Soc.*, 1 (1955), 101-17.
17   J.P. Hill and C.J. Martin, 'On a platypus embryo from the intra-uterine egg, *Proc. Linn. Soc. NSW*, 9 (1894), 738-9; idem, 'On a platypus embryo from the intrauterine egg', ibid., 10 (1895), 43-74.

18 J.T. Wilson and C.J. Martin, 'Observations upon the Anatomy of the Muzzle of Ornithorhynchus', *Macleay Mem. Vol., Linn. Soc. NSW*(1893), 179–89; idem, 'On the peculiar rod-like tactile organs in the integument and mucous membrane of the muzzle of Ornithorhynchus', ibid., 190–200.

19 J.T. Wilson and J.P. Hill, 'Observations upon the development and succession of the teeth in *Perameles*; together with a contribution to the discussion of the homologies of the teeth in marsupial animals', *Quart. J. micr. Sci.(NS)*, 39 (1897), 427–588.

20 G. Elliot Smith, 'A preliminary communication upon the cerebral commissures of the Mammalia, with special Reference to the Monotremata and Marsupialia', *Proc. Linn. Soc. NSW*, 2nd series, 9 (1894), 635–57; idem, 'The brain of a foetal *Ornithorhynchus*', *Quart. J. Micr. Sci.*, 39 (1896), 181–206; idem, 'The structure of the cerebral hemisphere of *Ornithorhynchus*', *J. Anat. Physiol.*, 30 (1896), 465–87.

21 J.P. Hill, 'J.T. Wilson, a biographical sketch of his career', *J. Anat. Lond.*, 76 (1941), 3–8.

22 W.R. Dawson, *Sir Grafton Elliot Smith: A Biographical Record by his Colleagues* (London, 1938); J.T. Wilson, 'Sir Grafton Elliot Smith', *Obit. Not. Fellows Roy. Soc.*, 2 (1936–38), 323–33.

23 Russell, op. cit. (n.3).

24 C.J. Martin and A.W. Pierce, 'Studies on the phosphorus requirements of sheep. I. The effect on young merino sheep of a diet deficient in phosphorus but containing digestible proteins and vitamins', *Bull. C.S.I.R.*, No. 77 (Melbourne); C.J. Martin. 'Observations on *Myxomatosis cuniculi* made with a view to the use of the virus in the control of rabbit plagues', *Bull. C.S.I.R.*, No 96 (Melbourne).

25 J.P. Hill, 'Contributions to the morphology and development of the female urogenital organs in the Marsupialia, I. On the female urogenital organs of *Perameles*, with an account of the phenomenon of parturition', *Proc. Linn. Soc. NSW*, 14 (1899), 42–82; idem, 'Contributions to the morphology and development of the urogenital organs in Marsupialia, II. On the female urogenital organs of *Myrmecobius fasciatus*, III. On the female genital organs of *Tarsipes rostratus*, IV. Notes on the female genital organs of *Acrobatus pygmaeus* and *Petaurus breviceps*, V. On the existence at parturition of a pseudovaginal passage in *Trichosurus vulpecula*', *Proc. Linn. Soc. NSW*, 15 (1900), 519–32; idem, 'Contributions to the embryology of the Marsupialia, II. On a further stage in the placentation of *Perameles*, III. On the foetal membranes of *Macropus parma*', *Quart. J. Micr. Sci.*, NS, 43 (1900), 1–22; J.P. Hill and J.T. Wilson, 'Observations on the development of *Ornithorhynchus*', *Phil. Trans. Roy. Soc.* Series B, 199 (1907), 31–168; J.T. Wilson and J.P. Hill. 'Observations on tooth development in *Ornithorhynchus*', *Quart. J. Micr. Sci.*, 51 (1907), 137–65.

26 Young, Sefton and Webb, op. cit. (n.5).

27 H.H. Dale, 'Fifty years of medical research', *Brit. med. J.*, 2 (1963), 1287–90.

28 A. Landsborough Thomson, 'Origin and development of the Medical Research Council', *Brit. med. J.*, 2 (1963), 1290–2; idem, 'Half a century of medical research', *Origins and Policy of the Medical Research Council (UK)*, Vol. 1 (London, 1973).

29 H. Himsworth and J.F. Delafresnaye, *The Support of Medical Research: A Symposium Organized by the Council for International Organization of Medical Sciences* (Oxford, 1956).

30 C.R. Harington, 'The place of the research institute in the advance of medicine', *Lancet*, 1 (1958), 1345–51.

31 A.A. Lendon, *The University of Adelaide Jubilee of the Medical School, 1885–1935* (Adelaide, 1935).

32 W.E. Le Gros Clark, 'Frederic Wood Jones, 1879–1954', *Biogr. Mem. Fellows Roy. Soc.*, 1 (1955), 119–34.

33 C.H. Kellaway, 'The Walter and Eliza Hall Institute of Research in Pathology and Medicine', *Med. J. Aust.*, 2 (1928), 702–8; F.M. Burnet, *Walter and Eliza Hall Institute 1915–1965* (Melbourne, 1971); V. de Vahl Davis, 'Sir Harry Allen and the foundation of the Walter and Eliza Hall Institute of Medical Research', *Hist. Rec. Aust. Sci.* 5 (4) (1983), 31–8; G.J.V. Nossal, 'The Walter and Eliza Hall Institute of Medical Research: 1915–1985', *Med. J. Aust.*, 143 (1985), 153–7.

34 C.J. Martin and F.E. Williams, 'I. Types of dysentery bacilli isolated at No. 3 Australian General Hospital, Cairo, March–August 1916, with observations on the variability of the mannite-fermenting group, II. The value of agglutination in the identification of members of the mannite-fermenting group of dysentery bacilli', *Brit. med. J.*, i (1917), 479–80; C.J. Martin, C.H. Kellaway and F.E. Williams, 'III. Epitome of the results of the examination of the stools of 422 cases admitted to the No. 3 Australian General Hospital, Cairo, for dysentery and diarrhoea, March–August 1916', *Brit. med. J.*, i (1917), 480.

35 H.H. Dale and P.P. Laidlaw, 'The physiological action of β-iminazolylethylamine', *J.Physiol.*, 41 (1910), 318–44 and 43 (1911), 182–95.

36 N.H. Fairley, 'The present position of snake bite and snake bitten in Australia', *Med. J. Aust.*, i (1929), 296–313; idem, 'The dentition and biting mechanism of Australian snakes', Ibid., 313–27; idem, 'Venom yields in Australian poisonous snakes', Ibid., 336–48; idem, 'Criteria for determining the efficacy of ligature in snake bite', Ibid., 377–94.

37 C.H. Kellaway, 'The Charles E. Dohme Memorial Lectures: Snake venoms', *Bull. Johns Hopk. Hosp.*, 60 (1937), 1–17, 18–39, 159–77.

38 W. Feldberg and C.H. Kellaway, 'Circulatory effects of the venom of the Indian cobra (*Naia naia*) in

cats', *Aust.J.exp.Biol.med.Sci.*, 15 (1937), 159-72; idem, 'Circulatory effects of the venom of the Indian cobra (*Naia naia*) in dogs', ibid., 441-60.

39   C.H. Kellaway, H.F. Holden and E.R. Trethewie, 'Tissue injury by radiant energy and the liberation of histamine', Ibid., 16 (1938), 331-42; C.H. Kellaway and E.R. Trethewie, 'Photodynamic action and liberation of histamine', Ibid., 17 (1939), 61-76; C.H. Kellaway and W.A. Rawlinson, 'Studies on tissue injury by heat, I The influence of anoxia', Ibid., 22 (1944), 63-8; idem, 'II The liberation of enzymes from the perfused liver', ibid., 69-82; idem, 'III Isolated limb preparation', ibid., 83-94.

40   C.H. Kellaway, A.H. Tebbutt and P. MacCallum, *Report of the Royal Commission of Inquiry into the Fatalities at Bundaberg* (Canberra, 1928).

41   J. Boyd, 'Neil Hamilton Fairley', *Biogr. Mem. Fellows Roy. Soc.* 12 (1966), 123-45.

42   C.L. Oakley, 'Gordon Roy Cameron', Ibid., 14 (1968), 88-116.

43   Landsborough Thomson, op.cit. (n.28).

44   Sir Macfarlane Burnet, *Changing Patterns: An Atypical Autobiography* (Melbourne, 1968).

45   F.M. Burnet and J. Macnamara, 'Human psittacosis in Australia', *Med. J. Aust.*, ii (1936), 84-8; F.M. Burnet and M. Freeman, 'Experimental studies on the virus of Q-fever', Ibid., ii (1937), 299-305; F.M. Burnet and S.W. Williams, 'Herpes simplex: a new point of view', Ibid., i (1939), 637-42; F.M. Burnet and A.V. Jackson, 'Poliomyelitis: IV The spread of poliomyelitis virus in cynomolgus monkeys with particular reference to infection by the pharyngeal-intestinal route', *Aust.J.exp.Biol.med.Sci.*, 18 (1940), 361-6.

46   H.H. Dale, 'C.H. Kellaway', *Obit. Notices Fellows Roy. Soc.*, 8 (1953), 503-21.

47   R.E. Richards, 'The National Health and Medical Research Council of Australia', in *The Support of Medical Research*, H. Himsworth and J.F. Delafresnaye (eds), (Oxford, 1956), pp. 53-9.

48   C.H. Kellaway, 'The Richard Stawell Oration', *Med.J.Aust.*, i (1938), 365-74.

49   T.E. Lowe, *The Thomas Baker, Alice Baker and Eleanor Shaw Medical Research Institute: The First Fifty Years* (Melbourne, 1974); P.I. Korner, 'The Baker Medical Research Institute', *Med.J.Aust.*, 143 (1985), 296-9.

50   F.W. Gunz, *The Kanematsu Memorial Institute, April 20 1933 to April 20 1973*; Brochure written to commemorate the 40th anniversary of the Institute (Sydney, 1973), F.C. Courtice, 'The Kanematsu Memorial Institute of Pathology: The Inglis era, 1933-1960', *Hist. Rec. Aust. Sci.*, 6(2) (1985), 115-36; P.C. Vincent, 'The Kanematsu laboratories', *Med. J. Aust.*, 143 (1985), 502-7.

51   J.C. Eccles, 'My scientific odyssey', *Ann. Rev. Physiol.*, 39 (1977), 1-18.

52   Ibid.

53   D.S. Nelson, 'The Kolling Institute of Medical Research', *Med. J. Aust.*, 143 (1985), 97-101.

54   M.R. Lemberg, 'Chemist, biochemist and seeker in three countries', *Ann. Rev. Biochem.*, 34 (1965), 1-20.

55   J. Barrett and R.N. Robertson, 'Max Rudolph Lemberg', *Rec. Aust. Acad. Sci.*, 4(1) (1978), 133-56; M.R. Lemberg and J.W. Legge, *Haematin Compounds and Bile Pigments* (New York and London, 1949).

56   E.P. Abraham, 'Howard Walter Florey, Baron Florey of Adelaide and Marston 1898-1968', *Biogr. Mem. Fellows Roy. Soc.*, 17 (1971), 255-302.

57   E. Chain, H.W. Florey, A.D. Gardner, N.G. Heatley, M.A. Jennings, J. Orr-Ewing and A.G. Sanders, 'Penicillin as a chemotherapeutic agent', *Lancet*, ii (1940), 226-8.

58   Sir Howard Florey, Speech at opening of John Curtin School of Medical Research, Canberra, 27 March 1958.

59   F. Fenner, 'The history of the John Curtin School of Medical Research: A centre for research and postgraduate education in the basic medical sciences', *Med. J. Aust.*, ii (1971), 177-86; R. Porter, 'The John Curtin School of Medical Research', Ibid., 142 (1985), 205-13.

60   E.H. Derrick, 'The birth of the Queensland Institute of Medical Research', Ibid., ii (1972), 952-9; C. Kidson, 'The Queensland Institute of Medical Research', Ibid., 142 (1985), 355-9.

61   F.M. Burnet, *Principles of Animal Virology*, 2nd edition (New York and London, 1960).

62   F.M. Burnet, 'A modification of Jerne's Theory of antibody production using the concept of clonal selection', *Aust. J. Sci.*, 20 (1957), 67-9; idem, *The Clonal Selection Theory of Acquired Immunity* (Cambridge and Nashville, Tennessee, 1959).

63   F.M. Burnet and F. Fenner, *The Production of Antibodies*, 2nd edition (London, 1949).

64   J.C. Eccles, 'From electrical to chemical transmission in the central nervous system', *Notes and Records, Royal Society*, 30 (1976), 219-30.

65   L.G. Brock, J.S. Coombs and J.C. Eccles, 'Action potentials of motoneurones with intracellular electrode', *Proc. Univ. Otago Med. School*, 29 (1951), 14-15; idem, 'The recording of potentials from motoneurones with an intracellular electrode', *J.Physiol.Lond.*, 117 (1952), 431-60.

66   Eccles, op. cit. (n.51).

67   J.C. Eccles, *The Physiology of Nerve Cells* (Baltimore, 1957).

68   J.C. Eccles, *The Physiology of Synapses* (Berlin, Göttingen, Heidelberg, 1964).

69   Eccles, op. cit. (n.51).

13

Early years of Australian radio astronomy

Woodruff T. Sullivan, III

The undisputed leaders in post-war radio astronomy were Australia and England. While it is not surprising to find England at the forefront of a scientific field in the middle of the twentieth century, Australia's presence calls for more explanation. How was it that a small, isolated country succeeded so impressively in such an arcane field? The answers revolve around the course of the Second World War, the Australian government's policies toward its scientific laboratories, and the relationship with the mother country. First, a strong community of radio physicists developed in Australia in the 1930s, based on intimate ties with the ionospheric community in England (see Chapter 10). Second, Britain shared the secret of radar with its Dominions as the war began, nurturing intense radar research, development, and manufacture in Australia. Third, the team of scientists and engineers that grew out of that effort, primarily at the Radiophysics Laboratory in Sydney, remained intact at war's end, and soon put their new skills to use in developing peacetime research ventures. And finally, dynamic and skilful leadership was provided by E.G. Bowen and J.L. Pawsey — two men whose styles of science and complementary personalities produced a favourable mix for exploring and exploiting the most profitable avenues into the radio sky.

This chapter presents the history of the origins and early years of radio astronomy in the Radiophysics Laboratory, Sydney, and its post-war transition into a multi-faceted programme, which by 1949 was dominated by extraterrestrial radio noise. The first solar discoveries in 1945–46, and the 1946–47 pioneering work on radio stars are some of the highlights of this period of development of the laboratory up until 1952.

Other early radio astronomy

Radio astronomy, the study of the naturally emitted radio waves from extraterrestrial objects, did not take hold until after the Second World War, but there had been pioneer efforts before this. At the turn of the century several European physicists made unsuccessful attempts to detect 'Hertzian' waves from the sun. The first discovery of extraterrestrial radio waves, however, was not until 1931–32, and then only made accidentally by Karl G. Jansky of Bell Telephone Laboratories in the United States. While investigating the sources of static affecting trans-Atlantic shortwave radio-telephone circuits, Jansky found a 'steady, weak, hiss-type static' at 20 megahertz, which a year's monitoring allowed him to establish as coming from the Milky Way. In the ensuing decade, only one man, Grote Reber, followed up Jansky's discovery. Reber built a 31 foot dish antenna in his backyard in Wheaton, Illinois in 1937 and over the period 1939–46 made detailed maps of the Milky Way at the much higher frequencies of 160 and 480 megahertz.

Despite Jansky's and Reber's work, radio astronomy as a field of study does not really stem from them, but rather from the intensive wartime development of radar in Britain, America and Australia. As part of a military operations investigation in England in 1942 J.S. Hey accidentally discovered powerful radio bursts from the sun, and in America in 1942–43 G.C. Southworth purposely made measurements of the microwave sun. As radar receivers improved, it also became apparent that the galactic radiation was becoming a limiting factor to sensitivity in many cases. At war's end in England, three different men determined to follow up on this extraterrestrial noise from the sun and the Milky Way. Hey led an excellent and active Army group, but, after about two years, it was no longer able to pursue radio astronomy. Martin Ryle at Cambridge University and A.C. Bernard Lovell at Manchester also were soon leading strong groups. Ryle in particular worked on areas that placed him in a distant, but stiff, competition with the Australians. Until the early 1950s Lovell concentrated on meteor radar research, that is, bouncing radar off the ionized trails left by meteors in the upper atmosphere. Other smaller post-war efforts took place in France, Canada, the Soviet Union, and Japan, but these were dominated by the work in England and Australia. In the decade following the war, ironically, there was little productive radio astronomy in the United States, a situation discussed at the end of this chapter.[1]

The Radiophysics Laboratory, 1945–1952

Transition to peacetime

The Radiophysics Laboratory (RP) had been established in 1939 in the grounds of Sydney University as a secret branch of Australia's Council of Scientific and Industrial Research (CSIR). Its staff was largely drawn from the strong radio ionospheric community that had been built up in the 1930s by J.P.V. Madsen at Sydney University, T.H. Laby at Melbourne University, and the Sydney research laboratory of Amalgamated Wireless (Australasia), Ltd (AWA). During the Second World War RP both designed wholly new radar systems and adapted British radars to Australian needs, and by its end the staff numbered no fewer than 300,[2] of whom sixty were professionals and fourteen bore names that would later become familiar in radio astronomy. As the war closed, various memoranda began to circulate on potential peacetime roles for the laboratory, culminating in an agenda paper put together by E.G. 'Taffy' Bowen for a meeting of the CSIR Council in July 1945. Bowen, who was then acting chief of the division and would soon take over as chief, had been working on radar for over a decade. Born in Wales in 1911, he was trained in physics at the University of Wales and obtained his PhD under E.V. Appleton at the University of London in 1933, working on atmospheric physics at Slough Research Station. There he came under the eye of Robert Watson Watt and in 1935 was co-opted into the initial team of four, who developed the first operational military radar systems, systems that were vital in the defence of Britain against the Luftwaffe. He led the development of 200 megahertz airborne radars,[3] for which he flew thousands of hours. In 1940 Bowen was a member of the famous Tizard mission, which delivered radar secrets to the United States, including that of the cavity magnetron, the first source of power sufficient to make microwave radar a feasible proposition. He remained in the United States for three years, first as a radar liaison officer in Washington, DC and then developing airborne radar systems at the MIT Radiation Laboratory. In early 1944 he went to the Radiophysics Laboratory as its deputy chief, and, although still officially on loan from the British, he soon took a liking to RP and to Sydney and spent the remainder of his career (until 1971) there as chief.[4]

Bowen's proposals for RP's peacetime role were warmly received and quickly endorsed by his bosses in CSIR's Melbourne headquarters,[5] A.C.D. Rivett and F.W.G. White, a New Zealander who himself had been RP chief for three years. Rivett felt strongly that each CSIR division should achieve a roughly even balance between free-running basic research and applied research. This was a vital element in what Schedvin calls the 'culture of CSIRO'.[6] So RP's proposed programme emphasized new scientific possibilities as well as areas where Australian commerce and industry would more immediately benefit. Bowen and his staff were clearly as excited about the potential of radar techniques in peacetime as they were weary of applying them to warfare. For them the new radar techniques were 'perhaps as far-reaching in themselves as the development of aircraft [during the First World War] or the introduction of gunpowder in a previous era'. They laid out a long shopping list of possible projects, the main categories of which were radio propagation, vacuum research (directed toward generating power at millimetre wavelengths), radar aids to navigation and surveying, and radar study of weather. These topics, together with the production of *A Textbook of Radar*, which incorporated RP's knowledge and was edited by Bowen (1947),[7] were to form the initial post-war programme.

Radiophysics was CSIR's glamour division, arguably containing within its walls the densest concentration of technical talent on the continent, and CSIR was eager to keep this winner intact. As F.J. Kerr, one of RP's early staff members, recalls:

> [Basic radio research] was thought of as a good subject for the Lab to get into, partly in order to keep the Lab in being because it was a collection of good people, well trained in the arts of radio. Especially at that time there was a feeling that it had been a great national value to have had the Lab, and so it was possible to sell the idea to the authorities that the group should be kept in existence as a 'national asset'.[8]

Keeping the best of the research staff at RP was also immeasurably helped by the fact that research in physics and engineering at Australian universities after the war was minimal;[9] government funds for CSIR scientific research were thirty times greater than for universities.[10] In contrast to the situations in England and the United States, the young cadre of RP researchers saw their wartime laboratory as the best place to continue their peacetime careers.

There they had an unbeatable combination of expert colleagues, support staff, camaraderie, laboratory space, workshops, and the latest radio electronics.

While RP's continued existence was assured, its direction changed from developing military hardware to a mixture of fundamental research and applications of radio physics and radar to civilian life. Rivett felt strongly that classified military research should be removed from CSIR once the war was over and this in fact did largely happen. In retrospect, Bowen[11] names this policy as one of the key ingredients of Radiophysics' post-war success, but even in his 1945 document he strongly affirmed this principle, saying that peacetime military work in such laboratories 'stifles research and seldom produces effective assistance to the Armed Forces'.[12] Rivett's policy, however, came not without political repercussions. In the post-war atmosphere of a deepening Cold War and its attendant espionage episodes in a variety of countries, the loyalties of many prominent figures were called into question. Rivett was attacked by right-wing politicians, in particular over the loyalty of CSIR workers and over his policies on classified research. This, plus his opposition to a major administrative reorganization of CSIR (which he felt would substantially reduce its independence from the government), eventually led to his resignation. In 1949 CSIR became CSIRO, the Commonwealth Scientific and Industrial Research Organisation, and its number-two man became White, who over the next two decades was a major force in fostering the growth of his old division.[13]

Overall research programme

The major programmes at RP waxed and waned over the years 1946–53, as shown in the graphs below, which are based largely on annual reports and lists of publications[14] issued by the Division of Radiophysics. Vacuum physics work died away within two years and work on radar applications steadily lessened over the first five years. The two research programmes that grew were radio astronomy (although this term was not used until the 1949 report) and rain and cloud physics. Between 1946 and 1949 these increased their share of the professional staff from 6 per cent to 63 per cent. Because the total staff grew by only one-quarter over this same period, there were clearly many reassignments of personnel. In terms of papers published in the scientific literature, radio astronomy and rain and cloud physics also dominated, accounting for 71 per cent of the papers by 1949 and 65 per cent over the eight-year span. The radio astronomy staff, however, produced more than double the number of papers per person.

Rain and cloud physics, in which Bowen himself specialized and which he personally oversaw, was centred on attempts to understand the way clouds and rain form and behave. Microwave radar measurement, often from aircraft, became a central technique. Buoyed by one of the first successes in seeding clouds (as early as February 1947),[15] the RP group hoped that rainmaking for the dry Australian climate would ultimately become a reliable and economic proposition. Although this never happened, Bowen's group became one of the international leaders in the field. This effort, as well as the development of radar systems for commercial aviation, such as a distance-measuring equipment allowing airliners to locate themselves relative to beacons, were important as practical areas balancing off fundamental research in astronomy, fast becoming RP's most visible sector. But even the radio astronomy work sometimes found itself 'shoe-horned' into the role of being practical.[16,17] From the *1949 Annual Report*:

> Radio astronomy has already made important contributions to our knowledge and, like any fundamental branch of science, is likely to lead to practical applications which could not otherwise have been foreseen. For example, attempts to explain how certain types of radio waves arise in the Sun are already leading to new techniques for the generation and amplification of radio waves.[18]

Other projects included a mathematical physics section, a largely unsuccessful vacuum physics section, and (after the late 1940s) a group developing an early

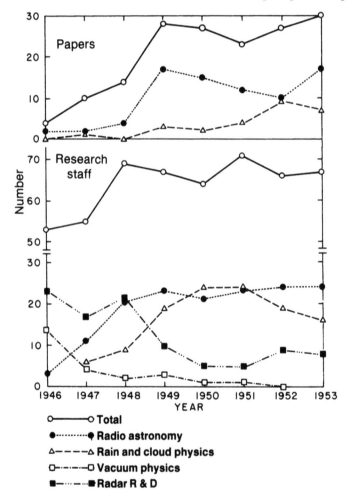

The growth and decline of different research areas at RP over the period 1946–53, as gauged by the number of published papers and the number of research staff. Papers are plotted for the year of publication and staff for the year of issuance of the annual report published in the middle of each calendar year. Staff levels not plotted include ionosphere (which fell in a similar manner to vacuum physics) and mathematical physics and electronic computing (which rose to a level of about 10 by 1951–53).

electronic computer (CSIRAC).[19] And there were always a few ionospheric radio projects going on.

Growth of research on extraterrestrial radio noise

Buried in the twenty-four pages describing Radiophysics' post-war plan is a fraction of a page under 'Radio propagation', subheading 1.2 (ii), 'Study of extra-thunderstorm sources of noise (thermal and cosmic)':

> Little is known of this noise and a comparatively simple series of observations on radar and short wavelengths might lead to the discovery of new phenomena or to the introduction of new techniques. For example, it is practicable to measure the sensitivity of a radar receiver by the change in output observed when the aerial is pointed in turn at the sky and at a body at ambient temperature. The aerial receives correspondingly different amounts of radiant energy (very far infrared) in the two cases. Similarly, the absorption of transmitted energy in a cloud can be estimated in terms of the energy radiated to the receiver by the cloud. None of these techniques is at present in use.[20]

It is this enigmatic paragraph, with its heading designed primarily to indicate that it was *not* talking about thunderstorm noise (atmospherics), that would develop into RP's radio astronomy programme! Surprisingly, it did not explicitly mention *solar* noise, but seems to propose an exploratory programme of radiometry wherein antennas would be pointed to different parts of the sky.

Triggered by reports of anomalies from radar stations (see below), J.L. Pawsey and his colleagues started solar observations in October 1945 and never turned back. Pawsey by this time had become the linchpin and recognized leader of Radiophysics' fundamental investigations through his Propagation Research Group. He had studied physics under Laby at Melbourne and obtained his PhD in 1934 under J.A. Ratcliffe at Cambridge. His dissertation involved a study of the intensity variations suffered by radio waves reflected off the abnormal E layer of the ionosphere. For five years he then worked for Electronic and Musical Industries, Ltd (EMI) developing the equipment needed to make television a commercial reality. This was at the famous BBC station at Alexandra Palace and Pawsey's main contributions, which involved no fewer than twenty-nine patents, were in designing the transmission lines and antennas necessary for television's broad bandwidth. After the outbreak of war he hastened home and joined the Radiophysics staff early in 1940. Pawsey was viewed as the local wizard on antennas and transmission lines, and by war's end he had also gained new skills working on receivers and operational aspects of radar systems. More importantly, his abilities to lead scientific research teams had been nurtured and honed in the intense wartime environment.[21]

Work on extraterrestrial noise started off small, but Bowen as Chief and Pawsey as his right-hand man in charge of most of the research activities of the laboratory were willing to shift resources into any programme showing superior results or great promise. The flexibility was the CSIR style,[22] largely moulded by Rivett, who believed that research programmes should be based on people, not topics — getting the right people and then letting them loose. As Bowen recalls:

> We tried many things, but the criterion for going on with any programme was, of course, success. And the things that Pawsey was trying on the Sun and Bolton on point sources were so outstandingly successful that that's the way we went . . . With our first-rate staff as a handout from the war, we had the freedom and the encouragement to find new projects.[23]

Or as Pawsey put it:

> [Scientific directors must] very quickly make decisions and supply facilities for the really promising developments. In all too many cases elsewhere the energies of scientists are taken up in advertising the potentialities of their prospective investigations in order to obtain any support at all.[24]

As the years passed, work on solar and cosmic noise grew in importance at RP and there emerged a circle of group leaders, all of whom were world-class in their scientific contributions. Besides his overall supervision, Pawsey led a large group studying numerous aspects of the radio sun. In 1947 J.G. Bolton began his pioneering work on discrete radio sources and soon had an active group around him. J.P. Wild arrived in 1947 and, after a year languishing in the laboratory's instrument test room, moved into research on solar radio bursts with a swept-frequency receiver. B.Y. Mills started off working on a linear accelerator in the short-lived vacuum physics section, then shifted to several other projects, and in 1948 permanently switched to radio astronomy, briefly on the sun and then into his own programme on discrete sources. W.N. 'Chris' Christiansen arrived at Radiophysics in 1948 from AWA and immediately plunged into his own solar research programme. He was unique among this group in that, despite his career as an antenna engineer, he had long wanted to be an astronomer.[25] Wartime RP veterans J.H. Piddington and H.C. Minnett in 1948 began a programme of microwave research, in particular on lunar radiation, and Kerr and C.A. Shain started on lunar radar in 1947. Finally, S.F. Smerd and K.C. Westfold complemented all of the observational work by working on the difficult theory of solar radio emission. Of all this work, however, there is only space here to cover the science from the earliest important results, those from the first year of Pawsey's and Bolton's groups.

Among all these successes, however, the RP archives also give evidence, in retrospect, of at least one important missed opportunity, that of the 21-centimetre spectral line arising from interstellar hydrogen. The line had been predicted in 1944 in Holland, and its 1951 discovery at Harvard and Leiden universities was to be one of the major turning points in early radio astronomy. Pawsey first heard of the idea in early 1948 while on a tour of the United States. Over the next year or two, several Radiophysics researchers considered the problem (and Wild wrote up a report), but none of them were willing to construct the equipment necessary to search for the line. The decision not to pursue the hydrogen line can hardly be called a managerial mistake, since the researchers involved were all doing excellent work on other projects. Nevertheless, given its resources and technical expertise, the fact remains that RP surely would have soon succeeded in detecting the interstellar 21-centimetre line if it had ever made a serious effort.[26]

* * *

Early solar work

Wartime

As radar receivers during the war became more sensitive and moved to higher frequencies, concepts of receiver noise, background noise, and antenna temperature gained currency:

> Receivers were getting more and more sensitive and we were concerned with the whole thermodynamic theory of their noise level and its relationship, through the antenna, to space — if the antenna were in an enclosure at three hundred degrees, what would be the noise level? This was different from the purely circuit approach that had been worked up by Nyquist and others ... And it obviously occurred to Ruby Payne-Scott and Joe Pawsey that radiation from objects might possibly be seen. I remember that Ruby had a small paraboloid poking out a window at certain objects in the sky to see how the noise level varied.[27]

Ruby Payne-Scott, the only woman to make a substantial contribution to radio astronomy during the post-war years, was able to do so only because she kept her marriage secret from 1944 to 1950, when CSIRO changed its policy forbidding married women on permanent staff. But the following year she resigned from RP in order to raise a family, and never again participated in research. She had been trained before the war as a physicist at the University of Sydney, worked on cancer radiology, spent two years at AWA, and from 1941 on at RP mainly worked on display systems and calibration of receivers. She soon became known around RP for her considerable intellectual and technical prowess, forthright personality, and 'bushwalking' avocation.

It was in March and April 1944 that Pawsey and Payne-Scott[28] looked at the microwave sky. In their subsequent RP report they discussed various contributions to the noise power measured by a receiver–antenna combination and cited Jansky's and Reber's work on cosmic static. But their operating wavelength of 10 centimetres was far shorter than that of earlier reported work on noise from either terrestrial or extraterrestrial sources. They used a 20 x 30 centimetre horn connected to a receiver with a system temperature of approximately 3500 K, one person pointing the horn around the room or out of the window in varying weather, the other taking readings from a meter. Changes of 20–300 K (in antenna temperature) were noted, and they were particularly struck by the apparently low absolute temperature of the sky, less than 140 K. Moreover, they noted the 'most unusual' consequence of this, that, as they demonstrated for themselves, inserting attenuation between the horn and receiver actually *increased* the output!

They also tried to detect microwave from the Milky Way with the same receiver and a 4 foot dish pointing first in the vicinity of Centaurus and then away. There was no detectable difference, that is, less than 0.25 per cent (less than 10 K) — 'very much less than that observed by Jansky and Reber'. This they ascribed to the fact that the material in space must be at a very low temperature, and they appealed to Eddington's work,[29] about which they undoubtedly learned from Reber's citation.

These Milky Way results were accompanied by a single sentence stating that they did not try for any solar radiation. It would seem that they were then unaware of either Hey's or Southworth's secret reports on the sun, but given that they mentioned the sun at all, why did they not try for it? If they had, it can be calculated that they probably would have easily detected a change in power output.

These kinds of ideas were thus in the air around Radiophysics and therefore, as already discussed, merited a short paragraph in Bowen's proposed post-war programme. But the archival evidence indicates that what really galvanized Bowen and Pawsey into working on solar noise was not this preliminary experiment, nor reports from overseas, nor *ab initio* calculations, but the 'Norfolk Island effect', referring to the tiny island about 1000 miles north-east of Sydney. This was where the commander of a New Zealand Air Force radar station reported that during 27–31 March 1945 his operators observed on 200 megahertz a large increase in noise on their 'range tubes' each morning and evening for about 30 minutes as the sun rose and set. The noise level was maximum at the azimuth of the sun and followed a pattern close to that of the antenna pattern.

This report was referred to the Radio Development Laboratory of the Department of Scientific and Industrial Research in Wellington, New Zealand, where its director, Ernest Marsden, became interested. He assigned Elizabeth Alexander to investigate the effect further, and she co-ordinated observations over the period 11–20 April at Norfolk Island and four similar stations on the northern coast of New Zealand. Each station managed to observe the effect at least twice, although sometimes only slightly, and in general they found the azimuth of peak noise very close to the sun's position. But there were nagging exceptions where the noise peak seemed to be well away from the sun, or where the weather or local terrain seemed to influence its strength. And why was the effect never observed once the sun was more than 8 degrees above the horizon? And why had the radar stations never previously seen the effect?

The measurements in April were crude. Only one station used as much as a meter; the rest were simply estimating the amount of increase in the 'grass' on a cathode ray tube display. But in her report Alexander[30] was reasonably confident that the effect was real, that it represented radio radiation emitted by the sun at a level far above that expected from black-body theory, and that it deserved further study. She next arranged for a more quantitative watch, and some further observations were carried out, but the work ended when the radar stations were shut down at the end of 1945.[31]

Bowen learned of the Norfolk Island effect in July 1945 and was entranced. In a letter to White, he wrote:

> These results are remarkable in that while one would expect to receive solar noise radiation on S. or X. band equipment [10 or 3 cm wavelength], a C.O.L. antenna and receiver at 200 Mc/s is quite unlikely to do so. I have heard rumours of the same thing happening in England, but as far as I am aware, the subject has never been followed up. We are therefore going to attempt to repeat the observations here in Sydney to see if we can track down the anomaly.[32]

This letter testifies that in August 1945 Bowen and Pawsey knew about thermal, microwave radiation from the sun, presumably from Southworth's

1942–43 work, but were not specifically aware of Hey's 1942 and 1945 restricted reports about low-frequency solar bursts. Instead, their first investigations were triggered by the New Zealand work. Furthermore, the thrust of these investigations was toward monitoring the sun for non-thermal bursts of radio waves, unlike what was stated in Bowen's proposed programme.

Solar bursts and the sea-cliff interferometer

Pawsey swung into action and mounted an observing programme on a frequency of 200 megahertz using existing Air Force radar installations along the coastline near Sydney. Working with him on this were Payne-Scott and L.L. McCready, a receiver expert, pre-war AWA engineer, and Radiophysics veteran who at the time was Pawsey's number-two man and who eventually became the head of all engineering services at RP. The first observations were on 3 October from Collaroy, fifteen miles north of Sydney on a hilltop half a mile inland. The antenna was an array of about forty half-wave dipoles and observations were carried out by Air Force as well as RP personnel. After only a week or two of data, Pawsey noticed that the general level of 'this noise effect' was highly variable and seemed to correlate with the number of visible sunspots.[33] For the latter information he had made contact with C.W. Allen, a long-time solar astronomer at the Commonwealth Mt Stromlo Observatory near Canberra. After three weeks of monitoring, Pawsey, Payne-Scott and McCready sent a letter to *Nature*[34] pointing out the close correspondence between the total area of the sun covered by sunspots and the average daily radio noise power from the sun. Because the antenna's elevation angle could not be changed, observations were only possible at dawn or dusk and various corrections had to be made for ground and sea reflections, but it was nevertheless clear that the daily values of solar noise varied by as much as a factor of thirty over the three weeks. They also pointed out that, for a thermally emitting disk the size of the optical sun, their detected levels implied 'equivalent temperatures' ranging from $0.5–15 \times 10^6$ K, much higher than the sun's 'actual temperature' of 6000 K. Such incredible signals, they reasoned, could not come from atomic or molecular processes, but more likely from 'gross electrical disturbances analogous to our thunderstorms'.

With such a promising start, Bowen and Pawsey decided to increase their efforts on solar noise, and continued monitoring for another ten months. Gradually Air Force personnel and equipment were phased out as RP took over. Pawsey's group made measurements at a variety of frequencies, mostly at metre wavelengths and with antennas (such as those shown in the illustration below) at a variety of coastal radar sites around Sydney, including Collaroy, North Head, Georges Heights, and Dover Heights.

The climax of this initial period came in early February 1946 when by good fortune the largest sunspot group of the century chose to make its appearance. When Allen announced that this huge group of sunspots had appeared (covering about 1 per cent of the sun's visible disk), the RP group intensified their monitoring and realized that they now had the opportunity to take advantage of a property of their antenna system that had previously been more a bother than a help. A single antenna situated on the edge of a cliff or a hilltop, looking near the horizon over a relatively smooth terrain or over the sea, in fact acts as an interferometer and can achieve far better angular resolution than would otherwise be possible. The interference in this case is between that portion of a wavefront directly impinging on the antenna and that reflected from the sea, which must travel an additional length equal to twice the cliff height multiplied by the sine of the source's elevation angle. In classical optics this arrangement is known as 'Lloyd's mirror' and the fringes obtained are equivalent to those with a conventional interferometer consisting of the antenna and an imaginary mirror image located under the base of the cliff. With the antennas at Dover Heights and Collaroy located 85 and 120 metres above the sea, the respective fringe lobes were spaced by 30 arc minutes and 21 arc minutes at 1.5 metre wavelength. In principle, then, one could locate objects with an accuracy of approximately 10 arc minutes, far better than the 6 degree beam of the antenna considered by itself. This phenomenon was not new to those who had been developing radar systems, for during the war fringes had often been observed when pointing a radar beam near the horizon, as with search radars on a ship or a coastline.

So Pawsey and his colleagues used this sea-cliff interferometer[35] to advantage as the bespotted sun rose over the Pacific. The general level of solar emission was far above normal for several days and often interspersed with bursts. As before, they found that the solar signal appeared at sunrise and gradually faded out as the sun rose beyond where the antenna's beam responded, but now there were striking oscillations, the interferometric fringes, superimposed on this. The exciting thing was that the very presence of these oscillations implied that the source of the solar signal was a good bit smaller than the spacing of the fringes (20–30 arc minutes), and therefore a good bit smaller than the 30 arc minutes size of the optical sun. Exactly how much smaller the emitting region was, as well as its location, could be worked out through details of the amplitudes and phases of the oscillations. This led to the sketches shown below (p. 339), where they inferred that the emitting region on any given day had a width of 8–13 arc minutes and coincided with the giant sunspot group being carried along by the sun's rotation. Even though the fringes of the sea-cliff interferometer were orientated parallel to the horizon and thus could give no information about the azimuth of the emitting region, it seemed eminently reasonable that the sunspot group itself was the source of the enhanced solar radiation. They had thus directly confirmed what Hey[36] and Appleton[37] had earlier surmised.

In a paper submitted to the *Proceedings of the Royal Society* in July 1946, McCready, Pawsey and Payne-Scott[38] reported the above results and much more. They expanded on their first results in *Nature* and now characterized the solar radiation as consisting of two components — a slowly changing type that could vary by a factor of 200 in intensity over many days; and intense bursts, lasting from less than a second up to a minute, that could be tens of times more powerful than the general level on a given day. These results were so unexpected that they worried at length that the bursts might somehow be induced by the ionosphere, but various arguments, principally the fact that separate sites observed the bursts at the same time (to within a second), convinced them that this indeed was an extraterrestrial, 'and presumably solar', phenomenon. As in their previous letter to *Nature*, they pointed out that the equivalent brightness temperatures were extraordinarily high — the sunspot group seemed as high as 3×10^9 K.

This seminal paper also explained many basics of the sea-cliff interferometer, considering effects such as refraction (the worst uncertainty), the earth's curvature, tides, and imperfect reflection from a choppy sea. Again, however, they were hardly starting from scratch in working on this problem; for instance, a 1943 Radiophysics report by J.C. Jaeger had covered many of these points.[39] Pawsey's group, however, also introduced a vital new principle, namely that their interferometer was sensitive to a single Fourier component (in spatial frequency) of the brightness distribution across the sun, and that in principle a complete Fourier synthesis could be achieved if one had enough observations with interferometers of different effective baselines:

> Since an indefinite number of distributions have identical Fourier components at one [spatial] frequency, measurement of the phase and amplitude of the variation of intensity at one place at dawn cannot in general be used to determine the distribution over the Sun without further information. It is possible in principle to determine the actual form of the distribution in a complex case by Fourier synthesis using information derived from a large number of components. In the interference method suggested here . . . different Fourier components may be obtained by varying the cliff height h or the wave-length λ. Variation of λ is inadvisable, as over the necessary wide range the distribution of radiation may be a function of λ. Variation of h would be feasible but clumsy. A different interference method may be more practicable.[40]

Much of the subsequent technical development of radio astronomy was to be concerned with this method of making high-resolution cuts across sources, and eventually complete maps. By the early 1950s their suggested type of Fourier synthesis was indeed regularly carried out and results on the sun and on discrete radio sources were fundamental. But the last two sentences of this quotation were prophetic, for it was not sea-cliff interferometry, but the more tractable and flexible conventional interferometry with two separate antennas, as developed first by Ryle at Cambridge University, that made such mapping a reality.

The million-degree corona

Some time toward the middle of 1946, Pawsey extracted another jewel from his plethora of data. He noticed that his large set of daily values of the 200 megahertz solar flux density had a peculiar distribution, with a sharp lower limit corresponding to an equivalent brightness temperature for the solar disk of about 1×10^6 K. This was drawn from the same data presented earlier, but looked at in a new way: first, with a histogram of values over the entire period (about 150 values over seven months) rather than a plot against time, and second, using single-day values rather than three-day averages. Pawsey had earlier argued that three-day averages were necessary because the solar bursts frequently vitiated individual half-hour daily observations, but now he saw that this averaging had also tended to mask the marked lower limit of intensity, since about two-thirds of all days exhibited enhanced levels.

At this same time D.F. Martyn, Australia's leading ionospheric physicist and RP's first chief,[41] who was now at Mt Stromlo and very interested in this new field, suggested a theory that could explain a million-degree base level for the solar radiation. He learned, undoubtedly from discussions with Allen and Richard Woolley, observatory director, that recent studies of spectral-line widths and ionization states strongly suggested that the solar corona had a temperature of about 1×10^6 K.

Why the corona was so hot was not at all understood, but the evidence was there. Martyn then realized he could apply standard techniques in ionospheric theory to calculate the expected radio emission from the sun. Once he had adopted likely values for the electron densities in the corona, he found that the corona was opaque at Pawsey's kind of frequencies. The observed radio waves were therefore emanating not at all from the 6000 K optical surface (photosphere) of the sun, but from well above the photosphere out in the million-degree corona. When the sun was quiet, this coronal thermal emission constituted the entire solar signal; when active, the coronal emission was dwarfed. Furthermore, at shorter wavelengths, the observed emission came from deeper in the corona, and eventually even from the chromosphere. This powerful idea thus explained why the measured brightness temperature of the quiet sun always seemed greater than 6000 K and sharply increased at longer wavelengths. It also meant that the corona could now be studied without the inconvenience of having to chase a total eclipse. As it turned out, in the Soviet Union a few months earlier Vitaly L. Ginzburg[42] had independently made similar calculations while considering the possibility of reflecting radar off the sun. The basic ideas were again independently presented, yet a third time, in the Russian literature in a late 1946 paper by Iosif S. Shklovsky.[43] But Martyn had access to better confirming data and was positioned more in the mainstream of post-war radio astronomy. His paper, in *Nature* for 2 November 1946,[44] had far more influence.

Pawsey's and Martyn's work as described above seems fairly well established, but there is controversy over whether Martyn first predicted the million-degree corona and then suggested to Pawsey to seek it in his data, or whether Pawsey first found it empirically and so instigated Martyn's working on the problem. In fact, Pawsey and Martyn at first planned a joint publication, but this went sour, resulting in two adjacent notes in *Nature*: Martyn's did not mention Pawsey's base-level data at all, while Pawsey's[45] acknowledged his indebtedness to Martyn for 'pointing out to me the probable existence of high-level thermal radiation'. The archival evidence indicates that Pawsey first noticed the lower bound corresponding to an effective temperature of approximately 1×10^6 K and then Martyn brought in the previous astronomical evidence of a million-degree corona, pointing out that the million-degree 'effective' or 'apparent' temperatures cited by the RP group could actually represent *thermal* emission from the solar atmosphere. Pawsey and his colleagues had calculated these temperatures, but thought of them only in a formal sense. In fact to them these incredibly high values were prima facie evidence of *non*-thermal phenomena.

Mt Stromlo and the Radiophysics Laboratory
It is remarkable that the Commonwealth Observatory at Mt Stromlo worked

so closely with Radiophysics right from the start — such active collaboration between astronomers and radio investigators occurred nowhere else in the world in the first few years after the war. Since 1930, however, staff at Mt Stromlo (in particular A.J. Higgs, who after the war was RP's technical secretary) had been doing a small amount of ionospheric research. Moreover, during the war C.W. Allen had worked on the effects of sudden solar disturbances on ionospheric conditions and optimum communications frequencies.[46] As noted above, Allen from as early as October 1945 was feeding optical solar data to RP and over the years his ties with RP remained strong.[47] Since Mt Stromlo was already in the solar monitoring business at optical wavelengths and RP did not want to maintain a strict daily patrol, the idea soon developed of Radiophysics installing a radio system at Mt Stromlo. This happened in early 1946 and from April onwards Allen oversaw regular 200 megahertz solar monitoring with a steerable array of four Yagi antennas. This led to a paper in which Allen[48] presented a year's worth of radio data and correlated it with optical and ionospheric activity. And in early 1949 he used the same array to make a complete map of the galactic background radiation.

In addition to Allen, there was Martyn, who worked on extraterrestrial noise as a sideline to his ionospheric research. Besides his important work on the million-degree corona, he also pointed out in his 1946 *Nature* note[49] that at wavelengths of 60 centimetres or less the quiet sun should appear brighter at its edges than in the centre. This prediction of 'limb-brightening' turned out to be qualitatively correct, although it was more than five years before observations of sufficient detail could settle the question. When a large sunspot group appeared in July 1946, Martyn had Allen's four-Yagi array modified so that it could search for polarization in radio bursts.[50] This led to detection of a high percentage of circular polarization and the discovery that the handedness of the polarization neatly flipped when the sunspot group crossed the solar meridian. Similar and simultaneous results were also obtained in England and were valuable for elucidating the influence of the sunspot group's magnetic field on the emitted radio waves.

This radio activity work could not have flourished without the encouragement of the observatory director and Commonwealth Astronomer, Richard Woolley, a stellar and dynamical theorist who had come to Australia to take over Mt Stromlo in 1939 and who would leave it in 1955 to become Astronomer Royal in England. He wrote an early paper on the theory of galactic noise and other papers on solar models incorporating radio data. In late 1946 Woolley suggested that Bolton should check for radio emission from the nebulosity near the bright star Fomalhaut, and in 1947, after Martyn had speculated that the Cygnus source (see next section) might be a distant comet, Woolley searched for such an object.[51] Moreover, relations between Woolley and RP were cordial enough so that Bowen first checked with Woolley before sending off the first RP paper on the Cygnus source.[52] Woolley was also elected in 1948 as the first chairman of the International Astronomical Union's new Commission 40 on Radio Astronomy and shortly thereafter became vice-chairman of Australia's national organization for the International Union of Radio Science (URSI).

Despite these fruitful exchanges of ideas, data, and know-how between the astronomers and the radio physicists, there was always tension between Woolley and Martyn on the one hand and Bowen and Pawsey. Much of this stemmed from Martyn and his status as an 'exiled' Radiophysics staff member, seconded to Mt Stromlo from Sydney. Martyn had been removed as RP chief late in 1941 after two years of continual problems — despite his scientific excellence, he simply did not have the managerial skills or temperament needed to run a large organization developing new technology under the threat of Japanese attack. By 1941 his relations with the military, industry, and his own staff were abysmal. On top of this, in early 1941 it became feared that he was a security risk because of his liaison with a German woman recently emigrated to Australia.[53] With this background, one can understand that his post-war relations with RP were often less than smooth. Things were not helped either by the emergence of the idea of an independent department of radio astronomy at Mt Stromlo. The matter culminated in 1951-52, after the departure of Allen to take up a professorship at the University of London. Mark Oliphant, head of physical sciences at the new Australian National University in Canberra, and Woolley made a major thrust to acquire a large radio telescope, but were beaten down by White at CSIRO headquarters and by Bowen and Pawsey.[54]

Radio stars

In August 1946 Bowen, then visiting England, excitedly sent Pawsey a reprint of a recent letter in *Nature* by Hey, S.J. Parsons and J.W. Phillips[55] of the British Army Operational Research Group. While mapping the general distribution of galactic noise, they had accidentally discovered that the noise from one particular spot in the constellation of Cygnus fluctuated in intensity on a time scale of minutes. Although they could directly measure only that the size of the fluctuating region was less than two degrees in size, they argued that such rapid changes must originate in a small number of discrete sources, perhaps only one. These were considered as stars, by analogy with the sun and its radio bursts. Pawsey jumped on this. As he soon wrote: '[Within a few days of receiving Bowen's letter] we immediately made some confirmatory measurements on 60 and 75 Mc/s, obtaining similar fluctuations, of the same form as the "bursts" observed in solar noise. We have no hint of the source of this surprising phenomenon'.[56]

This early success, however, was apparently followed by a period of conflicting observations, during which the reality of the Cygnus fluctuations came into question. In the end they gave up, no longer knowing what to make of Hey's claim.[57,58]

Cygnus work then lay dormant for several months until resumed by John Bolton, who had joined the RP staff as its second post-war recruit in September 1946. Bolton was a Yorkshireman who had studied undergraduate physics at Cambridge before joining the Royal Navy, where he first did radar development and then served as a radar officer on the aircraft carrier HMS *Unicorn* before his demobilization in Sydney Harbour. Assigned to the solar noise work at Dover Heights, Bolton built two 60 megahertz Yagi aerials to follow up on Martyn's earlier detection of circular polarization, and was soon joined by technician O.B. Slee, a former Air Force radar mechanic who had also just joined RP.[59] The sun was not co-operating with much activity, however, and so Bolton decided he would check for radio emission from the positions of various well known astronomical objects, as listed for instance in *Norton's Star Atlas*. His inattention to solar monitoring, however, got him into trouble: 'After a week or two our efforts were cut short by an unheralded visit from Pawsey, who noted that the aerials were not looking at the sun. Suffice it to say that he was not amused and we were both ordered back to the Lab for reassignment.[60]

Notwithstanding this setback, Bolton a few months later managed to return to Dover Heights, where he was joined by electrical engineer G.J. Stanley, a New Zealander who had come to Radiophysics on leaving the infantry three years earlier. This time the goal was to follow up work of the previous year by Payne-Scott and D.E. Yabsley on simultaneous solar-burst observations at widely spaced frequencies. On 8 March 1947 the sun obliged with a remarkable burst, exhibiting delays of a few minutes between signals arriving first at 200 megahertz, then 100 megahertz, and finally 60 megahertz. The behaviour of this and earlier bursts was taken to arise from emission at various critical frequencies as successively higher coronal layers were excited; with a model of electron densities in the corona, it was even possible to infer a speed of about 600 kilometres per second for the ejected material.[61] Here indeed was a dramatic confirmation of Martyn's model of different coronal levels effectively emitting different radio frequencies.

Bolton again grew tired of solar monitoring, however, and in June 1947 he and Stanley returned to the Cygnus phenomenon. The antenna was nothing more than a pair of 100 megahertz Yagis connected to a converted radar receiver and operated as a sea-cliff interferometer. This allowed a quick survey of the southern sky over three weeks, but they found only the Cygnus source and hints of two weaker ones.[62] So they spent several months checking out the Cygnus source, culminating at year's end in papers submitted to *Nature* and the very first issue of the *Australian Journal of Scientific Research*[63,64] Cygnus usually gave a workably strong set of fringes as it rose, but it never got more than 15 degrees above the northern Sydney horizon and observations were continually harassed by the intense fluctuations that had led to Hey's discovery in the first place. Bolton and Stanley's analysis broke up the signal into a constant component (which they estimated as 6000 Jy) and a variable component, which added (never subtracted) amounts that fluctuated over times of 0.1–1 minutes.

The heart of their study was concerned with the size and position of the source. Size came from the technique worked out on the sun over a year before, namely to measure the ratio of the intensity of fringe maximum to minimum. As the 'equivalent radiating strip' became broader, then the fringes would wash out in a predictable manner. But Cygnus gave difficulties with (1) subtracting off a considerable baseline slope caused by strong galactic noise in the vicinity; (2) isolating the constant component from the variable; and (3) determining a proper upper limit for the fringe minimum, for it appeared that the best records in fact showed minima that were not distinguishable (given the noise) from zero. They estimated that the maximum-to-minimum ratio was at least 50, implying that the source size was less than 8 arc minutes (about one-eighth of the fringe separation). Hey had inferred that the Cygnus fluctuations must arise from a small discrete source or collection of sources, but here was evidence more direct and quantitative. In fact they thought the source even smaller than their published limit:

> Careful examination of the records suggests a much smaller source size than stated above [8']. Further experiments using improved receiver stability and greater aerial height will probably substantiate the authors' belief that the source is effectively a 'point'.[65]

With a source size in hand, they moved on to the even trickier task of a position. This involved measurement and interpretation of the timing and spacing of fringes in terms of sky geometry and radio wave propagation. One had to find the sidereal time when the source was highest in the sky (culmination) and the length of the arc travelled by the source between rising and culmination. But Bolton and Stanley were forced to tie together observations at three different cliffs around Sydney (including ones facing north) in order to get reliable data; furthermore, the necessary corrections for refraction were large and, as it turned out, uncertain. In the end, the derived position was $19^h58^m47^s\pm10^s$, $+41°47'\pm7'$. With this first well defined position for the Cygnus source (Hey's group had been able to give its position to within 5° only!), their next step was of course to consult the optical catalogues and photographs. But this was disappointing:

> Reference to star catalogues, in particular the Henry Draper Catalogue, shows that the source is in a region of the galaxy distinguished by the absence of bright stars and objects such as nebulae, double and variable stars, i.e., the radio noise received from this region is out of all proportion to the optical radiation . . . The determined position lies in a less crowded area of the Milky Way and the only obvious stellar objects close to the stated limits of accuracy are two seventh magnitude stars.[66]

They did, however, request Woolley to have a special photograph of that portion of the sky taken for them, and this appeared as a plate in their paper, together with a tracing-paper overlay indicating their source position and error box. It certainly did appear to be a nondescript patch of sky.[67]

Given that there was no optical counterpart, could one nevertheless put any constraints on the distance to the object? Since they had been observing the source for three months, the changing position of the orbiting earth might have caused an apparent shift in position if the object were nearby. But they had detected no shift greater than 2.5 arc minutes (corresponding to their 10 second accuracy in timing the sudden appearance of the source at rising), and this meant the source was at least ten times the distance to Pluto (50 light hours) that is, well outside the solar system. But how far? Bolton and Stanley could only suggest that the farthest imaginable would be if somehow the radio object were a star with total power output similar to that of the sun, but all channelled into the radio spectrum. That distance came out to be 3000 light years. But no matter what its distance within those broad limits, it was not at all understood what caused the radio radiation. They could only say it had to be a non-thermal mechanism, for the measured effective (brightness) temperature was more than 4×10^6 K.

Just as Bolton and Stanley were writing up these results, they received an interesting communication from Pawsey, who was then on the first part of an around-the-world tour. He had visited Mt Wilson Observatory in Pasadena and there found Rudolph Minkowski and S.B. Nicholson 'intensely interested' in the Cygnus results and willing to undertake observations directed toward finding an optical counterpart. Pawsey then described optical objects that Minkowski had showed him in the region of the Cygnus position, and he ended with several suggestions from Minkowski for possible places to look for radio noise:

> The Magellanic Clouds [are] the nearest external galaxies, abnormal with much dust and blue stars ... If we are interested in interstellar dust, etc. the 'Crab Nebula', NGC 1952, is a good sample. If white dwarfs are of interest, the companion of Sirius is a convenient sample. The Orion region is a region of emission nebulae. [But] I do not think these ideas get us very far. I should recommend the method of empirical searching; our tools are not too fine to prevent this.[68]

With the Cygnus case temporarily closed, Bolton and Stanley indeed set out in November 1947 to search the sky in Pawsey's 'empirical' fashion. Stanley had made significant improvements to their receiver's short-term stability, in particular through constructing power supplies able to provide voltages stable to one part in a few thousand. Very weak fringes could now be reliably detected. They methodically took records at different points along the eastern horizon, and fringes for several sources appeared and were confirmed in the next few months. As it became clear that the sky had a lot more to offer than just the Cygnus source, Bolton introduced the nomenclature (still used today) that the strongest source in a constellation would be called A, the next B, etc. And so their second source became Taurus A, one-sixth as strong as Cygnus A, followed by Coma Berenices A at a similar level. The uncertainties of this work can be appreciated by noting that Taurus A appeared clearly on one November night, but it took another three months before its existence was satisfactorily confirmed.

By February the group had surveyed about half of the southern sky (man-made interference made daytime observations nearly worthless) and had good cases for six new discrete sources. Bolton wrote up a short note to *Nature*[69] announcing that a new class of astronomical object existed: Cygnus was not unique, either in its existence or in its lack of association with 'outstanding stellar objects'. Upper limits on the sizes of the new sources were no better than 15–60 arc minutes, but Bolton was becoming convinced that all these discrete sources were truly stellar, 'distinct "radio types" for which a place might have to be found in the sequence of stellar evolution'.[70] Since even the most powerful solar-style bursts would not do the trick, he appealed to either pre-main sequence, collapsing, cool objects or to old, hot objects related to planetary nebulae.[71]

After this survey Bolton chose next to improve his source positions, in particular to eliminate systematic errors by observing source *setting* as well as rising. A high westward-facing cliff was needed and so Bolton and Stanley headed off to New Zealand in the southern winter of 1948. As Bolton recalls:

[Just before the New Zealand trip] I remember Taffy Bowen asking me what I really thought of the positions of my sources, and I said, 'Well, they're the best I can do at the moment, but I'd like to be the first to correct them'. And indeed the corrections were absolutely massive when they came in.[72]

The superior observations in New Zealand put an even tighter limit on Cygnus A's size (<1.5 arc minutes), and, in co-ordination with simultaneous observations by Slee in Sydney, provided strong evidence that most of the intensity fluctuations originated in the earth's atmosphere, not in the source itself. Many new sources also turned up and it became apparent that incorrect refraction corrections and other problems had thrown most previous positions 5–10 degrees off. Some even changed names, as when Coma Berenices A migrated into Virgo. But Cygnus A was still vexing, as neither its new position (shifted about 1 degree south from earlier) nor its old one agreed with that measured by Ryle at Cambridge and privately communicated in June 1948. For a while it seemed that the source might actually be moving, but after six months both groups admitted earlier errors and were able to agree on a common position.

The really beautiful outcome of the new positions of approximately 10 arc minutes accuracy was that for the first time tentative optical counterparts could be identified. And these were no ordinary objects. Taurus A was associated with the Crab Nebula, the expanding shell of a supernova known to have exploded 900 years before; Centaurus A was found to coincide with one of the brightest and strangest nebulosities in the sky, so peculiar that astronomers were not even sure whether or not it was part of our galaxy; and Virgo A was a bright elliptical galaxy five *million* light years away.[73] These identifications infused new life into the study of discrete sources and caused several optical astronomers, among them Minkowski, to take serious note.

Overview of the early years of Radiophysics

The isolation factor

Almost any analysis of things Australian must consider the geographical isolation of Australia from the other centres of Western culture. Geoffrey Blainey[74] speaks of 'the tyranny of distance', that is, the overwhelming importance of distance, isolation, and transport in moulding the general history of Australia. Given that observational earth sciences are intrinsically an international enterprise, we would expect here also to find evidence of this tyranny.

The above account of the early RP years is rife with examples of things that would have gone differently if RP had not been located 10 000 miles from its sister institutions, but instead 100 miles, or even 1000. The best airline connections to Europe required a gruelling three days (or a civilized week) and the more common passage by ship took about four weeks; moreover, the cost of a ship's berth amounted to one or two months' pay for an RP staff member. The inability to have frequent contact with colleagues from other institutions, the long interval before learning about research conducted elsewhere, the delays in publishing Australian results in the prestigious overseas journals, and the lack of foreign readership of Australian journals — these circumstances constantly bedevilled the RP staff. One counterforce was the maintenance of Australian Scientific Research Liaison Offices in London, Washington, and Ottawa. These had been originally set up during the war to co-ordinate radar research, and served as scientific embassies to increase the flow of information to and fro. But a far better solution was to send an RP researcher on an extended jaunt through North America and Europe. In the six years after the war the primary overseas stays or trips of importance for the development of radio astronomy were taken by Bowen (1946), R.N. Bracewell (1946-9), Pawsey (1947-8), Westfold (1949-51), Bolton (1950), and Kerr (1950-1). The RP correspondence files resulting from these trips are particularly good sources for understanding the influence of the isolation factor on RP's work. What emerges is that such trips served four primary purposes: intelligence (in the military sense), education, publicity, and establishment of personal contacts.[75]

The first purpose of the overseas trips was simply to find out what was going on. The RP visitor to an overseas laboratory typically sent back a detailed report on recent and ongoing research, and this report (as evidenced by multitudinous initials on the original documents) was widely circulated back home. Pawsey and Bowen in particular were masters at picking up what was being done elsewhere and analysing its effects on RP's current research programme and future plans. To give but two examples, in August 1946 Bowen cabled back that British work on the sun was ahead of RP's at observing frequencies less than 200 megahertz and that therefore RP should concentrate on higher frequencies. And in April 1950 Bolton sent word home that the solar work by Hey's group lagged Wild's by at least eighteen months.[76]

Lack of knowledge of overseas research led of course to frequent duplication of experiments. In the first few years after the war, Pawsey often thought that poor communications were leading to fruitless repetition among British and Australian observers. Somewhat idealistically, he therefore tried to co-ordinate radio noise work on opposite sides of the earth, but this worked little in practice. As an example, in September 1946 Pawsey wrote to his mentor Ratcliffe at Cambridge:

> I got rather a shock when I received Ryle's note enclosing a copy of the letter to *Nature* contributed by himself and Vonberg[77] ... The Cavendish and Radiophysics Laboratories have unfortunately succeeded in duplicating a very considerable part of the work ... I do not know what we can do about this duplication or how we can avoid it in the future. My only suggestion is that you have a talk with E.G. Bowen, Chief of this Division, who is at present in England.

Ratcliffe replied with a summary of Ryle's plans for solar observations and noted that not all duplication should be avoided. He also stated that 'now that the Air Mail works so quickly [1-2 weeks], we will make a special attempt to keep you fully in touch'.[78] Other attempts by Pawsey at co-ordination took place in 1948-49. But eventually it became clear that Australian science, like so many other aspects of post-war Australian society, was becoming an independent entity, not just an extension of the mother country.

A second purpose of the overseas trips was education. Sometimes this was in the formal sense, as when Bracewell took a PhD at Cambridge and Westfold at Oxford, and Kerr a Master's degree at Harvard; but more often it was simply the wealth of knowledge to be garnered from overseas contacts. The background of the RP staff was of course far weaker in astronomy than in radio physics, and thus it was the visits to observatories that were particularly valuable. As Kerr recalls:

> Bolton and also Pawsey did some touring at that time and learned something of what generally were the interesting problems in astronomy, acquiring some of the attitude of astronomers toward astronomy, instead of just the electrical engineers' and physicists' attitudes.[79]

On the other hand, Mills points out that the paucity of astronomers in Australia may have helped more than it hindered: 'Our isolation did help us develop with an independent outlook. We had no famous [astronomer] names to tell us what we should believe, and to some extent we just went ahead following our noses.'[80]

The overseas trips also served to spread the word about RP research. The RP archives of this era are full of instances where Bowen and Pawsey sent reprints, complained about Australian work being neglected in reviews overseas, and urged people to subscribe to the *Australian Journal of Scientific Research*, started by CSIR in 1948 and a further sign of the growing independence of Australian science. RP sent thirty full articles to the journal in its first four years, but only eight to British journals (plus eight letters to *Nature*). Although this corpus in the end probably lent more stature to the journal than did any other single field, it took a while for a world readership to develop.

Direct word-of-mouth, when possible, was of course also important. After attending a 1948 URSI meeting in Stockholm, Pawsey wrote back: 'Martyn and I, to put the matter rather bluntly, attempted to put Australia on the map, and I think were fairly successful.[81] And Bracewell[82] recalls that while he was at the Cavendish as a post-graduate student, Ryle's group thought Sydney work way behind, but when he returned to Radiophysics he found that they thought the same of the Cambridge work — his conclusion was that each side was acting on dated information. Preprints were not common in those days, and in any case were sent by sea mail (as were journals, even *Nature*), taking two to three months for the passage. Bracewell also remembers wanting to act as a link between the two groups: 'Being young and idealistic, I felt that I should try to close the gap, that a freer flow of information was a blow struck against entropy, as well as my duty.' As he wrote to Bowen in early 1948:

> Publication [of Australian work] is slow and the diffusion of advance news by word of mouth does not occur. It results that ideas of priority are fixed before Australian work filters through. This is the case with solar noise. The attitude in the Cavendish Lab. is that nothing much of value is done elsewhere . . . [Since] I am in an effective position for informal dissemination of news from Radiophysics, I recommend for your consideration the transmission of this news.[83]

It is important to note that although Radiophysics as a laboratory was

Radio astronomers at Sydney University for the 1952 URSI General Assembly. **Left to right, ground level:**
*W.N. Christiansen, F.G. Smith (England), J.P. Wild, B.Y. Mills, J.L. Steinberg (France), S.F. Smerd,
C.A. Shain, R. Hanbury Brown (England), R. Payne-Scott, A.G. Little, M. Laffineur (France), O.B. Slee,
J.G. Bolton. First step: C.S. Higgins, J.P. Hagen (US), J.V. Hindman, H.I. Ewen (US), F.J. Kerr, C.A.
Muller (Netherlands). Second step: J.H. Piddington, E.R. Hill, L.W. Davies. (*Courtesy of Radiophysics
Division — Archives*).

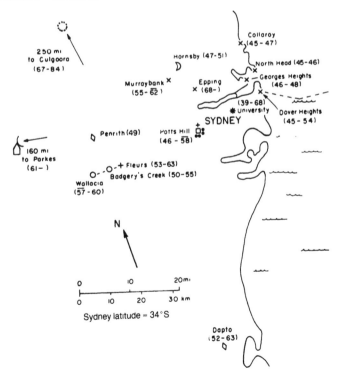

*Sites of chief Radiophysics
field stations and
headquarters at Sydney
University and at Epping.
Each station has the years
of operation indicated
(years with a bar overhead
are uncertain).*

isolated, its individual researchers were not. There were enough first-rate radio physicists in one place that they did not suffer from lack of intellectual interchange. RP thus avoided the phenomenon that Home[84] has called 'the isolation of the élite', referring to scientists (such as W.H. Bragg while in Australia) whose careers are debilitated by lack of peers within hundreds of miles.

One of the grandest opportunities for interchange and to advertise RP's work was the URSI General Assembly that met in Sydney in August 1952. This was a milestone for Australian radio research and a feather in the cap for all Australian science for it marked the first time that any international scientific union had met outside Europe or North America. In 1948 URSI had created a new Commission V on Extraterrestrial Radio Noise with Martyn as its first president and Pawsey as secretary. Martyn in particular engineered the General Assembly coming to Sydney and masterminded the organization and funding.[85] Sir Edward Appleton was the patriarch among the fifty foreigners in attendance, of whom about a third were active in radio astronomy. At last the RP staff could associate faces with names like Jean-Louis Steinberg from France, Robert Hanbury Brown from Jodrell Bank, F. Graham Smith from Cambridge, C. Alexander Muller from Holland, and H.I. 'Doc' Ewen from the United States. RP of course put on its best face for the guests, with a detailed, glossy 'Research Activities' booklet and full tours of all field stations. The visitors went home impressed.[86]

The field stations

The RP radio astronomy work took place at individual field stations, some as far as 30–50 miles from home base in the grounds of Sydney University. These sites provided sufficient land and isolation for observations free from man-made electrical interference. But since one or two sites well removed from Sydney could probably just as well have served for most of the experiments, the many sites also provided freedom from a second type of 'man-made interference': the RP staff simply preferred to spend most of their time alone at the field stations, not in a central laboratory, and management too found it to be a productive style of operation. By the late 1940s RP's research in radio astronomy was divided into many teams of two or three: leaders and sites about 1948–50 were Piddington and Minnett (University grounds), Kerr and Shain (Hornsby), Bolton (Dover Heights), Wild (Penrith), Mills (Badgery's Creek), Payne-Scott and Christiansen (Potts Hill), and Yabsley (Georges Heights). Christiansen has evoked the atmosphere of these stations:

> Each morning people set off in open trucks to the field stations where their equipment, mainly salvaged and modified from radar installations, had been installed in ex-army and navy huts . . . The atmosphere was completely informal and egalitarian, with dirty jobs shared by all. Thermionic valves were in frequent need of replacement and old and well-used coaxial connectors were a constant source of trouble . . . During this period there was no place for observers who were incapable of repairing and maintaining the equipment. One constantly expected trouble.[87]

Although groups had little day-to-day contact, Pawsey's skill as roving monitor and co-ordinator gave cohesion to the laboratory's radio noise work. This was achieved first through meetings every two to four weeks of the 'Propagation' committee (changed to 'Radio Astronomy' in 1949), which Pawsey chaired. These meetings provided a forum for progress reports, discussion of astronomical results and technical problems, floating of new ideas, co-ordination of experiments, and arguments about priorities. They also served to counter the danger that isolated groups would develop too narrow scientific or organizational perspectives. Several of those involved have commented on the value of these sessions, for instance Christiansen:

> Despite the fact that we were independent groups, we used to have these sessions, sort of what Americans call 'bull sessions', thinking of every conceivable sort of aerial . . . Á really good one would last all day. Joe Pawsey was one to stimulate that.[88]

Pawsey's second device for holding the radio noise research together was frequently to visit the field stations to see for himself what was happening and to give his advice. As Wild recollects:

> On some days he would arrive unexpectedly at one's field station, usually at lunch time (accompanied by a type of sticky cake known as the lamington, which he found irresistible), or else infuriatingly near knock-off time. During all such visits one had to watch him like a hawk because he was a compulsive knob-twiddler. Some experimenters even claimed to have built into their equipment prominent functionless knobs as decoys, especially for Pawsey's benefit[89] . . .[But] when one ran into problems, half-an-hour's discussion with Joe tended to be both soothing and rewarding.[90]

These visits, however, sometimes led to Pawsey seeing things he didn't like:

> Pawsey was in direct linkage with the little isolated groups. He'd try and make sure they didn't clash. And he stopped us working at times when Jack [Piddington] had had some idea and we'd started in a new direction . . . For instance, one day he found me [working on a radio analogue to a Fabry-Perot interferometer] and I was stopped. He said there are other people already there, and they've got a prior claim.[91]

But although Pawsey in general assigned exclusive turf to each small group, he sometimes encouraged two groups to plough ahead on the same problem if he felt their approaches differed enough. For example, Mills and Bolton for many years simultaneously worked on discrete sources with different types of interferometers.

Management of the radio noise work
Bowen turned over scientific leadership for the radio noise work to Pawsey, who was the division's number-two man from the start (although the office of Assistant Chief was not created until 1951). Pawsey thus had a free hand in running the radio noise side of things while Bowen took on the general administrative burden and concentrated on the rest of RP's programmes, taking a particular interest in the rain and cloud physics work to which he himself made several contributions. Bowen, however, minimized the number of his collaborations and so through 1951 published only seven papers. His career had seen more than its share of scientific directors who claimed credit for too much of what happened in their laboratory:

Pairs of 100 MHz (left) and 60 MHz Yagi antennas used for monitoring the sun and measuring its circular polarization at Dover Heights (1947). The same pairs, with the Yagi elements parallel and pointing toward the horizon, were used for the first studies and surveys of discrete sources by John Bolton (pictured) and his group (Courtesy of CSIRO Division of Radiophysics — Archives).

> When I became Chief, I was going to be quite certain of one thing . . . I was not going to jump in and claim credit when somebody else did the work . . . My previous experience of some pretty hard cases was that the best way to get first-class work out of people was to give them the credit.[92]

Along this line, Bolton recalls: 'Bowen was on our side in terms of letting people have their head — giving you a pat on the back when you did well and commiserating with you when something failed.[93]

The style of Bowen's and Pawsey's leadership very much fitted in with Rivett's philosophy discussed earlier — get the best people possible, give them the needed resources, and then let them run free.[94] But there were bounds to this freedom, as we have seen, leading to a creative tension between tight control of the laboratory's work, as it had necessarily operated during the war, and the kind of individual freedom one might find in a university department. This delicate balance is well illustrated by the juxtaposition of allowing workers to be scattered all over the countryside, while still closely supervising what they did. Other strong limits existed. For example, most scientific correspondence of staff members during these years was routed through either Pawsey or Bowen. More significantly, RP maintained a system of rigid internal reviews of all proposed publications, involving one or more of Bowen, Pawsey, and Arthur Higgs (technical secretary). The RP archives are replete with internal memoranda shuttling drafts back and forth between authors and management (and sometimes anonymous third-party RP referees), often to the frustration of the authors. As Bowen wrote in 1948: 'As I keep on telling the chaps, there is no doubt about the excellence of the work being done in the Laboratory, but the writing up is awful'.[95] But once a paper surmounted this first hurdle, a journal's referees almost seemed easy. The extent of Pawsey's influence on the radio noise papers can be gauged by the fact that half of them from the 1946–51 era specifically acknowledge his assistance with either preparation of the paper itself or the project in general. Yet he himself, like Bowen, published only seven papers through 1951.

Bowen and Pawsey agreed on the basic policies needed to run Radiophysics, but their differing temperaments led to differing contributions to RP's success in radio astronomy. As Christiansen recalls:

> [Pawsey's scientific style] set the tone completely . . . but he was a very, very unworldly fellow . . . Bowen was the man who got the money, the tough businessman, while Joe was the rather academic scientist. And it was an excellent combination.[96]

Bowen knew how to deal with the CSIRO hierarchy, how to pull off the necessary balance of applied and fundamental research, and how to manage Radiophysics as a whole. On the other hand, numerous RP staff members have indicated in interview that Pawsey by nature was not suited for such things. For instance, he abhorred (and avoided) making managerial decisions that he knew would cause upset.

Pawsey played a vital role for the radio noise researchers, however, as scientific father figure and mentor. He was about ten years older than most of the staff, who averaged only about thirty years of age, and he quickly gained their respect and confidence. The words of his protégés speak for themselves:

He had the ability to develop the latent powers in other people. All of the people who came out of that group — Christiansen, Mills, Wild, and so on — I also count myself in it — were made independent and skilful in their subject, experienced and self-reliant, quite largely because of Pawsey's way of drawing people out. He was not the kind of research leader who'd insist on claiming everything himself. But he fed in the ideas that other people developed — he was a teacher as much as anything. In the written record you don't find his name on many papers, but he was the inspiration behind an awful lot.[97]

It was in his research direction that Pawsey made his greatest contribution: patient, kindly, selfless direction; always probing for the simple vital question and significant experiment; and subtly transforming inexperienced research workers into leading contributors to their respective fields.[98]

Pawsey's style of science grew out of his training in the Cavendish Laboratory of Ernest Rutherford. He inherently loved the simple, inexpensive experiment and distrusted anything coming from complex set-ups. He also had an innate distrust of theory and mathematics,[99] complemented by a faith in experimentation. As he himself wrote in 1948 (regarding the possibility of solar bursts at frequencies less than a few hundred hertz):

My present guess is that the theory is wrong in general, and consequently I do not advise any time-consuming observations which are based on the theory. On the contrary, the observation of low-frequency noise is a fundamental scientific observation which is of value independent of the theory. Positive or negative results are of use. Hence this investigation is in order, and it is up to the experimenters to decide how far they go.[100]

The experimental style that Pawsey inculcated was particularly striking to 'Doc' Ewen, accustomed to much larger American budgets, when he visited Sydney for the 1952 URSI meeting:

Their equipment was shoestring stuff, but there were a lot of cute tricks . . . They didn't waste much time with hardware where it wasn't all that important, [or with] trying to make it look pretty. But wherever a part was critical to the operation of a device, they spent a lot of time thinking about it.[101]

And from the other perspective Christiansen recalls how Ewen reacted upon seeing his 21-centimetre hydrogen line receiver:

Ewen came out and said he had to see how these damn Australians did in three weeks what took other people eighteen months to do. And when he saw our gear, lying all over the room and on the floor, he just about passed out.[102]

The words of his colleagues capture Pawsey's scientific style better than anything else:

He had an enormous enthusiasm. It was always a delightful experience to bring to Pawsey some new idea or some interesting new observation. His immediate reaction would be one of intense interest, followed by suspicion as he looked for some mistake or misinterpretation, or what he called 'the inherent cussedness of nature'. Finally, if convinced that all was well, his face would shine with boyish pleasure . . . He never forced his opinions on a younger colleague; if the matter was open to doubt he was willing to leave it to experiment. He was, in fact, the arch-empiricist. 'Suck it and see' was one of his favourite expressions.[103]

One of the more active RP field stations, Potts Hill (October 1953), adjacent to a large Sydney water reservoir. Antennas visible include a 97 MHz Yagi element of Alec G. Little and Payne-Scott's solar interferometer (upper left), a 97 MHz 'model' (120 x 120 ft) Mills cross (lower left), a 10 ft dish used for 50 cm observations of solar eclipses in 1948–49 by Christiansen (right centre), a 16 x 18 ft rectangular paraboloidal section used for solar work and in 1951 by Christiansen and Hindman for 21 cm hydrogen line observations (top right), and a 36 ft dish used by Kerr and Hindman for later hydrogen line work (top centre) with a small dish near it used for testing and as the reference signal of a switched receiver.
(Courtesy of CSIRO Division of Radiophysics — Archives)

> You were always expected to keep your main programme going, but you could also do more speculative things on the side — 'on the wrong job number' as Pawsey put it. He was a strong believer in devoting part of one's efforts on 'long shots' or 'wildcats'.[104]
>
> Pawsey had a childlike simplicity about him, a childlike curiosity. He was not a sophisticated man in the least. I find this is a talent that a lot of people who are truly great have in common — retaining a feeling that science is not a business, that it's a game . . . If Joe had been a businessman, you would have called him a sucker, but [for science] I think that's actually an important characteristic.[105]

And a somewhat dissenting view has been given by F.F. Gardner, an ionospheric colleague of Pawsey's during these years:

> The impression of Joe as a naive, unworldly type is misleading. To some extent this was a pose, which contributed to his ability to 'draw people out' . . . Nor was he opposed to theory . . . In discussions he was able to grasp immediately what was said to him, even if poorly expressed, and he also was able to concentrate one's attention on the problem under consideration. Occasionally he would suggest solutions to some degree with tongue-in-cheek. His suggestions might not be appropriate, but enabled others to see the solutions.[106]

Why so successful?

When a laboratory is created for one specific mission and then, because of changed circumstances, tries to adapt to a different role, the results are often not successful. RP's shift from war to peace, however, is a striking counter to this maxim. Through skilful leadership, scientific expertise, and good fortune (for instance, how might the fledgling solar noise efforts have gone if the 'sunspot group of the century' had not shown up in February 1946?), RP put Australia at the forefront of post-war radio astronomy. In no other natural science did such international stature come to Australia during these years — perhaps the closest was the immunology research led by MacFarlane Burnet at Melbourne's Walter and Eliza Hall Institute of Medical Research. Many of the factors important in this achievement have already been discussed, but others deserve mention. One was the sheer size of the radio noise group, far larger than other institutions in the field — with so many projects going on simultaneously, one is much more likely to have at least one winner at any given time. The radio physicists were also supported by invaluable assistance from the large staff of technicians for electrical and mechanical work. There were also distinct advantages from a mild climate for research involving outdoor construction and experimentation.[107] One possible factor can be dismissed, however, namely that the Australians had the southern sky to themselves and therefore had no competition and only needed to mimic northern observers. The evidence of this chapter shows that this is untenable, if for no other reason than that the same sun is shared between north and south.

Post-war developments in Australian radio astronomy form an interesting contrast to those in England and in the United States. The high degree of inter-connectedness of the development of radar in the Allied nations meant that all three nations found themselves at the end of the war with giant radar laboratories and large numbers of scientists skilled in radar and in the practicalities of getting a job done quickly and under trying circumstances. In each nation, too, the relationship between government, the military, and science had been revolutionized, and physicists had gained a new prestige because of their contributions to winning the war. Why then did things develop so differently? In England and Australia, similar factors acted to foster the new science, with the major difference that the British leaders of radio research, unlike their Australian counterparts, had prestigious university posts to which they wanted to return. Meanwhile in America, little happened for a decade. After his tour of the US in 1948, Pawsey sent this assessment back to Bowen:

> Since my arrival I have been struck by an anomaly. Astronomers and physicists have displayed a great interest in our work but have not undertaken similar work themselves ... The astronomers of the US ... have now become thoroughly interested in the implications but have not yet taken the plunge of tackling a totally new technique. Meanwhile, the physicists, who at the close of the war had the skill and the inclination to undertake the radio side, but failed to interest the astronomers then, now have other interests. The result is that we have a first-class opportunity to establish the lead which we at present hold.[108]

With its world dominance in (optical) astronomy, its pioneering work by Jansky and Reber, and its much greater funding for science, why indeed did the United States not go anywhere with radio astronomy? First, the leadership in optical astronomy is of little relevance, for the early radio noise researchers in all countries were not part of astronomy, but came out of a tradition of ionospheric research and radar development. Second, the work of Jansky and Reber had little influence on the development of the post-war field, although in retrospect it can be seen to be the intellectual start of what later became radio astronomy. Neither man was influential in the post-war community and the war meant that the course of radio noise studies would undoubtedly have proceeded much as it did even if Jansky and Reber had never made their discoveries. Third, the bounteous funding for science in the US came primarily from the Atomic Energy Commission and the military, and most funds went to projects that fulfilled a 'requirement', that is, had some military bearing. This meant, for example, that nuclear physics and related fields were heavily supported, and indeed many of the best American wartime radar researchers, such as E.M. Purcell and R.H. Dicke, went this way. Fourth, those who did stay in radio and ionosphere work went after the large contracts available for research related to military communications and radar needs, and these were seldom ideal for obtaining byproducts of use to radio astronomy. For example, the military was intensely interested in pushing most radar systems to ever higher frequencies, and so groups such as that at the US Naval Research Laboratory concentrated their efforts there. But, as it turns out, the sky emits only very weak signals in the microwave region and receivers of that era were in general not of sufficient sensitivity to make decent measurements. Progress in radio astronomy at that time, where the low-frequency sky had all the action, involved the much cheaper and easier-to-handle equipment at low frequencies. With little military influence and limited by their budgets to low-frequency work, the British and the Australians found themselves discovering the radio universe far more efficiently than could those with big budgets in the US.[109]

Furthermore, radio physics in the British Empire was traditionally done in physics departments, where the conceptual leap from the ionosphere to the solar atmosphere and then on out to the galaxy was more likely made than it would be in electrical engineering departments (such as at Stanford University), where American radio physics work was carried out.[110] Finally, the greater intensity of the war as experienced in England and in Australia led to a camaraderie among the post-war groups that was not found in the United States. Ewen recollects his meeting the RP researchers in 1952:

> My first impression was of a very gung-ho group . . . eager minds delighted with the opportunity to get together and chat about what was going on in the field. The [URSI] meeting was not typical of what you would have found in the US at that time. . .there was an excitement and camaraderie.[111]

As work on radio noise developed over the years in the Radiophysics Laboratory, there was a gradual integration of the research into astronomy proper and a transformation of radio physicists into radio astronomers. Even from the beginning Pawsey recognized that this new radio technique was fundamentally altering astronomical knowledge. As he stated during a talk to the August 1946 ANZAAS meeting in Adelaide:

> This [solar noise] work is a new branch of astronomy . . . New observational tools [in astronomy and astrophysics] have an unusual importance. The last outstanding development in solar instruments was probably the spectro-heliograph (developed at the turn of the century). Consequently it is reasonable to expect that the discovery of this radiation will come to be recognised as one of the fundamental advances in astrophysics.[112]

Yet although the RP staff realized that they were essentially doing astronomy, albeit of a wholly different type and not well understood by astronomers, their astronomical education proceeded in a chequered manner. Whereas Bolton[113] chose methodically to plough through volume upon volume of the *Astrophysical Journal* during long observing nights, most just picked up what they deemed necessary to interpret their observations as they went along. Books such as George Gamow's *The Birth and Death of the Sun* were read, and Bolton undertook a partial translation of Max Waldmeier's 1941 treatise on the sun. The contacts with Mt Stromlo, including joint colloquia held now and then, were also important as an exposure to astronomical knowledge. But RP had no staff astronomer and its orientation during the first post-war years was as often toward the techniques as the astronomy:

> We were simply radio people trying to provide another tool for detecting what these astronomers said was likely to be there . . . We didn't consider ourselves to be astronomers — our primary interest was in the equipment. In fact we'd just left a wartime situation and we knew that our success in radar stemmed from having people who were very well trained in the techniques.[114]

Things evolved, however, over the years, and by the early 1950s overseas trips, increasing contacts with astronomers, and the gradual accumulation of astronomical knowledge had caused a clearer picture to emerge of how the radio work fitted into astronomy as a whole.

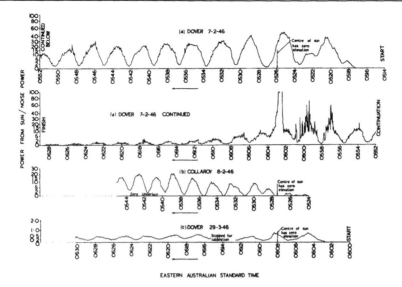

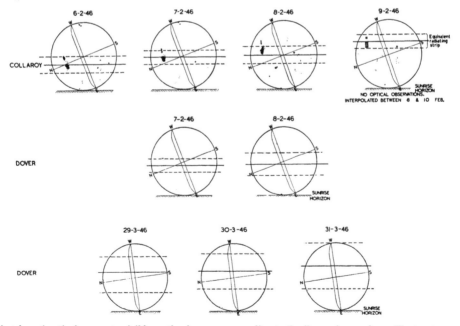

Sea-cliff interference patterns obtained at 200 MHz in February and March 1946. When the sun rose, the fringes suddenly appeared (note that the 'radio sunrise' came earlier than the optical sunrise) and then gradually faded away about an hour later as the sun moved above the beam of the antenna. Note the greater ratio of fringe maximum to minimum for the top observations, indicating that the radiation was originating from a smaller region of the sun. The very fast variations and intense signals recorded at 0600 on 7 February represent a solar outburst. Note the closer spacing of the fringes for the Collaroy observations taken from a higher elevation above the sea. (McCready, Pawsey and Payne-Scott, 1947 (n.38))

Sketches of optical sunspots visible on the days corresponding to the figure shown above. The top two rows are dominated by a large sunspot group, while the March observations show much less activity. 'N–S' indicates the rotation axis of the sun. The three horizontal lines on each sketch indicate the centre and estimated width of the 'equivalent radiating strip' causing the radio fringes. (McCready, Pawsey and Payne-Scott 1947 (n.38))

The early 1950s represent a watershed from several other perspectives. For the first time work on solar noise was overtaken in quantity by that on 'cosmic' (non-solar) noise — the percentage of solar papers dropped from around 70 per cent before 1951 to approximately 40 per cent during 1952-54. At this time, also, Pawsey and Bracewell wrote a masterful monograph, *Radio Astronomy* (mostly written in 1952, but not published until 1955),[115] that formed a capstone to the first stage of the field's development and was to remain the definitive textbook for a decade. And of course the 1952 URSI meeting also happened at this juncture.

The most important change that began during the early 1950s was that from a large number of relatively small experiments to a smaller number of projects on a large scale. This was the start of the transition from Little Science to Big Science — or, as Wild had pungently described it, moving from trailers 'with a characteristic smell' to air-conditioned buildings.[116] This story is beyond the scope of this chapter, but can be very briefly summarized as follows. Scientific progress now largely demanded huge antennas and arrays and many of these were beyond the capacity of RP to produce in-house. For example, in 1951 bids for antennas were sought outside for the first time (for 50 ft and 80 ft dishes).[117] In 1953 RP funded its last major antenna from its own resources: the 1500-foot Mills cross array at Fleurs for £2500.[118] More costly ventures did not come easily, however, for the government and CSIRO were not willing to support large capital projects.[119] [120] First thoughts about a 'Giant Radio Telescope' and its funding went as far back as 1948. Bowen at that time tried to convince the Royal Australian Air Force to build a huge radar antenna that could be sometimes used for radio astronomy.[118] Several designs were studied over the next few years, some as large as 500 feet in dimension, but the funding never materialized. By 1951 the search for funds shifted to non-military sources and eventually the key money came from American foundations. In the end Radiophysics finished the transition to Big Science only with the commissioning of a resulting 210-foot dish at Parkes in 1961.

The maturation of the science was of course also mirrored in the growing seniority of the RP staff. Pawsey had had great success in scientifically rearing his junior colleagues, but RP was not like a university department with its steady stream of students — the RP 'students' had nowhere to go and there were no positions for new ones. Already in 1951 Pawsey was saying that the outstanding defect in the radio astronomy group was its lack of the 'research student' type[122] with which he worked so well. Through the 1950s the various group leaders became strong-willed, confident individuals, arguing their own particular visions of how radio astronomy at RP should be done. Major disputes centred on two questions. Should the focus continue on small technique-oriented groups or shift to a single major facility? And, which types of antennas will pay off best? Furthermore, Bowen and Pawsey began to work less well as a team and developed their own significant differences on these issues. In sum, the tale of the later 1950s turned out to be anything but a simple extension of the pioneering period covered in this chapter.

Australia's history was linked with astronomy from the start — Captain Cook's first voyage was as much to observe the 1769 transit of Venus across the solar disc in Tahiti as it was to explore for *Terra Australis Incognita*, leading to the discovery of Australia's eastern coast the following year. But 175 years would pass before Australians became part of the first rank of world astronomical research. And when this happened, from cliff edges only a few miles removed from Cook's landing site at Botany Bay, it was in a most unlikely manner, for they did their astronomy not with glass lenses, but with rods of metal.[123]

Notes

This article is an adaptation and shortening of a chapter of a forthcoming book, to be published by Cambridge University Press, on the worldwide history of radio and radar astronomy until the early 1950s. The tapes and transcripts of the interviews cited below (conducted over the period 1971-85) will eventually be deposited at the Center for the History of Physics of the American Institute of Physics in New York City.

1   For further details about the overall development of early radio astronomy, see the following: J.S. Hey, *The Evolution of Radio Astronomy* (London, 1973); D.O. Edge and M.J. Mulkay, *Astronomy Transformed* (New York, 1976); W.T. Sullivan, III (ed.), *Classics in Radio Astronomy* (Dordrecht, 1982); W.T. Sullivan, III (ed.), *The Early Years of Radio Astronomy* (Cambridge, 1984), and W.T. Sullivan, III, 'Early radio astronomy', in *Astrophysics and Twentieth-century Astronomy to 1950*, O. Gingerich (ed.) (Cambridge, 1984), pp. 190-8.

2   The subsequently most famous member of the wartime staff was certainly eighteen-year-old Joan Sutherland, who worked as a typist for £2 per week, before resigning to pursue her career as one of the premier sopranos of this century [April 1944 to January 1945, file G23/11, CSIRO Archives, Canberra].

3   S.S. Swords, 'A technical history of the beginnings of radar', Tech. Rept. MEE1 (PhD thesis, Dept of Microelectronics and Electrical Engineering, Trinity College, Dublin, Ireland, 1983), pp. 539-57.

4   Bowen's career particularly well illustrates the extremely close cooperation among the Allies on the wartime development of radar. For his memoirs of the pre-1944 period see his book *Radar Days* (Bristol, 1987).

5   E.G. Bowen, 2 July 1945, 'Future programme of the Division of Radiophysics'; White to Bowen, 15 June 1945; Rivett to Bowen, 3 July 1945; G.A. Cook to Bowen, 26 July 1945 — all in file D1/1, Radiophysics Division Archives, Epping [RPS].

6   C.B. Schedvin, 'The culture of CSIRO', *Australian Cultural History*, 2 (1982), 76-89.

7   E.G. Bowen (ed.), *A Textbook of Radar* (Sydney, 1947).

8   Interview with author, 1971.

9   R.W. Home, 'Origins of the Australian physics community', *Historical Studies*, 20 (1983), 383-400.

10 D.P. Mellor, *The Role of Science and Industry* (Series 4, Vol. 5 of *Australia in the War of 1939-1945*) (Canberra, 1958).

11 E.G. Bowen, 'The origins of radio astronomy in Australia', in Sullivan (ed.) 1984, op. cit. (n.1.)

12 Bowen, op. cit. (n.5), p.23. Bowen did not, however, propose a complete break from military ties. He suggested that RP personnel should serve as scientific consultants to the various services, and that research work on *defensive* radar against ballistic missiles ('likely to form a great bulk of the offensive weapons in the next war') would be proper and important.

13 Bowen, op. cit. (n.11), p.110.

14 File D2, RPS.

15 E.B. Kraus and P. Squires, 'Experiments on the stimulation of clouds to produce rain', *Nature*, 159 (1947), 489-91.

16 Bowen, op. cit. (n.11), pp.105-9.

17 Interview with H.C. Minnett, 1978.

18 CSIRO Division of Radiophysics *Annual Report 1949*.

19 CSIRAC, which was used at RP from 1951 on, employed 2000 tubes, had a 2 msec cycle time, and stored 1K 20-bit words in mercury acoustic-delay lines. See M. Beard and T. Pearcey, 'The genesis of an early stored-programme computer: CSIRAC', *Annals of the History of Computing*, 6 (1984), 106-15.

20 Bowen, op. cit. (n.5), p 6.

21 Biographical information on Pawsey comes from the obituaries by A.C.B. Lovell (*Biog. Mem. Fell. Roy. Soc.*, 10 (1964), 229-43, including a section on wartime work by H.C. Minnett) and by W.N. Christiansen and B.Y. Mills (*Australian Physicist*, 1 (December 1964), 137-41).

22 Schedvin, op. cit. (n.6)

23 Interview with author, 1978.

24 J.L. Pawsey, 'Australian radio astronomy', *Australian Scientist*, 1 (1961), 181-6.

25 Interview with author, 1976.

26 Further details on this missed opportunity (as well as on the Pawsey-Martyn controversy discussed briefly in the section 'The million degree corona') will be in the forthcoming book mentioned in the introduction to these notes.

27 Interview with H.C. Minnett, author, 1978.

28 J.L. Pawsey and R. Payne-Scott, 'Measurements of the noise level picked up by an S-band aerial', Rept. No. RP 209, CSIR Radiophysics Laboratory, Sydney, 11 April 1944.

29 A.S. Eddington, 'Diffuse matter in interstellar space', *Proc. Roy. Soc.*, 111 (1926), 424–56, 430.

30 E. Alexander, 'Report on the investigation of the "Norfolk Island effect"', R.D. 1/518, Radio Development Lab., D.S.I.R., Wellington, New Zealand, 1945.

31 The primary source for the story of the Norfolk Island effect is the restricted report by Alexander, ibid. (n.30). This, as well as a sheet by Alexander titled 'Long wave solar radiation' (17 December 1945), can be found in the Appleton papers in the Library of Edinburgh University; I have also deposited a copy of the report in the Radiophysics Division Library, Epping, NSW. Relevant correspondence of Bowen and Pawsey is in file A1/3/1a, RPS.

32 Bowen to F.W.G. White, 1 August 1945, file A1/3/1a, RPS. A 'C.O.L. antenna' was one for 'Chain Overseas Low-flying', in contradistinction to the original Chain *Home* system in England.

33 Pawsey to C.W. Allen, 15 October 1945, file A1/3/1a, RPS.

34 J.L. Pawsey, R. Payne-Scott, and L.L. McCready, 'Radio-frequency energy from the sun', *Nature*, 157 (1946), 158–9.

35 In this article I use the term 'sea-cliff interferometer', although in the period under discussion the arrangement was called either a 'sea interferometer' or a 'cliff interferometer'.

36 J.S. Hey, 'Solar radiations in the 4–6 metre radio wave-length band', *Nature*, 157 (1946), 47–8.

37 E.V. Appleton, 'Departure of long-wave solar radiation from black-body intensity', *Nature*, 156 (1945) 534–5.

38 L.L. McCready, J.L. Pawsey, and R. Payne-Scott, 'Solar radiation at radio frequencies and its relation to sunspots', *Proc. Roy. Soc.*, A190 (1947), 357–75.

39 J.C. Jaeger, 17 March 1943, 'Theory of the vertical field patterns for R.D.F. stations', Report No. RP.174.

40 McCready et al., op. cit. (n.38), pp. 367–8.

41 A full obituary of Martyn is given by H.S.W. Massey in *Biog. Mem. Fell. Roy. Soc.* 17,(1971), 497–510.

42 V.L. Ginzburg, 'On solar radiation in the radio-spectrum' [in English], *Dokl. Akad. Nauk. SSSR*, 52 (1946), 487–90.

43 I.S. Shklovsky, 'On the radiation of radio-waves by the galaxy and by the upper layers of the solar atmosphere' [in Russian], *Astron. Zh.*, 23 (1946), 333–47.

44 D.F. Martyn, Temperature radiation from the quiet sun in the radio spectrum', *Nature*, 158 (1946), 632–3.

45 J.L. Pawsey, 'Observation of million degree thermal radiation from the sun at a wavelength of 1.5 metres' *Nature*, 158 (1946), 633–4.

46 Mellor, op. cit. (n.10), pp. 502–9.

47 Interview with S.F. Smerd, 1978.

48 C.W. Allen, 'Solar radio-noise of 200 Mc./s. and its relation to solar observations', *Mon. Not. RAS*, 107 (1947), 386–96.

49 Martyn, op. cit. (n.44).

50 D.F. Martyn, 'Polarization of solar radio-frequency emissions', *Nature*, 158 (1946), 308.

51 Martyn to Bowen, 16 April 1947 and 25 June 1947, file A1/3/1a, RPS. Further information on Woolley and non-radio work at Mt Stromlo during his era may be found in S. Davies, *Historical Records of Aust. Science*, 6 (1984), 59–69, and in S.C.B. Gascoigne, *Proc. Astron. Soc. Aust.*, 5 (1984), 597–605.

52 Bowen to Woolley, 11 November 1947, file D5/4/35, RPS.

53 C.B. Schedvin to author, 27 May and 1 October 1986; idem, *Shaping Science and Industry: A History of Australia's Council for Scientific and Industrial Research, 1926–49* (Sydney, 1987), pp. 256–9.

54 Woolley to Bowen, 12 April 1947; Martyn to Bowen, 16 April 1947; Woolley to White, 12 February 1951; White to Woolley, 23 February 1951; Pawsey to Bowen (in London), 30 March 1951; Pawsey to Oliphant, 10 September 1951 — all file A1/3/1n, RPS. I also thank R. Peter Robinson for information on this proposal.

55 J.S. Hey, S.J. Parsons, and J.W. Phillips, 'Fluctuations in cosmic radiation at radio-frequencies', *Nature*, 158 (1946), 234.

56 Pawsey to Woolley, 11 September 1946, file A1/3/1a, RPS.

57 Interview with J.G. Bolton (1976).

58 Interview with G.J. Stanley (1974).

59 Although Slee did not join RP until November 1946, he had made an independent discovery of the radio sun while operating a radar set near Darwin in March 1946, a discovery which he duly reported in detail to RP [J.N. Briton to Slee, 18 March 1946, file A1/3/1a, RPS].

60 J.G. Bolton, 'Radio astronomy at Dover Heights', *Proc. Astron. Soc. Aust.*, 4 (1982), 349-58.

61 R. Payne-Scott, D.E. Yabsley and J.G. Bolton, 'Relative times of arrival of bursts of solar noise on different radio frequencies', *Nature*, 160 (1947), 256-7.

62 Bolton, 'Summary of Cygnus results — 4th-25th June, 1947', 1 page, undated, file A1/3/1a, RPS.

63 J.G. Bolton and G.J. Stanley, 'Variable source of radio frequency radiation in the constellation of Cygnus', *Nature*, 161 (1948), 312-3.

64 J.G. Bolton and G.J. Stanley, 'Observations on the variable source of cosmic radio frequency radiation in the constellation of Cygnus', *Aust. J. Sci. Res.*, A1 (1948), 58-69.

65 Ibid., p. 64.

66 Ibid. p. 68.

67 As it turned out, Bolton and Stanley's first Cygnus A position was a full degree north of the correct position and so there was no chance of finding an optical counterpart. But even positions measured to accuracies of a few arc minutes a few years later were unable to disclose an optical identification. Not until 1' accuracy was achieved by F. Graham Smith at Cambridge in 1951 did it become possible to identify an extremely faint, peculiar galaxy as the source of the radio emission.

68 Pawsey to Bowen and Bolton, 11 November 1947 (from Lincoln, Nebraska), file A1/3/1a. RPS.

69 J.G. Bolton, 'Discrete sources of galactic radio frequency noise', *Nature*, 162 (1948), 141-2.

70 Ibid., p. 141.

71 It was only fitting that an 'aleing' star observed from Australia should radiate its swan song in the form of Cygnus-type radio noise!

72 Interview with author, 1976.

73 J.G. Bolton, G.J. Stanley and O.B. Slee, 'Positions of three discrete sources of galactic radio-frequency radiation', *Nature*, 164 (1949), 101-2.

74 G. Blainey, *The Tyranny of Distance: How Distance Shaped Australia's History* (Melbourne, 1968).

75 Groundwork was also being laid for a fifth purpose of overseas trips, namely fundraising for large antennas, but here the payoffs only began in the mid-1950s.

76 Bowen (from London) to Pawsey, 23 August 1946, file A1/3/1a, RPS; Bolton (from London) to Bowen, 14 April 1950, file F1/4/BOL/1, RPS.

77 M. Ryle and D.D. Vonberg, 'Solar radiation on 175 Mc./s.', *Nature*, 158 (1946), 339-40.

78 Pawsey to Ratcliffe, 10 September 1946; Ratcliffe to Pawsey, 17 September 1946 — both file A1/3/1a, RPS.

79 Interview with author, 1971.

80 Interview with author, 1976.

81 Pawsey to Bowen, 12 August 1948, file F1/4/PAW/1, RPS.

82 Interview with author, 1980.

83 Bracewell to Bowen, 19 January 1948, file F1/4/BRA, RPS. Despite his request, it appears that Bracewell himself remained little better informed than others in Cambridge. Upon his return to Australia in late 1949, he was 'surprised at the magnitude of the projects that had already been completed in Australia without his having heard of them in England' (H. Newton, *Observatory*, 70 [1950], 56). This induced Bracewell over succeeding years to send three short papers to *Observatory* for the purpose of advertising RP's work.

84 R.W. Home, 'The problem of intellectual isolation in scientific life: W.H. Bragg and the Australian scientific community, 1886-1909', *Hist. Recs. Aust. Sci.*, 6 (1986), 19-30.

85 For details, see R.N. Bracewell, 'Early work on imaging theory in radio astronomy', in Sullivan, (ed.), op. cit. (n.1), pp. 167-90; at pp. 169-74

86 Interview with F.G. Smith, 1976.

87 W.N. Christiansen, 'The first decade of solar radio astronomy in Australia', in Sullivan (ed.) 1984, op. cit. (n.1), pp. 113-4.

88 Interview with author, 1976.

89 Pawsey was in good company with his compulsive knob-twiddling. Philadelphia Orchestra conductor Leopold Stokowski was so notoriously unable to resist turning knobs of audio equipment during recording sessions in the 1930s that the engineers finally provided him with an entire control booth that was nothing more than an electronic placebo [R.E. McGinn, *Technology and Culture*, 24 (1983), 38-77 (p. 45)].

90 J.P. Wild, 'The beginnings of radio astronomy in Australia'. *Records Aust. Acad. Sci.*, 2(3) (1972), 10.

91 Interview with H.C, Minnett, 1978.

92 Interview with E.G. Bowen, 1973.

93 Interview with author, 1978.

94 Bowen, op. cit. (n.11), p. 110.

95 Bowen to Pawsey (in England), 23 July 1948, file D5/7, RPS.

96 Interview with author, 1976.

97 Interview with F.J. Kerr, 1971.

  98 J.P. Wild, 'The exploration of the sun by radio', *Australian Physicist*, 5 (1968), 117–22.
  99 Interview with K.C. Westfold, 1978.
 100 Pawsey (from Ottawa) to Bowen, 13 February 1948, file A1/3/1a, RPS.
 101 Interview with author, 1979.
 102 Interview with author, 1976.
 103 Christiansen and Mills, op. cit, (n.21), p. 139.
 104 J.P. Wild to author, 30 July 1986.
 105 Interview with G.J. Stanley, 1974.
 106 F.F. Gardner to author, 12 November 1986.
 107 Interview with J.G. Bolton, 1978.
 108 Pawsey to Bowen, 15 April 1948, 'Solar and cosmic noise research in the United States and Canada', file A1/3/1b, RPS.
 109 If RP had had access to US-level funds in the late 1940s, one can speculate that its high-energy accelerator work might have thrived to the point of dominating the entire laboratory's programme. In such a case it would seem unlikely that RP's staff would have achieved such a distinguished international stature as they in fact did with radio astronomy.
 110 Madsen's Electrical Engineering Department at the University of Sydney was somewhat of an exception to this rule, but even in this case almost all of the undergraduate degrees (for those who eventually became researchers in radio astronomy) were combined engineering *and* physics.
 111 Interview with author, 1979.
 112 Pawsey, 'Solar radiation at radio frequencies and its relationship to sunspots: interpretation of results', talk delivered to meeting of the Australian and New Zealand Association for the Advancement of Science, 19 August 1946, file A1/3/1a, RPS.
 113 Interview with author, 1978.
 114 Interview with J.V. Hindman, 1978.
 115 J.L. Pawsey and R.N. Bracewell, *Radio Astronomy* (Oxford, 1955).
 116 J.P. Wild, 15 October 1965 lecture delivered to CSIRO Division of Plant Industry, 'Origin and growth of radio astronomy in CSIRO', file D12/1/5, RPS.
 117 'Tentative specifications for 80-foot and 50-foot radio astronomy aerials', 8 August 1951, file A1/3/1h, RPS.
 118 B.Y. Mills, 31 March 1953, 'Large aerial for meter wavelengths', file A1/3/10, RPS.
 119 Interview with E.G. Bowen, 1978.
 120 E.G. Bowen, 'The pre-history of the Parkes 64-m telescope', *Proc. Astron. Soc. Aust.*, 4 (1981), 267–73.
 121 Bowen, op. cit. (n.11), pp. 98–9.
 122 Pawsey to Oliphant, 10 September 1951, file A1/3/1n, RPS.
 123 I thank the US National Science Foundation's Program in History and Philosophy of Science for substantial support for this project. My research into the Australian history has also been immeasurably aided by Sally Atkinson, Archivist of the CSIRO Radiophysics Laboratory. Her knowledge of the RP archives is minute and her assistance to me over the past decade has been indefatigable. I have greatly profited from detailed comments on a draft of this article by Sally Atkinson, John Bolton, Taffy Bowen, Ron Bracewell, Chris Christiansen, Frank Gardner, Mott Greene, Jim Hindman, Rod Home, Frank Kerr, Bernie Mills, Harry Minnett, Jim Roberts, Peter Robertson, Boris Schedvin, Bruce Slee, Gordon Stanley, Kevin Westfold, Paul Wild, and Don Yabsley.

14

Australian astronomy since the Second World War

S.C.B. Gascoigne

The end of the Second World War saw a remarkable upsurge in Australian astronomy. Within a few years a series of brilliant discoveries had placed Australia in the front rank of the world's astronomical nations, a position it has maintained ever since. Once before, in the latter part of the nineteenth century, astronomy had been important in Australia. The colonial observatories were prominent in the scientific and cultural life of the day, their directors were acknowledged as the principal scientific advisers to their governments, and the world's largest telescope, a 48-inch reflector, was part of Melbourne Observatory. But at turn of the century a combination of factors, economic, political and scientific, saw the observatories go into a long decline. In particular, they had little connection with the post-war resurgence: that event was initiated by different people, at different institutions, with different objectives, and in a quite different social and economic climate. Rather than in the state observatories, its beginnings are to be found in two Commonwealth bodies, the Commonwealth Solar Observatory (hereafter Mount Stromlo Observatory or MSO), and the Radio Research Board (RRB).

The precursors

That the MSO was established at all was due almost entirely to the efforts of W.G. Duffield,[1] an Australian physicist then living in England. Duffield conceived the idea at a scientific conference in 1905, while still a graduate student. He enlisted some powerful support, and though no great scientific achiever himself, had the spirit and the tenacity of purpose to overcome many setbacks and finally bring his vision to reality in 1924. The observatory was built on Mount Stromlo, west of Canberra, with Duffield, naturally enough, as its first director.

Duffield died early, in 1929, and because of the depression his position was not filled until Richard Woolley[2] was appointed in 1939. The choice of Woolley was critical. He was an astronomer, of the classical British school of Cambridge and Greenwich, and his avowed purpose was to change completely the direction of the observatory, by discontinuing its existing programme of solar and geophysical work and concentrating exclusively on stellar astronomy. The change had momentous consequences.

Woolley's plans were delayed by the onset of the Second World War, during which the observatory became a munitions factory, employing seventy people at peak and engaged mostly in the design and manufacture of precision optical instruments for the forces. As it happened, more useful experience for a young observatory-to-be could hardly have been imagined. The return to peace found the observatory with a staff of thirty-odd, a three-fold increase over the prewar figure; a workshop inherited from the war, the envy of astronomers everywhere; a 30-inch reflector, the largest working telescope in the country; and, of course, access to the relatively unexplored southern sky. But the staff had still to find their way into an old and well-established subject, and even more to master for themselves its particular observational techniques. This took time, and it was some years before the MSO programmes were fully under way.

The Radio Research Board was founded in 1926, following approaches by Professors Madsen (Sydney) and Laby (Melbourne) to CSIR, the PMG, and industry.[3] Strong support was forthcoming, and laboratories were set up in Sydney, for work on the ionosphere, and in Melbourne, for work on atmospherics. The RRB was a most successful organization, both as a research group and as a training-ground for young Australian physicists. Many notable people passed through its ranks, and D.F. Martyn, who developed into its chief scientist, achieved a considerable international reputation. (With Woolley, he was a leading contender for the directorship of Mount Stromlo in 1938.)[4]

Early in 1939, word was passed to Australia about the then new and secret system of radar. Steps were at once taken to set up within CSIR a suitable radar laboratory; the Radio Research Board was a natural starting-point. From the beginning it was agreed that because of the outstanding work of the CSIR–RRB physicists, the new laboratory would not merely duplicate existing British equipment, but would also carry out its own research and development. Later, a second equally important decision was taken: after the war the laboratory would remain in being as a permanent division of CSIR, devoting itself to peace-time applications of radio science. Thus the Radiophysics Division (RP), which was to bring such lustre to Australian science, and of which no fewer than five members were to be elected to the Royal Society.[5] (It is to be noted that after the war the corresponding body in the USA, the Radiation Laboratory at MIT, was disbanded.)

Radiophysics

The early discoveries[6]

The end of the war found the Radiophysics Division with a brilliant young staff completely at home in the most advanced radio techniques of the day. They had a mass of now surplus radar equipment, and in J.L. Pawsey a scientific director of the very first rank.[7] There was not an astronomer in sight. Evidence had been accumulating for the existence of sources of 'cosmic static' in the sky, some of them associated with the sun, and under Pawsey's leadership several groups turned to this field. Corresponding British groups did likewise, and within a few years a series of striking discoveries had laid the foundation for the new science of radio astronomy.

Pawsey himself worked on the sun. He and his group had found that at metre wavelengths, solar radiation was subject to rapid and violent variations that correlated in a general way with the presence of sunspots. To investigate these more closely required an antenna with a beamwidth small compared with the diameter of the sun (30 arc minutes); this need for high resolution was already urgent in radio astronomy. Pawsey met it by observing from an old radar site at Dover Heights, Sydney, where the aerials looked out over the sea from clifftops high above it. By using them to detect the direct rays from a source rising above the sea, simultaneously with the rays reflected by the sea, he constructed an interferometer, the first used in radio astronomy. It was the radio analogue of Lloyd's mirror, well known in physical optics, and is described more fully elsewhere in this volume.[8] It enabled a point source to be distinguished unambiguously from one, say, the size of the sun, and also allowed its altitude and hence its position in the sky to be measured to within a few arc minutes.

Two major lines of work were initiated with the cliff interferometer, as Pawsey called it. The first concerned the sun, which at the time (1946) was being crossed by a huge spot. Its radio emission was abnormally strong, and Pawsey showed conclusively that the excess radiation came from the immediate vicinity of the spot; it was so intense — a thousand times more so than the 'quiet' sun — that it could hardly be explained by the physics known at that time. Sensing the importance of this discovery, he and his group embarked on a full-scale investigation of solar radio emission and its relation to optical activity. In 1949 Wild built his first dynamic spectrograph and began his classification and analysis of solar bursts, and in 1952 Christiansen built his first grating interferometer and produced the first detailed pictures of the radio sun, on the way inventing earth rotation synthesis. These lines of work transformed solar physics, and we shall return to them later.

Using the same cliff interferometer, Bolton and Stanley discovered a number of 'radio stars', and having determined reasonably accurate positions for them, proposed in 1949 the identification of three with known optical objects. The first was the Crab Nebula, the well-known remnant of a supernova that had exploded in AD1054; the second, M 87 or Virgo A, was a galaxy with a peculiar and now famous jet; the third, NGC 5128 or Centaurus A, was another peculiar object, the true nature of which as an external galaxy was not established until 1954. This was a great discovery. It revealed for the first time the existence of a quite new astronomical species, the radio galaxies, with properties at once completely unexpected and of the greatest significance.[9] The work was followed by Mills and others. A succession of different antenna systems, culminating in the famous Mills cross, enabled observations to be pushed to increasingly fainter limits and more accurate positions, and to achieve better discrimination between point and extended sources. Mills made the first systematic source survey of the southern sky, and showed that, as foreshadowed by Bolton, there were two types of source. One proved to be associated with supernovae within our own galaxy, the other with remote galaxies outside it, though it took time to establish this. He built the first Mills cross, and was among the first to apply radio astronomy to cosmological problems.

The third main development followed the discovery, made in 1951 in the Netherlands and in the USA, of the 21-centimetre line of atomic hydrogen. Like other radio lines discovered later, this one had two properties that made it a powerful tool for exploring galaxies. First, if the gas is moving along the line of sight its velocity can be measured quickly and accurately from changes in its wavelength (the Doppler effect). Second, radio waves can travel freely through regions of interstellar dust that are opaque to visible light; such regions are common in galaxies like ours. The discovery of the 21-centimetre line was immediately taken up in Australia, and, in what grew into a major collaborative programme with the Dutch, the large-scale spiral structure of our galaxy was revealed for the first time. A second application, the first to an external galaxy, was to confirm that the Large Magellanic Cloud is a flat rotating galaxy not unlike our own, to make the first estimate of its total mass, and to explore its radio structure, which was to reveal most unexpected features.

The Parkes telescope

These discoveries reasonably defined the radio astronomy of the 1950s. Many new observatories were established throughout the world and the subject advanced rapidly. The Australian contribution had been an essential one, nowhere more so than in the development of high-resolution radio telescopes. Australians had built the first radio interferometer, the first linear array of steerable paraboloids, and were the first to combine the output of two crossed arrays to give a pencil beam — the Mills cross. They had made major contributions to aperture synthesis, to earth rotation synthesis, and to the mathematics of image reconstruction. These techniques refer to the mapping in fine detail of extended areas of sky, using observations made not with single dishes but with arrays of spaced antennae. Difficult to explain in simple terms, they are, as R.H. Frater has said, the absolute foundation of modern radio astronomy.[10] And, to repeat, they are to be attributed not to astronomers, but to engineers and physicists.

Against this background the next step at the Radiophysics Division was not obvious, but after a number of alternatives had been considered, Bowen decided it was to be a large steerable paraboloid, the radio analogue of an optical reflector. This was a major decision. Not only was the cost considerable — $1 500 000 in 1960 — but also it meant that the several individual groups engaged in cosmic research would have to amalgamate and redirect their resources to the new telescope. It meant, too, a complete break with the Pawsey style of single-purpose, inexpensive equipment, built for the short rather than the long term. Limits on funding allowed only one other major instrument to be built; this decision went to the Wild radioheliograph.

These changes, and the accompanying reorganization, triggered other effects. Before the telescope was commissioned Mills and Christiansen had left for the University of Sydney, and Pawsey for the USA. Pawsey was to have been the first director of the US National Radioastronomy Observatory at Greenbank, but a fatal illness intervened and in 1962 he died in Sydney. He was a much-loved man, and nobody had contributed more to physical science in this country.[11]

Over the years other people of course left Radiophysics, many of them to make considerable names elsewhere.[12] But the advent of the Parkes telescope was certainly the end of an era. With it RP entered the realm of 'big' science, and if some of the magic seemed to have gone, perhaps it was a sign that the pioneer days were truly over. At the same time the subject itself was changing: radio and optical astronomy were merging into one, and the new generation already coming up would hardly distinguish between the two.

The telescope was commissioned at Parkes, NSW, in October 1961, with John Bolton, back from Caltech, as director. Much of the credit for the whole telescope project belongs to Bowen, who organized the funding virtually single-handed, and played a major part in both the design and construction. It is a huge, splendidly engineered instrument, and was pre-eminent in its class for years.[13] It served, for instance, as the prototype for the three almost identical space communications telescopes for the NASA Deep Space Network, one of them located at Tidbinbilla, ACT.

The CSIRO radio telescope at Parkes, NSW. The diameter of the dish is 64 m, its area 0.32 hectares or 0.79 acres. When the telescope is pointing vertically upwards the cabin at the top of the tripod is 56 m above ground level. (Courtesy of CSIRO Division of Radiophysics, Epping, NSW)

Perhaps the outstanding feature of the Parkes telescope is the quality of its huge dish — 64 metres in diameter and approaching an acre in area. It was found possible to upgrade it, in successive stages, so that by 1986 the central 44 metre diameter could be used at wavelengths as short as a centimetre, and the central 17 metre at wavelengths of a few millimetres; this innermost fraction does not depart from a true paraboloid by more than 0.3 millimetres. None of this would have been of any avail without matching receivers, built to cover the appropriate frequency ranges, and sensitive and stable enough to measure the minute signals emitted by astronomical sources (optical astronomers have of course the same problem, but towards the other end of the electromagnetic spectrum, where the technology is different). Building receivers of this quality means working at the frontiers of electronics, with exotic solid-state devices operating at temperatures close to that of liquid helium to reduce the noise level. The shorter wavelengths are more difficult to work at, but they offer technical advantages such as better directivity, while the great prize is that this region is so rich in molecular and other spectral lines, among them, for example, the highly important 2.6 millimetre line of carbon monoxide.

An early major programme at Parkes, one in which Bolton took the leading part, was to prepare a new source catalogue. By 1969 it included 2133 sources in the region south of +27 degrees and had become a standard reference, distinguished for its accuracy, homogeneity, and wavelength coverage, sources being measured at 75, 21 and 11 centimetres. (The number of sources is now about 8000.) A similar large-scale project was the mapping of the southern galaxy and of the Magellanic Clouds in the 21-centimetre neutral hydrogen line. A beam as narrow as that of the Parkes dish (15 arc minutes half-width at 21 centimetres) was unprecedented at the time, and the detail in the resulting maps transformed current ideas, particularly about the Magellanic Clouds. The quality of the velocity data was similarly enhanced by a newly developed sixty-four channel receiver, which could measure a whole line profile in one operation, no matter how complex. This meant at least a fifty-fold gain in efficiency, and made practicable programmes that otherwise would have been too time-consuming to consider. It also brought about, in Australia, the first direct use of computers in the observing process.

Bolton has described the Parkes telescope as 'one of the most successful research instruments ever built',[14] and in its heyday papers flowed from it at the rate of fifty or sixty a year. It was nothing if not versatile, and to the programmes already mentioned can be added work on polarization and magnetic fields; on hydrogen lines in emission and absorption, and the phase structure of the interstellar medium; on supernova remnants; on the periods, dispersion measures, and other properties of pulsars; on complex molecules in interstellar space; and, through the 21-centimetre line, on the dynamics and masses of remote galaxies. In the early 1960s an 18-metre dish was added. It was mounted on rails so that its distance from the big telescope could be varied, and the two together could then be combined as an interferometer pair, used for example to measure the structures and sizes of many of the sources in the Parkes catalogue.

The individual discovery for which the telescope is best known was made in

1963.[15] An observation at Parkes of the source 3C 273 as it was being occulted by the moon yielded a very accurate position, good to a second of arc. With this position it was identified unambiguously with an apparently nondescript thirteenth magnitude star. But a spectrum taken at Palomar showed the 'star' to be a remote galaxy receding at about a sixth of the velocity of light, therefore at a great distance, and the source was revealed for the remarkable object it was, the first quasar, or QSO (quasi-stellar object).

QSOs have come to occupy a central position in modern astronomy, for two reasons. They are the most distant and most luminous objects known, and therefore the best available tools for exploring the outer reaches of the universe and cosmological problems generally. Second, it follows from the short time-scale of their variability that relatively they are exceedingly small. How such a prodigious amount of energy can be generated within such a small volume, or, more simply, what makes quasars shine, has been a central problem in contemporary physics. Current ideas incline to one of Einstein's massive black holes, located at the centre of the quasar. Matter falling into such a hole would create energy according to his $E=mc^2$ equation, and indeed this and other evidence make the existence of black holes almost certain. So after decades in the scientific doldrums, Einstein's general relativity theory has at last entered the mainstream of physics and is acquiring the familiarity of everyday experience.

It is not generally known that it was from a transmission relayed through the Parkes telescope that the world's TV audiences saw man's first steps on the moon. It is even less well known that the telescope was involved in most of the lunar missions, on two occasions playing a critical role, and that it was an essential communications link in the Giotto and Voyager observations of Halley's Comet. With the Harbour Bridge and the Opera House, the telescope has become one of the most familiar of Australian national symbols. As W.F. Evans has put it: 'This magnificent scientific instrument emerged to become ... in a subtle way ... a visible symbol of Australia's intellectual "coming of age" '.[16]

In 1971 Bowen retired, after 25 years as chief of the Division of Radiophysics. He was succeeded by Paul Wild, who later became chairman of CSIRO. Radio astronomy had been by no means the only activity at RP. Two of Bowen's chief personal interests were rain and cloud physics, and air navigation. Wild in his turn became deeply involved in the Interscan aircraft landing system, and after a tremendous amount of development had gone into it, RP had the satisfaction of seeing it adopted internationally in 1978, to be installed in all the world's airports in the 1990s. As Radiophysics staff like to point out, Interscan was a direct spin-off from radio astronomy techniques.

The radio sun

During the 1950s and 1960s our knowledge of the radio sun grew apace, Australians largely pioneering the way. Solar radio emission has two main components, quasi-constant thermal emission from the 'quiet' sun, and a variable part, which becomes increasingly dominant at longer wavelengths. At the decimetre wavelengths where Christiansen worked the radiation is mostly thermal, originates not far above the 'optical' surface, and its variation is relatively long term and minor. It conveys information about the sun's temperature, the structure of its atmosphere, and about small regions (radio 'plages') found above sunspots. The latter are important because they can influence the earth's ionosphere and through it our radio communications.

What was more important was the means Christiansen devised to map these features. He produced first, in 1952, his celebrated grating interferometer. Referred to earlier, this instrument was a linear array of thirty-two paraboloids steerable and equally spaced, which worked at 21 centimetres and produced a fan beam 3 arc minutes wide. With it, repeated slow scans were made across the sun's disc, their direction in the sky changing as the day wore on; from the accumulated scans, in what was the first application of earth rotation synthesis, the first two-dimensional maps were built up. The computations were heavy, and had then to be done by hand, and it was largely to avoid them that the Fleurs or 'chris-cross' was built. This was two grating interferometers, intersecting at right angles with their outputs combined electronically as in the Mills cross. It had a pencil beam 3 arc minutes across, and following its completion in 1957 it was used to map the sun daily for several years. The pendulum had since swung back completely; the data are now handled entirely by computers, and earth synthesis has become the most widely used technique in the subject.[17]

Metre-wave radiation originates in the corona, and is characterized by bursts and noise storms like those first studied by Pawsey — sporadic, violent, and highly variable. It was soon found that for many bursts the shorter wavelengths reached us earlier than the longer. This suggested that bursts originated in disturbances that travel outwards from the sun. As they rise through the corona they excite radiation, the wavelength of which increases with height because it goes inversely as the electron density, which decreases with height: the shorter wavelengths are therefore generated earlier than the longer.

This was the genesis of Wild's swept-frequency or dynamic spectrograph, the instrument that first demonstrated the extraordinary complexity of burst and storm phenomena, and enabled him to classify them into his Type I, Type II and so on, now recognized universally. One could differentiate between Types II and III, for example, because in Type III the delay between the arrival of the short and long waves is reckoned in seconds whereas in Type II it lasts for minutes; in addition, the Type II pattern is often duplicated at the second harmonic. Both types were soon associated with solar flares. These are violent electromagnetic eruptions at the solar surface, which are the cause of two immediate effects: a stream of electrons is shot out through the corona at speeds up to a third that of light; and simultaneously a shock wave is generated, also travelling outward, much more slowly. Both agencies excite, by different mechanisms, the radiation we record as bursts: Type III are attributed to the electron streams, Type II to the shock waves. Theory and observation have gone hand in hand in arriving at these explanations, which could hardly have been reached without the development of a much deeper understanding of physical processes in the coronal plasma, and of plasma physics generally. Given this stimulus the corona has more and more taken on the aspect of an immense high-temperature plasma physics laboratory, with vast magneto-hydrodynamic experiments running continually.[18]

The success of these investigations, and the need to confirm and elaborate on the coronal model they suggested, led to the construction of the radio observatory at Culgoora, NSW, at its time the most advanced solar radio observatory in the world. Its centrepiece was the radio heliograph, ninety-six paraboloidal dishes equally spaced around the circumference of a circle 3 kilometres in diameter. Every second they formed high-resolution images of the sun, one left-hand, the other right-hand circularly polarized, on which one could follow in real time the passage of a solar eruption as it travelled from the sun's surface outwards through the corona. Such short time resolution was necessary to study events that varied as rapidly as those on the sun. The installation was completed by a much upgraded dynamic spectrograph.[19]

The ensuing investigations, by this time pursued in many parts of the world besides Australia, transformed our perception of the solar corona, which we now see as configured almost entirely by the sun's magnetic field, and as highly non-uniform and continually varying. One of the first achievements of radiophysics had been to confirm the million-degree temperature already proposed for the corona by the optical astronomers. With this high temperature the corona is a prolific source of X-rays, and as soon as space observations became possible they were studied intensively. When in 1980 NASA ran their special Solar Maximum Mission satellite, the Culgoora group was a co-investigator with the High Altitude Observatory at Boulder, Colorado, in a fundamental programme on flares. Our understanding of surface and coronal activity on the sun has grown enormously, and now that X-ray and radio observations have shown that solar-type phenomena are common attributes of ordinary stars, solar studies have attained a new importance, and are entering the mainstream of stellar astronomy with a vengeance.

Also at Culgoora, and under the leadership of R.G. Giovanelli,[20] the CSIRO Division of Physics (now part of the Division of Applied Physics) undertook a programme of optical observations of the sun. It aimed to study in fine detail the complex and violent events that occur on the sun's surface — much more violent than any weather on the earth's surface — and was notable for its advanced instrumentation. This included a 30 centimetre telescope with a 'seeing monitor' that permitted photographs to be taken only when the atmosphere was very steady, and a tunable optical filter with a bandwidth of only one eighth of an angstrom, the narrowest achieved to that time. Pictures taken with it at a series of wavelengths closely spaced within, say, the hydrogen-alpha line have made it possible to study the point-to-point kinematics of an area that might surround a sunspot or a flare. Some of the most exciting photographs of the sun have been obtained by this group.

Before the war, C.W. Allen, one of the original staff members of the Commonwealth Solar Observatory, had established an Australian presence in solar physics with his celebrated photometric atlas of the solar spectrum. His results for the mean electron density of the corona, determined at the 1940 total eclipse, were invaluable for the solar radio astronomers. He initiated a half-century during which Australia remained consistently in the forefront of solar research, and he was one of the few links between pre-war and post-war Australian astronomy. This half-century ended in 1984, when lack of funds caused the radioheliograph to be closed. Only the spectrograph is still being operated, by the Ionospheric Prediction Service.[21]

Sydney University

Christiansen and Mills went to the University of Sydney in 1960. They were joined a short time later by Hanbury Brown, a British radio astronomer who, like Bowen, had been one of the pioneers of British radar. He planned to measure stellar diameters by a completely new method, devised and tested with another colleague, Richard Twiss, and he went to Sydney because an excessive proportion of the bright blue stars for which the method worked best are to be found in the south.

Christiansen's main objective was to extend his 'chris-cross' so that it could be used as a rotational synthesis telescope for sources other than the sun. The new telescope, the Fleurs synthesis telescope or FST, has provided the highest resolution so far available in the south (20″).[22] It has been used for observing complex and extended radio sources, especially supernova remnants, and for finding accurate positions for special objects, notably the Vela pulsar. Earlier, in the Netherlands, Christiansen had been closely associated with the design considerations that led ultimately to the Westerbork synthesis telescope. The commissioning of this telescope in 1973 was a landmark in radio astronomy.

Mills built a new version of his cross at the Molonglo Observatory near Canberra. Its two arms, NS and EW, were 1570 metres long by 11.6 metres wide, and the resulting large collecting area and high resolution made it the most powerful instrument in the south for source surveys. The Molonglo Reference Catalogue of more than 12 000 sources remains the largest and most homogeneous in either hemisphere. The cross also proved ideally suited for pulsar searches, the first pulsar having providentially been found in 1967, just as the cross went into operation. More than half the known pulsars have been discovered at Molonglo. One of the first was that in Vela, important on two counts: its location at the heart of the Gum nebula *(q.v.)* was the first real evidence that pulsars were linked with supernovae; and its short period of 89 milliseconds — eleven pulses a second — pointed decisively to neutron stars, as opposed to white dwarfs, as the correct source of the pulsar phenomenon.

In 1978, the main survey complete, it was decided to convert the cross to a rotational synthesis telescope, which became known as the MOST (Molonglo Observatory synthesis telescope).[23] The NS arm was abandoned and the EW arm made steerable with a phased feed system. Extensive new electronics had to be incorporated, and the operating wavelength was changed from 73 to 36 centimetres. The longer integration time, now up to 12 hours, meant an increased sensitivity, and the shorter wavelength, a better angular resolution, while the fast time resolution was retained. The telescope has been used for many programmes — mapping radio galaxies, counts of faint sources, searches for short-period pulsars and for supernova remnants, especially in

The University of Sydney's intensity interferometer at Narrabri, NSW. The picture shows the two 7.6 m diameter mirrors, the circular track, diameter 190 m, around which they moved, and in the foreground the garage that housed them. The building at the centre of the track is the control cabin. This interferometer is being replaced by another of different design. (Courtesy of the School of Physics, University of Sydney, NSW)

the Magellanic Clouds, and generally as a precursor for the Australia Telescope *(q.v.).*

Normally expressed in milliseconds of arc, the diameters of even the nearest stars are so minute that to measure them at all might seem beyond the power of human ingenuity. The instrument with which Hanbury Brown and his group succeeded in doing so was called a stellar intensity interferometer. It was the optical version of a radio method devised originally to measure the diameters of radio sources, and while the principle on which it works was not entirely new, this application certainly was, and it broke new ground in optical science. So the interferometer is important historically as well as scientifically.[24]

Located at Narrabri, NSW, it consisted of two rather rudimentary telescopes, which were notable for their 7.6 metre diameter multiple-segmented mirrors, and which ran on a circular track of 190 metre diameter so that their separation and orientation could be varied continuously. Beams of light from the star under observation were received by each telescope, and what was measured was the extent to which the fluctuations in the two beams were correlated, that is, the extent to which the highs and lows of the intensity in one beam matched the highs and lows of the intensity in the other. When the mirrors were close, in principle overlapping, the correlation was perfect. As they were separated the correlation decreased, more rapidly for stars of larger diameter, and when the separation was found for which the correlation vanished the diameter of the star followed at once.

The diameters of thirty-two stars were measured, with accuracies of a few per cent; they made a significant quite independent contribution to the temperature scale for hot stars. The instrument was then closed down (1973), as further work, on necessarily fainter stars, would have been uneconomical. However, its success has led to the funding of another interferometer, a sophisticated and much more powerful version of the classical Michelson instrument. It will have a maximum baseline of 640 metres, appreciably longer than that of the Narrabri instrument, and will be 400 times as sensitive. With these factors it should produce large amounts of fundamentally new data, and it promises to have a major impact on stellar astronomy.[25]

Post-war Mount Stromlo

As noted above, Mount Stromlo Observatory came out of the war much strengthened, though its staff was still only a fraction of that of Radiophysics. But whereas RP moved into a subject newly created from wartime radar, MSO had to make its way into optical astronomy, the oldest and one of the most conservative of all the sciences. The task was not made easier by the fact that nobody on the staff had had any extended experience of stellar astronomy. Even Woolley's work had been mostly concerned with the sun.

The way ahead was made clearer by two seminal papers, published in 1938 and 1944. In the first, Bethe and Critchfield showed that the primary source of stellar energy was the thermonuclear conversion of hydrogen to helium. Their ideas have been developed, at MSO among other places, to the point where they can not only explain the life cycles of common stars in considerable detail, but also show how the primordial hydrogen and helium can be synthesized into heavier elements such as carbon, oxygen, magnesium and iron, each with its proper cosmic abundance.

In the second paper, Baade introduced his concept of stellar populations. What he had found was a correlation between the evolutionary properties of stars — their chemical compositions and their ages — and their dynamical properties. For example, in our own galaxy, Baade's Population I, the young and metal-rich stars, belongs to the flat spinning disc — the Milky Way —, while his Population II, old and metal-poor, is to be found in the central nucleus and in the slowly moving spheroidal halo that encompasses the whole. Baade's ideas have been most influential, and are now incorporated into the more general theories of the chemical and dynamic evolution of galaxies that have come to dominate this part of the subject.

A number of the earlier papers at MSO were written against this background, and introduced lines of work that have been followed ever since. Thus Colin Gum, a graduate student, used an improvised spectrograph camera to discover forty-four hydrogen emission regions. Following an idea of Baade's, he found for the first time that they delineated the nearest southern spiral arm of the galaxy, conforming well to a similar pattern just discovered in the northern hemisphere. One of Gum's discoveries was the great emission nebula in Vela that now bears his name. Quite the most spectacular object of its kind, it was shown by the radio astronomers to be the remnant of a relatively nearby supernova, and is referred to several times in this chapter.

The Magellanic Clouds, the two nearest galaxies of any consequence, were an obvious place to test Baade's population concept. In one such programme Gerald Kron and I found that some of the star clusters in the Magellanic Clouds were much bluer, and hence much younger, than their counterparts in the galaxy. This was the first clue that the evolution of the Clouds had taken a different path from that of the galaxy, an idea that thirty years later is being explored in detail. Kron, who with his compatriot Olin Eggen was on a visit from the Lick Observatory in California, was the leading photoelectric instrumentalist of his day. He was the first of a small group of American instrumentalists — others were Aller (photoelectric spectrum scanner), Dunham (coude spectrograph) and Kent Ford (image intensifiers) — who played a significant role in the development of MSO.

In a third of these early programmes Gerard de Vaucouleurs concluded, from deep, small-scale photography made with an old wartime aerial camera, that the Large Magellanic Cloud was a flat rotating system not unlike our own galaxy, a point of view distinctly at variance with that currently held. However, the 21-centimetre observations obtained almost simultaneously at RP not only confirmed de Vaucouleurs' proposal handsomely, but provided the first reasonable mass estimate of the Cloud as well.

A fine paper from an earlier period, a fitting *envoi* to the old Commonwealth Solar Observatory days, was one in which Woolley and Allen constructed a model of the solar chromosphere and corona that incorporated all the data known at the time, including the radio data, and was the 'last word' for some years.

While the impact of these (and other) papers was hardly commensurate with the brilliant output from RP, the new institute had made a start and was clearly on the way.

Besides being an astronomer, Woolley was also a public servant, and a good one. During the war he had been given additional duties that brought him into close contact with the upper echelons of the Public Service. It was invaluable experience and he made the most of it, establishing many useful personal contacts along the way (Bowen had had similar wartime experiences in Washington). So it was no surprise when in 1948 he obtained government approval for the purchase of a 74-inch telescope. He had a good case. All the big telescopes were well north of the equator, while many of the best astronomical objects were well south of it; he had the blessing of the Astronomer Royal, brought out in 1947 especially for the purpose; the observatory had done well in the war. The Prime Minister, J. B. Chifley, evidently thought so too.

Woolley said afterwards he got the telescope so easily he wondered if he ought not to have gone for a 100-inch.[26]

The telescope was commissioned in 1955, and with a sister instrument in Pretoria was for twenty years the largest optical telescope in the south. This was an important step for MSO, which with the 74-inch, became a force to be reckoned with, and much that was learnt with the 74-inch telescope later went into the Anglo-Australian telescope (AAT). In fact without the 74-inch there might not have been an AAT. That such a good case could be made for the AAT came directly from the experience gained in installing, maintaining and above all in observing with the 74-inch.

In 1948, Woolley acquired as scrap the Great Melbourne Telescope. It had never worked well and had been a source of trouble and embarrassment since it was first set up in 1869. Given a new set of optics and extensively rebuilt, it went back into service in 1955, at last functioning as it had always been hoped it would. For the next twenty years it was worked hard, but once the AAT was commissioned in 1975, demand for it dropped and it was retired. Woolley gave MSO two further windows on the astronomical world when he arranged for the Yale-Columbia Observatory to move their 26-inch refractor from Johannesburg to Mount Stromlo, and for the Uppsala Observatory to set up a new schmidt telescope there. The Uppsala schmidt was later moved to the darker skies of Siding Spring.

Woolley's final contribution, and possibly his best, was to transfer the observatory from the then Department of the Interior to ANU, clearly a more congenial home for a purely research organization. More particularly, the move made it possible to set up a graduate school, which has become an essential part of Australian astronomy. He concluded an eventful term of office when he left in December 1955 to become the eleventh Astronomer Royal and the director of the Royal Greenwich Observatory. He was succeeded by Bart Bok.

Bok and Eggen

Bok was an energetic, ebullient man whom nobody ever forgot.[27] He was born in Holland but had spent most of his astronomical life in the USA, at Harvard. In the early 1950s he had been a prime mover in re-establishing radio astronomy in his adopted country. He was an exceptionally talented popularist and public lecturer, and to many people, including not a few parliamentarians, he *was* Australian astronomy. Moreover his scientific reputation stood high. He established on Siding Spring Mountain what is now one of the world's best known observatories; he greatly extended and upgraded MSO's technical resources; he was the moving spirit behind a highly successful graduate school; and he was an articulate and effective advocate for the Anglo-Australian telescope. He left a deep mark on Australian astronomy, and when in 1966 he returned to the USA, after nine years as director, he had raised MSO to near the front rank of the world's observatories.

One of Bok's first steps was to bring in T.E. Dunham, celebrated for the spectrographs he made at Mount Wilson Observatory, to construct a similar instrument for the 74-inch. Dunham's spectrograph was built in the MSO workshops, going into operation about 1962. It is a fine instrument, and by giving MSO a point of entry into major new subjects like stellar atmospheres it broadened the observatory's scope in a most significant way.

Bok recognized early the extent to which Mount Stromlo was threatened by the growth of the Canberra lights, and as a result he instituted the extensive site survey from which Siding Spring was chosen. Situated in northern NSW, Siding Spring Mountain was a fortunate choice; its sky is suitably dark, and it has grown more rapidly than even its founder could have hoped. The first telescope, installed in 1964, was a new one-metre. Two smaller telescopes were acquired, to be followed within ten years by the AAT and the UK schmidt, then by ANU's 2.3-metre and the Uppsala schmidt.

He did much to foster relations between MSO and Radiophysics. Joint meetings were held most years, and students went regularly to the Radiophysics telescopes, some to write PhD theses on radio topics under joint RP–MSO supervision. One of them, Ron Ekers, became director of the VLA synthesis telescope in New Mexico, currently the leading institute in US radio astronomy. What Bok himself regarded as his most important achievement was the graduate school.[28] Given his wholehearted support, it has won an enviable reputation overseas, and in attracting able students who might otherwise have gone elsewhere it has injected a necessary degree of continuity into Australian astronomy. The school had an unexpectedly stimulating effect on the observatory, becoming the pivot about which much of its research effort turned; perhaps there is nothing quite like a group of clever competitive graduate students for extracting the best from a tenured staff.

These were times of rapid scientific growth. Previous work on the Magellanic Clouds, cepheids and emission nebulae, was pushed ahead, and new subjects appeared. The relevance of globular clusters, the ancient survivors of the earliest phase of our own galaxy, to theories of galaxy evolution and stellar evolution was becoming clear, as was their role as self-contained dynamical systems with evolutionary histories of their own. They continue to be a central part of the MSO research effort. In 1957, in a famous paper, the Burbidges, Fowler and Hoyle described in detail how the chemical elements are formed from primordial hydrogen and helium, each with its proper abundance. The determination of cosmic abundances became immediately fashionable, and Dunham's spectrograph, admirably suited for obtaining the appropriate high-dispersion spectra, had arrived in good time.

In one area, however, progress was slow — the identification of radio sources with their optical counterparts. Without such an identification no optical spectrum could be obtained, and optical spectra were all-important as the only source of the radial velocities and hence of the distances of sources. They also carried much other information. This field, which had become intensively competitive, was particularly frustrating for Australians, who, although they had pioneered the subject and discovered as many sources as anyone, found themselves 'outgunned' by the big telescopes in California. To make matters worse, the radio sources, especially the quasars, were providing some of the most exciting problems in the whole of physics. The situation provided one of the most compelling arguments for a new, large optical telescope.

In 1966 Bok was succeeded as director by Olin Eggen. Eggen was an

American who had worked with Woolley in England, a dedicated research astronomer whose over-riding objective was to maintain the research impetus and increase the flow of research papers. This he did, he himself producing the awesome total of ninety-nine papers in the eleven years he was at Mount Stromlo; but his term may be remembered more by other events, such as those that surrounded the building of the AAT, and the great changes that took place in the conduct and style of astronomy during the 1970s.

A foretaste of these changes was brought to MSO by another American visitor, Kent Ford, a leading expert in the new field of image intensifiers. At that time they were used, in essence, to shorten photographic exposures. The factors of ten or twenty that could be so achieved at last made it possible for the MSO telescopes to identify quasars. By 1972 eighty had been found south of declination –30 degrees, as opposed to one in 1970. Intensifiers are still an essential component of most of the advanced light detectors in astronomy.

Finally, two important results reflect the MSO interest in the structure and dynamics of 'normal' galaxies. The surface brightness of the discs of spiral galaxies was found to fall off exponentially with distance from the nucleus. The unexpected feature of this result is that it is so general, and though still not understood completely it has become the recognized starting-point for any discussion of disk structure. The other was a method of finding the total masses of star clusters. This made ingenious use of a Fourier technique to measure the Doppler-broadening of lines in their integrated spectra. The method not only yields good masses, but has become the essential tool for exploring the dynamics of elliptical galaxies, thereby breaking new ground in an important branch of astronomy.

The Anglo-Australian Telescope

Following several years campaigning, the decision to build a major optical telescope in Australia was reached in 1967. The telescope was to have an aperture of 150 inches, was to be located on Siding Spring Mountain, and was to be built and operated jointly by the British and Australian governments, costs and observing time being shared equally.

The scientific case was well known. Until the early 1970s the world's largest telescopes were all located well north of the equator. They were unable to reach the southern third of the sky, although that third held many of the most important objects in astronomy. These included the Magellanic Clouds, the galactic centre, the southern third of the Milky Way, most of the globular clusters, and many special objects like Centaurus A and the Gum nebula, to name but two. The southern galaxies were particularly poorly observed. The need for optical observations of radio sources has already been emphasized. There was also a need to acquire optical data to complement those coming from space observatories, especially data on ultraviolet and X-ray sources. At least three other groups, the European Southern Observatory (ESO), the Kitt Peak National Observatory in the USA, and the Carnegie Institution, which operated the Mt Wilson Observatory, had recognized the force of these arguments, and in due course all three established large telescopes in the south, in Chile.

*The Anglo-Australian telescope inside its dome at Siding Spring Observatory, Coonabarabran, NSW.
The telescope is pointing east, and the cabin for observing at the prime focus can be seen at the end of the
tube. The top of the dome, not shown in the figure, is 50.3 m above ground level. (Courtesy of the Anglo-
Australian Observatory, Epping, NSW)*

The first serious moves towards the AAT were made in Britain in 1959.
Woolley played a leading role, and the British continued to make the running
for several years. Negotiations were lengthy and at times difficult. As Sir
Bernard Lovell has written:

> The telescope has undoubtedly had a major impact on astronomical research in
> the southern hemisphere and in retrospect it seems remarkable that the idea of
> the instrument was nearly abandoned on more than one occasion ... The
> problems and difficulties that besent the concept until 1967 and almost led to its
> abandonment were not primarily financial. The issues were political ones
> coupled with protracted indecision amongst the astronomers and scientists both
> in the United Kingdom and Australia.[29]

Sir Leonard Huxley described a case in point. The Royal Society of London
had written to the Australian Academy enquiring about their interest in the
proposal to build a large telescope:

> This letter . . . elicited a surprising response. Seven Fellows eminent in biological sciences wrote to the Secretary (Biological Sciences) expressing their fears that such a 'gigantic' expenditure would adversely affect funds for research in other areas . . . The Academy's support for the telescoope project led directly to the establishment of the Flora and Fauna Committee, to advise Council on major biological projects.[30]

Agreement was of course reached, but there was more trouble in store. In 1972 a dispute arose as to whether the telescope was to be operated through ANU, or by an independent scientific and technical organization set up for that specific purpose. The ANU case was that they had the expertise, they could do the job more economically, and they owned the mountain.[31] The AAT Board considered that the operating body must be answerable directly to them, as representing the astronomical communities of the two nations. The differences could not be reconciled, and exacerbated by personal and institutional rivalries, created a rift in Australian astronomy deep enough to cause anxiety.[32] Matters were not improved by an inconclusive meeting — 'confrontation' might be a better word — between Malcolm Fraser, then Minister for Science in Australia, and Margaret Thatcher, his opposite number in the United Kingdom. Finally, the board's view prevailed, though not without an appeal to the Prime Minister.

In the finally adopted arrangement, only the telescope and its immediate ancillaries were located on Siding Spring; the headquarters proper, where most of the staff normally worked, was established in the Radiophysics grounds in Epping, NSW. Once the telescope went into regular operation it worked so well that the differences were swept away on a wave of euphoria; only vestiges remain.

That was in mid-1975. The telescope was accepted almost at once as a technological *tour de force*. The mounting and optics were clearly of the highest standard, but what created the real impression was the computer control system, which was comprehensive, versatile and efficient to a degree beyond anything previously contemplated. For a while, indeed, visiting astronomers were nonplussed by it. For example, the AAT could be pointed as a matter of routine to an accuracy of 2 or 3 arc seconds, an unheard-of figure; it could execute scan patterns, in grids or spirals or whatever else one cared to programme, and it lent itself generally to a speed and ease of operation that came as a revelation.[33]

A comparable impression was created by a new spectrograph built by Joe Wampler, the telescope's first director. It was a copy of one he and Lloyd Robinson had made for the 120-inch Lick telescope. Its key element was a new electronic detector in which the signal was recorded not on a photographic plate, but picked off TV-style by a scanning beam and stored in a computer memory. At any time the observer could see how he was progressing by displaying the signal on a video screen, and this, coupled with the linearity that enabled the sky background to be subtracted accurately, conferred enormous advantages. This detector and its associated computer inaugurated a full-scale instrumental revolution that could raise the efficiency of existing telescopes ten or twenty-fold. And the new control systems made observing so much easier that suddenly the traditional observers' skills had become redundant, and the telescope was open to a much wider circle of users. So it is not surprising that it has had such a decisive effect on both British and Australian optical astronomy.

The Anglo-Australian observatory

In 1976 Don Morton succeeded Wampler as director. The task was now to maintain the leading position of the AAT, and in particular to keep it ahead in the development and exploitation of new instruments. Increased grants from the two governments and a supportive board were a great help. The observatory's own workshop space being limited, much of the instrumental development is carried out through design contracts with universities and observatories, a practice that has worked well. The achievements of the first ten years, instrumental and observational, are summarized in the AAT's *1984/85 Annual Report;* just a few items are described here.[34]

The first is a technique of multi-object spectroscopy the observatory's staff have pioneered themselves. The light from up to fifty objects, reasonably close to each other in the sky, is channelled along optical fibres to the spectrograph slit so that the spectra of all fifty are obtained at one time. This is an immense gain in efficiency, and, given a computer with a big enough memory, up to a hundred objects could be so observed. The technique is made possible by a two-dimensional detector developed at University College, London (the IPCS), which can record many spectra simultaneously whereas the Wampler detector could record only one.

Though not located at a particularly favourable infrared site, the AAT has been used with good effect in the near infrared. One factor has been the development of sensitive detectors in the 1 to 5 micron range. Extended to a sixteen element array and incorporated in a cooled grating spectrometer built at MSO, this has produced the most sensitive infrared spectrometer available. A similar detector, used in conjunction with the AAT's unique ability to make accurate and repeatable raster scans, has enabled small areas — the galactic centre, the dark side of Venus, the centre of the 30 Doradus nebula are examples — to be mapped with a precision and depth unequalled elsewhere.

While electronic devices like CCDs have taken over much of the field, there remain programmes that can be carried out in no other way than by direct photography. Here the AAT has again excelled. David Malin, the observatory's scientific photographer, has pioneered new techniques to make a series of brilliant pictures that have not only led to valuable scientific discoveries, but have been reproduced throughout the astronomical world and publicized the Anglo-Australian telescope as probably nothing else has.

The detection in 1977 of the optical counterpart of the Vela pulsar was a spectacular achievement, and gave clear notice that the new telescope had 'arrived'; it was the faintest optical object to have been measured up to that time. The pulsar was discovered at Molonglo and then intensively studied at Parkes. An accurate position, an essential prerequisite, was found by combining observations from Molonglo, Fleurs, and the NSW State Observatory, and the final observations were made by a multi-observatory team using a sophisiticated TV technique.

Perhaps the AAT has been nowhere more successful than when observing radio sources, particularly in conjunction with the UK schmidt, and particularly when observing very remote quasars; it has several times held the record for the most distant quasar. The current example was found with the schmidt and recognized by its peculiar energy distribution relative to normal stars. The data from which this was found were five plates taken in five different colour bands, from the ultraviolet to the infrared. The source (0046-293) is receding at 0.923 times the velocity of light(z=4.01).

There can be no denying the impact of the AAT. Its technical influence has been widespread and, together with the UK schmidt, it has not only been the prime mover in the great current revival in British optical astronomy, but it has come to dominate Australian optical astronomy in much the same way. Had it not been for the Anglo-Australian Telescope it is doubtful whether the British 4.2–metre telescope would ever have been built, or for that matter the ANU 2.3 metre at Siding Spring. As the *New Scientist* put it:

> Foremost of these, and the undisputed star of the south . . . the AAT was the greatest single stimulus to the resurgence of British optical astronomy . . . The excellence of the AAT stems as much from its comprehensive instruments and dedicated support staff as from its fine optics and precision engineering.[35]

In turn, the AAT owes much to the Parkes telescope and perhaps even more to 'Taffy' Bowen. Bowen was the first chairman of the AAT Board. He set up the AAT project office (on much the same lines as that at Parkes), guided the project through its first years, and had much to do with resolving the political problems of the early 1970s. At its peak the project office employed about twenty-five people, mostly engineers. Unlike other similar organizations it was the engineers who ran it, in the sense that it was they and not the astronomers who had the executive and financial authority — an eminently workable arrangement. The two astronomers 'permanently' attached, Professor R. O. Redman, of the University of Cambridge, and myself, acted more as project scientists, and as a link between the project office and potential users. Working with that talented and dedicated group, one could feel only a step away from those earlier craftsmen who built the great cathedrals of medieval Europe. Or, as Sir Fred Hoyle put it, 'a large telescope is a good example of the things which our civilisation does well'.[36]

After the AAT

With the completion of the AAT in 1975 Australian astronomy found itself on something of a plateau in that the ensuing decade was largely one of coming to terms with the new telescope. Directorships changed: at MSO Eggen was succeeded by Don Mathewson, who had spent the early part of his career at Radiophysics, while at RP, Wild was followed first by Harry Minnett, a major architect of both the Parkes telescope and the AAT, then by Bob Frater, a product of the Christiansen school at Sydney University. The Molonglo synthesis telescope was completed in 1981, and the Sydney University prototype interferometer passed its key tests in 1986, by which time the new MSO 2.3 metre optical telescope was well on the way to regular operation. With its alt-azimuth mounting, fast primary, and rotating building this telescope is thoroughly contemporary in feel, and has taken good advantage of the experience gained with the AAT and with the multiple mirror telescope in Arizona. It was designed at MSO, funded by ANU, and except for specialist components was largely built in Australia.

This period also saw most of the astronomical 'old guard' step down. Bolton, Wild, Mills, Christiansen and Hanbury Brown all retired, not to mention a number of their contemporaries. They had seen Australian astronomy through a time of spectacular growth. In place of at most six 'professional' astronomers before the war there were now some 130, with corresponding increases in the numbers of engineers, support staff, and especially of graduate students. Many universities and other institutions had entered the subject, and it had for some time been the strongest branch of physics in the country.[37]

It was a measure of the standing of astronomy that in 1976 the Commonwealth government set up an inter-departmental committee, assisted by a sub-committee of senior astronomers, to enquire into the 'co-ordinated development and rationalised use of observatory facilities.' Endorsed by ASTEC (The Australian Science and Technology Council), their report reached the Prime Minister in 1978. Following a thorough survey of the subject, it recommended that Australia's research effort be maintained at the current level in real terms, that priority for the next major instrument should go to a new radio synthesis telescope, the Australia Telescope, and that future government-funded instruments be regarded as national facilities.[38] With its acceptance, astronomy seemed secure at least until the end of the century.

The funding of the Australia Telescope was duly authorized in 1982, for completion in 1988. It will be located at Culgoora, on the site of the old radio heliograph, and its inner 'compact' array will consist of six 22-metre paraboloids on an east-west line, five of them movable along 3 kilometres of broad-gauge railway track. There will be a seventh 22-metre dish at Siding Spring, 120 kilometres to the south-west, and it will be possible to incorporate also the Parkes telescope and telescopes at Tidbinbilla and Hobart so that the whole can be used as a long baseline array. Special features of the AT will be its broad wavelength coverage, its high spectral and time resolution (high spatial resolution goes without saying), and some very advanced technology.[39]

This has been a fruitful period in Australian astronomy, and only a few research topics can be described here — a few among a great many. Those few are chosen to illustrate some current trends and modes of operation.

Australian work on supernovae had its beginnings in Gum's pioneering search for hydrogen emission regions, and in its successor, the MSO hydrogen-alpha atlas.[40] The first survey made with the original Mills cross found a number of extended continuum sources near the galactic plane, though their true nature as supernova remnants was not revealed for some time. This work was followed with the Parkes telescope, very effectively because it proved easier to find supernova remnants at radio than at optical wavelengths, where they are often obscured by interstellar dust. Many new objects were found, and it became clear that a similar search of the Magellanic Clouds would pay handsomely. One obvious reason was that this would provide a sample — a rich one as it happened — of relatively unobscured objects, all at the same distance. This search was made at the Molonglo Observatory. It was accompanied by a systematic programme of optical spectroscopy, followed by another that recorded the images formed at certain critical wavelengths; both were carried out on the AAT. A third source of data was a series of X-ray observations made from an American satellite by a group at Columbia University, New York. Discussed in terms of theories largely developed at MSO, the whole made a substantial addition to our knowledge of supernovae, incidentally illustrating a favourite maxim of Pawsey's — 'clues add up non-linearly'. The part played by theory in constructing models and suggesting critical observational tests is not to be underestimated. Note also the pooling of data from three widely separate parts of the spectrum, each carrying unique and particular information.

The next example is purely 'optical', with no input at all from the radio or X-ray spectral regions. It concerns element abundances in stars. The relative abundances of the heavier elements — oxygen, nitrogen, silicon, calcium, iron and so on — normally mimic those on earth (or on the sun). But in 1960 it was discovered at Mt Stromlo that the iron peak elements in the star HD 101065 are underabundant by orders of magnitude, and the star's atmosphere is dominated by lanthanides, especially by the rare earth holmium. This is an extraordinary result and it makes the star, in this sense, the most peculiar known. It was named after its discoverer, Antoni Przybylski, a most unusual distinction. In another remarkable but quite different star, CD -38° 245, calcium and iron were found to be about 100 000 times less abundant than in the sun, and at least ten times less than in the next most metal-deficient star. Przybylski's Star is an enigma, the true significance of which has yet to be found; but the CD star, because its composition may well reflect that of the primordial medium from which our galaxy formed, has a most important place in the stellar hierarchy. The observations of Przybylski's Star were made with the 74-inch at MSO, of the CD star with the AAT.

The Magellanic Stream is a long wake of neutral hydrogen stretching from the Magellanic Clouds almost half-way across the sky.[41] It is regarded as a result of interaction between the Clouds and the galaxy, partly tidal and partly the drag on the Clouds of the hot gaseous halo that surrounds the galaxy. It is important because of the light it throws on the nature of interactions between galaxies and on the recent history of the Galaxy-Magellanic Clouds complex. The Magellanic Stream was discovered by MSO observers in the course of observations made at Parkes of the 21-centimetre hydrogen line. Subsequent observations, especially of its cepheid variables, showed that the Small Cloud is actually two separate bodies lying along the line of sight, the system thus acquiring an unexpected complexity. Note that however dominant the Anglo-Australian telescope may sometimes appear to be, life without it could still go on: these observations, at least in their initial phase, made no use of it. The results have been widely discussed, and the subject has been taken up at many overseas observatories.

The final example comes from the amateur ranks. The Reverend Robert Evans, now of Lawson, NSW, works from his back yard with a 10-inch reflector, more recently with a 16-inch, looking for supernovae in the several hundred galaxies he observes on a regular basis. He scans his galaxies by eye, comparing what he sees with what appears on a photograph. By the end of

1986 he had discovered fifteen 'new stars', four before maximum light, a remarkable achievement and an important one. With their key roles in chemical evolution, the physics of interstellar matter and cosmic ray astronomy, supernovae lie at the heart of modern astronomy, and Evans' discoveries are followed up at once on our largest telescopes. He has shown, too, how much can still be done with the unaided hand and eye — together with a little dedication.[42]

Other instruments, other institutions

Astronomy in Australia is pursued seriously at some eighteen institutions, only four of which have so far been discussed. But while those four operate most of the larger instruments and so set the stage for most of Australian astronomy, the other fourteen include groups that have made major contributions and have long-standing commitments to the subject, so much so that it will hardly be possible to do them all justice.

The decision to 'nationalize' major facilities greatly enhanced the ability of various institutions, especially the smaller ones, to carry out effective observing programmes. Virtually all large instruments are now open to virtually all comers. Observing time is awarded on scientific merit, and it is no longer necessary for an institution to possess a telescope of its own. The decision was the more important because the large instruments now operate in a manner and with an efficiency that smaller establishments hardly have the resources to match.

This change in style goes deep. While observers now work in warm, well-lit control rooms, watching their observations unfold on video screens, they are distanced from the telescope and instruments — spectrographs or whatever — by computers and resident technicians, and not only the old-time observers' skills but the observers themselves are steadily becoming redundant. They are already so at the UK schmidt and the Molonglo telescope. More than ever the bulk of the astronomer's work is done at his home institution, much of it at a computer with the help of one of the massive, many-faceted programs that are so characteristic of contemporary astronomy. These will have taken thousands of hours to write, and be culled from the work of many people in many countries; while an astronomy department can do without a telescope, it certainly cannot do without a computer.

We begin with the UK schmidt at Siding Spring, which though owned and operated by the UK Science, Engineering and Research Council has become almost an integral part of the Australian scene. A schmidt is a specialized form of telescope that can be used only for wide-angle direct photography of the sky, but does that very well. This one was erected in 1974 as the southern counterpart of the Palomar schmidt, and was built in the first instance to extend to the south the sky survey made with the Palomar instrument. It closely follows the Palomar design, but better optics give it a wider wavelength coverage, and it has objective prisms that allow it to take spectra of whole fields of objects at once. It is widely used by Australian groups, and its success in identifying quasars, and radio sources generally, is one reason why the AAT has contributed so effectively to this field.[43]

The NASA Deep Space Station, at Tidbinbilla outside Canberra, has

64 metre and 34 metre radio telescopes, which when not needed for satellite tracking are made available for radio astronomy. In this mode they have been used almost exclusively for interferometry. Initially they worked as an interferometer pair to measure accurate positions (good to 3″) for compact Parkes sources, and in conjunction with the UK schmidt 'sky survey' they enabled identifications to be made by high-grade radio-optical positional coincidences. Their many successes included Pks 2000-330, for several years the most remote and most luminous object known. The programme was then extended to very long baseline interferometry (VLBI), in which the NASA telescope is used together with other large telescopes such as its sister instrument at Goldstone, California, or telescopes in South Africa, Japan, and elsewhere, all at distances of many thousands of kilometres. The first objective was to set up an accurate grid of positional reference standards, but the programme naturally expanded to include identifications and radio and optical source morphology. Such interferometers allow radio sources to be delineated and their positions found with accuracies of milliarc seconds, far in excess of anything achievable by ordinary optical means. Simultaneously the baselines — the intercontinental distances — are measured to centimetres. Changes in these distances, and in the positions of tectonic plates, can now be determined, bringing into being a new branch of geophysics.

In a promising development, even longer baselines were achieved by using an orbiting spacecraft as one station of the network, the other two being the 64 metre telescopes at Tidbinbilla and Usuda, Japan. A maximum projected baseline of 1.4 earth diameters was achieved (17 800 km) and resolutions of a tenth of a milliarc second seem well within sight. Other plans involve observations of OH and water masers, thought to be the sites of newly forming stars. The Parkes telescope is to play a major part in some of these programmes, thereby giving it a new lease of life; in recent years it has been rather overshadowed by the capabilities of the big arrays in the northern hemisphere.

Because of the high magnetic latitude of Hobart, the ionosphere there is unusually transparent to long radio waves. The University of Tasmania's Physics Department has therefore concentrated on longer wavelengths, up to 100 metres, and has run successful programmes on Jupiter, the Milky Way, and on pulsars. More recently NASA donated to the university a 26 metre radio telescope previously operated at Orroral Valley near Canberra; one of its main uses will be to extend the pulsar work. The department built their own one-metre optical telescope, and have co-operated with Imperial College, London, among other institutions, in X-ray observations made from high-flying balloons, and of observational cosmic-ray physics.

The University of Adelaide also has a long-standing interest in gamma and cosmic rays, and is upgrading its cosmic ray air shower array to enable it to detect radiation from discrete sources. Previously the university had made radar observations of meteor showers. Encouraged by proximity to Woomera, the Adelaide staff were early workers in space astronomy, made some of the first ultraviolet observations above the atmosphere, and flew X-ray packages in UK Skylark rockets.

At Monash University the Mathematics Department has carried out theoretical work on pulsating variables, radiation theory, turbulence, and planetary cosmogony. A recent promising development is a new computer method (called SPH, smoothed particle hydrodynamics) of simulating colliding or otherwise interacting gas masses, specifically interstellar or protostellar masses. Since 1970 a group in the Monash Chemistry Department, headed by R. D. Brown, has been measuring the wavelengths of lines in molecules likely to be found in interstellar space. Initially the Parkes telescope was invoked to search for these molecules, and four were found — thioformaldehyde, formamide, methanimine and methyl formate. The most effective observations have been made with the NRAO microwave dish at Kitt Peak. It includes the identification of three isotopic versions of hydrogen cyanide, and the 'invention', for want of a better word, and detection in space of a previously unknown molecule, tricarbon monoxide(C_3O). The group is regarded as one of the world's leading microwave spectroscopy laboratories.

The Australian Defence Force Academy at Canberra (previously the RAAF Academy in Melbourne) has had a long interest in balloon flights, partly in collaboration with Hobart, but its most important work has been in the infrared, especially at longer wavelengths. ADFA observers have also taken part in several high-altitude flights on the American Kuiper Airborne Observatory (KAO).

Since the early 1980s a Department of Astrophysics and Optics has been operating at the University of NSW, with emphasis on higher technology. Work is well advanced on a patrol telescope that will utilize a converted Baker-Nunn satellite tracker (donated by the Smithsonian Astrophysical Observatory) and a CCD detector. Infrared arrays (64 x 32 pixels) have been successfully made for operation in the 1 to 5 micron region. Macquarie University has taken over most of the astrograph material of the Sydney Observatory *(q.v.)*, and will continue the programme with emphasis on proper motions. The main lines at Wollongong are Wolf-Rayet stars and long-period variables, mostly with material from Siding Spring, with a new programme on star formation regions. The University of Queensland department continues work on stellar atmospheres.

Since the 1890s the four state observatories had been burdened with an ambitious international programme for the photographic mapping of the whole sky (the *Astrographic Catalogue)*. The Melbourne and Adelaide Observatories were closed during the war, and Sydney was asked to complete both its own and the unfinished Melbourne part of the catalogue. Thanks to the leadership of its director, Harley Wood, the immense task — Sydney had to measure the positions of more stars than any other of the twenty-odd participating observatories — was at last completed, in 1964. It was followed by a collaboration with Yale Observatory on a more manageable star-mapping project, and an extensive programme on minor planets. Harley Wood was a much-respected figure in Australian astronomy. In 1966 he was elected foundation president of the newly formed Astronomical Society of Australia, and in 1973 organized in Sydney a highly successful General Assembly of the International Astronomical Union, the first held in the Southern Hemisphere. The decision to close the observatory, greatly regretted by the astronomical community, was announced in 1982.[44]

Perth, the only survivor of the state observatories, completed its Astrograph zone with the help of the Royal Observatory, Edinburgh. Then in 1967 the Hamburg Observatory in Germany sent out a six-man expedition with a new automated transit telescope to observe accurate positions for some 26 000 stars. The programme occupied four years, and the resulting *Perth 70 Catalogue* has become the standard reference in its field. When the expedition returned home the telescope was left in Perth.

Astronomy in Australia

Why did astronomy become so strong in this country? First, a unique opportunity was created by the tremendous expansion of the subject in the aftermath of the Second World War. Wartime radar gave us radio astronomy, wartime rocketry satellites and space astronomy, and between them they extended the astronomical spectrum by a factor of 10^{12}, from 100 metre radio waves at one end to X-rays and gamma rays at the other. It was this extension, from 2 to 40-plus octaves, and the glittering discoveries it led to, that gave astronomy such impetus and have kept it so long in the forefront of the physical sciences.

But nothing would have happened without the high-level decisions that created the Radiophysics Laboratory early in the war, continued it in being after the war, and redirected Mount Stromlo into stellar as opposed to solar astronomy — these decisions, together with the generous support that went to both bodies. As Bowen has written

> we received tremendous support from Sir Frederick White and successive chairmen of CSIRO . . . Above all, we owe a tremendous debt of gratitude to Sir David Rivett, whose philosophy was very simple — to appoint the very best men available, give them the facilities, point them in the right direction and let them go for their lives.[45]

RP and MSO went into the post-war period with no outside commitments, clean slates to write on, and a great enthusiasm to prove themselves in open company. They were well placed to take advantage of whatever turned up, and plenty did. Two such institutions, at once rivals and collaborators, were probably necessary to achieve the progress that was made. It helped too that both were well led. Lovell[46] has written of the new breed of scientists produced by the war, men used to authority, who knew about money, and who moved easily in the upper echelons of government. Bowen and Woolley were two such, able and decisive, and their institutions benefited accordingly.

Australia's location in the southern hemisphere, though often quoted and certainly useful politically, was perhaps less important. Good advantage has been taken of it by the 21 centimetre observers, as witness their success with the Magellanic Clouds and southern Milky Way programmes, and the Magellanic Clouds have become steadily more valuable to the optical astronomers. But astronomy being the competitive business it is, there are astronomers who prefer to compete on level terms with their northern counterparts, finding that the programmes from which they gain most satisfaction are those that could have been pursued equally well from either hemisphere.

Probably the most important reason, especially in the long run, was that success bred success. Good people attract good people, and they also attract money; money means better instruments; good instruments mean further success. Not for nothing the old astronomical adage that medals should be given to telescopes, not to astronomers. It was only the early achievements of Radiophysics that allowed Bowen to attract so much US money for the Parkes telescope, Mills for the Molonglo cross, and Wild for the radioheliograph. And it was the good use to which the Parkes telescope and the MSO 74-inch were put that paved much of the way for the Anglo-Australian telescope.

The last word on the subject may well be left with Pawsey:

> Australia with its relatively less congested radio spectrum and its view of the southern skies is favourably situated for the study of radio astronomy but these factors have not played a decisive part. Research in this field has been very generously supported by the Australian authorities, but financial support, though essential, must be supplemented by something else. The reasons must involve personalities, methods of organisation, historic background, and such complex factors, and all of these appear to have been particularly favourable in Australia.[47]

As pointed out earlier, Australian astronomy seems secure for some time yet. The subject appears to be heading in two main directions. One, not new, is the continued development of space observations of all kinds, from X-rays and gamma-rays to the radio spectrum. Australia has already a substantial interest in communication, meteorological and other satellites, and while research satellites like those for astronomy are not cheap, so many advanced countries are now investing in them that it cnnnot be long before Australia follows suit. In the early 1980s strenuous efforts were made, jointly with the USA and Canada, to build an ultraviolet orbiting satellite known as Starlab. The project came to nothing, but a similar venture, a share with NASA and the European Space Agency in the orbiting ultraviolet telescope Project Lyman, promises better.

The other main direction — almost a fashion — is the construction of very large optical telescopes, their mirrors with diameters of 8 metres or more. Major efforts are being made to keep the costs within bounds, and it is possible that a future 8-metre may be built for little more than the combined cost of the Australian Telescope and the Anglo-Australian Telescope. If Australia is to maintain its position in the astronomical world, serious consideration will have to be given to both these possibilities.[48]

Notes

1   R. Love, *Historical Records of Australian Science*, 6(2) (1986), 71.
2   R. v.d. R. Woolley, *Records Australian Academy of Science*, 1 (3) (1968), 53; S. Davies, *Historical Records of Australian Science* 6(1) (1984), 55; S.C.B. Gascoigne, *Proc. Astronomical Society of Australia*, 5(4) (1984), 597. Woolley died in 1986 in South Africa, the country of his birth.
3   F.W.G. White and L.G.H. Huxley, *Records Australian Academy of Science*, 3(1) (1975), 7.
4   Cf W.F. Evans, *History of the Radio Research Board* (Melbourne, 1973), p. 127.

5   The five fellows of the Royal Society were: Pawsey (1954), Mills (1963), Wild (1970), Bolton (1973) and Bowen (1975). Martyn (1950) and F.W.G. White (1966) could be included as the first and second chiefs of Radiophysics. Woolley (1953) should also be mentioned. Hanbury Brown(1960) had been elected a Fellow before he came to Australia.

6   There is already a formidable bibliography for the origins and first years of radio astronomy, in Australia and elsewhere. See: J.P. Wild, *Records Australian Academy of Science*, 2(3) (1972), 52; E.G. Bowen, *Proc. Astronomical Society of Australia*, 4(2) (1981), 267; J.G. Bolton, *Proc. Astronomical Society of Australia*, 4(4) (1982), 349; P. Robertson, *Australian Physicist*, 21(8) (1984), 178; R.H. Frater, ibid., 21(9) (1984), 203; W.T. Sullivan, *The Early Years of Radio Astronomy* (Cambridge, 1984).

7   J.L. Pawsey was assistant chief and head of the radio astronomy group. The chief of the Radiophysics Division was E.G. Bowen.

8   See W.T. Sullivan, Chapter 13 of this volume.

9   Because of uncertainties in the cliff interferometer radio positions not all astronomers accepted the Bolton and Stanley identifications as proven, but it was not long before other means had yielded better positions, since when Bolton and Stanley have been universally credited with the first correct identifications of radio sources. Meanwhile Graham Smith in the UK had determined much more accurate positions for Cygnus A, which Baade and Minkowski could then identify, in 1954, with a pair of faint apparently interacting galaxies, indubitably extragalactic. The disparity between the brightness of the radio source and the faintness of the optical source created a tremendous impression. R. Minkowski, *Stars and Stellar Systems, VIII* (1968), 177; F.Graham Smith and B. Lovell, *J. Hist. Astron.*, 14 (1983), 155.

10   R.H. Frater, 'Radio science in Australia 1932-1982', pp 15-18, Golden Jubilee Publication of the IREE Australia (1982).

11   For biographies of Pawsey see: W.N. Christiansen and B.Y. Mills, *Australian Physicist*, 1 (1964), 137-41; A.C.B. Lovell, *Biographical Memoirs of the Royal Society*, 10, (1964), 229-43; see also J.P. Wild, cited in note 6.

12   They included Bolton (who returned later), Bracewell, Christiansen, Gum, Kerr, Little, Mathewson, Mills, Piddington, Stanley, Warburton, Watkinson, Westfold.

13   For accounts of the Parkes telescope see: H.C. Minnett, *Sky and Telescope*, 24 (1962), 184; F. J. Kerr, Ibid, p. 254; E.G. Bowen, *Proceedings Astronomical Society of Australia*, 4(2) (1981), 267-73.

14   J. Bolton, *Nature*, 246 (1973), 282.

15   For an authentic first-hand account of this episode see C. Hazard, *Active Galactic Nuclei*, J.E. Dyson (ed.), (Manchester, 1985).

16   W.F. Evans, op. cit (n.4), p.231.

17   See W.N. Christiansen, in Sullivan op. cit. (n.6). p. 113.

18   For a comprehensive account of the radioheliograph and its work see: D.J. McLean and Labrum (eds), *Solar Radiophysics* (Cambridge, 1985). Briefer accounts are given in: J.P. Wild, *Proc. Astronomical Society of Australia*, 1(8) (1970), 365; D.J. McLean, *Proc. Astronomical Society of Australia*, 4(2) (1981), 132; D.B. Melrose, ibid., p. 139; Wild's Matthew Flinders Lecture, *Records Australian Academy of Science, 3 (1)* (1975), 93 is well worth reading.

19   J.P. Wild, *Proc. Astronomical Society of Australia*, 1(1) (1967), 38.

20   For a semi-popular account of this work see: R.G. Giovanelli, *Secrets of the Sun (Cambridge, 1984)*. Giovanelli died in 1984. For obituaries see: J.H. Piddington, *Historical Records of Australian Science*, 6(2) (1985), 223; W.R. Blevin and R.E. Loughhead, *Australian Physicist*, 21(3) (1984), 72.

21   This body was largely created by Allen in the Second World War, for the purpose of supplying the armed forces with optimum frequencies for radio communications. These frequencies are mostly controlled by the sun's influence on the ionosphere.

22   M.J. Batty et al. *Proc. Astronomical Society of Australia*, 6(3) (1986), 346.

23   B.Y. Mills, *Proc. Astronomical Society of Australia*, 4(2) (1981), 156; ibid., 6(1) (1985), 72.

24   R. Hanbury Brown, *The Intensity Interferometer* (London, 1974). Prof. Hanbury Brown is the only President of the International Astronomical Union to have come from Australia.

25   J. Davis and W. Tango, *Proc. Astronomical Society of Australia*, 6(1) (1985), 34, 38.

26   Personal communication.

27   Bart Bok died in 1983. Volume XIII, number 2 (1984), of *Mercury* contains recollections of him and assessments of his work by people who knew him.

28   *MSO Annual Report for 1965*, p.1: 'I feel personally that our most important success over the past nine years has been the establishment of a first class graduate school of astronomy and astrophysics, with comparable emphasis on optical and radio astronomy.'

29 A.C.B. Lovell, *Quarterly Journal Royal Astronomical Society,* 26(4) (1985), 393.

30 *The First Twenty-five Years* (Canberra, 1980), p. 48.

31 ANU based much of their case on the wording of the Agreement, Article 4 of which read (in part): 'The arrangements for the provision by the University of facilities and services for the purposes of construction, operation and maintenance of the telescope shall be such as are agreed upon by the Telescope Board and the University, and the Commonwealth Government shall accord its good offices as appropriate in the negotiations and the putting into effect of these arrangements'.

32 In 1968 a disagreement arose among the Canadian astronomers as to whether the telescope they were building, comparable in size with the AAT, was to be located in Chile or Western Canada. Although the blank for the primary mirror had been paid for and delivered, the Prime Minister (Pierre Trudeau) resolved the conflict by cancelling the whole project. AAT people were of course well aware of this possible precedent. See G.J. Odgers and K.O. Wright, *Journal Royal Astronomical Society Canada,* 62 (1968), 392-4.

33 See E.J. Wampler and D.C. Morton, *Vistas in Astronomy,* 21 (1977) 191-207 for a general account of the telescope.

34 *Anglo-Australian Telescope 1984–1985* (Canberra, 1985).

35 *New Scientist,* 13 February 1986, p. 20.

36 From a speech at the official opening of the telescope in 1974.

37 See *Physics in Australia 1981: a Review by the National Committee for Physics* (Canberra, 1981). The Astronomical Society of Australia was formed in 1962. It has a membership of about 250, and publishes a regular professional journal.

38· *The Next Generation of Australian Telescopes* (Canberra, 1979); see also *Review of Observatories: Report of Expert Sub-Committee* (Canberra, 1978) p.22. The Expert Sub-Committee was chaired by Professor K.C. Westfold.

39 See R.H. Frater, op. cit. (n.6); J.B. Whiteoak, *Proc. Astronomical Society of Australia,* 6(3) (1986), 290.

40 A.W. Rodgers, C.T. Campbell and J.B. Whiteoak, *Monthly Notices of the Royal Astronomical Society,* 121 (1960), 103.

41 D.S. Mathewson, *Proc. Astronomical Society of Australia,* 6(1) (1985), 104; G.R. Meurer et al, *Proc. Astronomical Society of Australia,* 6(2) (1985), 195.

42 *Sky and Telescope,* 73(6) (June 1987), 596.

43 The UK schmidt is to be amalgamated with the AAO early in 1988. The AAO will then be responsible for two telescopes, the schmidt and the ATT.

44 See Harley Wood, *Proc. Astronomical Society of Australia,* 5(2) (1983), 273 for a history. The observatory is to be maintained as an astronomical museum. Harley Wood died in 1984. See W.H. Robertson, *Proc. Astronomical Society of Australia,* 6(1) (1985), 111.

45 W.T. Sullivan, op. cit. (n. 6), p. 110.

46 A.C.B. Lovell, *Journal of the History of Astronomy,* 8 (1977), 15.

47 J.L. Pawsey, *The Australian Scientist* (April 1961), 181.

48 Thanks are due to Dr Dick McGee, editor of the *Proceedings of the Astronomical Society of Australia,* for essential assistance, and to many others who supplied information, encouragement and criticism. They included 'Taffy' Bowen, Ron Brown, Robert Hanbury Brown, 'Chris' Christiansen, Harry Hyland, Dave Jauncey, Agris Kalnajs, Bernie Mills, Kevin Westfold, and Paul Wild. A few short passages have been taken from the entry on Astronomy in the *Australian Encyclopaedia,* and from a paper, 'The Woolley Era', referred to in note 2.

15

The shaping of contemporary scientific institutions

Ron Johnston and Jean Buckley

The Second World War marked a turning point in the standing and organization of Australian science. The need to produce military equipment and to become self-sufficient with respect to many goods previously imported required the establishment of new industries employing increased numbers of scientifically trained staff. Secondary industry production rose from less than £200 million prior to the war to £362 million in 1944/45, for the first time overtaking the value of production of primary industry.[1] The number of scientists and employment opportunities for them expanded considerably.

In addition, the contribution of science and technology to the war effort raised its visibility and perceived significance in the eyes of governments and the general community. In this climate, expenditure of large sums of public funds on research was seen as quite appropriate, and not requiring detailed justification.

CSIR was pre-eminent in the performance of research, and the quality of its staff and facilities. In 1939 it had a budget of £350 000 and a total staff of 400 (of which about half were scientists); by 1946/47 this had risen to £1.4 million and 2300 respectively.[2] In addition it also constituted the major and frequently the only source of advice to the Commonwealth government on scientific matters. Research was strongly focused on agricultural problems, through the operations of CSIR and the six State departments of agriculture. CSIR, however, had also diversified, with the establishment of three non-agricultural divisions — Radiophysics, Industrial Chemistry and Aeronautics — and the National Standards Laboratory.

The one area of notable strength outside CSIR was medical research, which was located primarily in independent medical institutes such as the Walter and Eliza Hall Institute of Medical Research[3] and the Kanematsu Memorial Institute of Pathology.[4] Following the formation of the National Health and Medical Research Council in 1937, the Medical Research Endowment Fund had been established with an annual appropriation of approximately £250 000.

In the more than forty years since the war, the institutions of science have become firmly established and indeed familiar elements of the Australian social and economic structure. There has been a very large increase in the resources devoted to science. Moreover, the nature and role of the science institutions have been transformed from one of essentially 'service science', meeting national needs in exploration, mapping, public health and agriculture. Over this period science became recognized as a major source of economic growth, an important national resource, an activity of relatively high status, and the basis of a rewarding career. At the same time, with a few marked exceptions, science did not find itself on the public political agenda; rather it was bureaucratic politics that provided the arena for the shaping of the scientific enterprise.

This chapter seeks to chart the changing structure and influence of Australian science institutions over the period 1944-1980, in the context of the major economic and political influences of the day. The major focus will be on the period 1965-1980, when the scientific community systematically sought greater influence and financial support. During this same period bureaucratic control over science was steadily increased, often as a direct trade-off for greater financial support, as explicit science policy machinery was developed.[5]

This period was preceded however, by a period of institution building, characterized by Peres as being one of apolitical stability, with minimal interaction between science and politics:

> The simple fact of the matter is those who participate in the policy processes, and society generally, have not seen Australian science as having a great instrumental role in tackling the range of national problems that, elsewhere, have provided the major justifications for investment in science and technology. Instead of turning to science and technology to relieve our national anxieties, it has been our habit to seek out economic, political and diplomatic bargains to gain access to the instrumental science and technology we have needed or to block threats from innovating foreign technology.[6]

Institution building

The establishment and growth of science in Australia has been achieved largely through government support. The first available comprehensive data for expenditure on research, for 1958/59, show governments to be responsible for more than 80 per cent of the national research expenditure of £37 million.[7] The Department of Supply had the highest research expenditure (£11.7 million) largely through its support of the long-range weapons project, followed by the Commonwealth Scientific and Industrial Research Organisation (CSIRO) (£8.5 million).[8]

A major boost to post-war defence research came with the Joint Long Range Weapons Project with the United Kingdom in 1945, and the setting up of the Weapons Research Establishment at Salisbury, South Australia, in 1949. The research focus was on electronics, high-speed flight research and propulsion,[9] and significant achievements included the Jindivik pilotless drone plane, the Malkara rocket, and advances in computer-aided weapons and flight simulation, underwater engineering, new alloys and metal ceramics, and microelectronics. However, little of this research output flowed into the civilian sector, and there was only a small return to the Australian economy from this public investment.

For CSIR/O the post-war period was marked by early trial, but there followed a flowering into a position of command and esteem in the Australian, and indeed the international, science scene.[10] The growth of this Commonwealth research organization in part reflects the federal structure of Australia, whereby the states had responsibility for education and science, but in the latter, at least, relatively little interest. It is also a consequence of the strenuous efforts of successive leaders of CSIR/O, and in particular Sir David Rivett, the first full-time Chief Executive Officer of CSIR.[11]

In 1948 CSIR had to weather its first major political controversy. In the 'cold war' climate then prevalent, members of the Liberal–Country Party Opposition seized on Rivett's public opposition to secrecy in science to attack the government, through CSIR, for its employment of professed communists in areas of strategic significance. This led, in 1949, after a considerable bureaucratic struggle, to CSIR's reconstitution as CSIRO, under increased Ministerial, Public Service Board, and public control and accountability.[12]

However, through a series of scientific and technical achievements in radio physics,[13] in animal nutrition through the addition of trace elements,[14] and apparently in rain-making,[15] the new CSIRO was able to regain its image of success. In particular, the decimation of the rabbit population by myxomatosis[16] provided a dramatic demonstration of the 'magical' power of science, which captured the public imagination and was 'a signal triumph for CSIRO and served to blot out the memory of the spy stories of the 1940s'.[17]

Throughout the 1950s, CSIRO's budget grew from £5.9 million to £19.5 million. A number of very large research facilities were constructed, including the Parkes radio telescope,[18] a radio heliograph, and a phytotron,[19] at a cost of £0.8, 0.6 and 0.5 million respectively. Over this period CSIRO also had the support of a committed and influential Minister — R.G. (later Lord) Casey.[20] It also had a high level of autonomy, and a dominant position as adviser to the government on scientific matters.

The third major Commonwealth research organization was the Australian Atomic Energy Commission (AAEC), established in 1953. As the results of a co-operative agreement with the United Kingdom, the AAEC was set up primarily as an R&D organization with laboratories and a DIDO-type research reactor at Lucas Heights near Sydney.[21] A considerable in-house expertise and annual budget was built up, but after the failure of the push to establish a full-scale nuclear power reactor in 1972, the AAEC became what has been described as 'a body in search of a mission'.[22]

Probably the greatest extent of science institution building occurred in higher education. The prime task of the six universities and two university colleges after the war was the rehabilitation of service personnel. The emphasis was entirely on teaching. By 1950, the total expenditure on research activities in universities, including post-graduate training, was only £350,000, compared with CSIRO's budget for that year of £2.35 million.[23] The contrast between CSIRO, with its high-quality staff and good salaries and conditions, and the overcrowded, overworked and underpaid conditions of the universities was sharply felt.[24]

The first move to improve the lot of the universities and at the same time the supply of post-graduates to CSIR, was the establishment of the Australian National University (ANU) in Canberra in 1945, as a post-graduate institution. The emphasis on attracting outstanding expatriate scholars led to the return of M.L. (later Sir Mark) Oliphant, who was to have a profound influence on the organization of science and its relation with governments over the following twenty years.[25]

A very different tradition, building on the strength of technical education in New South Wales and modelled on the Massachusetts Institute of Technology (MIT), led to the establishment of the New South Wales University of Technology as the second Sydney university in 1949.[26]

Neither of these moves improved the lot of the existing universities. It required a substantial personal campaign mounted by Sir Ian Clunies Ross, Chairman of CSIRO,[27] and lobbying by the Australian Vice-Chancellors' Committee (AVCC)[28] and the Australian National Research Council,[29] to create the climate for government action. The final trigger was a 1956 report of the Joint Committee of Public Accounts on the Commonwealth Office of Education, which urged a wide-ranging inquiry into universities in Australia.[30]

Subsequently, a Committee of Inquiry was established by Prime Minister R.G. Menzies, chaired by Sir Keith Murray, Chairman of the British Universities Grants Commission. The ensuing report[31] called for a large increase in funding for the universities, and the establishment of an Australian Universities Commission (AUC) to provide advice on the rational and coherent development of the universities. The massive intrusion of the Commonwealth into an area of states' responsibility was justified by the Prime Minister in terms of national importance and inadequacy of state funds.[32]

However, the concomitant of expansion in the national interest was greater government co-operation, if not control. Thus, Menzies asserted the AUC's functions to be:

> to co-ordinate the work of the universities, to avoid avoidable overlapping, to push somebody here and hold somebody back there, and to get the overall pattern of University training in Australia brought into the most useful shape.[33]

While funding for the universities was substantially increased, the extent of the growth in student numbers did not lead to a great increase in the support for research. The first chairman of the AUC, Sir Leslie Martin, argued for an earmarking of funds to support special research activities at the post-graduate level.[34] An alternative proposal was made by Oliphant on behalf of the Australian Academy of Science, for the establishment of a National Science Trust 'to give financial support to specific research projects in the natural sciences which would otherwise not be carried out'.[35] It was the latter model that appealed to Minister John Gorton, and the Australian Research Grants Scheme was established in May 1965 to fund excellent research on a national competitive basis.

The last of the major science institutions established in this period was the Australian Academy of Science,[36] in 1954. The major proponents were Oliphant and D.F. Martyn, who argued for an 'Australia-wide body, representative of all the sciences, membership of which would become a privilege and mark of achievement'.[37] An unashamed commitment to élitism was necessary, it was argued, to endow the body with sufficient prestige to enable it to speak for science as a whole.

The early history of the Academy was marred by fierce battles over eligibility for election, particularly with respect to officers of CSIRO. However, there emerged a fellowship that was able to achieve considerable influence, at least in some areas. Important reports were produced in its early days on the issues of scientific manpower[38] and the future of the Kosciusko Summit Area.[39]

Perhaps the most significant of the academy's initiatives was the establishment of the Science and Industry Forum in 1967. It followed a symposium in 1964 that called for regular meetings between academicians and industrialists.[40] The forum was made up of about eighty scientists, industrialists and government officials and, as we shall see, through its reports and its provision of occasions for interaction played a significant role in the shaping of Australian science.

It can be seen that the period 1945–1965 was one of very considerable building of the institutions of science in Australia. Prior to that time science, in an organized and institutionalized form, had achieved only a fledgling existence. By the end of 1965 institutions, structures and values had been thoroughly embedded in a way that was to have considerable consequences for Australian life in the remainder of the century.

The development of these new institutions was, in general terms, not a particularly difficult process, though it undoubtedly would not have seemed that way to the protagonists at the time. However, the political and economic conditions were favourable; it was a time of stability characterized by great promise, modest performance and few demands.[41] There was an almost universal commitment on the part of the scientists to autonomy in their institutions. Decisions about science were regarded without question as being of a sufficiently special nature to be appropriately made by scientists only.

There was also a marked pattern of insider influence, or what C.P. Snow has called 'closed politics'.[42] The government's decision-making about key directions in the development of science were made with the advice and influence of a small élite behind closed doors; there was almost no discussion with the wider community of scientists, let alone the general public.

It was also a period of very considerable support from governments, particularly that of R.G. Menzies. However, Menzies' support for science was by no means unqualified. In a speech in 1958, just after the launching of Sputnik, he argued that science and wisdom have no necessary connection, and that there was a need for

> more scientists; more scholars; more scholarly scientists; more scientific scholars; more historians and philosophers and poets. For if science is truly to serve the cause of civilisation it will be because the people to use the advances of science have grown to use them in a civilised way.[43]

The growth of bureaucratic control

It is apparent that the considerable growth of the scientific institutions after the war was not achieved without growing pressure to reduce their autonomy. However, it was predominantly in the period following 1965 that bureaucratic control was significantly expanded. Some of the changes undoubtedly emerged as a consequence of the sheer growth of the enterprise. Thus, the size of the science institution alone forced a change from the cosy interaction of an élite with one or two Ministries (including the Prime Minister) and a handful of senior public servants, to a more formal public administrative structure and process.

Other changes, however, reflected a world-wide change in the perception and influence of science. No longer were the arguments of Vannevar Bush for unfettered scientific research as the source of economic growth[44] accepted without question. The memories of the scientific contribution to the war effort had faded and, in the face of the new and urgent national problems emerging in the 1960s and 1970s, there was a strong demand that science be more effectively organized and mobilized to produce solutions. Hence attention shifted towards mission-orientated, 'relevant' research.

Another major force for change was the growth of a new and widely publicized disenchantment with science and technology:

> Scientific research became associated in the minds of many with war (Vietnam); and with the environmental and social deterioration resulting from the large scale application of technology. For the first time in many years, the steady and occasionally spectacular growth of R&D began to falter, and scientists, whether deservedly or not lost some of their influence and credibility in government and before the public.[45]

A measure of the changing political status of science is provided by its ministerial representation. From 1950 to 1960 science, in the form of CSIRO, was represented by one senior and influential minister (Casey) and had the strong support of the Prime Minister. In contrast, over the period 1963 to 1982 there were eleven ministers (see the table below). While two of these (Gorton and Fraser) went on to become Prime Ministers, in general the ministry was lowly ranked.

Ministerial representation for science, 1960-1982

Minister	Responsibility	Period
Dr D.A. Cameron	Minister-in-Charge CSIRO	Feb. 1960–Dec. 1961
Mr R.G. Menzies	Minister-In-Charge CSIRO	Dec. 1961–Feb. 1962
Senator J.G. Gorton	Minister-in-Charge	Feb. 1962–Dec. 1963
Senator J.G. Gorton	Minister assisting Prime Minister in Education and Research	Dec. 1963–Feb. 1966
Senator J.G. Gorton	Education and Science	Feb. 1967–Feb. 1968
Mr J.M. Fraser	Education and Science	Feb. 1968–Nov. 1969
Mr N.H. Bowen	Education and Science	Nov. 1969–Mar. 1971
Mr D.E. Fairbairn	Education and Science	Mar. 1971–Aug. 1971
Mr J.M. Fraser	Education and Science	Aug. 1971–Dec. 1972
Mr W.L. Morrison	Science	Dec. 1972–Jun. 1975
Mr C. R. Cameron	Science and Consumer Affairs	Jun. 1975–Nov. 1975
Senator J.J. Webster	Science	Dec. 1975–Apr. 1978
Senator J.J. Webster	Science and Environment	Apr. 1978–Dec. 1978
Brig. D. Thompson	Science and Environment	Dec. 1978–Nov. 1980
Brig. D. Thompson	Science and Technology	Nov. 1980–Mar. 1982

Funding continued to grow, however, if more slowly than in the institution-building phase. Indeed, by drawing on data collected by the Department of Science (in its Project SCORE — 'Survey and Comparison of Research

Expenditure'), it can be seen that funding in real terms peaked early in the 1970s, and then declined to a lower level for the next eight years, as shown in the second table, below. The decline can be attributed almost entirely to a dramatic fall in industrial R&D.

Science funding over the period 1968–1982

Sector of performance	A$ million at current prices				
	1968–69	1973–74	1976–77	1978–79	1981–82
Commonwealth government	177	208	326	353	540
State Government	41	76	132	157	211
Private enterprise	119	191	160	206	288
Higher education	89	174	244	326	440
Private non-profit	2	6	11	13	17
Total	368	655	873	1055	1496
Total (1974/75 prices)	599	724	639	636	662

Source: Ronayne, (see n.5) and ABS publications

The increasingly dominant role of the government in supporting research is apparent. Given that most of the research performed in the higher education sector was funded by governments, it emerged that by the end of the 1970s more than seventy-five per cent of all research funds were provided by government, compared with 20 per cent from industry. Within the Commonwealth sector, about one-third of funds went to CSIRO.

Perhaps the most visible feature of this period was the campaign to establish an independent body to provide scientific advice to the Commonwealth government — a campaign that Ronayne traces back to the 1951 Jubilee Seminar at ANU, 'in order to overcome CSIRO's monopoly of the provision of scientific advice to the government'.[46] A detailed examination of this campaign provides the means to observe the emergence of explicit science policy in Australia.

In pursuit of a science advisory council.

The first formal proposal seems to have been produced in 1957, when the Academy of Science issued its first report on the subject of scientific and technological manpower in Australia. It called for the establishment of a committee of enquiry on science and technology, the terms of reference of which should include 'to report on the state of development of science and technology in relation to the present and future needs of the community'.[47] In November 1960, Prime Minister Menzies, accompanied by F.W.G. White from CSIRO, met with Sir Alexander (later Lord) Todd, then Chairman of the UK Advisory Council on Scientific Policy. Records show that the Prime Minister was sceptical about establishing a national body, largely because of difficulties arising from federalism, and that White expressed doubt about the 'need for a new scientific co-ordinating committee in Australia'[48] — a view he was to continue to put for many years.

However, in 1962 there was a flurry of lobbying on the need for a more effective government policy on science. Firstly, White, in the context of a letter to the Prime Minister concerning the choice of a new Minister-in-Charge of CSIRO, argued thus:

> The Government's attitude to science and the part the results of research can play in national affairs has of recent years been that of a "patron" of science ... What is missing is the expressed interest of the Government in the development of science in Australia and the Government's conviction that the application of the results of research can aid vitally in the agricultural and industrial development of this country.[49]

He called for the Minister-in-Charge of CSIRO to be charged with developing a 'more comprehensive policy of science in relation to national affairs'.[50]

Only a few weeks later Sir Mark Oliphant wrote to the Prime Minister proposing the establishment of an advisory committee on scientific policy. The basis for Oliphant's proposal was concern at the large number of costly proposals emerging from the scientific community: 'Nobody examines all major requests and recommends whether or not a given project should be initiated or expanded in the light of the needs of the country, the ability of the investigators or the state of the national economy.'[51]His solution was a committee that would advise on major new projects above a stated cost, and in addition review the national research effort each year.

White's response was highly critical, and raised so many interesting points it is worth examining at length. It should be noted that there was already a history of irritation on the part of CSIR/O with regard to Oliphant's many proposals, whether on the basis of their appropriateness or resentment at a competing influence. White rejected Oliphant's proposal firstly on the grounds that it rested on an excessive belief in the rationality of government decision-making: 'Whether the money is approved by the Government or not seems to me to depend much more on the Government's confidence in the institution putting forward the plan and its understanding of the status and merit of those concerned.'[52]

He also objected to a centralized approach that might threaten CSIRO: 'Surely it is of great importance for the Government not to set up an arrangement which undermines the authority and responsibility of existing institutions.'[53]

Finally, he made a strong claim as to the appropriate aims of any policy:

> My chief reaction to your paper is, however, that it tends to create the impression that the Government should set up some machinery to restrict and control an over-ambitious scientific community which this country cannot afford ... In my view, if the Government needs anything more in the way of organization, it would be for the purpose of assisting the Government itself to have a progressive policy for the development of science in the interests of our economy.[54]

Thus, for White, policy with regard to science was to advise government on how to nurture science, rather than to assist government with decisions on resource allocation.

The third submission in 1962 came from Sir Leslie Martin, chairman of the AUC, who proposed the establishment of an advisory committee to the cabinet on scientific policy, one of whose functions would be a stocktake of the Australian science capability.

In response Menzies met with Oliphant, Martin and White on 16 April, 1962. There was support for the Oliphant proposal from the Prime Minister's secretary but the emphasis was on keeping control in government hands and 'strong opposition to any notion that an outside body should have any advisory role direct to Government'.[55]

After the meeting, White wrote further to the Prime Minister, softening his original stance somewhat:

> There can be only one primary reason for creation of an Advisory Committee on Scientific Policy. This step would be an indication of the Government's determination further to stimulate and encourage the development of Australian science . . . Such a Committee could broadly survey the Australian scene. It could make recommendations to the Government for action to fill gaps in the existing structure of science or to improve and extend the present activities . . . It could also be charged with the responsibility to advise on the proper balance between the Government expenditure in different areas of activity, for example, the Universities, CSIRO, the Atomic Energy Commission and so on.[56]

But he remained adamant about the need for autonomy for government research institutions:

> It would be a very grave mistake, and detrimental to the advancement of science and to the application of science to the nation's problems to give to an Advisory Committee on Scientific Policy any function which made it an intermediary between the Government's appointed agencies and the Ministers to whom they report. To do so would be tantamount to the creation of a sort of Department of Science.[57]

The Academy also pursued its initiative. The president wrote to the Prime Minister on 30 June 1962 arguing that a strong prima facie case for the establishment of a national science trust, or foundation, existed, but that in order to advance the matter properly, the government should appoint a committee of inquiry into science and technology in Australia 'to inquire into the state of development of Science, and its application to Technology, in Australia in relation to the needs of our community'.[58]

Subsequently, at a meeting with the Prime Minister on 9 November, the Academy was encouraged to formulate detailed proposals for an Academy/ government-administered scheme. In response to this proposal the government formed a small committee on May 1963 to advise it on ideas for a national policy on scientific research. However, attention now shifted to the competing proposals from the AUC and the Academy for support for research. It was this issue that was to dominate government concern until the formation of the ARGS in 1965.

Meanwhile the 'campaign', not so much organized as promoted through the grapevine, continued. An article in the *Current Affairs Bulletin* one week before the opening of the 1962 ANZAAS Congress, by two anonymous chemists, apparently from Sydney University, argued for the 'appointment of an advisory council of strong and distinguished scientific-economic composition and with a suitable breadth of representation, reinforced by the nomination of a senior minister to convey its recommendations to Cabinet'.[59]

At the 1962 ANZAAS Congress, the Presidential Address concluded: 'The time has arrived when a man, or a small group of men, should be appointed in

this country to examine the implications of our scientific policy as a whole.'[60]
White, as President of the 1963 congress, returned to the theme:

> Australia cannot leave science to develop in the hope that its influence will spread like an infection. A more deliberate approach is imperative. This is easy to say but difficult to achieve. It is because there must be a clear understanding of the elements of such policy and the reasons for it that I find Tizard's strategy of science an attractive approach.[61]

This strategy was defined as 'the art of so directing the application or advance of science as to make the most rapid progress of society or of knowledge',[62] but the means for achieving it were not examined.

The campaign appeared to reach its culmination with the statement of the Prime Minister on the occasion of the tabling of the Martin Report in March 1965. In addition to providing the government's response, the Prime Minister announced

> that there ought to be a body to advise the Government on the most effective methods of co-ordinating, and achieving results from, expenditure on research through the universities, through the Government's own agencies, and through any other bodies to which grants are made by the Government. What we seek is a situation in which, the Government having decided what proportion of national income can go to scientific research, and having indicated its views on the general fields in which advances would, economically, most benefit the nation, some competent advisory body would recommend the allocation of available money among the various governmental research bodies.[63]

Rubenstein has argued that this proposal was shelved before the end of the year, probably because of

> the decision in September 1965 by the Menzies government to reject also the main recommendation of the Vernon Committee of Economic Inquiry for an Advisory Council on Economic Growth. Clearly it was felt that the mere existence of independent expert advisory councils on the economy — and implicitly on science policy — would create strong pressures to follow their recommendations, thereby severely curtailing the government's freedom of action.[64]

However, it is worth noting that White continued to work on a report to his minister, Senator John Gorton, on science policy in other nations, arising from an overseas trip from April to July 1965.[65] In addition, in his speech at the opening of the David Rivett Laboratory on 1 April 1966, Senator Gorton indicated the need for some form of co-ordination of the national research effort and invited suggestions as to how this might be achieved. Indeed, it appears there was action to establish a 'so-called Inter-Departmental Committee, to recommend appropriate machinery for continuing advice to the Government on science policy, made up of six people, three of whom will represent science, and three from the Commonwealth Service'.[66]

With the retirement in January 1966 of Prime Minister Menzies, who had displayed a strong interest in the place of science in Australian life, it was inevitable that the political agenda should be somewhat rewritten. The new leader of the government, Harold Holt, announced in his election policy speech in 1966 the intention of the government to form a Ministry of Education and Science.

There was considerable opposition to such a move on the part of CSIRO. The establishment of a ministry of science in Britain in 1959 had inevitably led to proposals for its imitation in Australia. However, the minister in 1960, Dr D.A. Cameron, took the advice of White that no case had been made for its effectiveness in a federal system, and that there was doubt whether government should become involved in direction and co-ordination of scientific activities.[67] Apparently, neither the fact that the limitations of federalism had been so effectively and recently over-ridden in implementing the Murray Report, nor that the Commonwealth government was responsible for the vast proportion of Australian research, was seen as providing an argument to the contrary.

However, in this new era CSIRO was not finding its relationship with its ministers as easy as in the days of Casey. For example, in February 1963, not long after taking up the responsibility of Minister-in-Charge of CSIRO, Senator Gorton noted:

> The practice of work being approved by the Minister-in-Charge of CSIRO before it is undertaken has, however, been honoured in breach rather than in the observance, and I gather that, over the years, a practice has arisen where it is 'assumed' that any work that is being carried-out by CSIRO has been approved, *ipso facto*, by the Minister-in-Charge.[68]

White's response, in defence, was to place responsibility squarely on the minister:

> Although the Minister may think it desirable, as indeed previous Ministers have, to depend largely on the Executive to develop its policy, nevertheless it is quite obvious that the Minister must spend a reasonable amount of time with us to understand what we are doing and whether our policy is in line with Government responsibility in science.[69]

With the installation of the Holt government in 1966, Gorton took over the new portfolio in education and science that incorporated responsibilities for the previous Education Division of the Department of the Prime Minister (which included the ARGC), the Commonwealth Office of Education, the AUC and CSIRO. Previously, in 1963, the Minister-in-Charge of CSIRO had been replaced by a minister assisting the Prime Minister in education and research, the consequence of which was that educational priorities displaced those of research.

A major force in the promotion of any explicit science policy at this time was the rapidly strengthening opposition Labor Party. At the August 1965 Federal Conference the ALP issued a platform statement on science and technology. In tune with the determination of the ALP to present itself as a party of the 'future', the platform stated: 'Australia desperately needs national scientific policies which will embrace not only planning for scientific R&D, but also enable the results of scientific research in Australia and elsewhere to be applied in every aspect of Australia's industry and its culture.'[70] To achieve this they committed a future Labor Government to the establishment of a ministry of science, an advisory science council, a parliamentary standing committee on science and technology, a national science foundation and tax incentives to encourage industrial R&D. To some extent this platform provided an agenda for the subsequent debate about the machinery of science policy.

During Gorton's ministry, the Department of Education and Science dealt almost exclusively with its education responsibilities. It was not until 1968, when Gorton became Prime Minister and Malcolm Fraser took over the Department of Education and Science, that a science branch was formed and the department began to become a force in science policy formation. However, its initial functions were merely to provide the secretariat for the ARGC and initiate an information-collection exercise entitled Project SCORE (Survey and Comparison of Research Expenditure), to provide for the first time detailed statistics on levels of research activity and cost.

Over the ensuing years there was intense activity, largely within the science bureaucracy, to establish an appropriate form of machinery for science policy. However, it was an oft-quoted off-the-cuff comment by Prime Minister Gorton to an American journalist, in contrast to much of his previous interest, that set the public image of his government's views:

> I don't know what a science policy is. The critics want an overall advisory committee to allocate funds, but I don't see the need for any advisory body. These committees are only a group of individuals pushing the barrow for their own disciplines.[71]

It should be noted that there were good grounds for these views. When the OECD examiner conducted a review of Australian science and technology in 1973, they noted there was little push outside 'a well-developed scientific community for a coherent science policy.'[72]

The establishment of a science advisory mechanism was promoted by the officials of the Department of Education and Science as a means of 'extending their limited resources.'[73] Given their small staff, and their commitments to provide the ARGC secretariat and conduct Project SCORE, there was concern about the ability of departmental staff to provide guidance to the minister on scientific and technical matters:

> Nor for that matter would any one Commonwealth Department ever be able to do so if objectivity and national needs were among the criteria on which advice should be based. Secondly there is already considerable pressure from universities, scientific bodies and the press for some expression of the Government's attitude towards what is loosely termed a "national science policy".[74]

They proposed a standing committee of departmental heads, to be known as the Australian Science Advisory Committee, with limited outside membership, to consider proposals with a 'science' element that came to cabinet. However, the proposals from academic circles, including the Academy of Science, for the establishment of an advisory council to assist the government in the formulation of science policy, were rejected:

> Such a Council would mean the involvement in Government matters of a large number of people extraneous to Government. In practice, Government decisions on scientific matters frequently involve large expenditures of public funds, there are security aspects to consider, and the various projects on which decisions have to be made may involve the setting of priorities between different elements in government itself. Such matters do not lend themselves to consideration by a Council type body.[75]

Indeed one early draft suggested that:

> In any event, receipt of advice from an outside body of this kind can well be embarrassing, especially since the avowed policy of many science academicians is largely to promote more funds for science rather than to assess priorities in the light of the national needs.[76]

The essential argument, encapsulated by Oliphant, was that Cabinet was being faced with a large number of increasingly expensive requests for support on which it was unable to make a balanced judgement. Technical advice, and a broader judgement, could be provided by a standing committee.

Throughout the development and extensive consultation processes associated with the preparation of a cabinet submission, CSIRO consistently sought not so much directly to oppose as to use the full range of bureaucratic techniques to divert, refer back, and seek wider consideration. White portrayed science policy developments overseas as chaotic, experimental and ineffective — a view that was, to say the least, idiosyncratic. While CSIRO accepted that a Minister for Education and Science might quite appropriately establish mechanisms to advise him, the idea of a centralized advisory body that was seen as potentially intruding on CSIRO's privileged relationship with government was strongly opposed. Thus:

> Does every proposal have to go through the Committee? CSIRO for example at every budget has proposals for changes in its scientific activities that are very carefully considered by the Executive and expert scientists. Surely it is not envisaged that these have to go through the Committee.[77]

In June 1968 the proposal went to cabinet but was rejected on the grounds of concern that the advisory committee might adopt an 'overlord' role. However, Minister Fraser was invited to consider the matter further, particularly in the light of information emerging from Project SCORE, and, if appropriate, to bring forward a new submission.

There were two specific stimuli for further action. The first was the production of the first report of the Science and Industry Forum entitled 'Science Policy Machinery for Australia'.[78] The report, produced by a working group chaired by Professor G.M. Badger, later to become the first Chairman of ASTEC, argued the need for a national science policy:

> To foster and develop science; to ensure that basic scientific research is encouraged and that research in fields of applied science and technology is carried on at adequate levels in all areas relevant to the development of Australia and the achievement of our economic goals; and to encourage the implementation of scientific knowledge.[79]

The order of objectives clearly implies that the commitment of policy for science was still pre-eminent.

The report argued, on the basis principally of comparison with the Canadian model, that these goals could best be achieved by the establishment of an independent advisory committee on science and technology, served by a full-time secretariat within the Department of Education and Science.[80] The committee should report to the Cabinet through the Minister for Education and Science, and its composition should be about fifteen members from the scientific community, private enterprise and the public service.

The Minister, Malcolm Fraser, responded cautiously to this report at the meeting of the forum in Thredbo in March 1969. There is evidence that he was sympathetic to the proposal but was restricted by the recent Cabinet rejection. Nevertheless he did express caution about the value of a centralized approach:

> there is no easy answer and there is no simple solution to the problems of priority and that the machinery specially established to help determine these problems does not seem to have worked particularly well ... We may then be wisest to continue our pragmatic evolutionary approach seeking advice from different people as different projects arise.[81]

It is worth noting that Fraser appeared to be somewhat sceptical about the inevitability of benefits from research. As one example of this attitude, he challenged the chairman of the ARGC, Rutherford Robertson, that national interest should be more explicitly included in the criteria for allocation of grants. Robertson successfully defended the principle that excellence alone should be the measure — a principle that still holds in 1987, and that may be partly responsible for the limited government funding for this scheme over the years.

The second stimulus came from the public pronouncement of the leader of the Opposition, E.G. Whitlam, particularly at the ANZAAS Congress in 1970. In his speech on this occasion he set out the basic aims of a national science policy:

> First and foremost such a policy should be aimed at guidance and management rather than direction and control.
> Second, it must envisage means by which the government can obtain guidance for the continuing formulation of science policy ...
> Third, the policy must aim at getting value for money spent ...
> Fourth, the policy would need to ascertain the order of priorities to be afforded the major areas of research ...
> Fifth, a policy should assist to maintain and improve the effectiveness of scientific activity by minimising unwarranted duplication ...
> Sixth, the policy should aim to deploy available scientific and technological resources so as to bring about planned innovation.[82]

Whitlam went beyond the ALP platform that had been adopted in 1965 to argue that under his government, the Minister for Science, together with an advisory science council, 'would formulate policy recommendations for submission to Cabinet and would seek through the operational arm of his portfolio — the CSIRO and comparable organizations — and through the advisory council, the implementation of policy'.[83]

Thus it was that in May 1970, under a new minister, N. Bowen, a further attempt was made to produce a cabinet submission on science policy. At this time J.R. Price took over from White as chairman of CSIRO, following White's retirement. The new draft was essentially the same as that of 1968, except that the Standing Committee of Permanent Heads had been upgraded to a Ministerial committee.[84] The opposition from CSIRO was also almost a facsimile of previous arguments.[85]

To provide a better basis for the proposal Bowen, together with Price from

CSIRO and Rutherford Robertson, president of the Academy, visited the USA and Canada to examine their science policy machinery.

In September 1971, on his return to the portfolio of Education and Science, Fraser sent Price and A.H. (later Sir Hugh) Ennor, his Departmental Secretary, to Europe to report jointly on science advisory machinery and procedures. This may well have been a device not only to gather information, but also to commit these two antagonists to some common view and terminate the continual flow of critical correspondence between them.

Fraser was also keen to involve industry more in the development of machinery for science policy, and as a consequence met in November 1971 with a senior representative group from the Manufacturing Industry Advisory Committee, led by Sir James Vernon. In the same period the Liberal Party's Secondary Industry Consultative Committee carried out a detailed examination of the proposed Advisory Council on Science and Technology. From both these groups there emerged a new proposal for independent membership, with permanent heads excluded but available as 'assessors'. Thus the original proposal for a single standing committee of Cabinet had now bifurcated, into a new independent advisory committee, reporting to a new inter-ministerial committee of Cabinet.[86]

By the time the Cabinet submission was prepared in February 1972, however, discussion with officials of other departments, and in particular of the Prime Minister's Department, led to abandonment of the ministerial committee tier, it being replaced by the reporting of the Advisory Council to the Minister of Education and Science.

On 27 April 1972 the then Prime Minister, W. McMahon, announced the government's intention to establish an Advisory Committee on Science and Technology (ACST) 'to make recommendations to the Government on Australian efforts in civil science and technology'.[87] The Opposition response was to complain that 'the proposal is almost a direct steal from the science and technological policy of the Australian Labor Party'.[88] In September an eleven-man membership under the chairmanship of Sir Colin Syme and including four scientists, an economist, and six members of the industrial and pastoral community was announced.

The major terms of reference were to report on the development and application of science and technology to national needs; on new areas of science and technology of importance to Australia, including fields of industrially and commercially orientated R&D; on the balance, adequacy and effectiveness of national efforts in various fields of science and technology; and on the relative importance of efforts in those fields that may contribute to national economic development and welfare.

Thus, after a campaign of almost twenty years, the scientific community finally achieved a so-called 'independent advisory body'. But the 'success' was short-lived. The Labor Party was elected to power in December 1972 and, in line with the party platform, a new Ministry of Science was established under W.L. Morrison, who also held the portfolio of External Territories. In the Whitlam government this placed science inside the Cabinet, though it was lowest ranked at 27 — a far cry from the earlier direct representation by the Prime Minister.

The Whitlam years

One of Morrison's first acts, having consulted the Prime Minister, was to call a meeting of the ACST, 'give us a good lunch, and tell us we were sacked'.[89] At the time of the suspension, the major reason given was the inappropriate membership of the ACST.[90] The fact that, as we shall see, it took approximately two-and-a-half years for an interim ASTEC to be established, with terms of reference very similar to those of the abolished ACST, justifies this action being described as 'unnecessary and damaging'.[91]

The years of the Whitlam government were in general very difficult years for the Australian scientific community. There was a disenchantment with science and technology by many in the community. At the same time, governments were pressing for more accountability for, and payoff from, public expenditure on research.

Initial expectations had been high. With its platform on science, which had been prepared primarily by scientist party members in 1964, many in the scientific community had hopes of a new golden age. However, this platform had been revised in 1971 with the aim of linking science more effectively with environmental, energy and economic goals; specific proposals for a fuel and energy policy, environmental and technology assessment organizations, and extended scientific education had been added.[92] But the machinery proposals for science itself were essentially unchanged. It was thus perhaps inevitable that in 1972 they should be somewhat out of step with, and placed lower than, other priorities.

Other difficulties emerged from the style and operation of the minister. Firstly, in a manner that was common in the Whitlam government, the minister appointed personal advisers who were given considerable authority and operated not only between the minister and interest groups, but also between him and his department. Partly as a consequence, there was poor co-ordination and occasionally outright antagonism between the ministry and the department. A third factor was the committed anti-elitism of the Labor government ministers and staffers, who were much less likely to be influenced by traditional arguments about the essential value of science. Morrison complained bitterly about the lack of support from scientists: 'One of the disappointments I had was that I received precious little assistance from scientists on our early efforts to enunciate a science policy... From the comments I received, it seemed that a science policy to them merely meant more money for scientists'.[93]

After the dissolution of the ACST, the minister expressed a wish to proceed as quickly as possible to set up a science council. The first draft proposal was produced by August 1973. Gone was Whitlam's proposal for an executive and policy-formulating body, which presumably would have displaced the Department of Science. The proposal for elected representatives had also been dropped. Even the full-time chairman's position was changed to part-time, at the suggestion of Oliphant.[94] In this form, little different from the ACST, a lengthy consultation procedure with all other ministers was initiated at the Prime Minister's suggestion.

At the same time, following Australia's admission as a member of the Organisation for Economic Cooperation and Development (OECD) and perhaps also to obtain an independent, international source of advice, the minister initiated a request to the OECD to conduct a review of Australian science and technology along the lines carried out already in most other OECD member countries. In November, the department pointed out to the minister the possible conflict between an independent review and the establishment of science advisory machinery.

This led to the idea of the publication of a 'green paper', or discussion paper, a practice new to the Australian parliamentary process but borrowed from Britain, where it had been widely used during the days of the Wilson Labour government. The 'green paper', which was to be prepared by the minister's personal assistants, would be tabled in parliament, made available to the OECD, and widely distributed for comment. The OECD review was seen as assisting the proposed science council to 'come quickly to grip with its tasks'.[95] Thus the establishment of the advisory council was to await the OECD review.

The 'green paper', entitled 'Towards an Australian Science Council',[96] was tabled on 21 March 1974, accompanied by a statement in which the minister placed emphasis on the fact that

> in recent years public disenchantment has risen with the part played by science and technology in developments of dubious value to mankind... In an increasingly environment-conscious society, it is now apparent that science and technology have themselves created problems. It is my belief that these problems can be solved only by the application of a re-directed scientific and technological effort.[97]

Thus a new focus emerged for the science council.

The OECD Panel of Experts, chaired by Alexander King, visited Australia in April 1974, and an assessment meeting was held in Canberra under the chairmanship of Oliphant. The final report,[98] which was not issued until October 1974, recommended the establishment of a ministerial committee for science, an advisory council for scientific and technological policy, and a parliamentary committee — in effect endorsing the ALP platform.

It was not until January 1975, almost two years after the 'suspension' of the ACST, that the Minister for Science issued a 'white paper' announcing the intention to form a ministerial committee to co-ordinate government programmes on science and technology, a statutory body to be known as the Australian Science and Technology Council (ASTEC), and a parliamentary science and technology forum. The foreword to the white paper announced it as flowing 'from the most intensive and extensive review ever conducted of science and technology in this country'.[99]

Perhaps this was accurate; but the return from this process appears to have been meagre. An interim membership was duly appointed in May 1975 but the Council 'was promptly ignored by the government'[100] on a variety of important matters, including the disposition of CSIRO divisions and the funding of research.

The CSIRO ethos of autonomy and independent research did not sit well with a Labor Government, and there were many areas of conflict. Morrison in particular considered CSIRO to be too big and insufficiently attuned to the needs of industry and the wishes of the government. Discussions were held about transferring some defence laboratories to CSIRO, but with the bitter 1948 experience still in mind, CSIRO was not interested. Morrison then canvassed the establishment of a new CSIRO-type organization for engineering research and development through a combination of some divisions with parts of the Department of Supply laboratories. The proposal generated a great deal of activity within the administrative echelons, much of it directed successfully to oppose any such move.

There were particular pressures on CSIRO in the energy area. Mr Rex Connor, Minister for Minerals and Energy, conducted a strong campaign for energy research to be linked to, and directed by, policy and needs established by his ministry. His first success was gaining control in September 1974 over research into the direct conversion of solar energy to electricity. This was followed in June 1975 by an announcement, without consultation with CSIRO, that their divisions concerned with minerals and energy matters — the Minerals Research Laboratories and the Solar Energy Unit — would be transferred to the control of the Department of Minerals and Energy. This created an enormous furore, with the executive and staff of CSIRO conducting a strong and highly public campaign to protect the Rivett ideal of independence. The campaign attracted a great deal of public support and eventually a backdown compromise was achieved, with the divisions in question staying with CSIRO but being required to report to both the Minister of Science and the Minister for Minerals and Energy.

Another indication of the relatively low standing of science with the Labor government occurred in the Budget of 1975 when, under budget pressures and a new Minister of Science, Clyde Cameron, who professed no interest in science, the allocation to the Australian Research Grants Scheme was cut by 66 per cent and that for the National Health and Medical Research Council by 47 per cent. In response to frantic lobbying by the Australian Vice-Chancellors' Committee (AVCC), the Academies, the interim ASTEC and the scientific community, a tacit admission was extracted that a 'mistake' had been made. A strong suspicion was expressed by many that the cuts resulted from a confusion between annual and financial years. With the connivance of the AVCC and the chairmen of the two research-funding bodies, a face-saving scheme was devised whereby an extra allocation of $7.2 million would be made available for 1976 in the 1976/77 budget, and the universities would produce $2.1 million of their own moneys for the ARGC in the period January-June 1976, to be reimbursed in the next financial year.

The final element of 'extensive review' of science policy by the Labor Government was initiated in August 1975, with the establishment of a science task force of the Royal Commission on Australian Government Administration that had been set up in 1974 to inquire into the Commonwealth Public Service. The science task force was 'to examine the conduct and co-ordination of scientific work carried out, financed and/or supported by the government'.

Ronayne has noted that the interim ASTEC was 'neither consulted by the Royal Commission about the appointment of the task force and its terms of reference, nor by the science task force itself during its deliberations'.[101] In effect it was yet another independent review covering much of the same ground already so tortuously examined in the two-year lead-up to the interim ASTEC. The report of the Science Task Force under the chairmanship of J.R. Philip, then a member of the CSIRO executive, stressed the importance of autonomy to science: 'Creative, productive science depends on the autonomous operation of self-imposed values and controls and it is ultimately self-defeating for a society or government to erode the autonomy of the scientific community.'[102] Hence it argued that most departmental research organizations should obtain the same independent statutory corporation status as CSIRO, 'providing the maximum diversity and adaptability possible, and conferring as much autonomy at as low a level as could be responsibly achieved.[103]

In addition, it recommended that ASTEC become responsible to the Prime Minister, the abolition of the Department and Ministry of Science, and that CSIRO should report to one minister without responsibilities for a specific science or technology field. It was, in effect, a thinly disguised attempt to re-establish CSIRO and the view of science it had espoused for so long as pre-eminent in Australian science, and to dismantle much of the bureaucracy that had developed in the previous eight years to organize and direct science.[104] The response of the Labor Government remains a matter of conjecture, as it was dismissed by the Governor-General on 11 November 1975, only two weeks after the Science Task Force Report was presented.

Throughout these two years of intensive review, what was achieved in science policy and organization itself? Most of the action was in the newer mission-orientated aspects of policy. The Australian Institute of Marine Sciences was established at Townsville; reviews of the Australian Defence Scientific Service, the Industrial Research and Development Grants Scheme, and research in Antarctica were initiated; the Biological Resources Council was set up; and the 154-inch Anglo-Australian telescope was commissioned at Siding Spring Observatory. A strong emphasis on environmental research and 'people science' was reflected in a reorganization of divisions within CSIRO to commit greater resources to ecological research[105] in the promotion of energy policy and funding for research on alternative energy sources, particularly solar power; in the initiation of the 'Botany Bay' project with a $1 million grant over five years for a multidisciplinary environmental study; and in the establishment of a council for consumer standards by the Minister for Science.

The Fraser years

In contrast to the tumult of the Whitlam years, the science institutions went through a relatively calm period of scrutiny, reorganisation and consolidation under the Fraser government. The emphasis was much more on administrative efficiency than ideological views about the place of science or its contribution to the economy. Funds continued to be provided for research, primarily by the Commonwealth government, but without the substantial growth of earlier times. In addition, it was a period for refinement of the newly established science policy machinery.[106]

The Fraser government was quick to reject the approach of the Science Task

Force, in as much as it described the Department of Science as the 'scientific nerve centre' of the Australian community. Members of the Academy of Science also moved quickly and met the Prime Minister to urge the new government to build on the interim ASTEC rather than start afresh yet again. This led to the establishment of a small advisory group in February 1976,[107] indicating a return to the 'cosy influential élite' characteristic of earlier Liberal Country Party administrations.

The group recommended that the Council should continue with some changes in membership and functions. Subsequently a second interim ASTEC was announced, which continued for about a year until the permanent Council was set up by executive powers in March 1977, and finally established as a statutory body, reporting to the Prime Minister, by an Act of parliament that came into effect on 28 February 1979. Depending on the starting point chosen, it had taken between seventeen and twenty-two years to establish this body.

The major thrust of activity under the Fraser government was justified in terms of ensuring that the considerable amount of public funds invested in research, particularly in the Commonwealth sector, should be efficiently directed towards national objectives. Whatever the rhetoric, there followed a long series of enquiries, a number of them conducted by ASTEC itself.

The first and perhaps most significant of these enquiries was commissioned in October 1976 into CSIRO and was entrusted to an independent committee rather than to ASTEC. Its significance lay not only in its consequences for the future of CSIRO, but also in the model it provided for subsequent enquiries into government research establishments. The committee of inquiry, under Professor Arthur Birch, was required to report on the objectives of the organization, the extent to which its research programmes filled these objectives, and the adequacy of existing arrangements and procedures.

In essence, the report confirmed the appropriateness of CSIRO's role, organization and funding, and made recommendations for internal restructuring, greater external influence on decision-making through an independent advisory council, and a significant increase in commitment to working with and assisting manufacturing industry.[108] The new Act specifically confirmed CSIRO's role in undertaking basic research, and led to the establishment of the Institute of Industrial Technology as one of the five CSIRO institutes and, within it, a Division of Manufacturing Technology.

These various changes can be considered fairly marginal, though there was a recognition that the environment in which CSIRO operated was changing:

> CSIRO has entered a new phase of its existence, a phase marked by the exigencies of zero growth, greater redeployment of resources, a much more public stance and greater involvement of outside advice and reviewers, a determination greater than ever before to ensure that its research is relevant to the [sic] adopted by industry.[109]

However, CSIRO was to fall into considerable disfavour with the public and the government in the early 1980s, and to be held at least partly responsible for the low level of technology commercialization in Australia. Another inquiry only ten years after that chaired by Birch was to find that there was a need for significant restructuring of the decision-making in CSIRO, to enable the organisation to interact more effectively with commercial interests.[110] These findings suggest that CSIRO, and the authors of the Birch Report, may not have appreciated just how much the environment was changing, nor the growing dissatisfaction with CSIRO's view of the place of science.

Most other government research establishments were also reviewed in the following years. These included the Bureau of Meteorology, the astronomical observatories, the Bureau of Mineral Resources, the Rural Industry Research Funds, the Australian Atomic Energy Commission and the Defence Science and Technology Organisation.[111] In various ways the objectives and structure of each institution were modified. Some lost significant resources, as in the transfer of non-nuclear energy research from AAEC to CSIRO. Perhaps the single common feature of all the reviews was a clearer reporting and accountability requirement, in an attempt to ensure that the research establishments were under more direct control and more effectively committed to appropriate national interests.

Finally, in 1981 and 1982, government support for science began to come under a new form of scrutiny. For almost the first time in Australia, under conditions of strong budgetary constraint, the question of whether the government should be involved in supporting research at all was raised. The massive Commonwealth government funding of R&D (63 per cent of the national total) was held to distort market forces to such an extent that it inhibited the development of adequate levels of industrial R&D. This led to the Review of Commonwealth Functions (colloquially known as the 'Razor Gang'), a report by the Industries Assistance Commission on Budgetary Assistance to Industry, and the establishment of reviews of productivity and innovation programs (the Kirby Report) and of Commonwealth laboratories (the Ross Report).

Conclusion

The history of the establishment of the 'independent science advisory body', ASTEC, described as 'the longest running show in the history of Australian science',[112] is remarkably revealing of the contrasting, and competing, aims of the scientific community and the scientific bureaucracy, that is, those officials and areas of government with a responsibility for science.

The original hopes of the scientific community, to establish a body of scientists with a position of official influence with the government, who would also be able to ensure a flow of resources to research, were never fulfilled.

That the campaign for the establishment of an advisory council took more than a quarter of a century to succeed is itself a clear indication of the low political priority accorded to institutionalizing scientific advice. Even more revealing, the haste and pre-election pressure associated with the announcement of the first ACST by McMahon, the long years of 'exhaustive review' leading to the almost identical ASTEC under Morrison, and the three years of interim councils before ASTEC was established as a statutory body under Fraser, all reveal the mixture of hesitancy and lethargy with which successive governments moved.

By the time ASTEC was finally promulgated, the passage of time had produced an entirely new political and economic climate. The mystique of science and scientists had diminished. In addition, both science and the technical needs of government had become far more specialized and fragmented. Interest was now far less in science as symbolized by CSIRO, and far more in the technical dimensions of energy, or the environment, or health. As a consequence, the charter of ASTEC emphasized the provision of advice about the more effective employment of science and technology in the national interest. It was a body to control and direct the scientific community, through its advice to government, rather than one to represent the interests of scientists. In hindsight, at least some scientists must wonder whether the twenty-five years of struggle have been worth it.

However, while the establishment of ASTEC may have been the most visible vehicle of explicit science policy, far more was actually going on within the bureaucratic arm of governments over that period. In the period of institution building between 1945 and 1965, as funds for science were increased, so was government control. CSIRO waged an unflagging battle to protect its autonomy and to maintain its position as the major science institution in Australia, but its influence on areas beyond its direct purview steadily diminished as new institutions were established. Each of these new institutions brought additional funding to the scientific community outside CSIRO, principally in the universities, and was warmly welcomed. There appears to have been little recognition that the institutions were purchased at the price of increased government influence and control, principally through the Department of Education and Science and the Australian Universities Commission, and their various successors.

In the period 1965–1980, the institutional structure of Australian science apparently changed little. However, bureaucratic control was substantially increased and refined. The very scale of the scientific enterprise, the success of the arguments of the apologists for science concerning the economic returns to be expected, and changing patterns of world trade, together made science both more important at the national level and less 'special'. It became just another factor of economic performance, to be appropriately organized and directed.

A new phase of the organization of sciences and science policy has emerged in the 1980s. Where previously science had been regarded as a potential solver of national problems, the emergence of technology as the major determinant of patterns of world trade, the high levels of growth of investment and trade in 'high technology' products and the sharp decline in commodity-based economies have cast a new role for science. It is now seen as the prime source of technological opportunity.

This does not represent a simple return to the Vannevar Bush[113] argument of science as the source of economic growth. Rather, the management and organization of the development of new technology has advanced enormously, particularly in Japan and the USA, in such a way that fundamental research and commercial exploitation are closely linked.[114] In this climate science has become far more visible to the commercial sphere and to the community generally. At the same time it has become clearly too important to be left to the scientists. Science in the 1980s is perhaps most clearly characterized by its management.[115]

Notes

1 D.P. Mellor, *The Role of Science and Industry* (Canberra, 1957), p. 705.

2 S. Encel, 'Financing scientific research in Australia', *Science*, 134 (1961), 260-6, at p. 263.

3 G. Nossal, 'The Walter and Eliza Hall Institute of Medical Research: 1915-1985', *Medical Journal of Australia*, 143 (19 August 1985), 153-7.

4 F.C. Courtice, 'The Kanematsu Memorial Institute of Pathology: the Inglis era, 1933-60', *Historical Records of Australian Science*, 6(2) (1985), 115-36.

5 J. Ronayne, *Science in Government* (Melbourne, 1984), p. 146. Other articles exploring the emergence of explicit science policy in Australia include S. Encel, 'Science and government policy — Australia', *Public Administration*, 27(2) (1968); and A. Moyal, 'The making of Federal government's science policy', in R. Lucy (ed.), *The Pieces of Politics* (Melbourne, 1975).

6 L. Peres, 'The changing politics of Australian science', in A.T.A. Healy (ed.), *Science and Technology for What Purpose? An Australian Perspective* (Canberra, 1979), p. 56.

7 S. Encel, op. cit. (n.2).

8 Ibid.

9 Mellor, op. cit, (n.1). pp. 690-2; W.A.S. Butement, 'Science and defence', in M.L. Oliphant (ed.), *Science in Australia* (Melbourne, 1952), pp. 93-114.

10 'Science in Australia', *Nature*, 172 (1953), 969-71, p. 970.

11 A. Moyal, 'The effect of institutional evolution on science policy', in Healy, op. cit. (n. 6), pp. 67-83; p. 70.

12 For an insider view, F. White, 'CSIR to CSIRO — the events of 1948-1949', *Public Administration*, 34 (1975), 281-93. Other more general versions can be found in Ronayne, op. cit. (n.5), pp. 162-3; Mellor, op. cit. (n.1), pp. 683-5; and A.I. Clunies Ross, 'Ian Clunies Ross', *Records of the Australian Academy of Science*, 3(3/4) (1977), 98-100.

13 J.P. Wild, 'The beginnings of radio astronomy in Australia', *Records of the Australian Academy of Science*, 2(3) (1972), 52-61; J.L. Pawsey and R.N. Bracewell. *Radio Astronomy* (London, 1955); W.T. Sullivan, this volume, Chap. 13.

14 H. Marston, 'Cobalt in animal nutrition', *Australian Journal of Science*, 18 (1956), 161; E.J. Underwood, 'The cobalt story', *Records of the Australian Academy of Science*, 2(2) (1971), 23-8.

15 E.B. Kraus and P. Squires, 'Experiments on the stimulation of clouds to produce rain', *Nature*, 162 (1974), 489; Mellor, op. cit. (n.1), pp. 685-6.

16 F. Fenner and F.N. Ratcliff, *Myxomatosis* (Cambridge, 1965).

17 Clunies Ross, op. cit. (n. 12), p. 101.

18 F.W.G. White, 'Casey of Berwick and Westminster, Baron Richard Gardiner Casey', *Records of the Australian Academy of Science*, 3(3/4) (1977), 55-83.

19 Letter from Sir Ian Clunies Ross to R.G. Casey, quoted in White, op. cit. (n. 12), p. 73: 'The phytotron is a large series of cabinets or rooms in which plants can be grown under any predetermined intensity and duration of light, temperature and humidity or, in short, under conditions simulating the climatic and seasonal characteristics of any environment'.

20 White, op. cit. (n.18).

21 A. Moyal, 'The Australian Atomic Energy Commission — a case study in Australian science and government', *Search*, 6(9) (1975), 365-81.

22 Moyal, op. cit. (n.11), pp. 74-5.

23 Oliphant, op. cit. (n.9), p. vi.

24 Ibid., p. vii.

25 S. Cockburn and D. Ellyard, *Oliphant* (Adelaide, 1981).

26 'Technical Education and University of Technology Act, 1949', NSW Government Printers, 1949.

27 I. Clunies Ross, *The Responsibility of Science and the University in the Modern World* (Sydney, 1952), p. 12.

28 AVCC, *A Crisis in the Finances and Development of the Australian Universities* (Melbourne, 1952).

29 Letter from Professor A.P. Elkin to R.G. Menzies, 3 March 1954, CSIRO Archives.

30 Joint Committee of Public Accounts, *Enquiry into the Office of Education* (Canberra, 1956).

31 *Report of the Committee on Australian Universities* (Murray Report) (Canberra, 1957), p. 5.

32 Parliamentary Debates, House of Representatives, No. 20, 26-28 November 1957, pp. 2699-701.

33 Quoted in A.P. Gallagher, *Co-ordinating Australian University Development* (Queensland, 1982), p. 62.

34 *Second Report of the Australian Universities Commission on Australian Universities 1960–1963* (Canberra, 1963), p. 27.

35 'A national science trust for Australia', 20 April 1959, AAS Archives, p. 1.

36 For example, see D.F. Martyn, 'Personal notes on the early days of our Academy', *Records of the Australian Academy of Science*, 1(2) (1967), 53-72; F. Fenner and A.L. Rees, *The First Twenty-Five*

Years (Canberra, 1980); A. Moyal, 'The Australian Academy of Science: the anatomy of a scientific elite', *Search* 11 (7 and 8) (1980); G. Taylor, 'History of the events leading to the foundation of the Australian Academy of Science', mimeo, Academy Archives; Cockburn and Ellyard, op. cit. (n. 25), pp. 202-16.
37 Oliphant, op. cit (n. 9).
38 Australian Academy of Science, *Scientific and Technological Manpower in Australia* (Canberra, 1957).
39 Australian Academy of Science, *A Report on the Condition of the High Mountain Catchments of New South Wales and Victoria* (Canberra, 1957).
40 F.C. Courtice, 'Arnold Hughes Ennor', *Records of the Australian Academy of Science*, 4(1) (1978), 117.
41 Peres, op. cit. (n.6).
42 C.P. Snow, *Science and Government* (New York, 1960).
43 R.G. Menzies, 'Modern science and civilization', *Australian Journal of Science*, 20 (1958).
44 For a summary of these arguments and their influence, see Ronayne, op.cit. (n.5).
45 H. Brooks, *Science, Growth and Society* (Paris, 1971); for a more detailed description of the growth and effects of the 1960-70s disenchantment with science in Australia, see R. Johnston, 'Social responsibility of science: the social mirror of science', in R. MacLeod (ed.), *The Commonwealth of Science* (forthcoming, Melbourne, 1988).
46 Ronayne, op. cit. (n.5), p. 147. An account of the development of ASTEC has also been written by S. Encel, 'Pushing the barrow uphill: the establishment and re-establishment of ASTEC', in S. Encel and P. Wilenski, *Decisions* (Melbourne, 1981), pp. 21-40.
47 Australian Academy of Science, op. cit. (n.38).
48 Briefing for the Prime Minister, Department of Education and Science Archive 68/2374.
49 Attachment to letter from F.W.G. White to R.G. Menzies, 17 January 1962, CSIRO Archives.
50 Ibid.
51 Memorandum from M. Oliphant to R.G. Menzies, 19 February 1962, CSIRO Archives.
52 Letter from F.W.G. White to M. Oliphant, 16 March 1962, CSIRO Archives.
53 Ibid.
54 Ibid.
55 E.J. Bunting, note for file, 2 April 1962, Department of Education and Science Archives.
56 Letter from F.W.G. White to R.G. Menzies, 24 April 1962, CSIRO Archives.
57 Ibid.
58 Letter from Professor T.M. Cherry to R.G. Menzies, 30 June 1962, Academy Archives.
59 Anon., 'Science in Australia', *Current Affairs Bulletin*, 30(7) (1962), 111.
60 N.S. Bayliss, 'The chain reaction of science', *Australian Journal of Science*, 25 (1962), 134.
61 F.W.G White, 'The strategy of Australian science', *Australian Journal of Science*, 26 (1964), 194.
62 Ibid., 191.
63 Hansard, House of Representatives, 24 March 1965, p 274.
64 C.L. Rubenstein, 'Changes in Australian science and technology policies: from ends to means', *Australian Journal of Public Administration*, 37 (1978), 243.
65 This report was finally released in public form, with recommendations for action in Australia deleted, as *Science and Government*, a report to the Advisory Council of CSIRO, November 1966.
66 Memorandum from Senator Gorton to Secretary, Prime Minister's Department, undated, Department of Education and Science Archives 68/2374.
67 Briefing for the Prime Minister, ibid.
68 Notes on CSIRO, Senator J. Gorton, January 1963.
69 Letter from F.W.G. White to Senator J. Gorton, 1 February 1963, CSIRO Archives.
70 Australian Labor Party, *Platform, Constitution and Rules* (Canberra, 1965), p. 15.
71 B. Nelson, 'Australian education and science are looking up down under', *Science*, 160 (1968), 173.
72 OECD, *Science and Technology in Australia* (Paris, 1984), p. 6.
73 Proposal for a Draft Cabinet Submission, Department of Education and Science Archives 68/2374.
74 Ibid.
75 Ibid.
76 Ibid.
77 Memorandum attached to letter from F.W.G. White to H. Ennor, 24 April 1968, Department of Education and Science Archives 68/2374.
78 'Science policy machinery for Australia', Science and Industry Forum Report No. 1 (Australian Academy of Science, Canberra), 1968.
79 Ibid., p 6.
80 Ibid., p 8.
81 M. Fraser, 'Government approaches to science', address to Science and Industry Forum, 1 March 1969 (Canberra, 1969), p 12.

82 E.G. Whitlam, 'A national science policy', *Search*, 1 (1970), 134–5.

83 Ibid., p 135.

84 Memorandum from Secretary to Minister, undated (approximately March 1971), Department of Education and Science Archives 68/2374.

85 Memorandum for file by J.R. Price, 'The role of the proposed advisory committee on science and technology', 7 September 1971, CSIRO Archives.

86 Attachment to letter to J.M. Fraser from Sir C. McGrath, 16 December, 1971, Department of Education and Science Archives 68/2374.

87 Parliamentary Debates, House of Representatives, 27 April 1972, p. 2122.

88 Ibid., p. 2123.

89 Interview with R. Robertson, August 1985.

90 Letter from W.L. Morrison to E.G. Whitlam, 7 February 1973, Dept of Science Archives 73/1052.

91 S. Encel, quoted in Ronayne, op. cit. (n.5), p. 149.

92 Platform of the ALP, Section VII, 'Science and technology', 1971.

93 Address by Minister of Science to CSIRO Executive, 9 October 1974, reported in *Canberra Times*, 23 October 1974.

94 'Establishment of Science Council', Minutes of Meeting between A.F. Rainbird and T. Conlon, 19 June 1973, Department of Science Archives 72/1052.

95 Letter from W.L. Morrison to all Ministers, 14 August 1973, Department of Science Archives 73/1052.

96 *Towards an Australian Science Council* (Canberra, 1974).

97 Hansard, House of Representatives, 21 March 1974.

98 OECD, op. cit. (n.72).

99 Ibid., Foreword.

100 Ronayne, op.cit. (n.5), p. 150.

101 Ibid., p.151.

102 Science Task Force, *Report to the Royal Commission on Australian Government Administration* (Canberra, 1975), p. 6.

103 Rubenstein, op. cit. (n.64), p. 247.

104 A debate on the basis of the Science Task Force Report was conducted in the pages of *Minerva*, between J.R. Philip, 'Towards diversity and adaptability: an Australian view of governmentally supported science', *Minerva*, 16 (1978), 397–415, and J. Ronayne, 'Further thoughts on diversity and adaptability in Australian Science Policy', *Minerva*, 18 (1979), 445–58.

105 P. Pockley, 'Science under an Australian Labor Government', *Nature*, 245 (November 1973), 260.

106 R. Johnston, 'Australian science policy: Now we can steer, where do we want to go?', *Current Affairs Bulletin*, 59 (1982), 20–30.

107 The group consisted of Sir Colin Syme (Chairman of the 1972 Interim Council), Professor Geoffrey Badger (Academician, Vice-Chancellor of the University of South Australia, member of CSIRO Executive, and author of the Science and Industry Forum Report on science policy), Professor L. Matheson (Vice-Chancellor of Monash University and Chairman of the 1975 Interim ASTEC), Professor Bob Street (Vice-Chancellor of the University of Western Australia and Chairman, ARGS) and Professor (later Sir) Rutherford Robertson; R. Robertson, 'Scientific advice and government decision making', University of Melbourne Seminar in memory of Leon Peres, September 1985, mimeo, p. 9.

108 *Report of the Independent Inquiry into the Commonwealth Scientific and Industrial Research Organisation* (Canberra, 1977).

109 P. Wild, 'Research in CSIRO', in A. Birch (ed.), *Science Research in Australia: Who Benefits?* (Canberra, 1983).

110 ASTEC, *Future Directions for CSIRO* (Canberra, 1985).

111 Johnston, op. cit. (n.106), 20–1.

112 P. Pockley, 'Under a cloud of committees', *Nature*, 267 (9 June 1977), 476.

113 V. Bush, *Science, the Endless Frontier* (Washington, 1945).

114 D. Bartels and R. Johnston, 'The sociology of goal directed science: recombinant DNA research', *Metascience*, 1/2 (1984), 25–37.

115 Considerable assistance in obtaining archival material was provided by Colin Smith and the staff of the CSIRO Archives, Rosanne Clayton of the Basser Library at the Australian Academy of Science, and Dr Roy Green and staff at the Department of Science. In addition much of the arduous archival research was very efficiently conducted by Marion Pemberton.

Notes on Contributors

GEORGE BINDON is a lecturer in the School of Science and Technology Studies and Sub-Dean of the Faculty of Arts at the University of New South Wales. He is interested in both the historical analysis of and current policy issues associated with industrial innovation in peripheral countries, particularly in his native Canada and his adopted Australia.

JEAN BUCKLEY is a lecturer in technology studies in the School of Science and Technology Studies at the University of New South Wales. She has an M Phil. in the history of science from Griffith University and is currently doing research on the history of Australian technology.

ROBERT L. BURT is a Principal Research Scientist with the Division of Tropical Crops and Pastures at the CSIRO Davies Laboratory, Townsville. A native of England, he received his PhD in crop plant physiology from the University of Nottingham before moving to Australia. At Townsville, he established the Pasture Plant Introduction and Genetic Resource Section. He has collected plant material extensively throughout South and Central America and the Caribbean and has been associated with its domestication for use in tropical areas; stemming from this are 'outside' interests in the domestication of tropical fruit species and the history of plant science in general.

BARRY W. BUTCHER is a graduate in history and philosophy of science from the University of Melbourne. He is currently a tutor in Social Studies of Science at Deakin University, Victoria, and is completing a PhD thesis at the University of Melbourne on the reception and impact of Darwinism in Australia, 1860–1914.

F.C. COURTICE, emeritus professor of the Australian National University, has devoted his career to research and teaching in the medical sciences, firstly in the Physiological Laboratory at Oxford and then in the Kanematsu Memorial Institute at Sydney Hospital and the John Curtin School of Medical Research in Canberra. His wide interests in the medical sciences span a period of well over fifty years, during which time he has written two books and many original and review articles. His main field of endeavour has been the structure and function of the lymphatic system in health and disease. He was elected a Fellow of the Australian Academy of Science in 1954.

S.C.B. GASCOIGNE was born in New Zealand and studied at the Universities of Auckland and Bristol, specializing in astronomical optics at the latter. During most of the Second World War he worked on various aspects of military optical instruments, from 1942 at Mt Stromlo. He stayed on at Mt Stromlo after the war as a staff member, engaged mostly in problems of stellar evolution and the astronomical distance scale. He was associated with the Anglo-Australian Telescope from the early 1960s, and was Commissioning Astronomer in 1974-75. He was elected a Fellow of the Australian Academy of Science in 1966 and an Associate of the Royal Astronomical Society in 1979.

HUGH HAMERSLEY is librarian at the Australian Radiation Laboratory, Melbourne. A science graduate of the University of Western Australia, he has recently completed an MA in history and philosophy of science at the University of Melbourne.

L.R. HIATT is Reader in Anthropology at the University of Sydney and a Fellow of the Australian Academy of the Social Sciences. After studying anthropology and philosophy as an undergraduate at Sydney University, he received his PhD from the Australian National University on the basis of field research among the Gidjingarli people of Arnhem Land. He has written extensively on traditional Aboriginal culture and is currently co-editor of the journal *Oceania* and editor of *Oceania Monographs*. From 1974 to 1982 he was President of the Australian Institute of Aboriginal Studies.

R.W. HOME studied physics and then history and philosophy of science at the University of Melbourne before completing a PhD in history and philosophy of science at Indiana University. He has been Professor of History and Philosophy of Science at the University of Melbourne since 1975. He has published extensively on the history of eighteenth-century physics and, more recently, on the history of science in Australia. He is editor of *Historical Records of Australian Science* and is currently preparing a history of the Australian physics community to 1945.

MIRANDA J. HUGHES is presently completing her doctoral dissertation in the Department of History and Philosophy of Science at the University of Melbourne, where she also completed her first degree. She has presented papers based on her thesis, which is concerned with the Société des Observateurs de l'Homme and the voyage of Baudin to Tasmania in 1802, to several meetings in Australia and at the International Congress of the History of Science at Berkeley, California, in 1985.

IAN INKSTER is Associate Professor and Head of the Department of Economic History at the University of New South Wales. His publications include *Japan as a Development Model?* (Bochum, 1980), *Metropolis and Province* (London, 1983) and *The Steam Intellect Societies* (Nottingham, 1985). He is presently writing a book on science, technology and economic development from the eighteenth century to the present.

SYBIL JACK was educated in London and Paris and took her degrees at Oxford. She worked as a research assistant at the History of Parliament Trust, taught school briefly on first coming to Australia and has subsequently been employed by the University of Sydney, first in the Department of Economic History and more recently in History. Her interests range from sixteenth and seventeenth century European history to Australian science and technology.

RON JOHNSTON is Foundation Professor in the Department of History and Philosophy of Science (recently re-titled Science and Technology Studies) and Director of the Centre for Technology and Social Change at the University of Wollongong. He has written extensively on science and technology policy; his most recent book, edited with Stephen Hill, is entitled *Future Tense? Technology in Australia*.

RHYS JONES is a Senior Fellow in Prehistory in the Research School of Pacific Studies at the Australian National University. He is a Fellow of the Australian Academy of the Humanities. After reading natural sciences and archaeology in Emmanuel College, Cambridge, he received his PhD at the University of Sydney in anthropology. His research has been into the prehistory of Australia, with main field work in Tasmania, Lake Mungo and Arnhem Land. He has also studied the foraging ecology of the Gidjingarli people of Arnhem Land. He has an interest in the history of archaeology and the European perceptions of hunting and gathering peoples.

A.M. LUCAS graduated in zoology and genetics from the University of Melbourne before teaching biology in Victorian high schools. After a period in the School of Biological Sciences at the Flinders University of South Australia he was awarded a Fulbright scholarship to Ohio State University, completing his PhD in science education. He is now Professor of Science Curriculum Studies at King's College, University of London.

DAVID PHILIP MILLER has lectured in the School of Science and Technology Studies at the University of New South Wales since 1981. A former professional Yorkshireman, he received his PhD in the history and sociology of science from the University of Pennsylvania. His primary research interest is the social history of British science in the late eighteenth and early nineteenth centuries.

ROBERT STAFFORD is a post-doctoral research fellow in the Department of History at La Trobe University, having previously held a similar appointment in the Department of History and Philosophy of Science at the University of Melbourne. Before this, he earned a DPhil in modern history at Oxford University (Balliol College). His dissertation, on the role of the geologist Sir Roderick Murchison in promoting the scientific exploration of the British empire during the nineteenth century, is soon to be published by Cambridge University Press.

WOODRUFF T. SULLIVAN, III was born in Colorado. He obtained his BS in physics at the Massachusetts Institute of Technology and PhD in astronomy at the University of Maryland. He is presently a professor of astronomy at the University of Washington in Seattle. His research interests include galaxies, the search for extraterrestrial intelligence, and the history of astronomy. He compiled *Classics in Radio Astronomy* (Reidel, 1982), edited *The Early Years of Radio Astronomy* (Cambridge University Press, 1984), and has just completed a monograph on the early history of world-wide radio astronomy.

JAN TODD holds degrees in chemistry and sociology and is presently a doctoral student in the Department of Economic History, University of New South Wales. Her research interests centre on the genesis of science policy and industrial research in Australia.

W.T. WILLIAMS, formerly a member of the Agricultural Research Council and Professor of Botany at the University of Southampton, England, first visited Australia to advise on programming for computer classification of plants. He returned in 1966 as a member of the CSIRO Division of Tropical Crops and Pastures, initially at Brisbane and then at the Davies Laboratory, Townsville. A world authority on pattern analysis, and with interests as diverse as music and scientific fiction, Dr Williams was elected a Fellow of the Australian Academy of Science in 1978 and in 1980 was awarded the OBE for his service to science.

Milton Keynes UK
Ingram Content Group UK Ltd.
UKHW032138181024
449640UK00018B/257